# 16th International Scientific Conference: Civil Engineering and Building Services (CIBv 2021)



IOP Conference Series: Materials Science and Engineering
Volume 1242

Online
4 - 5 November 2021

**Editors:**

**Ioan Tuns**
**Radu Muntean**
**Dorin Radu**
**Teofil-Florin Galatanu**
**Christiana Emilia Cazacu**

ISBN: 978-1-7138-5441-8
ISSN: 1757-8981

**Printed from e-media with permission by:**

Curran Associates, Inc.
57 Morehouse Lane
Red Hook, NY 12571

**Some format issues inherent in the e-media version may also appear in this print version.**



Institute of Physics
Dirac House, Temple Back
Bristol BS1 6BE UK

Phone: 44 1 17 929 7481
Fax: 44 1 17 920 0979

techtracking@iop.org

**Additional copies of this publication are available from:**

Curran Associates, Inc.
57 Morehouse Lane
Red Hook, NY 12571 USA
Phone: 845-758-0400
Fax: 845-758-2633
Email: curran@proceedings.com
Web: www.proceedings.com

# TABLE OF CONTENTS

# International Conference CIBv2021
# Civil Engineering and Building Services

**4th – 5th November 2021, Brașov, Romania**

Selected, peer reviewed papers from the International Conference Civil Engineering and Building Services CIBv2021

**Edited by:**

Ioan TUNS
Transilvania University of Brasov, Romania

Radu MUNTEAN
Transilvania University of Brasov, Romania

Dorin RADU
Transilvania University of Brasov, Romania

Teofil-Florin GĂLĂȚANU
Transilvania University of Brasov, Romania

Christiana Emilia CAZACU
Transilvania University of Brasov, Romania

IOP Conf. Series: Materials Science and Engineering 1242 (2022) 011001 doi:10.1088/1757-899X/1242/1/011001

# FOREWORD

The International Scientific Conference CIBv represents one of the most important and traditionally scientific events of Civil Engineering and Building Services in Romania, enjoying increasingly higher visibility at the international level.

The aim of the CIBv conference is to provide a platform for researchers, engineers, academicians as well as professionals from all over the world to present their research results and development activities. It also provides opportunities for the participants to exchange new ideas and application experiences, to establish business or research relations and to find other partners and even friends for future collaboration.

The Conference has been organized each year at Transilvania University of Brașov, since 2000. In November 2021, Faculty of Civil Engineering was honoured to organize, in partnership with Shenyang Jianzhu University from China and to host the 16th edition of "International Scientific Conference – Civil Engineering and Building Services CIBv2021" in Brașov. More than 100 participants from China and Romania, as well as from United States of America, Serbia, Japan, Germany, Russia and Greece submitted over 65 papers to the scientific board which were presented during the conference. After a peer-review process (which included for each article at least two reviewers), 47 papers were finally accepted for publication in the CIBv2021 proceedings volume.

We would like to address special thanks to all those who were involved in the success achieved by the CIBv2021 International Conference, as well as to our distinguished publisher.

Prof. Eng. Ioan TUNS, PhD

Dean of Faculty of Civil Engineering,
Transilvania University of Brașov, Romania

IOP Conf. Series: Materials Science and Engineering 1242 (2022) 011001 doi:10.1088/1757-899X/1242/1/011001

# International Conference CIBv2021
# Civil Engineering and Building Services

## Faculty of Civil Engineering, Transilvania University of Brasov, Romania in collaboration with Shenyang Jianzhu University, China

## Organizing committee

# Peer Review Statement

All papers published in this volume have been reviewed through processes administered by the Editors. Reviews were conducted by expert referees to the professional and scientific standards expected of a proceedings journal published by IOP Publishing.

- **Type of peer review:** Single Anonymous
- **Conference submission management system:** The submission management was an on website uploading type one. Each abstract was uploaded on the conference website and the papers were sent directly via conference e-mail.
- **Number of submissions received:** 65
- **Number of submissions sent for review:** 51
- **Number of submissions accepted:** 47
- **Acceptance Rate (Submissions Accepted / Submissions Received × 100):** 72.3
- **Average number of reviews per paper:** 2
- **Total number of reviewers involved:** 48
- **Contact person for queries:**
  **Name:** Dorin Radu
  **Email:** dorin.radu@unitbv.ro
  **Affiliation:** Transilvania University of Brasov

# Improving the mechanical and thermal properties of hemp concrete by treating the hemp hurds with sodium hydroxide solution

**L Adam[1*], D N Isopescu[1] and S G Maxineasa[1]**

[1]" Gheorghe Asachi" Technical University of Iasi, Faculty of Civil Engineering and Building Services, Iași, Romania

* laurentiu.adam@student.tuiasi.ro

**Abstract**. The objectives of the research were to analyse how the treatment with sodium hydroxide solution of hemp stems influences the mechanical and thermal behaviours of concrete when hemp stems are used in the composition of concrete. Tests were performed on hemp concrete samples to evaluate the density and measure the heat transfer coefficient, compressive strength and modulus of elasticity. The activities carried out aimed to identify the influence of the sodium hydroxide solution concentration on the variations of the measured parameters and expressing some conclusions regarding the need to treat the hemp stems before their use in concrete. The need for research is given by the nowadays demand for obtaining eco-sustainable construction materials, in which the possibility of using a material that is considered plant waste becomes an economic resource in a circular economy.

## 1. Introduction

Hemp is a crop that absorbs and fixes the highest amounts of $CO_2$ among other crops, even if compared to an area of 10,000 square meters of forest. Many studies [1-6], refer to the ability of hemp to absorb an average of 15 tons of $CO_2$ in six months of growing time, an estimated amount per hectare of cultivated surface.

**Figure 1**. Hemp stems after natural drying.

**Figure 2**. Hemp core after removing the fiber by peeling.

**Figure 3**. Sodium hydroxide.

The research described in the article refers to the use of hemp stems (which are considered waste in the harvesting of hemp crops), to the realization of eco-sustainable concrete that can be used in various components of construction. After harvesting the hemp stems, figure 1, they are allowed to dry in a natural atmosphere. Prior to use in concrete, the hemp core shown in figure 2 is treated with alkaline solution using NaOH (sodium hydroxide), figure 3. This process effectively removes pectin and lignin from the wood [7].

An increasing of the tensile strength was observed after the treatment, [8], and high mechanical properties were obtained. In figure 4 are shown different part of hemp stems treated with 10% solution of sodium hydroxide after 24 hours of immersion. After the treatment, the samplings were bended and the difference between how the samples reacting to the stretching effort is shown in figure 5.

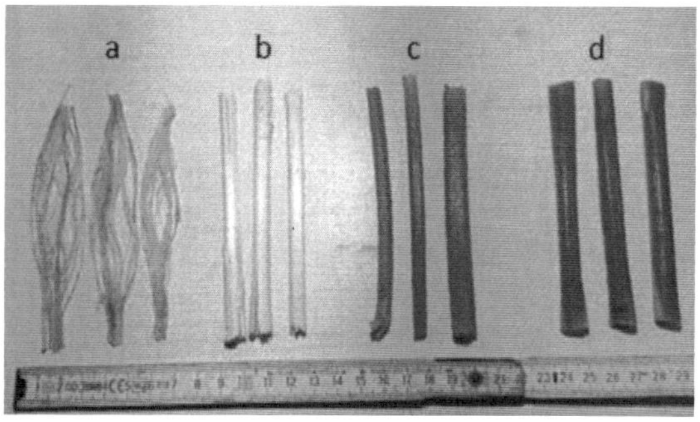


**Figure 4**. Hemp samples: a – hemp fibres untreated; b – hemp steam core untreated; c – hemp steam core treated with NaOH; d – hemp steams treated with NaOH.

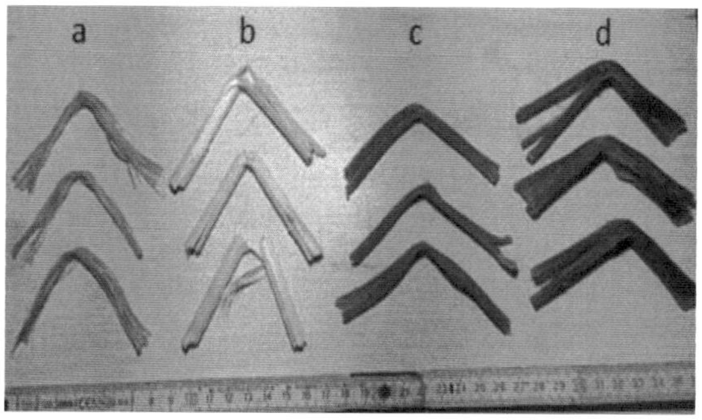


**Figure 5**. Bent hemp samples: a, b, c and d are the same material as in figure 4.

The process to treat hemp with NaOH solution has stimulated the individualization of the fibers shown in figure 6 and figure 7. Longer fibres are obtained after removing the pectin and lignin from the hemp stem core. The fibrous aspect of the hemp core treated with NaOH solution, figure 8 b, looks like those of natural fiber, figure 8 a.

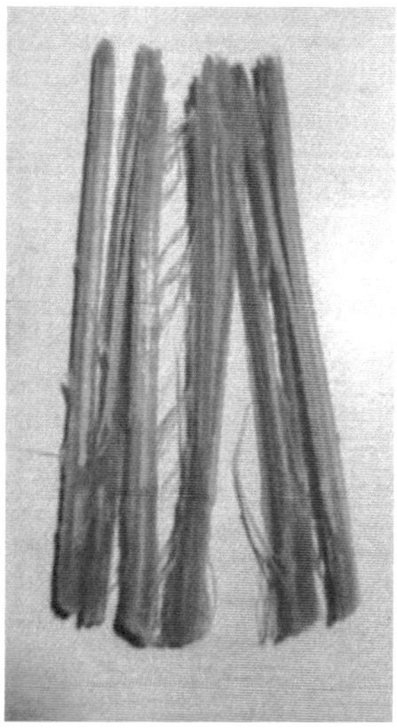

**Figure 6**. Hemp stem core after treated with alkaline NaOH solution.

**Figure 7**. The fibrous aspect of hemp stem core after treatment.

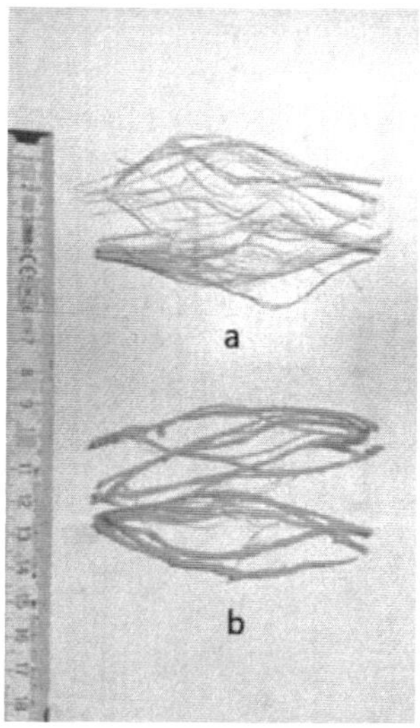


**Figure 8**. Hemp fibers: a. untreated hemp fibers, b. treated hemp fibers.

## 2. Materials

The materials used for the eco-sustainable concrete were hemp core from a "Zenit" variety crop species, figure 9, sodium hydroxide NaOH with 98% concentration, hydrated lime, Portland cement, sodium silicate solution and water which was used only for pre-treatment of hemp stemp core. In table 1 and table 2 are presented the values of the characteristic properties of the cement and the lime used for the eco-sustainable concrete.

**Figure 9**. Hemp core fibres with 0 – 4.5 cm length, 70 kg/m$^3$ bulk density.

**Table 1**. Performance specification for cement used.

| Components & property | U.M. | Values |
|---|---|---|
| Clinker (K) | % | 65 ÷ 79 |
| Furnace slag (S) and chalk (LL) | % | 21 ÷ 35 |
| Initial setting time | minutes | 60 |
| Resistance to initial compression | MPa | ≥ 20 |
| Standard compression strength | MPa | 42.5 ÷ 62.5 |
| Sulphate content in the form of $SO_3$ | % | ≤ 3.5 |
| Expansion stability | mm | < 3 |
| Chloride content | % | ≤ 0.1 |

**Table 2**. Performance specification for lime used.

| Components & property | U.M. | Values |
|---|---|---|
| CaO + MgO | % | ≥ 80 |
| Free lime Ca $(OH)_2$ | % | ≥ 65 |
| MgO | % | ≤ 5 |
| $CO_2$ | % | ≤ 15 |
| $SO_3$ | % | ≤ 2 |
| Free water | % | ≤ 2 |
| Penetration | mm | 10 ÷ 15 |
| Air content | % | ≤ 12 |

## 3. Methods

### *3.1. Absorption measurements of hemp fibres samples*
The first step of the research was to determine the absorption value of the hemp stem fibers immersed into the NaOH solution. A quantity of 10 grams of dried fibres was weighed and then it was immersed in water for 24 hours. The weight of immersed hemp fibres was four times its dry weight, therefore the absorption of water to fibres is 300%. This value was taken into consideration for the eco-sustainable concrete compositions.

Different concentration of NaOH solutions have been prepared to treat a 500 grams of hemp fibers for each case. To obtain 5%, 10% and 15% concentration of alkaline NaOH solution, were used 111 grams, 222 grams and 333 grams of NaOH dissolved in 2 litters of clean water. The solid NaOH pellets was subsequently added in water and mixed. Then the hemp fibres, in equal quantity of 500 grams, were immersed in the solutions. The aspects of immersed hemp fibres in the three solutions are presented in figure 10.

In figure 10, from left to right, the aspects of immersed hemp fibers in the cases with 5%, 10% and 15% of NaOH solution are presented. In the case of 5% concentration solution, the modification of fibres are minimum, the volume remains almost the same with the initially, and the brittle appearance of the immersed hemp fibres remains almost similar to the untreated fibers case. In the cases with 10% and 15% solutions concentrations it is observed a significant modification regarding the shrinkage of the treated fibres, an intense yellowish brown colouring and a softening of the fibres. This modification is better observed in W Fracz et al. [9] where are showed comparation between SEM photographs taken for samples treated with 2%, 5% and 10% of hydroxide solution that produced smaller diameter as the concentration of solution increased.

a. b. c.

**Figure 10**. Hemp fibres treated with sodium hydroxide solution: a. 5% NaOH concentration; b. 10% NaOH concentration; c. 15% NaOH concentration.

### *3.2. The mixing technology*
Each heap of 500 grams of treated hemp fibres (through immersion in NaOH solutions for 24 hours), was mixed with 375 grams of hydrated lime and 375 grams of Portland cement. No water was added because the hemp fibres were saturated in water during the treatment process. During the mixing, it was observed that the workability and compactness as improved as much the concentration of NaOH used in solutions increased. The fresh hemp concrete mixture is shown in figure 11, and the casted samples are presented in figure 12. The composition used are presented in table 3. For each composition two samples were made.

**Figure 11**. Fresh concrete.

**Figure 12**. Casted samples.

**Table 3**. The composition of concrete samples.

| | U.M. | | | | |
|---|---|---|---|---|---|
| The indicative of the samples set | | E | S1 | S2 | S3 |
| Hemp | kg | 0.500 | 0.500 | 0.500 | 0.500 |
| Lime | kg | 0.375 | 0.375 | 0.375 | 0.375 |
| Cement | kg | 0.375 | 0.375 | 0.375 | 0.375 |
| Concentration of NaOH solution | % | 0 | 5 | 10 | 15 |
| Water | l | 2.000 | 2.000 | 2.000 | 2.000 |
| Total weight of the mix | kg | 3.250 | 3.361 | 3.472 | 3.583 |

## 4. Results and discussion

The samples obtained with the compositions from table 3, are presented in figures 13, 14 and 15. A cylindrical mould of 200 mm height and with the diameter of 100 mm was used for samples.

**Figure 13**. The eco-sustainable concrete samples – the hemp fibres treated with 5% concentration of NaOH solution.

**Figure 14**. The eco-sustainable concrete samples – the hemp fibres treated with 10% concentration of NaOH solution.

**Figure 15**. The eco-sustainable concrete samples – the hemp fibres treated with 15% concentration of NaOH solution.

### *4.1. The density measurements*

After 28 days the density was determined, and the values are presented in table 4.

**Table 4**. Density values at 28 days.

| Sample | NaOH [%] | Mass 28 days [kg] | Density 28 days [kg/m$^3$] | Average density [kg/m$^3$] |
|---|---|---|---|---|
| E 1 | 0 | 0.606 | 385.99 | 371.34 |
| E 2 | 0 | 0.560 | 356.69 | |
| S 1.1 | 5 | 0.850 | 541.40 | 538.22 |
| S 1.2 | 5 | 0.840 | 535.03 | |
| S 2.1 | 10 | 1.125 | 716.56 | 714.01 |
| S 2.2 | 10 | 1.117 | 711.46 | |
| S 3.1 | 15 | 1.272 | 810.19 | 814.01 |
| S 3.2 | 15 | 1.284 | 817.83 | |

The values obtained for densities are below 1000 kg/m$^3$. The aspect of the samples S 3.1 and S 3.2 shown in figure 15, are significantly modified and improved when it is used the 15% concentration of sodium hydroxide to treat the hemp fibres. The compaction is better as the concentration of NaOH increasing. During the mixing, differences were observed between the hemp-concrete compositions, in terms of reducing the time required to mix and the ease of dispersion of binder throughout the hemp fibres heap. In the case of 5% NaOH concentration, the hemp fibres structure kept its brittleness appearance, while in the case of using 10% and 15% concentrations, the fibres have become more flexible turning into yarns whose malleability led to the loss of volume of the mixture and obtaining a higher density.

### *4.2. The heat transfer coefficient measurements*

The experimental measurements to determine the heat transfer coefficient were performed with Isomet 2114 Portable heat transfer analyser, figure 16. For each sample were taken 3 measurements within a time frame of one hour. The values obtained are shown in table 5. The values obtained for the sample E1, E2, S1.1 and S1.2 are comparable with the results for thermal conductivity measured by other outors [10] for 2 samples with appropriate density around 400kg/m$^3$. The difference ar that for a higher density the heat transfer coefficient measured are lower and this is because of the change in porosity made by hydroxide solution treatment.

**Table 5**. The heat transfer coefficient – λ.

| Sample | NaOH [%] | Mass [kg] | Density [kg/m$^3$] | λ [W/m*K] | λ Average values [W/m*K] |
|---|---|---|---|---|---|
| | | | | 0.0713 | |
| E 1 | 0 | 0.606 | 385.99 | 0.0720 | 0.0718 |
| | | | | 0.0721 | |

| | | | | | |
|---|---|---|---|---|---|
| E 2 | 0 | 0.560 | 356.69 | 0.0676<br>0.0681<br>0.0683 | 0.0680 |
| S 1.1 | 5 | 0.850 | 541.40 | 0.1463<br>0.1478<br>0.1481 | 0.1474 |
| S 1.2 | 5 | 0.840 | 535.03 | 0.1152<br>0.1167<br>0.1170 | 0.1163 |
| S 2.1 | 10 | 1.125 | 716.56 | 0.1916<br>0.1936<br>0.1944 | 0.1932 |
| S 2.2 | 10 | 1.117 | 711.46 | 0.2445<br>0.2471<br>0.2464 | 0.2460 |
| S 3.1 | 15 | 1.272 | 810.19 | 0.2576<br>0.2586<br>0.2595 | 0.2586 |
| S 3.2 | 15 | 1.284 | 817.83 | 0.2677<br>0.2707<br>0.2710 | 0.2698 |

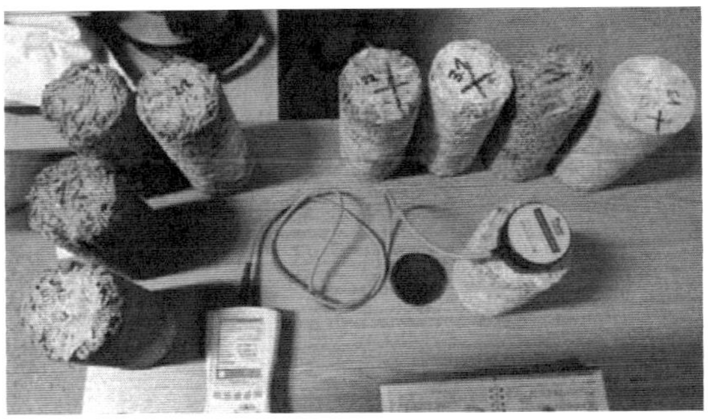

**Figure 16**. Experimental tests for heat transfer coefficient, λ.

An expected evolution is observed, so that the values obtained for the heat transfer coefficient increased with the increase of the density. Therefor the pre-treatment with NaOH of the hemp fibres heap gives a negative influence for the insulating properties compared to those of non-treated samples.

*4.3. The compression strength measurements*

Table 6 presents the experimental test results of compression strengths. Values obtained for samples S 2.1, S 2.2, S 3.1 and S 3.2 are comparable with the results obtained in a study [11], where was used MgO as a binder instead of aerated lime and Portland cement. The values obtained for samples poured using MgO are higher.

**Table 6**. The experimental compression strength values.

| Sample | NaOH [%] | Mass [kg] | Density [kg/m$^3$] | F at 5 mm displacement [kN] | F Average values [kN] | Compression strength [kN/m$^2$] |
|---|---|---|---|---|---|---|
| E 1 | 0 | 0.606 | 385.99 | 0.45 | 0.475 | 57.325 |
| E 2 | 0 | 0.560 | 356.69 | 0.50 | | 63.694 |
| S 1.1 | 5 | 0.850 | 541.40 | 0.90 | 0.825 | 114.649 |
| S 1.2 | 5 | 0.840 | 535.03 | 0.75 | | 95.541 |
| S 2.1 | 10 | 1.125 | 716.56 | 1.20 | 1.500 | 152.866 |
| S 2.2 | 10 | 1.117 | 711.46 | 1.80 | | 229.300 |
| S 3.1 | 15 | 1.272 | 810.19 | 2.88 | 2.665 | 366.879 |
| S 3.2 | 15 | 1.284 | 817.83 | 2.45 | | 312.102 |

**Figure 17**. Sample 3.2 before compressing.

**Figure 18**. Sample 3.2 after compressing test.

**Figure 19**. Sample E.2 after compressing.

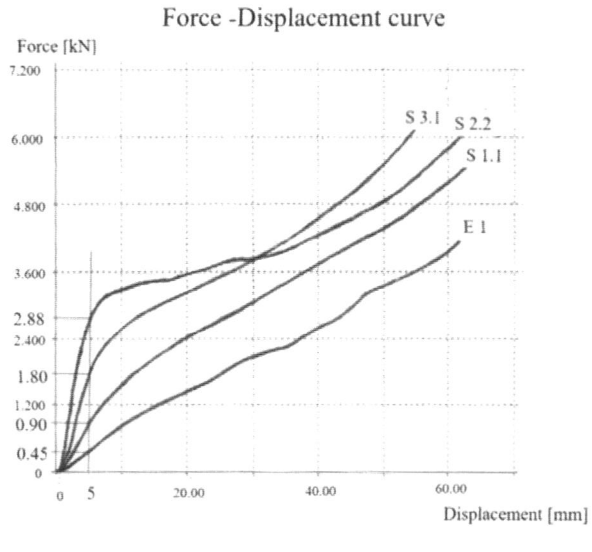


**Figure 20**. Force – Displacement curves for tested samples:
E 1 - 0% NaOH
S 1.1 - 5% NaOH
S 2.2 - 10% NaOH
S 3.1 - 15% NaOH

In figure 20 are presented the Force-Displacement curves for E 1, S 1.1, S 2.2 and S 3.1 samples. The shapes of the curves show that the elastic behaviour of the sample is depending by the type of treatment of the hemp fibres (concentration of NaOH solution), and it is in close connection with the density of the samples

### *4.4. The evaluation of longitudinal modulus of elasticity in compression*

The formula to calculate the longitudinal modulus of elasticity, E, when samples are subjected to compression is according to the Hook's law and presented in the equation (1). The values obtained for samples are shown in table 7.

$$E = \frac{\sigma}{\varepsilon} = \frac{FH}{A\Delta H} \quad [\text{Pa}] \qquad (1)$$

where:

$$\sigma = \frac{F}{A} \ [\text{N/m}^2]; \ \varepsilon = \frac{\Delta H}{H_i} \ [\%]$$

$$\Delta H = H_i - H_f \ [\text{mm}]$$

$$H = \text{the height of the samples, [mm]}$$

**Table 7**. Values for longitudinal modulus of elasticity at compression.

| Sample | NaOH [%] | F at 5 mm displacement [kN] | E [MPa] | E Average values [MPa] |
|---|---|---|---|---|
| E 1 | 0 | 0.45 | 22.929 | 24.203 |
| E 2 | 0 | 0.50 | 25.477 | |
| S 1.1 | 5 | 0.90 | 45.860 | 42.039 |
| S 1.2 | 5 | 0.75 | 38.217 | |
| S 2.1 | 10 | 1.20 | 61.146 | 76.433 |
| S 2.2 | 10 | 1.80 | 91.719 | |
| S 3.1 | 15 | 2.88 | 146.751 | 135.796 |
| S 3.2 | 15 | 2.45 | 124.841 | |

## 5. Conclusions

Alkaline treatment of hemp fibres has led to changes in their elasticity, which results in a more efficient compaction of eco-sustainable concrete compositions, directly proportional to the concentration of NaOH. The treatment of hemp fibres improved the mechanical strength properties of cylindrical samples. On the other hand, in terms of thermal properties, the pre-treatment of hemp fibres with NaOH solution resulted in a decrease in the thermal insulation capacity of the samples.

Research shows that, for eco-sustainable hemp concrete, density is also a determining factor in achieving the desired values for its characteristic properties, the influence being in a proportional relationship. The same composition of eco-sustainable concrete is influenced by the pre-treatment of hemp fibres and the type of product used to treat the fibers.

## References

[1] Aytac S 2018 An environmentally friendly plant in terms of oxygen supply: Hemp *ICOEST 4th Int. Conf. Environ. Sci. Technol. Sept.* 19–23
[2] Ruus A, Koosapoeg T, Pau M, Kalamees T and Põldaru M 2021 Influence of production on hemp

concrete hygrothermal properties: Sorption, water vapour permeability and water absorption *J. Phys. Conf. Ser.* **2069**

[3] Vosper J 2011 The Role of Industrial Hemp in Carbon Farming *GoodEarth Resour. PTY LTD* 1–6

[4] Wilson C 2019 The Future for Hemp An Environmental Perspective - What Can Hemp Offer? *16th EIHA HEMP Conf.*

[5] Hemp Technologies Global https://hemptechglobal.com/styled-17/page83/page83.html

[6] Adesina I, Bhowmik A, Sharma H and Shahbazi A 2020 A review on the current state of knowledge of growing conditions, agronomic soil health practices and utilities of hemp in the United States *Agric.* **10**

[7] Wang H M, Postle R, Kessler R W and Kessler W 2003 Removing Pectin and Lignin During Chemical Processing of Hemp for Textile Applications *Text. Res. J.* **73** 664–9

[8] del Rey R, Serrat R, Alba J, Perez I, Mutje P and Espinach F X 2017 Effect of sodium hydroxide treatments on the tensile strength and the interphase quality of hemp core fiber-reinforced polypropylene composites *Polymers (Basel).* **9**

[9] Frącz W, Janowski G and Bąk Ł 2021 Influence of the Alkali Treatment of Flax and Hemp Fibers on the Properties of PHBV Based Biocomposites *Polymers (Basel).* **13** 1965

[10] Fernea R, Manea D L, Plesa L, Iernutan R and Dumitran M 2019 Acoustic and thermal properties of hemp-cement building materials *Procedia Manuf.* **32** 208–15

[11] Stevulova N, Kidalova L, Cigasova J, Junak J, Sicakova A and Terpakova E 2013 Lightweight Composites Containing Hemp Hurds *Procedia Eng.* **65** 69–74

# Ventilated façade solutions

**I-R Baciu[1*], D N Isopescu[1], M L Lupu[1], S G Maxineasa[1], L Pruna[2] and S Dan[3]**

[1] "Gheorghe Asachi" Technical University of Iasi, Faculty of Civil Engineering and Building Services, Department of Civil and Industrial Engineering, Romania
[2] "Gheorghe Asachi" Technical University of Iasi, Faculty of Civil Engineering and Building Services, Department of Engineering Graphic, Romania
[3] SC FURNITURE VISION SRL, Ilfov, Romania

* ioanaroxana.baciu@yahoo.com

**Abstract**. Reducing the energy consumption needed for creating suitable indoor conditions has become a significant issue on a global scale. The building's envelope and service systems have the most important influence over the amount of energy consumed. This aspect is related to reducing the heat flux across the building envelope in summer conditions and preventing the condensation and infiltration risk in the winter period. The research regarding ventilated façades has advanced taking into account these advantages, which emphasized the need to study and create constructive solutions adapted to the conditions. In this context, this paper represents a brief introduction to ventilated façades solutions, taking into account the definition, their characteristics, the principal constructive elements, the main types of ventilated facades, and last, but not least, their advantages and disadvantages. The authors believe that the study is critical in fully comprehending the characteristics of these systems and their primary components, as well as designing and implementing them in accordance with current environmental needs. It is anticipated that comprehending this concept, as well as its evolution and trends, can contribute in the resolving of a number of ecological and societal issues.

## 1. Introduction

The construction industry in the European Union consumes 40% of global energy, with heating, ventilation, and air conditioning systems accounting for two-thirds of this consumption. To fulfill this growing energy demand, European regulation on building energy performance recommends that all EU member states implement energy policies that encourage the use of very low-energy and even zero-energy buildings [1].

In industrialized nations, energy conservation is a fundamental priority. Buildings use 40% of total energy in Europe [2, 3]. The building envelope is the structural component that has the largest influence on the building's total energy consumption [4, 5]. It's worth emphasizing that the façade serves as a link between the outside environmental components and the inside human needs. While the latter is rather steady, the former are ever-changing [6].

Recently, researchers have rekindled their interest in vented walls. It cannot be disputed that a considerable body of research and journal papers has been published on "ventilated envelopes" [7], with a concentration on Double Skin Façades [8, 9], Building Integrated Photovoltaic [9, 10], Solar chimney, Trombe, and Solar Walls [11], and Façade Solar Collectors [12]. On the other hand, Opaque Ventilated Façades have received less attention (OVFs).

Additionally, energy efficiency has emerged as a critical concern in the worldwide construction business. The energy efficiency of buildings, on the other hand, is often dictated by the building envelope and building services system.

The building industry divides thermal insulation choices for walls into two broad categories: continuous or wet systems and ventilated façade or dry systems. The wall pieces are cemented together, and the outside thermal insulation is applied using a wet technique to the support wall. Additionally, the insulation is thermally coupled to the outside plaster. The ventilated façade system, the second thermal insulation strategy, is fully dry due to the absence of adhesives or mortars. There has been much debate recently over residential thermal insulation, with the most intense debate centered on the thermal insulation of the façade. Thermal insulation work on the façade is often completed by plating with polystyrene, which adheres the insulation plate to the support surface. Apart from this kind of thermal insulation, authorities recommend others that are regarded to be more technically better. This time, though, we'll take a look at the vented façade.

The outer layer is often composed of modular panels and may include a range of tile types (e.g., metallic, ceramic, stone, or composite). Alternatively, the exterior paneling may be continuous or the tile seams may be sealed. In this scenario, ventilation is supplied through the cladding's bottom and top holes. Alternatively, the joints are left open (Open Joint Ventilated Façades (OJVFs)), allowing external air to enter and exit the cavity over the whole length of the wall; furthermore, the joints withstand temperature-related expansion [13].

This page addresses vented façade solutions, their features, structural components, the several varieties of ventilated facades, and the associated benefits and drawbacks. Additionally, the scientists employed this cutting-edge study to explore and understand the most effective processes for creating a new kind of ventilated façade composed of composite materials.

## 2. Literature research and gap identification

The ventilated façade concept, which has been widely used in Romania for less than two decades, was regulated two years ago by a normative that specifies the necessary materials, structures, and methods of execution, resulting in an energy-efficient and safe structure capable of withstanding the various challenges posed by nature or human activity (earthquakes, fires, weather conditions, pollution, various incidents, etc.).

Ventilated facades are structural sections of a structure that are coated in a cladding material and linked to the wall through supports, which are often metal (aluminum, steel). These components must ensure that the cladding material is properly connected while also supporting the exact loads generated by the weight and wind force of the plywood. Simultaneously, these structures must withstand the elements and perform in circumstances of thermal expansion.

Only in 2013 was the vented façade idea regulated in Romania via a normative that specifies the materials, critical installation structures, and execution techniques necessary to ensure that the resultant façade is both energy-efficient and safe. As described by the NP 135 – 2013 standard, a ventilated façade is a façade system that includes a naturally ventilated air blade as one of its components. The air blade is located between the vented air layer and the external environment, between the outer face of the heat-insulating layer and the interior face of the building element [14].

The major function of the ventilated air layer is to maintain a steady vapor pressure difference between the outside environment and the façade system.

The major function of the ventilated air layer is to maintain a steady vapor pressure difference between the outside environment and the façade system. The ventilated façade system is composed of the following primary components (figure 1): the support structure, which consists of all the profiles and supports applied to the façade in order to create the necessary space for mounting the insulation, the air gap, and attaching the façade finish; the thermal insulation, which can be mineral wool or polystyrene; and the finish [15, 16].

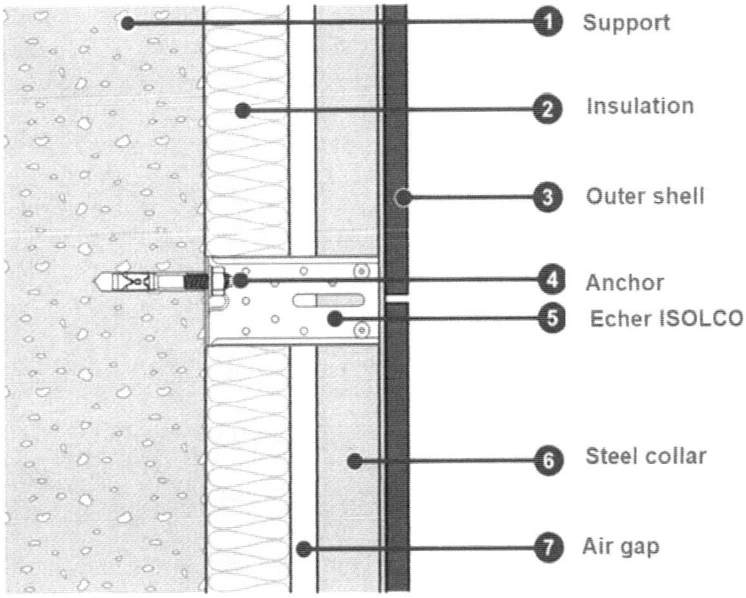


**Figure 1.** The structure of a ventilated façade [15].

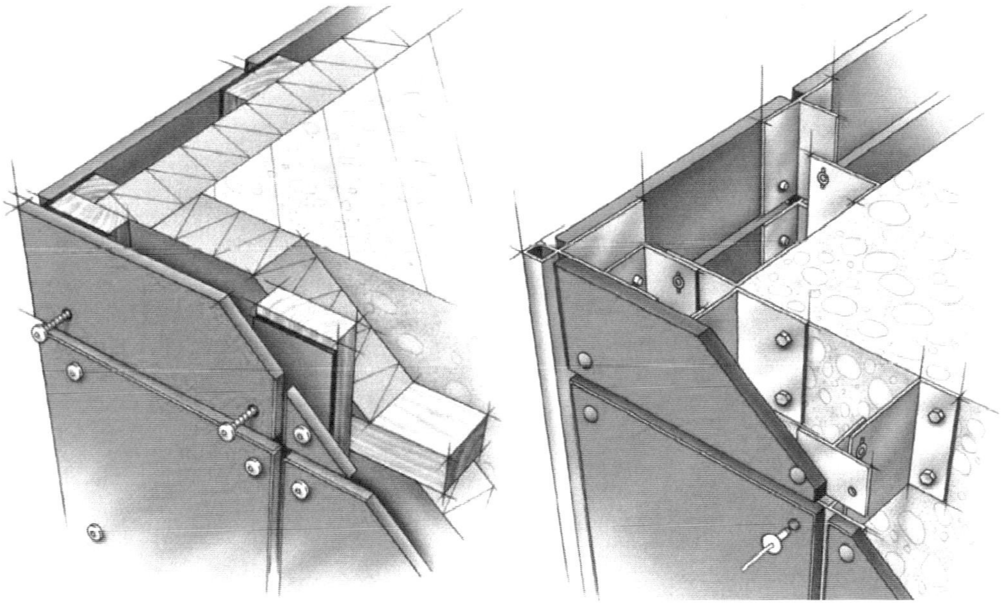

**Figure 2**. Detail ventilated façade TrespaMeteon [17].

Cladding materials are available in a broad range of compositions, physical and mechanical characteristics, chemical properties, and dimensions, and their selection is completely up to architects and designers. The supplier of cladding fastening systems will customize the mounting method for each application to meet the standards of the architectural project (figure 2). A cladding fastening system is made up of two major components: a substructure (the portion that connects to the building wall) and cladding elements (the portion that secures the plywood to the substructure), both of which may or may not be visible, resulting in visible or invisible cladding fastening systems [18].

*2.1. The substructure shall*
This substructure typically includes the following: consoles/model and size of application function with or without thermal insulation plates; fasteners for attaching consoles to the wall, such as dowels, expanders, or screws; fasteners for attaching support profiles (uprights) to consoles, such as rivets or stainless steel screws; and support profiles (uprights) that are not standardized profiles type T, L, Z, or other specialized profiles.

The attachment components for the cladding material are fairly various, including rivets and specialty screws made of non-embroiderable materials (with a head color-coordinated with the plywood), clamps, clips, unique profiles, specialty adhesives, bolts, and customized expands [19].

*2.2. The fastening systems for ventilated facades*
The systems are defined by their adaptability to a variety of cladding materials (natural and composite stone, ceramics, bond composite materials, aluminum hydroxide composite materials, metal in the form of flat and folded plates or cassettes, HPL, fibro cement, and similar materials, wood, glass, and plastic boards, etc.); adaptability to a variety of stereotypes in architectural projects; and lightweight, which facilitates manipulation.

*2.3. The structural skeleton*
This element may be built of wood, metal (steel or aluminum), or a combination of both (wood and metal). It is advised that this skeleton be separated from the support through the mounting manner on the local metal components, to provide thermal insulation behind it [18, 19].
The most frequently used systems for clamping plates to the supporting skeleton are as follows: with visible screws in the case of thin plates (aluminum plates, large cellulosic plates), or if the architectural plastic requires the rhythm of the field with point elements that may be shiny (brassed head, chrome, etc., or cover heads made of electrochemically treated metals); with masked screws in the case of complex plates or those profiled from sheet metal, with panel and h profiles.
Numerous firms specializing in the manufacturing of ceramic tiles might push the same fastening-assembly technology inside their own ventilated façade systems [20-21].

*2.4. The ventilated air layer*
The layer must be a reasonable thickness – minimum 40 mm, recommended 50 mm – except for masonry walls with air layer, where the air layer thickness may be up to 12 cm; have holes/slots connecting to the outside atmosphere located at the bottom and top of the façade, except in cases where the layer(s) located between the ventilated air layer and the outside of the façade have open joints or perforations; and if the vertical continuity of the ventilated air layer and the exterior [18-22].

*2.5. The component of external protection*
The most frequently used types of constructive-technological subassemblies for the realization of the external protection component are:

- continuous systems, typically in the form of wet plasters on perforated heat-insulating doubles or grooved plywood, finished with thin plasters;
- discontinuous systems, generally in the form of plywood mounted directly on the support wall; complex prefabricated, with thermal insulation and finish included ("vêtures" in French); thin stone plywood with mechanical fixation;
- exterior subassemblies of masonry with complex composition.

## 3. Types of ventilated façades
The facade is unique in its ability to blend aesthetic and practical characteristics. It is often the most significant aspect of the design since it sets the tone for the rest of the building. The term 'façade' refers to any predominantly vertical face of a building envelope, such as an outside wall. At times, the term 'façade' is used more specifically to refer to the outer faces of structures with a strong architectural

emphasis, such as an imposing design, ornamentation, or the building's main entrance. Advances in technology have led in the development of energy-efficient façade materials. These cutting-edge products harvest solar energy and are extensively employed as a backup source of energy in a range of commercial and residential constructions [23, 24].

The development of sustainable façade materials capable of sheltering buildings from adverse climatic conditions such as heat, wind, and rain is the main reason for their increased demand. The kind of façade you choose is critical in terms of making your building stand out both aesthetically and functionally. Product demand in commercial buildings is predicted to stay strong over the next eight years, due to the rising desire for enhanced security and ambiance in the corporate and hospitality sectors. The following sections discuss many forms of facades for building constructions [25].

The rest of this section will describe the most significant and frequently utilized forms of ventilated façades.

### *3.1. Veneer of masonry*

Brick is unmatched in terms of performance and durability. While many people expect brick as nothing more than a façade material, we compare it to other materials on almost every project in terms of cost, performance, and aesthetics. Brick often wins out over other materials owing to its durability, versatility, and familiarity (figure 3). A brick veneer, an air gap, three inches of stiff insulation, and a backup wall comprise our standard wall structure. While this design may seem simple, it offers a competitive price point in comparison to other wall systems, good durability, and excellent insulation, as well as a popular Midwest appearance. As with any material, bricks come in a variety of sizes, colors, and textures. Other types of masonry, such as stone and cement-based masonry units (CMU), may be used instead of or in addition with brick. Masonry is the most often utilized material for façades in our projects [26].

**Figure 3**. Masonry anchors are made with brick veneer [27].

### *3.2. Metal panels for walls*

Metal panels may be a viable alternative for the outside skin of a building - depending on the owner, the kind of project, and the budget (figure 4). These wall systems often spring to mind when envisioning a clean, contemporary style for a project. Metal panels provide a plethora of possibilities for achieving the desired aesthetic and performance. However, this material is often more costly than other materials and might impact the timeline of a project. Field verification and panel manufacture may have a significant influence on a project's timetable and the process of enclosing the building. It is critical to work within the manufacturer's specified panel sizes to keep costs down [26].

Insulated metal wall panels may give a better R-value than conventional encapsulated insulation wall construction. Supplemental framing may be necessary to support the façade, depending on the system and the backup wall structure. Although some manufacturers suggest that extra sheathing is not necessary, it is recommended that sheathing be included in your project to enable the structure to be enclosed and interior finishes to be put without delay.

**Figure 4**. Example of building with metal panels for walls [28].

*3.3. Exterior insulation and finish system (EIFS)*

Exterior insulation and finish system (EIFS) is a veneer system that insulates and can achieve a wide variety of looks. This material has been developed to replicate other materials in appearance. There are EIFS systems that even replicate brick and metal panels (figure 5). The system generically consists of a rigid insulation board adhered to back-up wall construction with a sprayed-on, or troweled-on, finish system. Since the material is applied as a liquid, architectural expression with details is easily achieved by carving the rigid insulation as desired. It is also very lightweight compared to other façade systems and can be used without much need for support. EIFS is the most economical system per square foot with the insulation performance it provides [26].

**Figure 5**. Example of building with EIFS [29].

The downside of EIFS is its durability. The EIFS shell over rigid insulation is thin and can be damaged easily. A puncture can occur from a flying rock from a lawnmower. Colors can sometimes fade or the surface can stain and get dirty over time. Due to these characteristics, we often use EIFS at higher parts of a building to keep it protected.

*3.4. Siding made of cementitious material (Fiber-Cement)*

Researchers are using this technique for residential construction as well as other situations when it seems to be suitable. Cementitious siding is becoming increasingly widespread in commercial buildings, and there are several producers of this material [26]. Although it most closely resembles wood lap siding, the material is also available in panels and panels that appear like lap siding or shakes. This façade technique is very durable and requires little maintenance. It is available in pre-finished or field-painted finishes. While it is lighter than stone, detailing this material with continuous insulation may be difficult. It is equivalent to EIFS in terms of cost and would give greater endurance. Insulation, on the other hand, should be considered.

**Figure 6**. Example of building with siding made of cementitious material (fiber-cement) [30].

## 4. Future research

The construction sector must embrace forward-thinking approaches that may pave the way for human-centered buildings equipped with energy and resource-efficient innovative façade technology. As such, this study intends to provide information on the state of the art in ventilated facades. This may help practitioners, traders, and industry participants better understand how to match their ideas and solutions with the social, environmental, economic, and technological realities and trends associated with ventilated facade technology.

To convert conventional facades to ventilated facades, substantial market penetration is required. This involves developing new approaches to low-impact and climate-responsive renovations via previously untapped partnerships involving sophisticated interdisciplinary design teams and creative façade engineering. Adopting a service-oriented business model is critical for the future growth of ventilated facade technology. It will benefit Europe in establishing a foundation early in the development of potential future ventilated facade technologies. The EU hopes to become the world's first climate-neutral continent by 2050 under the terms of the new Green Deal. Additionally, the European Energy Performance Building Directory promotes the relevance of comfort, smart preparedness, and high energy efficiency in the renewal of buildings. With climate change increasing the potential of overheating, the façade sector has a significant opportunity to re-establish the base for future competitiveness and development, as well as to accelerate the use of vented facade technology. The suggested classification and framework may assist players in the facade industry in comprehending the fundamental ideas of ventilated facades and concentrating their efforts on the development of important future facade technologies.

However, as shown in this paper, the study of ventilated facades is a relevant topic that will always need more research and study to properly comprehend all aspects. In this sense, the present work is a subset of wider research that is largely focused on composite materials and their application to ventilated facades.

## 5. Conclusions

The primary criterion for a building seems to be its ventilated façades, namely its outside insulation. As a result, issues about "thermal bridges" and "condensation" were avoided, resulting in a high degree of thermo-hydro-insulation for the façade. A vented façade is a "dual" façade in which one of the system's components is replaced with a naturally ventilated air blade (weak or heavily ventilated). This enables the structure to "breathe," allowing moisture to escape while preventing it from infiltrating the walls. The air blade is sandwiched between the vented air layer and the external environment between the outer face of the heat-insulating layer and the inner face of the building element (a structure that may or may not is thermally insulated, the outer face being opaque or glazed).

This study is a summary presentation with the technical requirements of a ventilated façade. As a consequence, the components of the system were highlighted and their attributes were documented. To aid in the grasp of the system's operation and design, numerous variations of ventilated facades and accompanying attributes were shown. The future researches into ventilated façades will enhance the opportunities of required market-driven development. The site-specific solutions will help this system to the nowadays challenges of climate changes.

**Acknowledgments**
This paper was elaborated with the support of the "Eco-innovative Products and Technologies for Energy Efficiency in Constructions – EFECON" research grant, project ID P_40_295/105524, Program co-financed by the European Regional Development Fund through Operational Program Competitiveness 2014-2020.

**References**

[1] European Commission 2010 The energy performance of buildings directive (EPBD) Bruxelles
[2] European Commission 2021 Proposal for a Directive of the European parliament and of the council on the energy performance of buildings
[3] International Energy Agency 2020 Global energy review ⟨http://www.iea.org⟩
[4] Manioğlu G and Yılmaz Z 2006 *Energy Build.* **38** (3) p. 266–172. http://dx.doi.org/10.1016/j.enbuild.2005.06.009
[5] Pacheco R, Ordóñez J and Martínez G 2012 *Renew. Sustain. Energy Rev.* **16** (6) p. 3559–3573. http://dx.doi.org/10.1016/j.rser.2012.03.045
[6] Badarnah L 2012 Doctoral thesis https://doi.org/10.4233/uuid:4128b611-9b48-4c8d-b52f-38a59ad5de65
[7] De Gracia A, Castell A, Navarro L, Oró E and Cabeza L F 2013 *Renew. Sustain. Energy Rev.* **22** 539–49. http://dx.doi.org/10.1016/j.rser.2013.02.029
[8] Barbosa S and Ip K 2014 *Renew. Sustain. Energy Rev.* **40** p. 1019–1029. http://dx.doi.org/10.1016/j.rser.2014.07.192
[9] Quesada G, Rousse D, Dutil Y, Badache M and Hallé S 2012 *Renew. Sustain. Energy Rev.* **16** 2820–32. http://dx.doi.org/10.1016/j.rser.2012.01.078
[10] Cerón I, Caamaño-Martín E and Neila F J 2013 *Renew. Energy* **58** p. 127–133. http://dx.doi.org/10.1016/j.renene.2013.02.013
[11] Harris D J and Helwig N 2007 *Appl. Energy* **84** p. 135–416. http://dx.doi.org/10.1016/j.apenergy.2006.07.001
[12] Leon M A and Kumar S 2007 *Sol Energy* **81** p. 62–75. http://dx.doi.org/10.1016/j.solener.2006.06.017
[13] Ibanez-Puy M, Vidaurre-Arbizu M, Sacristan-Fernandez J A and Martin-Gomez C 2017 *Renew. Sustain. Energy Rev.* **79** 180-191
[14] Al-Obaidi K M, Ismail M and Abdul Rahman A M 2014 *Front Archit Res* **3** p. 283-297 https://doi.org/10.1016/j.foar.2014.06.002
[15] Maciel A C F and Carvalho M T 2019 *J Build Eng* **25**
[16] Pujadas-Gispert E, Alsailani M., van Dijk K C A, Rozema A D K, ten Hoope J P, Korevaar C C

and Moonen S P G 2020 *Front Archit Res* **9** (3) p. 681-696
[17] Trespa 2020 https://www.trespa.com/en_GB/splash-screen?returnUri=%2Fen_GB%2Fproducts)
[18] Alonso C, Oteiza I, García-Navarro J and Martín-Consuegra F 2016 *Three Case Stud Madrid Energy Build* **126** p. 252-262
[19] Ascione F 2017 *Sol. Energy* 1**54** p. 34-100
[20] Krateur S, Salhi M J, Schroer S and Hanitsch R 2001 *International IBSPA Conference on Building Simulation*, Rion de Janeiro, Brazil
[21] Barabash A, Naumova E, Zhouvak O, Nemova D and Olshevskiy V 2016 *MATEC Web of Conference* **73** 02006
[22] Šadauskienė J, Ramanauskas J and Vasylius A 2019 *Thermal Sciences* **24** 299
[23] Sinclair R, Phillips D and Mizhibovski V 2009 *Ashrae Journal* **51** p. 16-27
[24] Bannier E, Catavella V, Pinazo V, Soto M and Moreno G S 2019 *QUALICER 2012,* Castellon, Spain
[25] Andaloro A, Gasparri E and Lucchini A 2013 *IAHS 2013*, Milano, Italy
[26] Schmidt-Arch 2018 (https://schmidt-arch.com/5-common-facade-systems-materials/)
[27] Constructionspecifier 2014 (https://www.constructionspecifier.com/durability-of-brick-veneer-a-deeper-look-at-masonry-anchors/)
[28] Bowles K 2019 (https://newtechmachinery.com/learning-center/what-are-metal-wall-panels/)
[29] Constrofacilitator 2021 (https://www.constrofacilitator.com/different-types-of-facades-for-commercial-building/)
[30] Architizer Editors 2021 (https://architizer.com/blog/product-guides/product-guide/eaktna-fiber-cement-cladding/)

# Study of the additives for polymeric materials by cheminformatics methods

**A Bărbulescu[1], L Barbes[2], and C S Dumitriu[1*]**

[1] Transilvania University of Brașov, Romania
[2] Ovidius University of Constanța, Romania

* cristian.dumitriu@unitbv.ro

**Abstract.** This article analyzes the structures of some antioxidants, UV stabilizers, and inhibitors utilized to improve the polymeric materials' functionalities using Cheminformatics tools implemented in the R software. Based on their properties, the dendrograms are drawn, and the molecules are classified in clusters, utilizing hierarchical clustering and the k-means algorithm.

## 1. Introduction

Plastics are essential materials for sustainable development and play an essential role in achieving ambitious strategies for harmonizing the sustainable development of contemporary society and various ecological projects. Developing competitive and high-performance products through efficient, long-term, and environmentally friendly solutions can be highly complex and challenging. Plastics represent a vast category of unique and versatile materials, widely used in industry and our daily lives, a fact that has led to a rapid increase in global production and market demand of 50.7 million tons only in Europe, recorded in 2019 until the onset of the COVID-19 pandemic crisis [1-3].

In the production process, the polymers are mixed with various additives and/or incorporated into technological adjuvants for increasing their performance and efficient use. The binding between these additives and the polymers is not usually covalent [4]. The additives can be antistatic agents, antioxidants, lubricants, plasticizers, photoinitiators, thermal stabilizers, UV stabilizers, etc.

Antioxidants can be synthetic or natural compounds capable of removing the formed free radicals and inhibiting the oxidation processes. UV stabilizers can absorb a significant part of the UV light and diminish the absorption of radiation in the matrix of the polymer. Photoinitiators contribute to improving the polymers' performances, for example, increasing the safety of the materials utilized for packaging the food.

Extreme temperatures, excessive humidity, exposure to UV radiation, mechanical stress contribute to the degradation of the characteristics of the polymeric material and can significantly decrease its performance. To increase the lifespan of the polymeric structures, different techniques have been proposed and implemented, like adding additives in the plastic or composite matrix. Still, it is necessary to improve the chemical compatibility and stability of these components' dispersion and increase the mixing efficiency when required.

Under the action of UV light, polymeric materials can suffer degradation or photooxidation, leading to increased yellowing or brittleness in the structure. Additives utilized for decreasing ultraviolet light effects include chromophores which absorb the UV. Photons absorbed by light are generally transformed to an excited electronic state and lose energy through phosphorescence or fluorescence.

Finding similarities of different molecules became a study field in the 90s [5], but they have been used mostly in drugs classification. Although many researchers investigated various chemical structures of antioxidants, UV stabilizers, or photoinitiators [3, 6-11] for polymeric materials, the classification of these molecules based on their components discovered using the new technologies attracted interest. Utilizing the capabilities of the latest software, this can be easily done. The rcdk, ChemmineR, and rpubchem packages of R are powerful tools in Chemoinformatics that help the oriented R user efficiently organize information [12-16].

This article proposes using the R software capabilities to emphasize the structural characteristics of the most important additives employed in the polymeric material composition. Clustering has also been done for assessing the molecules' degree of similarity.

## 2. Methodology

The structures of the molecules that were subject to the present study have been downloaded from PubChem (https://pubchem.ncbi.nlm.nih.gov.), where they are freely available. They are three antioxidants, nine UV stabilizers, and two photoinitiators, whose chemical names and abbreviations are presented in Table 1.

**Table 1.** Additives used in the composition of the polymeric materials.

| Type | Abbreviation | Chemical name |
|---|---|---|
| Antioxidant | BHA | Butylated hydroxyanisole |
| | BHT | 2,6-Di-tert-butyl-4-methylphenol |
| | Irganox 1010 | Pentaerythritol(tetrakis)3,5-di-tert-butyl-4-hydroxycinnamate |
| | Irganox 1076 | Octadecyl 3-(3,5-di-tert-butyl-4-hydroxyphenyl)propionate |
| UV stabilizer | UV-0 | 2,4-Di-hydroxybenzophenone |
| | UV-9 | 2-Hydroxy-4-methoxybenzophenone |
| | UV-24 | 2,2'-Dihydroxy-4- methoxybenzophenone |
| | UV-234 | 2-(2H-Benzotriazol-2-yl)-4,6-bis (1-methyl-1-phenylethyl) phenol |
| | UV-326 | (2'-Hydroxy-3'-tert-5'-methylphenyl)-5-chlorobenzotriazole |
| | UV-327 | 2-(2'-Hydroxy-3',5'-di-tert-butylphenyl)-5-chlorobenzotriazole |
| | UV-531 | 2-Hydroxy-4-n-octoxy-benzophenone |
| | UV-1577 | o-Toluidine |
| | UV-P | 2-(2'-Hydroxy-5'-methylphenyl)-benzophenone |
| Photoinitiator | UV-184 | 1-Hydroxycyclohexylphenylketone |
| | UV-651 | 2,2-Dimethoxy-2-phenylacetophenone |

The analysis has been performed employing the R software, which is free to download from https://cran.r-project.org/. ChemmineR, chemometrics, fingerprint, fmcsR, iqspr, rcdk, rgl, and vegan packages were installed for performing the structural analysis. The packages cluster, factoextra, ggplot2, gridExtra, NbClust were utilized for clustering and visualizing the structures and clusters. The readers interested in using these packages may download the vignettes from the mentioned website.

The characteristics of the molecules we are interested in are the molar weight (MW), the molar refractivity (MR), the atoms types and numbers, the types and number of functional groups, the Hydrogen Bond Acceptors and Donors (HBA, HBD), hydrogen bonds (logP), and the topological polar surface area (TPSA). The hydrogen bonds (logP) characterize the substrates (the compounds' lipophilicity). TPSA gives information on the hydrogen bond formation and is the area related to N, O, and polar hydrogen atoms [17-22].

The atoms-pair similarity was computed by using the Tanimoto index [23] implemented in the ChemmineR package. The compound similarities were assessed by a similarity metric implemented in the fmscR package. Finally, clustering the molecules has been performed by the hierarchical clustering and k-means algorithms implemented in the NbClust package.

## 3. Results and Discussion

Before starting the analysis of the structures, the ID numbers have been retrieved from PubChem using the command cid (sdfset), and they have been visually inspected by accessing them with the command plot(sdf[],print=FALSE). The IDs of the structures are 24667, 31404, 64819, 16386, 8572, 4632, 8569, 70355, 112412, 62531, 77470, 15797, 90571, 4992761, and 17113, respectively. The molecular structure for UV-326 (ID 112412) is shown in figure 1.

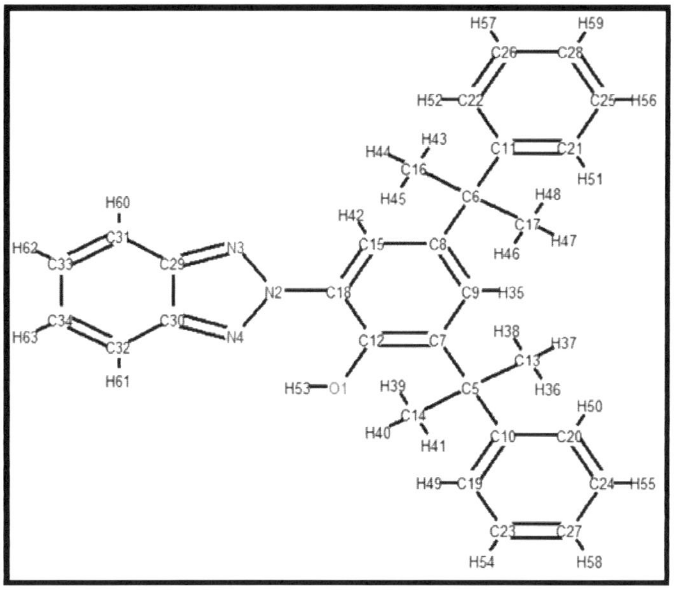

**Figure 1.** The structure of UV-326 (ID 112412).

The molecular weight (MW), the molecular formula (MF), the count of atoms, functional groups, and rings have been obtained and combined in one data frame using the propma command. Table 2 displays these data. Notice that only the functional groups with the number of elements different from zero for at least one molecule are displayed. Remark the difference between Irganox 1010 and Irganox 1076, on the one hand, and the other molecules, on the other hand (the first two molecules have a significantly higher number of carbon and hydrogen atoms).

Utilizing the CheminneOB library, one can compute Hydrogen Bond Acceptors and Donors, the molar refractivity, TPSA, and logP. The values for the study molecules are presented in Table 3.

The similarity of the atom pairs the cmp.search function. The highest atoms-pair similarities have been found (similarity value in the brackets) between the following molecules: UV-9 – UV-24 (0.7622), UV-9 – UV-0 (0.6954), UV-327 – UV-531 (0.6646). To determine all rings of the structure and the information on aromaticity, the ring command has been utilized.

**Table 2.** Molecules' characteristics retrieved by using *propma* command in R.

| Abbreviation | ID | MF | MW | C | H | O | N | Cl | ROH | RCOR | RCOOR | ROR | Rings | Aromatic |
|---|---|---|---|---|---|---|---|---|---|---|---|---|---|---|
| BHA | 24667 | C22H32O4 | 360.48 | 22 | 32 | 4 | 0 | 0 | 2 | 0 | 0 | 2 | 2 | 2 |
| BHT | 31404 | C15H24O | 220.35 | 15 | 24 | 1 | 0 | 0 | 1 | 0 | 0 | 0 | 1 | 1 |
| Irganox 1010 | 64819 | C73H108O12 | 1177.63 | 73 | 108 | 12 | 0 | 0 | 4 | 0 | 4 | 0 | 4 | 4 |
| Irganox 1076 | 16386 | C35H62O3 | 530.86 | 35 | 62 | 3 | 0 | 0 | 1 | 0 | 1 | 0 | 1 | 1 |
| UV-0 | 8572 | C13H10O3 | 214.21 | 13 | 10 | 3 | 0 | 0 | 2 | 1 | 0 | 0 | 2 | 2 |
| UV-9 | 4632 | C14H12O3 | 228.24 | 14 | 12 | 3 | 0 | 0 | 1 | 1 | 0 | 1 | 2 | 2 |
| UV-24 | 8569 | C14H12O4 | 244.24 | 14 | 12 | 4 | 0 | 0 | 2 | 1 | 0 | 1 | 2 | 2 |
| UV-234 | 70355 | C13H16O2 | 204.26 | 13 | 16 | 2 | 0 | 0 | 1 | 1 | 0 | 0 | 2 | 1 |
| UV-326 | 112412 | C30H29N3O | 447.57 | 30 | 29 | 1 | 3 | 0 | 1 | 0 | 0 | 0 | 5 | 5 |
| UV-327 | 62531 | C17H18ClN3O | 315.79 | 17 | 18 | 1 | 3 | 1 | 1 | 0 | 0 | 0 | 3 | 3 |
| UV-531 | 77479 | C20H24ClN3O | 357.87 | 20 | 24 | 1 | 3 | 1 | 1 | 0 | 0 | 0 | 3 | 3 |
| UV-1577 | 15797 | C21H26O3 | 326.42 | 21 | 26 | 3 | 0 | 0 | 1 | 1 | 0 | 1 | 2 | 2 |
| UV-P | 90571 | C16H16O3 | 256.29 | 16 | 16 | 3 | 0 | 0 | 0 | 1 | 0 | 2 | 2 | 2 |
| UV-184 | 4992761 | C27H27N3O2 | 425.52 | 27 | 27 | 2 | 3 | 0 | 1 | 0 | 0 | 1 | 4 | 4 |
| UV-651 | 17113 | C13H11N3O | 225.24 | 13 | 11 | 1 | 3 | 0 | 1 | 0 | 0 | 0 | 3 | 3 |

**Table 3.** The descriptors of the molecules

| Abbreviation | ID | HBA1 | HBA2* | HBD | logP | MR | TPSA |
|---|---|---|---|---|---|---|---|
| BHA | 24667 | 36 | 4 | 2 | 5.3966 | 108.4540 | 58.92 |
| BHT | 31404 | 25 | 1 | 1 | 4.2956 | 71.9710 | 20.23 |
| Irganox 1010 | 64819 | 120 | 12 | 4 | 15.8792 | 348.9130 | 186.12 |
| Irganox 1076 | 16386 | 65 | 3 | 1 | 10.7245 | 169.3960 | 46.53 |
| UV-0 | 8572 | 13 | 3 | 2 | 2.3288 | 60.3625 | 57.53 |
| UV-9 | 4632 | 15 | 3 | 1 | 2.6318 | 64.8315 | 46.53 |
| UV-24 | 8569 | 16 | 4 | 2 | 2.3374 | 66.8545 | 66.76 |
| UV-234 | 70355 | 18 | 2 | 1 | 2.5645 | 59.7573 | 37.30 |
| UV-326 | 112412 | 32 | 4 | 1 | 6.7779 | 138.7880 | 50.94 |
| UV-327 | 62531 | 21 | 4 | 1 | 4.3854 | 90.1340 | 50.94 |
| UV-531 | 77470 | 27 | 4 | 1 | 5.3745 | 104.4380 | 50.94 |
| UV-1577 | 15797 | 29 | 3 | 1 | 5.3625 | 98.4805 | 46.53 |
| UV-P | 90571 | 19 | 3 | 0 | 3.0151 | 72.7905 | 35.53 |
| UV-184 | 4992761 | 32 | 5 | 1 | 6.5373 | 128.6850 | 68.13 |
| UV-651 | 17113 | 14 | 4 | 1 | 2.4345 | 65.8540 | 50.94 |

*Note: HBA2 is computed as the sum of the elements in the column O and N in Table 2

For the structure in figure 1, the following result has been obtained:
ring1: "N_4" "N_2" "N_3" "C_29" "C_30" – aromatic
ring2: "C_19" "C_23" "C_27" "C_24" "C_20" "C_10" – aromatic
ring3: "C_21" "C_25" "C_28" "C_26" "C_22" "C_11" – aromatic
ring4: "C_12" "C_7" "C_9" "C_8" "C_15" "C_18" – aromatic
ring5: "C_29" "C_31" "C_33" "C_34" "C_32" "C_30" – aromatic
ring6: "N_4" "N_2" "N_3" "C_29" "C_31" "C_33" "C_34" "C_32" "C_30" – aromatic

The last step in collecting the molecules' characteristics was to assess their similarity, using the

Tanimoto and overlap coefficients as tools. The results of the similarity analysis for the first molecule (BHA) and the other molecules are presented in Table 4, where MCS denotes the maximum common structure. For example, there is a perfect similarity between that molecule and itself (first row). The similarity results of BHA and BHT are (the second row): the numbers of atoms are 26 and 16, respectively; the maximum common structure size is 16 atoms, the overlap percentage of these structures is 68.75%, while the Tanimoto coefficient is 0.3548. The higher the overlap and Tanimoto coefficients are, the higher the similarity is.

Based on the above results, hierarchical clustering has been used to classify the molecules. The dendrogram is presented in figure 2.

**Table 4.** Similarity descriptors for BHA and the study molecules.

| Molecule | Query_ Size | Target_ Size | MCS_ Size | Tanimoto_ Coefficient | Overlap_ Coefficient |
|---|---|---|---|---|---|
| BHA | 26 | 26 | 26 | 1.0000 | 1.0000 |
| BHT | 26 | 16 | 11 | 0.3548 | 0.6875 |
| Irganox 1010 | 26 | 85 | 11 | 0.1100 | 0.4230 |
| Irganox 1076 | 26 | 38 | 11 | 0.2075 | 0.4230 |
| UV-0 | 26 | 16 | 9 | 0.2727 | 0.5625 |
| UV-9 | 26 | 17 | 9 | 0.2647 | 0.5294 |
| UV-24 | 26 | 18 | 9 | 0.2571 | 0.5000 |
| UV-234 | 26 | 15 | 8 | 0.2424 | 0.5333 |
| UV-326 | 26 | 34 | 11 | 0.2245 | 0.4231 |
| UV-327 | 26 | 22 | 11 | 0.2973 | 0.5000 |
| UV-531 | 26 | 25 | 11 | 0.2750 | 0.4400 |
| UV-1577 | 26 | 24 | 9 | 0.2195 | 0.3750 |
| UV-P | 26 | 19 | 8 | 0.2162 | 0.4211 |
| UV-184 | 26 | 32 | 8 | 0.1600 | 0.3077 |
| UV-651 | 26 | 17 | 7 | 0.1944 | 0.4117 |

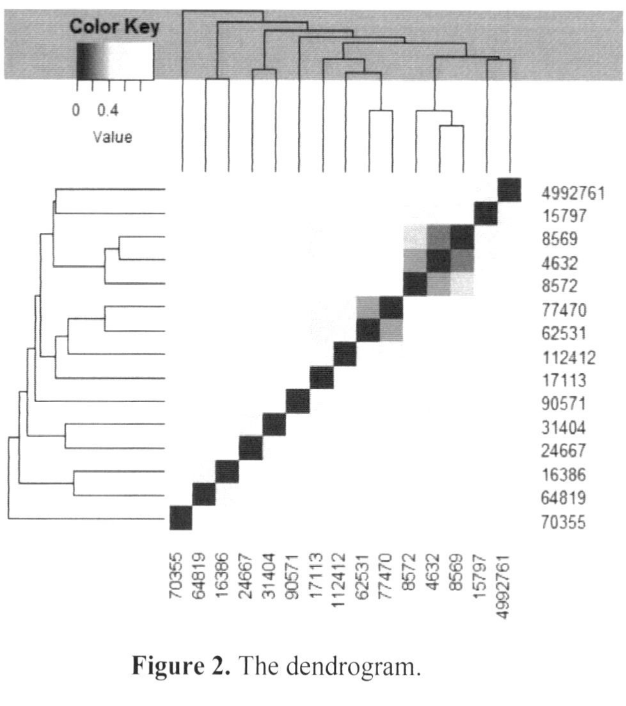


**Figure 2.** The dendrogram.

The darkest the color is, the highest the similarity is. Apart from the perfect superposition (the second diagonal), the highest similarities are those of the molecules with IDs 4632 and 8569 (UV-9 and UV-24); 77479 and 62531 (UV-531 and UV-327); 8572 and 4632 (UV-9 and UV-0).

The k-means algorithm has been used for grouping the molecules in clusters. The selection of the number of clusters has been done based on the minimum distance inside the clusters and the maximum distance between the clusters. The optimum number has been detected to be three. Within cluster sums of squares by clusters were 3.1994, 0.9880, and 3.760. The ratio between the sum of squares and the total sum of squares =39.2%. The clusters are represented in figure 3.

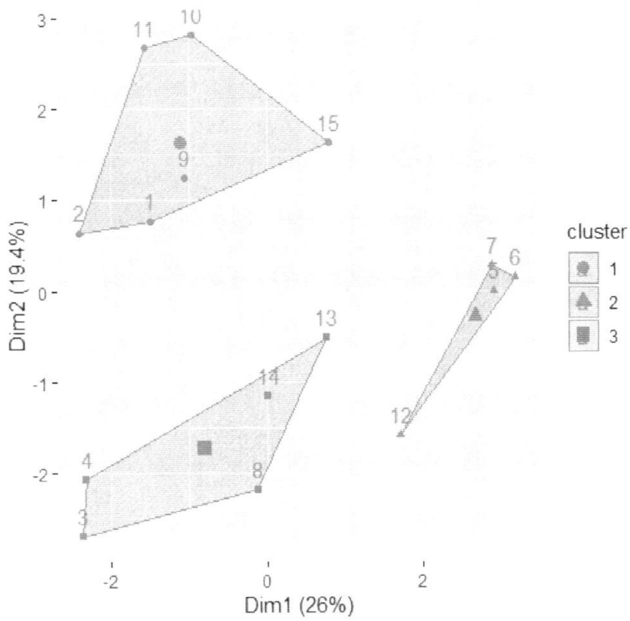


**Figure 3.** The clusters.

The first cluster contains the molecules BHA, BHT, UV-326, UV-327, UV-531, and UV-651; the second one - UV-0, UV-9, UV-24, UV-1577, whereas the last cluster is formed by the rest of the molecules. The clusters are non-overlapping, showing good clustering.

## 4. Conclusion

In this article, we determined the structural properties of 15 additives used to improve the qualities of composite polymeric materials. The structures can be successfully characterized by using the R packages specially designed for such analyses. Clustering the molecules is a first step in the analysis of the correlation between the chemical structure and their functional characteristics, which will be done as future work.

## References

[1] Barbeş L and Dumitriu C S 2021 *Polymeric Composite Materials* (Chisinau: Lambert Academic Publishing)
[2] Emoyan O O, Tesi G O, Ohwo E and Odali E W 2021 *Ovidius Univ. Ann. Chem* **32(1)** 53–62
[3] PlasticsEurope. Plastics-The Facts 2020: An Analysis of European Plastics Production, Demand and Waste Data, 2020. https://www.plasticseurope.org/en/resources/publications/4312-plastics- facts-2020 (accessed on 8 August 2021)
[4] Schyns Z O G and Shaver M P 2020 *Macromol Rapid Comm* **42(3)** 2000415
[5] Willett P 2011. Similarity Searching Using 2D Structural Fingerprints *Chemoinformatics and*

IOP Conf. Series: Materials Science and Engineering 1242 (2022) 012003 doi:10.1088/1757-899X/1242/1/012003

*Computational Chemical Biology, Methods in Molecular Biology* (*Methods and Protocols*), vol 672 ed J Bajorath (Totowa: Humana Press) pp. 133-158
[6] Hahladakis J N, Velis C A, Weber R, Iacovidou E and Purnell 2018 *J Hazard Mater* **344** 179-99
[7] Pérez-Albaladejo E, Solé M and Porte C 2020 *Curr Opin Toxicol* **20 –2**, 69-76
[8] Brostow W, Lu X, Gencel O and Osmanson A T 2020 *Materials* **13(7)** 16262020
[9] Chan-Park M B, Zhu A P, Shen J Y and Fan A L 2004 *Macromol Biosci* **4(7)** 665-73
[10] Mohammed A D and Hassan Y 2020 *Ovidius Univ. Ann. Chem* **31(1)** 9 – 13
[11] Bărbulescu A, Barbeș L and Dumitriu C Ș 2021 *J Chem* **2021** 4816970
[12] Cao Y, Charisi A, Cheng L C, Jiang T and Girke T 2008 *Bioinformatics* **24(15)** 1733-4
[13] Guha R and Cherto M R 2017 rcdk: Integrating the CDK with R. http://cran.nexr.com/web/packages/rcdk/vignettes/rcdk.pdf
[14] Guha R, Gilbert K, Fox G. Pierce M, Wild D and Yuan H 2010 *Curr Comput - Aid Drug* **6(1)** 50-67
[15] Mente S and Kuhn M *Curr Top Med Chem* **12(18)** 1957–64
[16] Wang Y, Backman T W, Horan K and Girke T 2013 *Bioinformatics* **29(21)** 2792 – 4
[17] Caron G, Vallaro M and Ermondi G 2018 *Drug Discov Today: Technol* **27** 65-70
[18] Cuesta S A, Mora J R and Márquez E A 2021 *Molecules* **6** 1100
[19] Ertl P, Rohde B and Selzer P 2000 *J Med Chem*, 3714 – 7
[20] Patrick G L 1995 *An Introduction to Medicinal Chemistry* (Oxford: Oxford University Press)
[21] Vistoli G and Pedretti A 2007) Molecular Fields to Assess Recognition Forces and Property Spaces *Comprehensive Medicinal Chemistry II* vol 5 eds J B Taylor and D J Triggle (Elsevier Science) pp. 577-602
[22] Voicu A, Duteanu N, Voicu M, Vlad D and Dumitrascu V 2020 *J Cheminformatics* **12(1)** 3
[23] Bajusz D, Rácz A and Héberger K 2015 *J Cheminformatics* 7 20

# Monthly precipitation field generation at Sulina (Romania)

**A Bărbulescu**[1*]

[1] Transilvania University of Brașov, Romania

* alina.barbulescu@unitbv.ro

**Abstract.** Nowadays, climate change is one of the most important threads for civilization. Having the origin mainly in the anthropic activity and intensive use of the environmental resources, climate change affects the ecosystems and the population lives. Changes in precipitation volume and cycles severely affect agriculture and food security. Therefore, building meteorological forecasting is important for planning agricultural works and water management. In this respect, this article attempts to create an image of the future precipitation evolution in the northern part of Dobrogea, a region more and more affected by extreme meteorological events - long drought periods, followed by high precipitation amounts.

## 1. Introduction

Precipitation is an essential factor in the water cycle in nature that impacts human life and activity and the ecosystems' existence and functioning. To make a realistic plan of water resources necessity and use, one should rely on precipitation forecast as correct as possible. Rainfall simulation can be done by stochastic procedures and numerical weather prediction models [1,2]. Multivariate time series analysis and precipitation generators are used for emphasizing the series characteristics and for the forecast at a single site or multiple locations [3-9].

This research used stochastic procedures since it requires much lower computational effort and fewer input data.

Since annual precipitation generation has lower practical importance, the majority of scientists focussed on the methods that synthetically generate the monthly precipitation series, of which: autoregressive processes (AR), Thomas-Fiering model (TF), fragment method (F), and its modified version (MF), two-tier model (TT), wavelets, etc. [10,11]. The fragment method, introduced by Srikanthan and McMahon [12], involves disaggregating the annual precipitation generated by an autoregressive model of the first order. Porter and Pink [13] proposed to obtain the monthly fragments from a monthly rainfall sequence, but the method does not keep the correlation between the first month of the current year and the last month of the previous one. Maheepala and Perera [14] modified the previous procedure, resulting in maintaining the monthly correlation over consecutive years. A significant drawback of this new approach is that the serial correlation of the first month of a year and the last month of the previous year is not maintained, which can be avoided considering that the hydrological year starts with a month with minimal serial correlation. Srikanthan and McMahon [15] observed the repetition of the same annual pattern when a limited number of annual data was available. To remove this inconvenience, Srikanthan *et al.* [16] introduced the synthetic fragments methods and applied the procedure to generate monthly precipitation at ten sites in Australia. It seems that it gives satisfactory results if there is no month without precipitation.

This article contains the results of generating monthly precipitation fields at Sulina, Romania, because such an investigation was never performed for the Dobrogea region (where Sulina is situated), even if the studies of climatic factors are presented in several scientific works [17-22]. We consider that this study will fill a gap in the knowledge on the weather pattern in the south-eastern part of Romania.

## 2. Data series and Methodology

The study series we are working with were recorded at the hydrometeorological station Sulina, Romania (figure 1), for 41 years. They are monthly precipitation series (figure 2) without gaps. The raw data were downloaded from https://www.ecad.eu/and processed according to obtain monthly series.

**Figure 1.** Map of Dobrogea region and Sulina (https://www.info-delta.ro/ghid-59/).

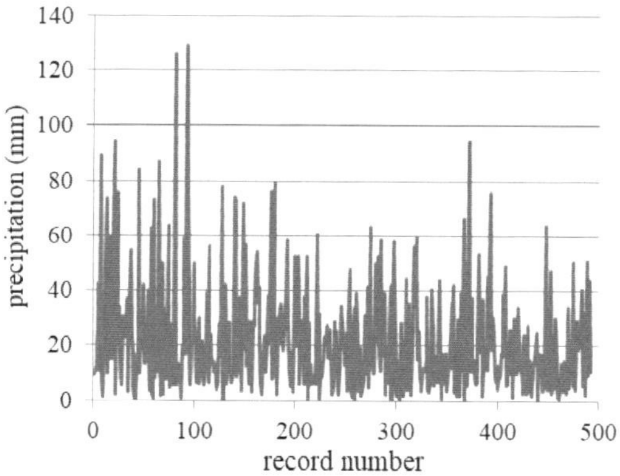


**Figure 2**. Sulina monthly precipitation series.

Sulina station has a particular position, in the Natural Reserve Danube Delta, 12 km off-shore (+ 45:04 latitude, + 29:39 longitude, and + 2:08 altitude). Maintaining an almost stable climate in the region is essential for species conservations and related economic activities and tourism.

The preliminary modeling step is the statistical analysis of the series properties, namely performing normality, correlation, homoscedasticity, stationarity, and breakpoint tests. For this aim, we performed the Shapiro-Wilks, runs, Levene, KPSS, and Hubert tests for data series [23].

We also tested the hypothesis of the existence of a monotonic trend by the Mann-Kendall, and seasonal Mann-Kendall followed by the Sen's slope computation [23]

The statistical distribution of the monthly precipitation series has been determined and tested by using the Kolmogorov-Smirnov goodness of fit test [23].

The tests were performed at a significance level of 5%, using the R software and Khronostat.

Next, the MMF algorithm was used to generate the precipitation field. The steps are the following:

a. Monthly data is generated using the Thomas-Fiering model [25], described by the equation:

$$x_{i+1} = \bar{x}_{j+1} + b_j(x_i - \bar{x}_j) + t_i s_{j+1}\sqrt{1-r_j^2}, \tag{1}$$

where $x_i, x_{i+1}$ are the rainfall generated for the months $i$ and $i$+1, $\overline{x_j}, \overline{x_{j+1}}$ are the average precipitations for the months $j$ and $j$+1, $b_j$ is the regression coefficient, $t_i$ is a Gaussian random variable with the mean zero and standard deviation 1, $s_{j+1}$ is the standard deviation of the precipitation in the month $j$+1, $r_j$ is the correlation coefficient between the precipitations recorded in the months j and $j+1$. Since rainfall is not always normally distributed, a random variable $t_i$ that follows a gamma distribution with three parameters could be used instead of the Gaussian one.

b. Annual data is generated. For this aim, it is possible to the Thomas-Fiering model by replacing the monthly data with the annual ones.

c. The monthly data are standardized, dividing the monthly precipitation from one year by the corresponding annual precipitation. Therefore, the sum of the standardized monthly precipitations in a year is 1. The result of this step is a set of fragments of monthly precipitations (equal to the number of years).

d. The annual data generated at the second step is disaggregated, employing the fragments generated at the first step.

The monthly fragments corresponding to a specific year, $k$, are selected considering how close the generated annual precipitation is to the monthly precipitation from the last month of the previous year of the already disaggregated data corresponding to the historical ones [14]. This is done by selecting the monthly fragments of the year $i$ in the generated monthly series that minimizes

$$\alpha_1 = \left(\frac{x'_k - x_i}{s_x}\right)^2 + \left(\frac{y'_{k-1} - y_{i-1}}{s_y}\right)^2 \tag{2}$$

where $x'_k$ is the annual precipitation generated for the year $k$, $x_i$ is the annual precipitation recorded in the year $i$, $s_x$ is the standard deviation of the annual precipitation, $y'_{k-1}$ is the monthly disaggregated precipitation from the last month of the year $k-1$, $y_{i-1}$ is the historical value for the last month of the year $i-1$ and $s_y$ is the standard deviation of the annual precipitation of the last month of the year.

The annual precipitations generated are disaggregated by multiplying the generated precipitation by each of the twelve fragments to obtain the monthly precipitations.

For modeling purposes, we utilized Stochastic Climate Library (SCL).

The tolerances used when generating the monthly models are listed in Table 1.

**Table 1.** Tolerances used in monthly models [11].

| Statistics | Tolerance |
|---|---|
| Average (%) | 7.5 |
| Standard deviation | 7.5 |
| Skewness | 0.75 |
| First order correlation coefficient | 0.2 |
| Maximum, minimum (%) | 10 |
| The proportion of the months without precipitation (%) | 5 |

Since the standard deviations of the monthly series were not always kept in the tolerance limits in the previous approach, we also used the procedure described in [26,27] to generate the precipitation field for the study data. The stages are the following:

a. Generating the annual series by using a symmetric moving average method;
b. Building the autocorrelation function for the annual series,

$$\gamma_k = \gamma_0(1 + hk\beta)^{-1/\beta}, \tag{3}$$

where $h$ and $\beta$ are parameters to be determined.

c. Generate the monthly series by a seasonal autoregressive model PAR(1), that has the equation

$$X^m = A^m X^{m-1} + B^m V^m, \tag{4}$$

where $X^m$ and $X^{m-1}$ are matrices that contain the monthly data, $V^m$ contains the covariances, $A^m$ and $B^m$ are the matrices of coefficients, depending on the month.

d. Computation of the monthly series by disaggregation.

A Monte Carlo simulation was performed to minimize the average of the sum of the absolute values of the variance coefficients.

Castalia software was used to run this algorithm.

## 3. Results and Discussion

The monthly precipitation varied between 0 and 129 mm during the study period, with an average of 21.80 mm, a median of 17.17 mm, a standard deviation of 19.55 mm, and a variation coefficient of 89.68%. The high variation coefficient and the right-skewed distribution of the monthly precipitation indicate a high variability of this series.

For the monthly data, the statistical tests provided the results presented in the following.

The Shapiro - Wilks test rejected the normality hypothesis. After a Box-Cox transformation with $\lambda = 0.34$, the null hypotheses couldn't be rejected.

The runs test rejected the null hypothesis of randomness. The autocorrelation existence was confirmed by the autocorrelation function. The Levene test rejected the homoskedasticity hypothesis. The Hubert test detected a change point in August 1982.

After performing the KPSS test, we rejected the hypothesis that the series is stationary in level, but we didn't find enough evidence that the series is not stationary in trend.

Performing the Mann-Kendall test for the entire series, we did not find enough evidence to reject the hypothesis that the series does not present a monotonic trend. The Sens's slope computation returned a value of -0.0145, with a p-value of 0.0009825. So the slope is significant, and the series trend is decreasing.

The same test has been performed, taking into account the data seasonality. The p-value returned was 0.0002608<0, so the existence of a monotonic trend can not be rejected. In this case, the seasonal Sen's Slope is -0.187451, a decreasing precipitation trend is noticed.

The simulated series, at the 2.5%, 50%, and 97.5% levels, and the monthly observations, are presented in figure 3.

Table 2 contains a part of the results related to the generated precipitation field at Sulina (mean, standard deviation, maximum and minimum precipitation) for each month. Yes/No in the last column means that the mean value (the fourth column) is or is not in the tolerance limits compared to the historical values (observation – the third column).

Figures 4-7 illustrate the output of generating precipitation fields by the second algorithm.

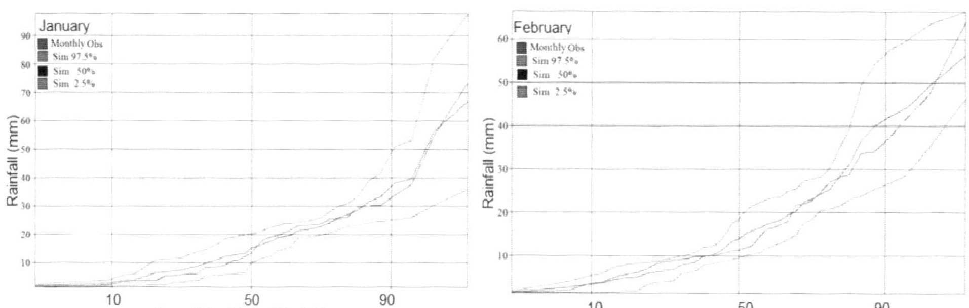


**Figure 3.** Simulated series

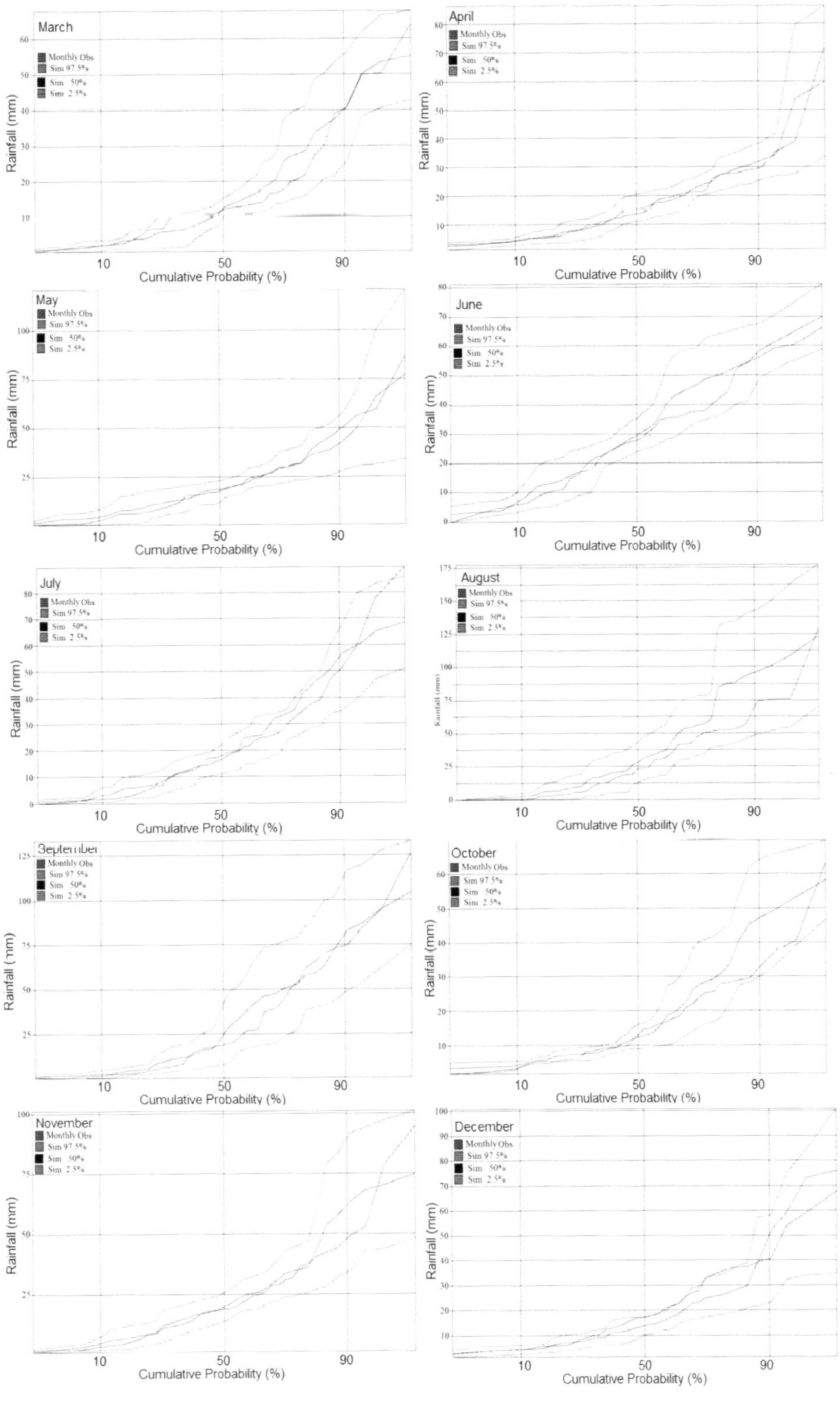


**Figure 3.** Simulated series (cont.).

**Table 2.** Results of generating precipitation field at Sulina

| Month | Stat | Observations | Mean | 2.50% | 50% | 97.50% | Yes/No |
|---|---|---|---|---|---|---|---|
| Jan | Mean | 17.193 | 17.245 | 13.298 | 16.904 | 23.587 | Yes |
| | St.dev | 14.996 | 13.358 | 86.60 | 123.30 | 308.10 | No |
| | Max | 73.7 | 58.51 | 34.097 | 52.24 | 125.38 | No |
| | Min | 1.5 | 1.515 | 1.269 | 1.446 | 2.864 | Yes |
| Febr | Mean | 16.937 | 17.171 | 13.723 | 17.014 | 22.173 | Yes |
| | St.dev | 13.619 | 14.297 | 11.019 | 14.326 | 19.074 | Yes |
| | Max | 63.9 | 59.128 | 44.44 | 60.845 | 66.7 | Yes |
| | Min | 1.5 | 1.582 | 1.427 | 1.513 | 1.992 | Yes |
| Mar | Mean | 15.883 | 16.561 | 11.378 | 16.576 | 21.958 | Yes |
| | St.dev | 14.36 | 15.903 | 11.049 | 16.117 | 19.797 | No |
| | Max | 63.9 | 60.214 | 41.564 | 62.769 | 67.792 | Yes |
| | Min | 0.3 | 0.52 | 0.253 | 0.298 | 2.035 | No |
| Apr | Mean | 17.022 | 17.24 | 12.916 | 17.13 | 21.368 | Yes |
| | St.dev | 12.778 | 12.509 | 8.005 | 12.789 | 18.953 | Yes |
| | Max | 71.9 | 61.483 | 32.163 | 74.366 | 88.376 | No |
| | Min | 3.4 | 3.288 | 0.914 | 3.566 | 4.254 | Yes |
| May | Mean | 22.163 | 20.945 | 14.34 | 20.488 | 28.668 | Yes |
| | St.dev | 18.069 | 14.696 | 9.212 | 12.659 | 28.326 | No |
| | Max | 86.7 | 68.421 | 33.303 | 53.715 | 130.359 | No |
| | Min | 0.7 | 1.034 | 0.269 | 0.659 | 5.796 | No |
| June | Mean | 28.988 | 31.087 | 24.469 | 30.86 | 40.929 | Yes |
| | St.dev | 16.853 | 18.401 | 13.942 | 18.23 | 23.823 | No |
| | Max | 66.1 | 69.845 | 57.95 | 69.072 | 88.802 | Yes |
| | Min | 0 | 1.5 | 0 | 0 | 5.721 | No |
| July | Mean | 23.512 | 21.227 | 15.425 | 21.126 | 27.496 | No |
| | St.dev | 21.143 | 18.319 | 13.014 | 18.13 | 23.031 | No |
| | Max | 89.5 | 77.018 | 50.227 | 81.114 | 86.195 | No |
| | Min | 0 | 0.311 | 0 | 0 | 1.707 | No |
| Aug | Mean | 29.01 | 36.007 | 21.069 | 33.838 | 61.716 | No |
| | St.dev | 27.695 | 34.559 | 18.07 | 34.164 | 56.195 | No |
| | Max | 129 | 138.41 | 69.743 | 143.194 | 194.16 | Yes |
| | Min | 0 | 0.006 | 0 | 0 | 0 | Yes |
| Sept | Mean | 29.366 | 31.98 | 16.055 | 31.364 | 47.675 | No |
| | St.dev | 29.837 | 31.472 | 18.37 | 31.82 | 42.326 | Yes |
| | Max | 126.3 | 117.517 | 73.925 | 121.616 | 160.6 | Yes |
| | Min | 0.1 | 0.284 | 0.095 | 0.103 | 1.6 | No |
| Oct | Mean | 16.176 | 18.147 | 13.518 | 17.907 | 24.423 | No |
| | St.dev | 12.813 | 16.123 | 11.09 | 16.135 | 20.924 | No |
| | Max | 63.2 | 64.684 | 43.537 | 66.765 | 70.831 | Yes |
| | Min | 1.7 | 2.794 | 1.727 | 2.192 | 4.997 | No |
| Nov | Mean | 23.471 | 24.724 | 16.763 | 24.001 | 33.866 | Yes |
| | St.dev | 19.913 | 20.446 | 11.629 | 20.448 | 28.834 | Yes |
| | Max | 94.6 | 87.56 | 46.556 | 94.167 | 101.28 | Yes |
| | Min | 1 | 1.61 | 0.939 | 1.821 | 2.097 | No |
| Dec | Mean | 21.915 | 17.842 | 12.571 | 17.4 | 26.518 | No |
| | St.dev | 18.122 | 14.171 | 8.423 | 13.729 | 30.999 | No |
| | Max | 75.8 | 65.09 | 34.411 | 68.426 | 128,956 | No |
| | Min | 2.7 | 2.978 | 0.726 | 3.15 | 4.311 | No |

Firstly, the annual autocorrelation function, based on which the annual series is generated, is shown in figure 4. Its mathematical formula is also given.

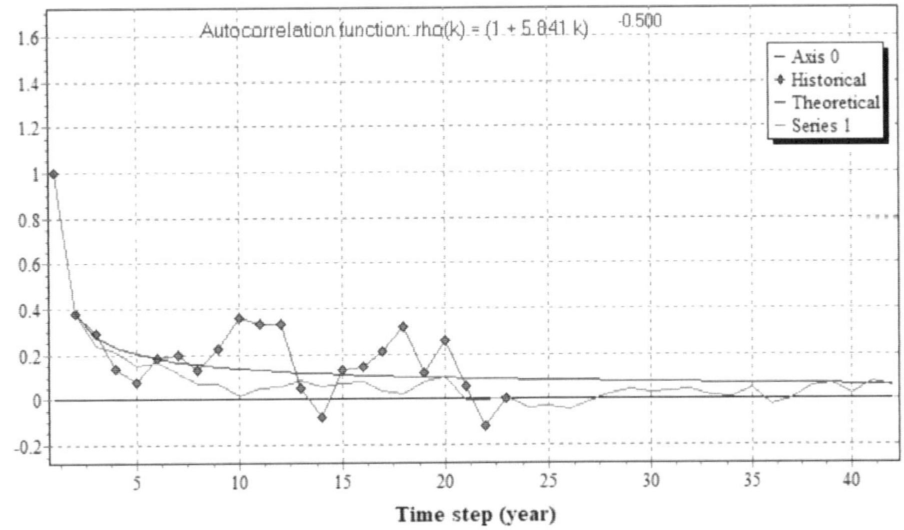


**Figure 4.** The annual autocorrelation function. Series 1 is the simulated series.

The estimated values of the parameters in equation (3) are $h = 2.9205$ and $\beta = 2$.

The average of the monthly generated series and the average values computed for the monthly series (based on the recorded data) are presented in figure 4, under the names Theoretical (computed) and Synthetic (simulated). The monthly averages are given under the chart. The annual averages computed utilizing the historical data and the simulated ones are represented by bars on the right-hand side of figure 5.

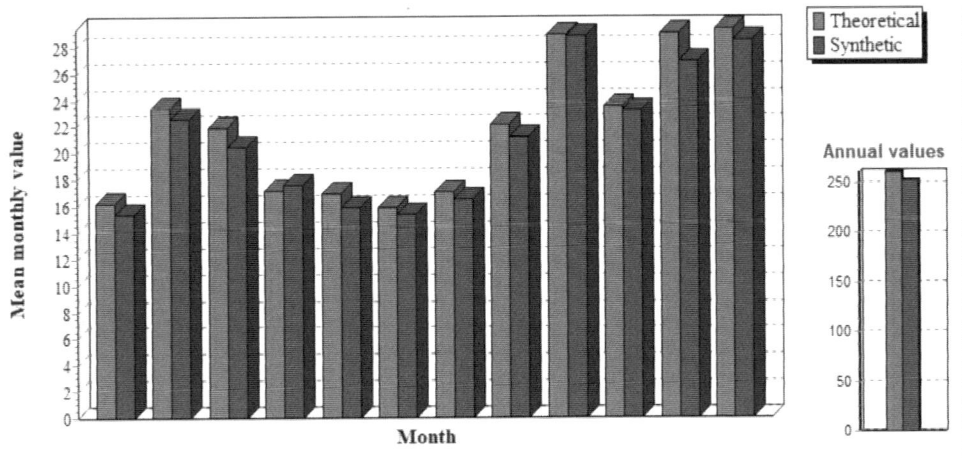


| Month | Oct | Nov | Dec | Jan | Feb | Mar | Apr | May | Jun | Jul | Aug | Sep | Annual |
|---|---|---|---|---|---|---|---|---|---|---|---|---|---|
| Theoretical | 16.18 | 23.47 | 21.91 | 17.19 | 16.94 | 15.88 | 17.02 | 22.16 | 28.99 | 23.51 | 29.01 | 29.37 | 261.27 |
| Synthetic | 15.39 | 22.57 | 20.48 | 17.64 | 15.91 | 15.34 | 16.56 | 21.19 | 28.84 | 23.26 | 26.90 | 28.53 | 252.47 |

**Figure 5.** Simulated and recorded mean monthly average of precipitation.

The mean monthly standard deviations of the historical and generated data are contained in figure 6, whereas the graphs of the annual and monthly autocorrelation functions are presented in figure 7.

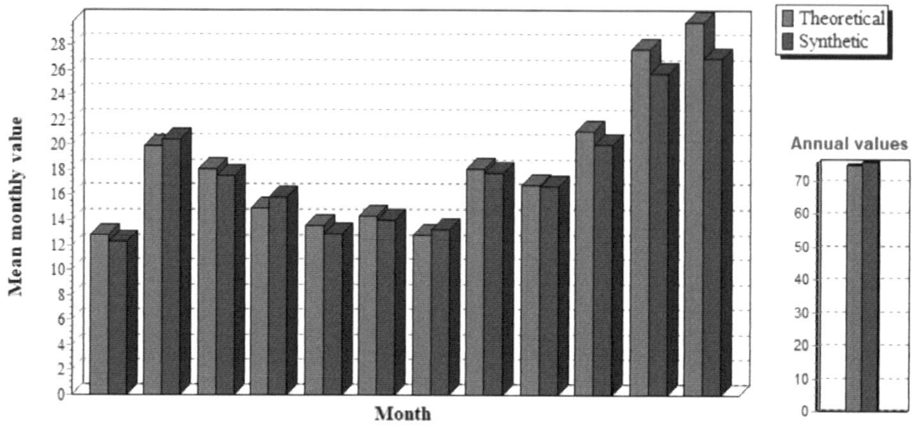


| Month | Oct | Nov | Dec | Jan | Feb | Mar | Apr | May | Jun | Jul | Aug | Sep | Annual |
|---|---|---|---|---|---|---|---|---|---|---|---|---|---|
| Theoretical | 12.81 | 19.91 | 18.12 | 15.00 | 13.62 | 14.36 | 12.78 | 18.07 | 16.85 | 21.14 | 27.70 | 29.84 | 74.72 |
| Synthetic | 12.33 | 20.50 | 17.53 | 15.88 | 12.90 | 13.98 | 13.26 | 17.75 | 16.67 | 20.04 | 25.71 | 26.90 | 75.66 |

**Figure 6.** Simulated and recorded mean monthly standard deviations of precipitation.

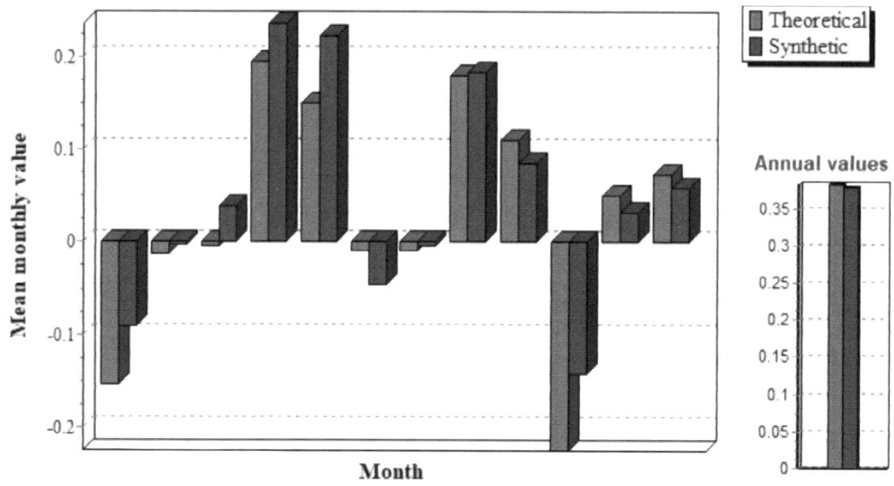


| Month | Oct | Nov | Dec | Jan | Feb | Mar | Apr | May | Jun | Jul | Aug |
|---|---|---|---|---|---|---|---|---|---|---|---|
| Theoretical | -0.1542 | -0.0132 | -0.0048 | 0.1949 | 0.1502 | -0.0102 | -0.0100 | 0.1800 | 0.1106 | -0.2251 | 0.0496 |
| Synthetic | -0.090 | -0.003 | 0.038 | 0.236 | 0.223 | -0.047 | -0.005 | 0.183 | 0.085 | -0.142 | 0.032 |

**Figure 7.** Monthly first-order autocorrelation.

The skewness, first-order autocorrelations, and the percentage of null precipitation have been preserved in the simulation of the monthly precipitation fields by the first algorithm; this is why we did not consider necessary to list their values in Table 2.

Analyzing the charts from figure 3, one remarks that the average values of the generated fields are generally close to the average of the recorded values. The best simulation results have been obtained for February (for which all the statistical parameters for the simulated series are in the admissible limits), November (for which only the minimum precipitation is not kept), and April (for which the maximum was not preserved). The worst results have been obtained for December and May.

For all the other months, two statistics for the simulated data were not in the tolerance range.

Standard deviations of the simulated series were not in the tolerance limits for the month January, March, May, June, July, August, October, December; the maximum values felt outside the tolerance limits for January, April, May, July, December, whereas the minima were outside these limits for March, May, June, July, September – December.

Analyzing the simulated series up to a 90% confidence level, the average of the simulated data (Sim 50%) are close to the historical ones for all but the August series. Therefore, the lack of fit in the tolerance limits is mainly due to a lower fit quality above the 90% level.

The series simulated by the second method satisfies the tolerance conditions as follows:

- The monthly average only for June;
- The standard deviations, only for November, December, March, April, and May;
- The autocorrelation function, only for May.

The first method had the best performances. Given the promising results obtained by other authors [26,27], we expected better results provided by the second method. The explanation that this did not happen could be the difference between the climates where the meteorological stations are situated.

## 4. Conclusion

In this article, we used two methods for generating monthly precipitation series a Sulina. Comparison of these approaches recommends the first method for this aim, even if some of the basic statistics are not preserved by the synthetic series, especially for July and December. Both methods build an initial synthetic series of first-order autoregressive model, which is not always the best choice for modeling a time series. Therefore, a possible approach to improving the simulation results would be selecting another model type at the first stage. Further research will be conducted in this direction. Still, the results provided by the first algorithm are promising and for the majority of monthly series, so they can be used for making decisions on the water management at Sulina.

## References

[1] Breinl K, Di Baldassare J, Girons Lopez M, Hagenlocher M, Vico G and Rutgersson A 2017 *Sci Rep* **7** 5449

[2] Khedhaouiria D, Mailhot A and Favre A-C 2018 *J Appl Meteorol Climatol* **57(10)** 2419

[3] Baigorria G A and Jones J W 2010 *J Climate* 23 5990

[4] Bărbulescu A and Dumitriu C S 2021 *Water* **13(8)** 1026, https://doi.org/10.3390/w13081026

[5] Mihăilescu M et al 2021 *Toxics* **9(5)** 111, https://doi.org/10.3390/toxics9050111

[6] Nazzal Y at al 2021 *Toxics* **9(5)** 95, https://doi.org/10.3390/toxics9050095

[7] Efstratiadis A, Dialynas Y, Kozanis S and Koutsoyiannis D 2014 *Environ Modell Softw* **62**139

[8] Forsythe N, Fowler H J, Blenkinsop S, Burton A, Kilsby C G, Archer D R, Harpham C and Hashmi M Z 2014 *J Hydrol* **517** 1019-34

[9] Khazaei M, Ahmadi S, Saghafian B and Zahabiyoun B 2013 *Climatic Change* **119** 631-64

[10] Srikanthan R and Pegram G G S 2009 *J Hydrol* **371** 142 -53

[11] Unal N E, Aksoy H and Akar T 2004 *Stoch Environ Res Risk A* **18**, 245-57

[12] Srikanthan R and McMahon T 1985 *Stochastic generation of rainfall and evaporation data*, AWRC Technical Paper No. 84 (Canberra: Australian Government Publishing Service) p. 10-32

[13] Porter J W and Pink B J 1991 A method of synthetic fragments for disaggregation in stochastic data generation. In: National Conference Publication - Institute of Engineers, Australia 187

[14] Maheepala S and Perera C J C 1996 Monthly hydrologic data generation by disaggregation, *J Hydrol* **178** 277 – 91

[15] Srikanthan R, McMahon T 2000 *Stochastic generation of climate data: A review*, Technical Report 00/16, Cooperative Research Centre for Catchment Hydrology (Canberra: Australian Government Publishing Service)

[16] Srikanthan R, McMahon T A and Sharma A 2002 *Stochastic generation of monthly rainfall data*, Technical Report **02/8**, Cooperative Research Centre for Catchment Hydrology

[17] Bărbulescu A, Dumitriu C S and Dragomir F L 2021 *Proc. Int. Conf. Electrical, Computer, Communications and Mechatronics Eng.* (ICECCME 2021), 7-8 October 2021, Mauritius, DOI: 10.1109/ICECCME52200.2021.9591085
[18] Bărbulescu A 2016 *Rom Rep Phys* **68**(2) 788
[19] Bărbulescu A, Dumitriu C S and Maftei C 2022 *Rom J Phys* **59(5-6)** 590
[20] Bărbulescu A, Postolache F and Dumitriu C S 2021 *Hydrology* **8(3)** 125
[21] Bărbulescu, C. Șerban, S. Caramihai 2021 *Rom J Phys* **66(3-4)** 806
[22] Șerban C, Bărbulescu A, Dumitriu C S 2021 *IOP Conf. Ser.: Earth Environ. Sci* **958** 012006
[23] Bărbulescu A 2016 *Studies on time series. Applications in Environmental Sciences* (Switzerland: Springer)
[24] Hubert P 2000 *Stoch Environ Res Risk A* **14**, 297
[25] Thomas H A and Fiering M B 1962 Mathematical synthesis of streamflow sequences for the analysis of river basins by simulation, *Design of water resource systems* ed A Maass et al. (London: McMillan) pp 459 – 93
[26] Koutsoyiannis D 2000 *Water Resour Res* **36**(6), 1519
[27] Koutsoyiannis D and Manetas A 1996 *Water Resour Res* **32**, 2105

# Structural design of the radar station of "Besna Kobila" Serbia

**M Besevic[1*], M Vojnic Purcar[1], L Kozaric[1] and S Bursac[1]**
[1] Faculty of Civil Engineering Subotica, Subotica, Serbia

* miroslav.besevic@gmail.com

**Abstract**. The location of the new radar station facility is located at the top of the Besna Kobila mountain in Serbia at an altitude of approximately 1904 m. The construction of the building is reinforced concrete. The building consists of ground floor and first floor with technical and other areas. From the 2nd to the 6th floor there is a communication central reinforced concrete core. At the level of the 6th floor, a platform was formed for storing equipment and supporting the antenna dome with a diameter of R = 15. The total height of the building with a dome is H = 37.0 m. Horizontal floor structures are in the form of reinforced concrete slabs supported on reinforced concrete walls 20 (30) cm thick, columns of different dimensions, and a system of reinforced concrete beams of different cross section. The vertical construction consists of load-bearing reinforced concrete columns, external and internal concrete walls. In the central part of the building, there is a staircase in the form of reinforced concrete slabs with a thickness of d = 16 cm, as well as reinforced concrete elevator core with a square base of approximately 560.0 cm. The calculation of the structure is performed for loads from its own weight, constant loads from the materialization of the object, loads from the equipment and for payloads: from snow and wind and seismic loads of intensity IX. The basic load of the dome is the wind according to the average speed of 70.0 m / sec (according to the measured data). The foundation of the building is on a reinforced concrete slab and reinforcements-capitals under the columns.

## 1. Introduction

The location, which was proposed for the construction of a new radar station facility based on previous detailed analyses, is located at the very end of the ridge that forms the peaks of *Besna Kobila*, north of its highest peak whose height is 1923 m, and which is otherwise occupied by PE Emission Techniques facilities, Telekom and other providers. The micro-location itself is a natural, saddle-flattened surface, which extends at altitudes from 1901 - 1904 m and with its natural characteristics meets the basic prerequisites for the construction of buildings with a suitable plateau. The planned plateau, whose approximate dimensions are about 55/30 m, in addition to providing space for setting up the facility, also provides the space needed for parking vehicles, as well as the required area necessary for specific technology when assembling or lowering the radar dome for installing or repairing antennas. The position of the building itself is conditioned by the already mentioned characteristics of the micro-location, as well as all functional-technological aspects, due to which the building is pulled to the highest point of the flattened part of the ridge and the border of the plot, in order to provide the best visibility with a minimum number of obstacles.

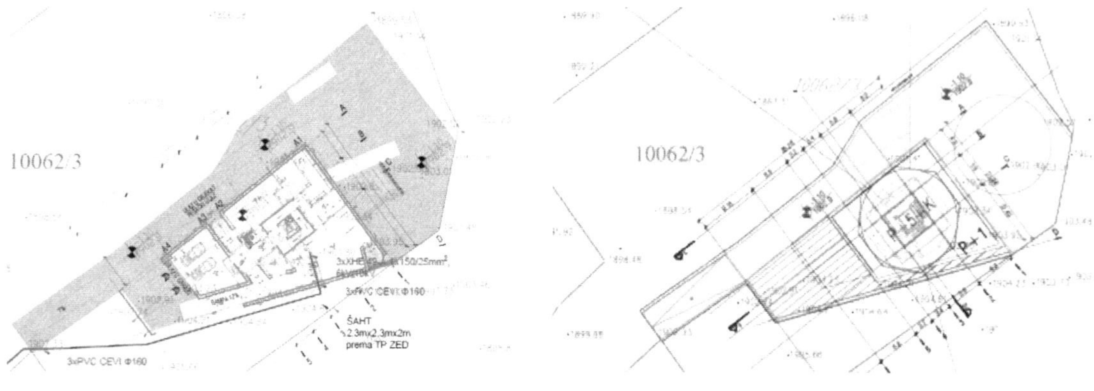


**Figure 1**. Ground floor and roof base.

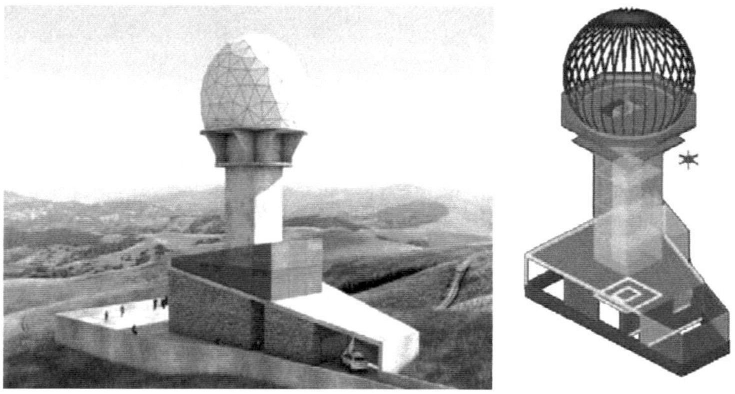

**Figure 2**. 3D model of the structure.

## 2. Structure of the building

The construction of the building is reinforced concrete [1]. The building consists of ground floor and first floor with technical and other spaces. From the 2nd to the 6th floor, the communication central reinforced concrete core continues. At the level of the 6th floor, a platform has been formed for storing equipment and supporting the antenna dome. The vertical construction consists of load-bearing reinforced concrete columns, external and internal concrete walls. In the central part of the building, there is a staircase in the form of reinforced concrete slabs with a thickness of d = 16 cm, as well as a reinforced concrete elevator core (cargo platform) with a square side base of approximately 560.0 cm. The antenna dome was adopted for this project with a diameter of R = 15.0 m, a circle with a diameter of 12.0 m is formed at the intersection with level 6. The dome is prefabricated, plastic, according to the supplier. The dome has no contact with antennas and radars inside and outside its contour. The building is based on a basic, healthy rock, in all respects according to geotechnical conditions - study. According to the bases from the geotechnical study and static calculation, the foundation will be made on a foundation slab with a depth of approximately 3.00 m from the elevation of zero (1904.00) terrain. This foundation level was adopted so that the foundations of the slabs would be performed in well-bearing soil, in order to avoid the zones of the surface layer and decomposed parts of the base rock, as well as the effects of freezing.

The part of the building from axis 6 to axis 7 is founded on foundation strips and connected by connecting beams in accordance with the plan of foundation positions. In the part between axes 5 and 6, i.e. from axis A to axis D, due to the installation of the water tank, the foundation slab is cascaded at an elevation of -4.95 m. The steel structure for the service platform for maintenance needs at elevation

+21.80 m is designed from hot-rolled steel profiles with a typical raster floor of galvanized sheets. A payload of 4.00 kN/m$^2$ was taken into account.

At elevation +24.0 m, an antenna mechanism is installed. The aforementioned equipment with a support is supported on a construction in the form of a steel grill. The equipment girder is calculated on the load from the base of the antenna mechanism of intensity (20 kN), as a constant load. The load from the equipment of intensity (70 kN) was taken into account. The load from the own weight of the structure is calculated in the Tower program. The load is applied to the circumference of the steel structure in the form of a line load. Considering the type of equipment during the structural analysis, the load is multiplied by the dynamic coefficient $k_d = 1.3$. The steel structure for supporting the antenna mechanism consists of two pairs of steel supports HEB 200 which are placed in two orthogonal directions at a mutual axial distance of 1.80 m. Each girder is supported on a reinforced concrete slab at elevation +24.0m via two anchors at the support location. The weight of steel 6t is also predicted, which refers to the antenna with motors and supporting structure. Static and dynamic analysis of the structure was performed in the analysis software *Tower 6*. Calculated displacements of the structure due to horizontal influences (wind and seismic forces) give maximum displacements at the platform level at + 24.00 m with a value of $u_{max} = 9.09$ mm, which is less than the allowable displacement, that is given by the project task and is 13.96 mm. The stiffness of the load-bearing reinforced concrete structure is in accordance with the design conditions according to the Eurocode standard, so that for a given horizontal load, the structure is not expected to move and affect the functioning of the equipment. Conducted calculations of the displacement of the structure show that the conditions regarding horizontal displacements and torsion are met. Based on the above, the rigidity of the structure fully satisfies all the necessary conditions prescribed for the functioning of the equipment. Due to the architectural requirements, it was necessary to design the protective construction of the plateau in the form of supporting reinforced concrete walls harmonized with the elevations of the terrain.

## 3. Analysis of the structure

Construction calculation is performed for loads from own weight, constant loads from materialization of the building, loads from equipment and payloads, loads from snow and wind (according to official data of competent meteorological services, which are part of the project task) and seismic loads for earthquake intensity IX, as given in the load analysis. The basic load of the dome is the wind. For the construction project, based on the data of the Investor, an average wind speed of 70.0 m/sec was adopted. The design of the dome shows the load of the dome according to SRPS and EC regulations-codes. The structure is calculated in the spatial system, in order to obtain realistic horizontal displacements and to determine whether their intensity affects the function of the structure. Local elements, external antennas, etc., will be calculated individually and as a whole, and this will be relevant for dimensioning the supports of these elements.

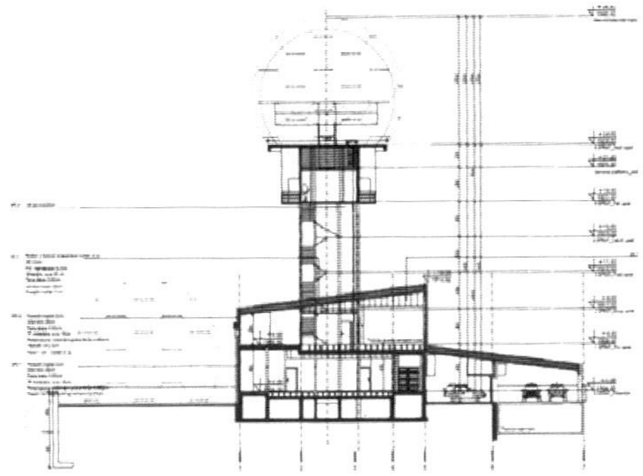

**Figure 3**. Cross-section.

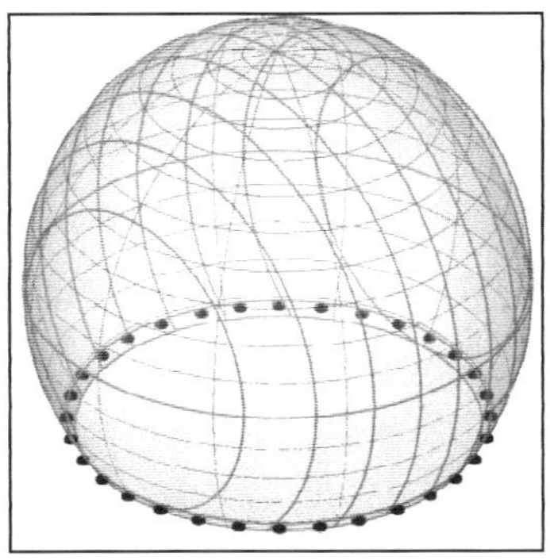

**Figure 4**. The dome.

## 4. Wind loads evaluation

Calculation of the basic wind speed was done according with the Serbian code design SRPS U.C7.110, and with the wind speed according to Eurocode 1 [2] for the location Besna Kobila, with the fundamental wind speed based on the project requirements $V_{bo} = 70$ m/s at a measured 10 m above the ground, for the roughness class B and an altitude of 1901 m and a total height of the structure of 37.1 m.

The air density was considered $\rho = 1.225 - \left(\frac{1901}{8000}\right) = 0.987\ kN/m^3$.

The basic wind speed was calculated according with [2] $V_b = C_{dir} C_{season}\, V_{b0}$, thus:

$$C_{dir} = 1\ C_{season} = 1 V_b = 1.0\ 1.0\ 70.0 = 70\ m/s$$

where mean wind pressure: $V_b = C_{dir} C_{season}\, V_b$

The wind gust pressure: $q_{p(z)} = \left[1 + 7 I_{v(z)}\right] \frac{1}{2} \rho\, V_{m(z)}^2 10^{-3} \left[\frac{kN}{m^3}\right] z_{max} = 200\ m$ for the terrain category I $\rightarrow z_{min} = 1.0\ m\quad z_0 = 0.01\ m$

Thus $k_r = 0.19 \left(\frac{z_0}{z_{0II}}\right)^{0.07} = 0.19 \left(\frac{0.01}{0.05}\right)^{0.07} = 0.17$ and $C_{r(z)} = k_r l_n \left(\frac{z_0}{z_{0II}}\right) \quad z_{min} \leq z \leq z_{max}$

considering topography coefficient $C_{0(z)} = 1.0$,

The turbulence intensity $I_{v(z)} = \frac{1}{ln\left(\frac{z}{z_0}\right)} = \frac{1}{ln\left(\frac{z}{0.01}\right)}$

The influence of wind on a part of the building up to the second floor is authoritative distribution of wind on the building by height, the height of the building from the base to the top being 37.1 m and the height of the tower with a sphere $h_t = 27.11$ m.

$$q_{p(z)} = [1+7\, I_v(z)]\ \frac{1}{2}\ \rho\ v_m^2(z) \tag{1}$$

**Table 1.** Wind load speed at different heights.

| z (m) | $C_{r(z)} = 0.17 \ln\left(\frac{z}{0.01}\right)$ | $V_b$ (m/s) | $V_m(z)$ (m/s) | $I_v(z)$ | $q(z)$ kN/m² |
|---|---|---|---|---|---|
| 9.99 - 15.59 | 1.2498 | 70.00 | 87.486 | 0.136 | 7.376 |
| 15.59 - 19.5 | 1.2878 | 70.00 | 90.149 | 0.132 | 7.719 |
| 19.5 -25.1 | 1.3308 | 70.00 | 93.154 | 0.128 | 8.115 |
| 25.1 - 37.1 | 1.3972 | 70.00 | 97.804 | 0.122 | 8.744 |

In order to calculate the wind forces acting on the structure, firstly it was calculated the wind pressure on surfaces with the relation:

$$fw = 2.1(Cpe - Cpi) q_p(z) \tag{2}$$

$B^2 = 1$ is adopted which is on the safe side.

$n_1 = \frac{\sqrt{b}}{0.09h} = \frac{\sqrt{5.6}}{0.09 \cdot 15.11} = 1.72Hz$ natural frequency

For the reference height = 37.1 m, $z_s = 0.6 \cdot 37.1 = 22.26\ m$ and $I_{v(z)} = 0.1297$ with $z_t$=200m and $L_T$ =300m.

$$\alpha = 0.67 + 0.05 \ln(z_0) = 0.67 + 0.05 \ln(0.01) = 0.44 \tag{3}$$

The value of $V_m$ at z height is $V_{m(zs)} = 111.18\ m/s$

The calculation of logarithmic decrement of construction damping was calculated as following

$$\delta = \delta_S + \delta_a + \delta_d \tag{4}$$

with $\delta_S = 0.03 \quad \delta_d = 0 \quad \delta_a = 0$

$$R^2 = \frac{\pi^2}{2\delta} S_L(z_s, n_1, x) \tag{5}$$

$$f_{l(z_s, n_1)} = \frac{n_1 L_{(zs)}}{V_{m(zs)}} = \frac{1.72 * 114.18}{111.18} = 1.776 \tag{6}$$

$$S_L(z,n) = \frac{6{,}8 f_L(z,n)}{(1+10{,}2 f_L(z,n))^{5/3}} = \frac{6.8\ 1.766}{(1+10.2\ 1.766)^{5/3}} = 0.089 \tag{7}$$

$$\eta_h = \frac{4.6h}{L(zs)} f_{L(z_S, n_1)} = \frac{4.6\ 37.1}{114.18}\ 1.766 = 2.64 \tag{8}$$

$$R_h = \frac{1}{\eta_h} - \frac{1}{2\eta_h^2}(1 - e^{-2\eta_h}) \tag{9}$$

Thus $R_h = \frac{1}{2.64} - \frac{1}{2\ 2.64^2}(1 - e^{-2\ 2.64}) = 0.307$

$R_h$,$R_b$ - aerodynamic access functions

$$\eta_b = \frac{4.6b}{L(zs)} f_{L(z_S, n_1)} = \frac{4.6\ 5.6}{114.18}\ 1.766 = 0.398$$

$$R_b = \frac{1}{\eta_b} - \frac{1}{2\eta_b^2}(1 - e^{-2\eta_b}) = 0.78$$

Thus $R^2 = 3.495$.

The impact coefficient $k_p = \sqrt{2\ln(\nu T)} + \frac{0{,}6}{\sqrt{2\ln(\nu T)}} \geq 3$ with the value of the frequency

$$\nu = n_{1,x}\sqrt{\frac{R^2}{B^2+R^2}} = 1.72\sqrt{\frac{7.968}{1+7.968}} = 1.517 \tag{10}$$

Thus T = 600s and $k_p = 3.921 > 3$

Coefficients of external wind pressure on the tower walls, e = b (dimension perpendicular to the wind) case e ≥ d.

**Figure 5**. Wind loads direction and areas.

$$\frac{h_k}{d} = \frac{15.11}{5.6} = 2.7 \tag{11}$$

Thus, the external wind pressure for zone A $c_{pe}$ = -1.2, for zone B $c_{pe}$ = -0.8, for zone D $c_{pe}$ = +0.8 and for zone E $c_{pe}$ = -0.585 and the coefficients for internal wind pressure on the tower walls are $c_{pi}$ = 0 and $c_{pi}$ = -0.23 (the building does not have a dominant side with openings - openings evenly spaced).

Wind load to the walls of the dome $fw = 2.1(Cpe - Cpi)q_p(z)$ (table 2).

**Table 2.** Wind load values coefficients.

| *z* *(m)* | qp(z) | *windward side D* fw (kN/m²) | | *windward side A* fw (kN/m²) | | *windward side B* fw (kN/m²) | | *windward side E* fw (kN/m²) | |
|---|---|---|---|---|---|---|---|---|---|
| | kN/m² | za Cpi=0 | za Cpi=-0.23 | za Cpi=0 | za Cpi=-0.23 | za Cpi=0 | za Cpi=-0.23 | za Cpi=0 | za Cpi=-0.23 |
| 9.99-15.59 | 7.376 | 12.392 | 15.954 | -18.588 | -16.530 | -13.633 | -9.714 | -9.969 | -6.050 |
| 15.59-19.5 | 7.719 | 12.968 | 16.696 | -19.452 | -16.530 | -13.633 | -9.714 | -9.969 | -6.050 |
| 19.5-25.1 | 8.115 | 13.633 | 17.553 | -20.450 | -16.530 | -13.633 | -9.714 | -9.969 | -6.050 |

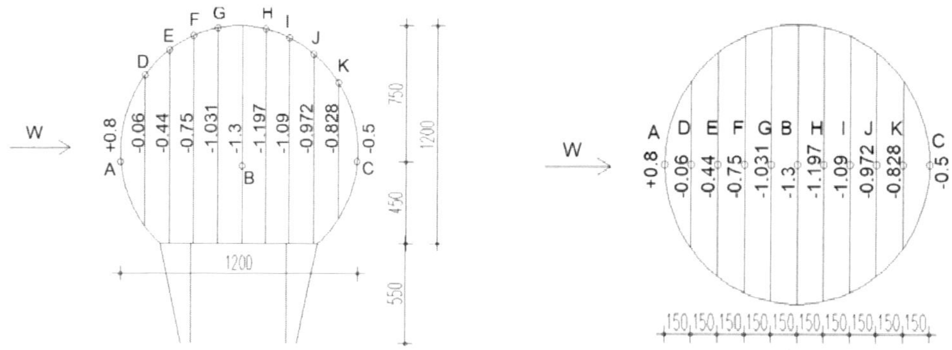


**Figure 6**. Coefficients of external wind pressure on the dome.

**Table 3.** Wind load.

| *z* (m) | $c_{pi}$ | $f_w$ (kN/m²) | | | | | | | | | | |
|---|---|---|---|---|---|---|---|---|---|---|---|---|
| | | A | D | E | F | G | B | H | I | J | K | C |
| 8.744 | 0 | 14.690 | -1.102 | -8.079 | -13.772 | -18.932 | -23.871 | -21.980 | -20.015 | -17.848 | -15.204 | -9.181 |
| | -0.23 | 18.913 | 3.122 | -3.856 | -9.548 | -14.708 | -19.648 | -17.756 | -15.792 | -13.625 | -10.981 | -4.958 |

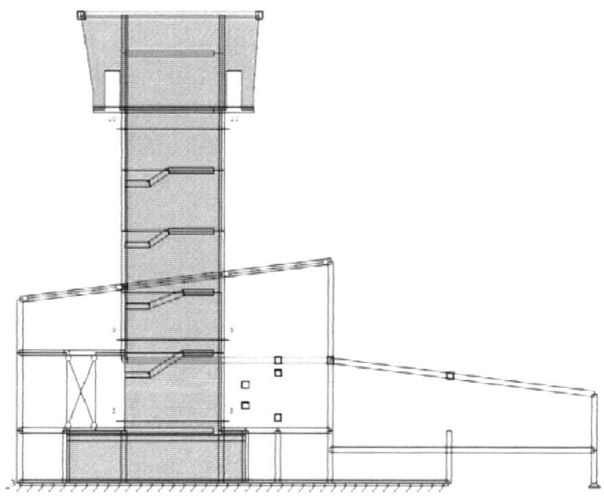

**Figure 7.** Cross section wall layout.

**Figure 8.** Design model of construction.

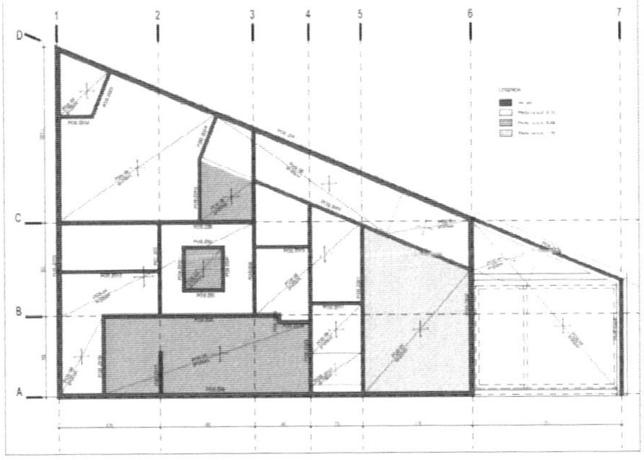


**Figure 9.** Ground floor layout.

The dimensioning of the structural bearing elements was carried out for the authoritative influences according to the load combinations, in everything as shown in the structural analysis [3].

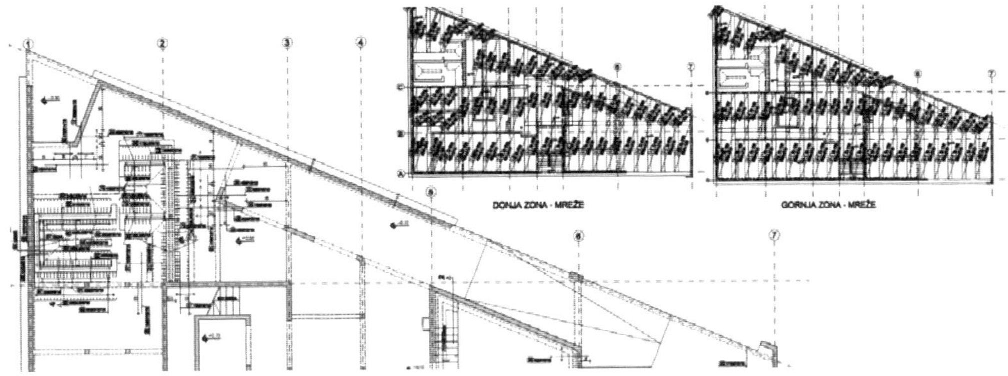


**Figure 10**. Plan of reinforcement.

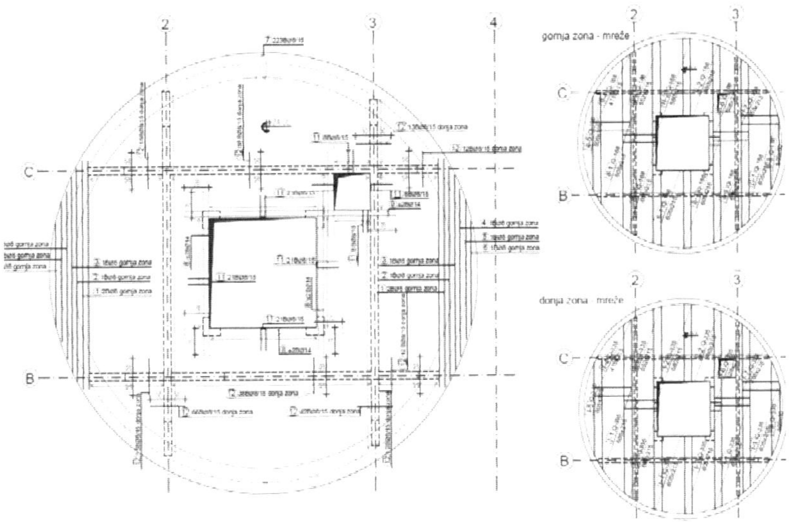


**Figure 11.** Reinforcement layout at elevation +24.00.

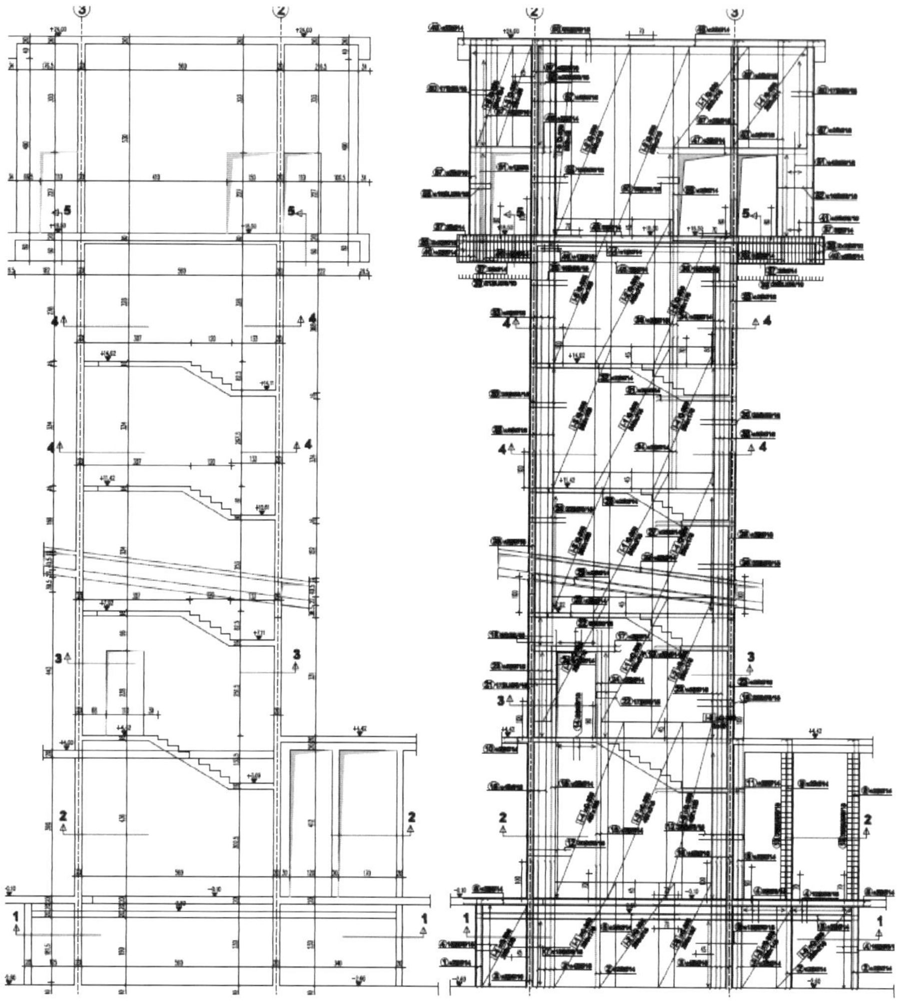

**Figure 12.** Reinforcement layout of the wall at elevation 2.60m to 24.00 m.

## 5. Conclusion

In this paper, the influences on the construction of the radar station are analysed in detail, and the optimal dimensioning of reinforced concrete and steel structural elements is performed according to the loads. The dominant load, determined for extreme forces and deformations, is the wind load. Wind as a load was analyzed in accordance with Serbian and European regulations. The detailed analysis of the effect of wind on the tower gave reliable results, which were further used as basic input data for the design of concrete and steel parts of the tower structure. All factors were analyzed: design wind speed, rigidity of the tower structure, the terrain of the tower location, aerodynamic coefficients and aerodynamic wind pressures. The calculated values of deformations were confirmed in practice after the construction of the facility, which proved that this approach and the way of calculating the wind load was correct and justified.

## References

[1] Design project of radar station "Besna Kobila"
[2] EN 1991-1-1 (2002): Eurocode 1: Actions on structures
[3] Besevic M, Landovic A, Prokic A, Tabakovic D, Tesanovic A 2015 Analysis of loads and calculation of steel chimney height H=40m according to EC and SPRS standards, *Journal of faculty of Civil Engineering Subotica*, 27, pp 7-20

# Analysis of up-to-date methods of non-destructive testing of performance properties of reinforced concrete building structures

**V Bredikhin[1*] and K Los[1]**

[1] Department of Real Estate Management, Mining, Southwest State University, Kursk, Russian Federation

* bvv001@mail.ru

**Abstract**. In the process of engineering survey of the operated reinforced concrete structures of buildings and erections, various methods are being used. According to their approach, they are divided into two groups: destructive and non-destructive testing methods. The main disadvantage of destructive testing methods is the violation of the "testing integrity", i.e. only the part of a structure is used in tests. When selecting core samples, for example, concrete cores from the "body" of a structure, a new, essentially independent, structural element is formed. In other words, the behavior of a sample that is the part of a reinforced concrete structure may differ from the behavior of the same part in the structure as a whole during experimental testing. The noted shortcomings can affect the accuracy of the research results, which means that the answers to the questions posed before the technical inspection of a facility cannot be adequate. Methods of reinforced concrete structures' inspection that partially or completely solve the problems of destructive testing are non-destructive methods. The paper concerns the system analysis of up-to-date methods of non-destructive testing of main performance properties of reinforced concrete structures.

## 1. Introduction

Currently operated buildings and structures are different in their design. At present, the most used material in the production of building structures is concrete being reinforced with steel rods, the so-called reinforced concrete. In the course of operation of such facilities, inevitably, there arises a need to monitor the technical condition of a facility as a whole and its structural elements in particular.

The main quality figures studied in the process of building structural survey of reinforced concrete structures are concrete density and compressive resistance. These parameters allow us to determine the current condition of concrete in the structure in order to consider the possibility of its further usage. The condition of the reinforcing component of reinforced concrete (reinforcing steel net, rolled steel bars, etc.) is evaluated not from the point of view of the qualities of reinforcement itself. It is assessed based on the behavior of an element under study as a whole. Therefore, during the building structural survey of reinforced concrete structures, namely reinforcement, two principal problems are considered: concrete reinforcement (compliance with the project specifications) and its condition.

As we said earlier, while examining the technical condition of the operated reinforced concrete structures, various methods can be used. According to their approach, they are divided into two groups: destructive and non-destructive testing methods. The first group of methods is based on the removal of a reinforced concrete structure or its separate part from the general structure of a building (structure) in order to conduct further tests under laboratory conditions. In its turn, sample taking in the application of destructive methods of quality control of reinforced concrete structures is carried out by two main methods:

- cutting out samples from a reinforced concrete structure “working” in the final design position;
- drilling out cores from the part of a structure

Methods of destructive testing make it possible to determine concrete qualities in the operated structure with high accuracy. However, these methods also have a number of disadvantages.

When “removing” a sample (or several samples) from a structural element, there always occurs an integrity disturbance of not only the element itself, but also a part of the facility’s structural layout, since the supporting structures are usually interconnected with each other. Therefore, when one of them weakens, the loads are redistributed (there appear beyond design loads).

Another disadvantage is the violation of the “experiment’s purity” when only part of the structure is used in the tests. In fact, when collecting cores or cutting out concrete samples from the “body” of a structure, a new independent structural element is formed. In other words, the behavior of a sample being a separate part of a reinforced concrete structure during experimental testing may differ from the “behavior” of that part in the structure as a whole under operating loads. The above noted shortcomings can affect the accuracy of the study results, which means that the answers to the questions posed before the building structural survey of a facility cannot be objective.

The methods of reinforced concrete structures survey that partially or completely solve the problems of destructive testing are non-destructive methods.

## 2. Non-destructive testing methods

Non-destructive testing methods of concrete quality include those that determine the strength characteristics of materials using local physical action [1-3]. These methods include:

- impact pulse method;
- detachment method;
- chipping method;
- pressing-in method;
- method of detachment with chipping;
- elastic rebound method

The above-mentioned methods are, in part, similar to destructive testing by their effect. However, since they have only a local application in the structure (i.e., do not imply any fracture of its integrity), they do not significantly affect its performance properties.

Another group includes non-destructive testing methods, being called indirect ones. By definition, indirect methods are those that allow us to determine the strength characteristics of concrete based on previously established dependencies. As a rule, these methods use the relationship between the compressive strength of concrete and the strength property which was previously determined by destructive or non-destructive methods.

However, it is much more convenient to classify methods of non-destructive testing by their physical impact on the reinforced concrete element, as well as the by the tasks that they perform during the building structural survey - figure 1.

Among other methods of non-destructive testing of concrete and reinforced concrete structures, methods of local destruction are more often used during the building structural survey. At the core of each of the three methods of local destruction is the identification of concrete strength characteristics through the internal resistance of the material to the detachment of its parts.

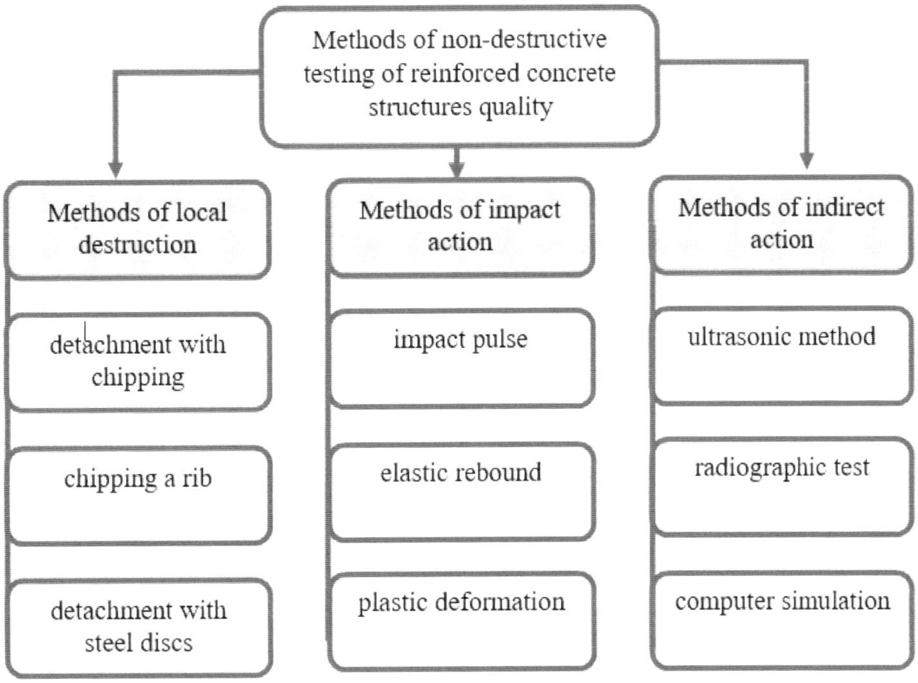


**Figure 1.** Classification of methods of non-destructive testing of reinforced concrete.

Out of the three methods of local destruction, the method of detachment with chipping is the most widely used. In practice, it determines the maximum force applied to tear off the part of concrete surface of the structure under study. Specifically, this method differs in the way of attaching a device to the surface of a test sample. To perform the study, a hole (anchor hole) is made in the prepared surface of the sample. Then, a special leaf-type anchor of the device is inserted. A spacer is provided between the supporting struts of the device installed on the edges of the studied area of a structure, and the movable part of the device connected to the anchor inside the structure. During mechanical action, the anchor is tensioned inside the structure that pulls its part. The monitoring device records an increase in the applied force. The process continues until the device shows a sharp decrease in the pull-off force, which will mean a partial destruction of the structural material in the anchor attachment area.

The method of detachment (same as the method of detachment of steel disc) is almost identical to the method of detachment with chipping. A distinctive feature of the method of detachment with steel disc is the absence of an anchor hole in the concrete structure. To conduct the study, a flat metal disk is used, on which an adhesive solution is applied. The disc is attached to the smooth surface of concrete (reinforced concrete structure) and is connected to the movable part of the device for measuring tensile forces. Further, the principle of the device operation is similar to the method of detachment with chipping. Despite the widespread application of the given method in foreign countries [4, 5], it is used less frequently in Russia. This is due to disadvantages, in particular, related to the dependence of adhesive composition on temperature and humidity conditions, the density of its application to the disk, etc.

Chipping a rib is the third method out of local destruction ones. The basis for determining the strength characteristics of concrete in this method remains the same as in the previous two – the determination of concrete tensile strength ($R_{bt}$), followed by its conversion to compressive strength ($R$). Currently, there are two ways to perform test in accordance with this method. They depend on the types of research devices. The first one is used for structures made of concrete (reinforced concrete) that have more than one adjacent face. In this case, the device is attached to a structure by the type of a clamp, being fixed on both sides of a structure. Then, through a manually-driven mechanical device element, a load is applied at an angle to the surface of the measurement area. The device determines the growth of the applied force until it sharply decreases, which, as in previous methods, indicates the detachment of the part of a structure.

Being more advanced, the other way makes it possible to determine the strength of concrete by chipping when there is a single face in the structure. The device is fixed on the face of a test sample using a device's hook with an anchor. Thus, the device is prevented from coming off when creating a force on the sample. Then, the test is carried out on the principle identical to the previous way discussed. At the same time, in both ways, in order to calculate the compressive strength of concrete, the amount of force (*P*) required for chipping the section of a structure is used.

The methods of local destruction have minor differences in the way of application, but they have their advantages and disadvantages. Besides, they have a different range of applications, depending on the ultimate strength of concrete of which the structure under study is made - figure 2.

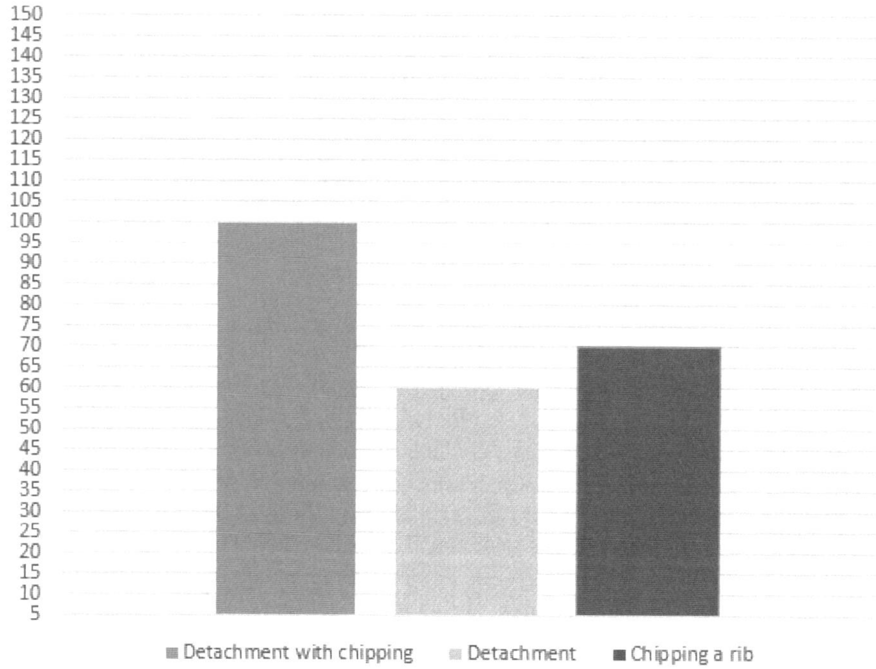


**Figure 2.** The interval of application of partial destruction methods for the ultimate strength of concrete [MPa].

One of the most common methods that are used to determine the strength characteristics of concrete are impact action methods. The methods of impact pulse, elastic rebound and plastic deformation are based on the principle of determining the concrete strength through impact-pulse action on it.

The basic principle for each of the above-mentioned methods of impact action is the principle of interaction of pulse being delivered by the impact of the device's striker and the structure's material. For example, rebounding velocity is higher in concretes of higher density and lower content of cavities in the structure [6, 7]. The conversion of the striker's kinetic energy into potential energy allows us to make calculations and build calibration dependencies between the readings of the device and the strength of concrete.

The essence of elastic rebound method is to measure the return velocity (rebound velocity) of the device's spherical striker and the height of rebound. Based on a number of measurements carried out by the device (sclerometer), the calibration dependence of the return position height of a striker and concrete strength is constructed – figure 3. When constructing the dependence, corrections for the position of the device are taken into account, since its deviations from normal to the surface of the structure affect the rebound velocity.

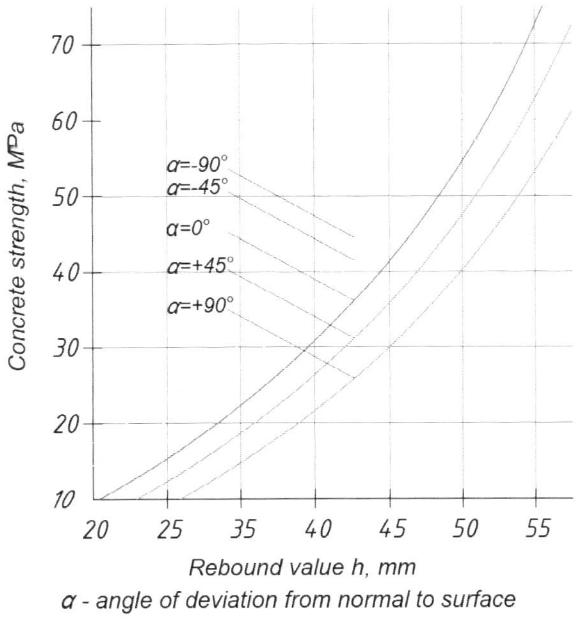


**Figure 3.** Calibration dependence of concrete strength on the height of elastic rebound.

A distinctive feature of the impact pulse method lies in the way of determining a relationship between the pulse and the strength of material that is subjected to mechanical impact. The electromechanical converter receives a signal about the energy of mechanical impact on the concrete surface. In other words, the potential energy of a striker is converted into kinetic energy, then is transferred to the particles of material and partially returned to the potential of the same striker during the rebound. In this case, the device directly measures the pulse energy at the moment of impact, obtained in the form of an electrical signal. The advantage of this method over the elastic rebound method is the possibility of allowing vertical deviations from the plane of the concrete surface, while maintaining the "purity" of the result.

The essence of the third impact action method – plastic deformation consists in constructing a relationship between the size of deformation imprint on the sample and its strength. With the help of a device, a spherical striker is struck on the material of a structure, as a result of which a deformation area is formed in the testing plane. Having an idea of the magnitude of the applied force on impact, the hardness of the striker material and the overall dimensions (diameter) of defect (dent) obtained on the plane of a structure, a relationship is built. It implies the following principle of strength determination: if the striker leaves a deeper and larger (in diameter) mark on surface from the impact, the less dense the material is. Based on this principle, the strength characteristics of concrete are calculated.

A separate area of research related to the non-destructive testing of concrete and reinforced concrete structures is determined by ultrasonic methods. They refer to the type of indirect impact methods [8]. The principles of ultrasound methods operation are determined by the tasks that they face. One singles out surface measurement, through measurement, and angular measurement. In surface measurement, the sensors of a device are located on one side of a structure, so the sound wave passes in the body of a structure in an arc - from one sensor to another. The angular and through measurement are almost similar to each other, with a difference in the location of sensors. In the first case, they are located in adjacent faces (at the corners) of a structure, in the second one – on opposite faces. The latter method of measuring the velocity of ultrasonic waves is more accurate, since ultrasound passes through the entire thickness of the material of a structure.

The application of ultrasonic methods for evaluating the strength of concrete is common in practice; however, it also requires the construction of calibration dependencies – figure 4. This is primarily due to the heterogeneity of concrete.

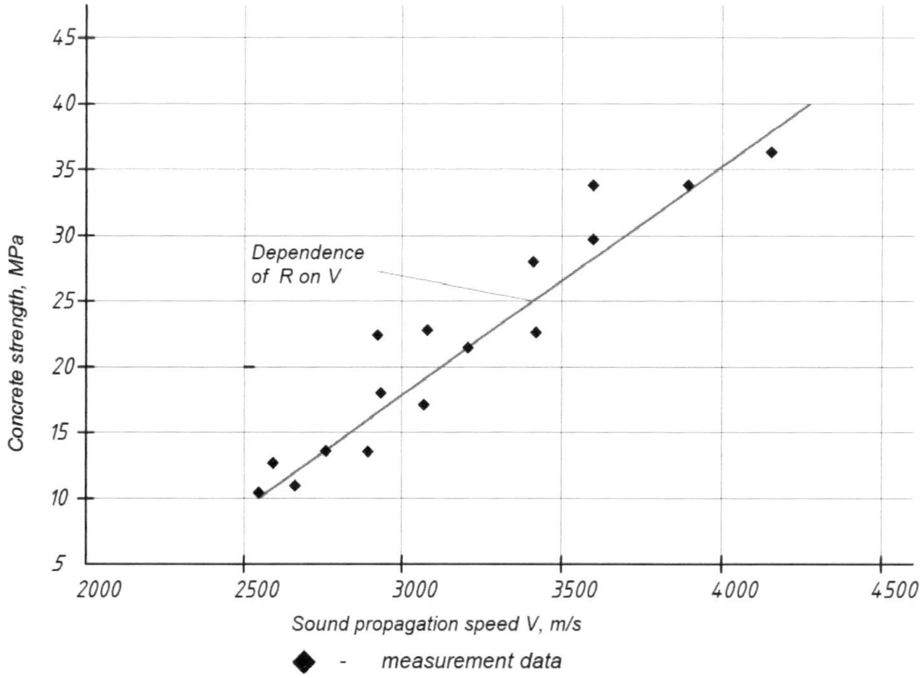


**Figure 4.** Calibration dependence of concrete strength on sound speed.

The use of ultrasonic methods is not limited to the tasks of determining the strength of concrete in building structures. The features of ultrasonic devices and the very principle of the methods' "work" allow us to solve such problems as:

- detecting voids and defects in the body of a concrete structure;
- determining the location of reinforcing bars;
- continuous monitoring of building construction state (increase or decrease in the strength of concrete in the structure)

In particular, one can distinguish such method of non-destructive testing as radiographic testing. The basis of the radiographic test method of reinforced concrete structures is the use of X-ray and gamma radiation (γ). This method makes it possible to reveal, among other things, the defects of a concrete structure. The main goals of this method are to control the quality of welded metal joints and to identify defects in reinforcing rods in concrete. The method of radiographic test can be attributed to research tests without direct physical contact with an object. In addition to that, the radiographic test is the most informative one, since according to the results of its implementation, specialists can obtain information about the places of reinforcement in concrete, about the dimensions of reinforcing rods and their condition, about defects and even their sizes.

## 3. Conclusions

The variety of non-destructive testing methods forms the scope of their application for solving individual problems, on the one hand, and allows to increase the overall accuracy of the conducted research, on the other hand. None of the non-destructive testing methods, direct or indirect, is universal for solving all the problems that building structural survey of working reinforced concrete structures faces. The accuracy of each of the described methods is also not absolute, due to the presence of shortcomings, including the underlying principles of measuring the characteristics of materials under study. Currently, the presence of shortcomings in non-destructive testing methods is solved through the complex application of several of them for the examination of a single object. Using indirect methods is another way to improve accuracy. The construction of calibration dependencies, repeated measurement and comparison of results with other methods (including destructive testing ones) - these are the ways to improve the accuracy of the results obtained in the examination of reinforced concrete structures in operation.

**References**

[1] INTERSTATE STANDARD 22690-2015 Concretes. Determination of strength by mechanical methods of nondestructive testing
[2] INTERSTATE STANDARD 17624-2012 Concrete. Ultrasonic method of strength determination
[3] INTERSTATE STANDARD 7512-82 Nondestructive testing. Welded joints. Radiography method
[4] Thomas H, Ede A 2019 Use of Non-Distructive tests for reinforced concrete damage assessment. Pp. 1-10
[5] Wiggenhauser H, Köpp C, Timofeev J, Azari H 2018 Controlled creating of cracks in concrete for non-destructive testing. *Journal of nondestructive evaluation*. Vol. **37**, no.3, p 67
[6] Chernova E S 2014 Application of elastic impact in quality control of materials and structures. National priorities of Russia. **4** (14) Pp 100-102
[7] Nesvetaev G V, Kolleganov A V, Kolleganov N A 2017 Features nondestructive control concrete strength exploited concrete constructions. vol. **9**, no.2. p 100
[8] Balayssac J-P, Garnier V 2017 Non-Distructive testing and evaluation of civil engineering structures. ISTEPress. Elsevier Pp 21-84

# How health influences the way we build

**Ch Cazacu**[1*]

[1] Transilvania University of Brasov, Faculty of Civil Engineering, Romania

*cazacu.christiana@unitbv.ro

**Abstract.** Knowing people's needs for a healthy life is a fundamental theme for those who design and execute constructions. The environment of a building must combine the most important factors for comfort and health. For a new construction, the hygiene and health of the occupants must become a priority strategy, and needs to be considered early on in the project because it is key to securing a safe construction. How do constructions make us sick and what are the main causes and phenomena, and what consequences do they have on our health are the principal subject from this study. The designer must use materials that are thermally efficient, healthy and comfortable that provides better constructions for the occupants.

## 1. Introduction

Human health is undoubtedly the most important subject when design buildings. In ancient times it was said that the architect must know, among many others, medicine in order to design the best possible constructions for people. From the design phase, we must understand that the relationship between people and the spaces, in which they carry out all their activities, is long and over time it can lead to worsening the health of occupants [1].

The purpose of this review study, is to show how an improperly design construction system can influence the health of the occupants in the long-term. The present study aims to present and analyses values of ambient temperatures, air movements due to heating systems in floors and walls, the quality and quantity of dust particles in the air, which inadequate can negatively influence the health of all of us.

But… how do our constructions get us sick? What are the main causes and phenomena, and what consequences do they have on our health?

The environment of a building combines the most important factors for comfort and health. It is directly influenced by temperature, humidity, purity or particle loading of the air, speed and movement of air currents, etc.

We can say that the state of health depends substantially on the air quality in the built spaces. In order to have a good quality of the environment of a room we must know very well aspects about ventilation, heating, refreshment, filtration, humidification [1].

From the air quality and interior comfort point of view, and in according to I5-2010 [2], the buildings are classified in four categories of environment:

- category I, High level, recommended for spaces occupied by very sensitive and fragile people (sick, disabled, young children, the elderly);
- category II, normal level, recommended for new or renovated buildings;
- category III, moderately acceptable level, recommended in existing buildings;

- category IV, level other than the above, recommended to be accepted for limited periods of time;

The built environment impacts our health through a variety of factors including inadequate ventilation, poor indoor air quality, chemical contaminants from indoor or outdoor sources, by making us feel too cold or too hot, traffic noise or poor lighting. The results are respiratory and cardiovascular diseases from indoor air pollution; illness and deaths from temperature extremes and inadequate energy access; anxiety and depression when buildings can't provide a sense of safety, as well as discomfort from less than optimal lighting conditions or irritability from noise levels [3, 4].

Unhealthy buildings even result in a distinct medical condition, known as Sick building syndrome (SBS). With one in six Europeans living in homes that make them sick, unhealthy buildings are a widespread problem that need political and public attention. The health impacts from living in energy-inefficient buildings and particularly in energy poverty have been studied extensively in some countries, for example the United Kingdom or New Zealand. According to a 2011 World Health Organization (WHO) report, the environmental burden of disease attributable to inadequate housing alone (not counting non-residential buildings) in Europe amounted to 100,000 premature deaths annually [5].

## 2. Interior temperature

We generally know that the thermal comfort conditions from a room, no matter if is for housing or various activities, are established according to the category of building, the destination of building, the destinations of indoor spaces, occupancy and type of activity, specified in the project [6]. Defining thermal comfort:

- from a sensory point of view: thermal comfort means the lack of unpleasant sensations of cold or heat;
- from a biological point of view: thermal comfort is achieved when the evacuation of internal heat from the body is done without the overload of the human thermoregulatory system;
- from a physical point of view: thermal comfort is achieved when the environmental parameters have the values that satisfy the equation of the thermal balance and which is at the same time the mathematical expression of the thermal comfort for man.

The "well-being feeling" of a person in a room (comfort level) depends on certain external factors that can influence it. For example, people are extremely sensitive to thermal radiation caused by cold walls or windows. We also feel thermal discomfort due to drafts in a room. The thermal comfort parameters characteristic of the heated space, as defined in SR EN ISO 7730 [7], are presented in figure 1:

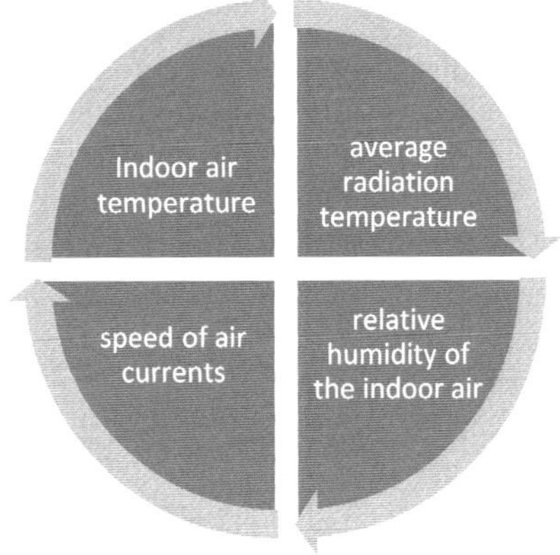


**Figure 1.** Thermal comfort parameters characteristic of the heated space.

The values of these parameters are established depending on the rooms type, the activity carried out and their effect on the perception of comfort of a person, depends all so on the activity carried out and the clothing of that person. The international standard ISO 7730 [7] combines all these parameters in the measurement of PMV / PPD ("Predicted Mean Vote / Predicted Percentage Dissatisfied").

- PMV is an index that shows the average value of the climate perception of a large group of people.
- The PPD index provides a quantitative prediction of the number of people who will be dissatisfied with a certain environmental atmosphere [7, 8].

The temperature of the indoor air must be adapted according to the destination of the rooms, according to ISO 7730 [7]. The heating systems must allow an adjustment of the heat flow yielded so that, during the activity of the rooms, the temperature of the indoor air can be modified according to the needs of the occupants.

Table 1 presents some of the indoor temperatures recommended for living, administrative and social cultural spaces, according to the regulations in force (SR 1907/2 of 2014 [9]):

**Table 1.** Recommended indoor temperatures for living, administrative and social cultural spaces [9].

| | Room type | Temperature (°C) |
|---|---|---|
| Living spaces | | |
| | rooms | 20 |
| | kitchens | 18 |
| | bathrooms | 22 |
| | stairs / corridors | 10 |
| | halls | 20 |
| | entrances to homes | 12 |
| Work spaces | | |
| | office | 20 |
| | libraries | 20 |
| | waiting rooms | 16 |
| | meeting / conference rooms | 18 |
| Administrative and social cultural spaces | | |
| | classes for students and pupils | 18 |
| | laboratory | 18 |
| | kindergarten halls | 20 |
| | workspaces or playgrounds | 18 |
| | reading room | 20 |
| | changing rooms | 22 |
| | exhibition halls | 18 |

An indoor ambiance can be characterized from a hygrothermal point of view taking into account the temperature of the indoor air, the temperature of the delimiting surfaces but also by the humidity and the speed of air movement. Experimental research based on large groups of people [10], maintained in environments with different characteristics, led to the conclusion that thermal comfort can be achieved for different combinations of values of microclimatic parameters, in correlation with the nature of the activity and clothing. For example, for production rooms, the comfort term can be replaced by the term of work efficiency, corresponding to the activity carried out.

Maintaining indoor temperature to a constant value is important for health. Temperature variations in rooms or values that do not fall within the standards, lead to health problems, as the body adapts to such changes within a few days.

Specialists in health believe that temperature fluctuations have a negative effect on the human body. According to them, lowering or increasing the temperature can affect the metabolism, and the risk of diseases such as asthma, various lung and respiratory diseases, allergies, depression and nutrition problems is quite high. Their studies have shown that the human body has begun to lose its ability to adapt to temperature fluctuations, and when the body is weakened, the impact on health can be quite serious. High temperatures have been associated with an increase in the number of patients affected by stress or nutrition problems, and low temperatures are responsible for respiratory diseases [10].

## 3. Heating systems

The human body is greatly influenced by the environment where it operates, and as previously mentioned air temperature is one of the most important factors. Carrying out daily activities in an environment unfriendly to the human body, can lead to problems of concentration, psychomotor performance or even behaviour problems.

In order to create the most favorable microclimate conditions in certain periods of the year, it is necessary to heat the rooms.

The purpose of heating is to maintain the environmental parameters necessary to regulate all the release heat by the body. Heating must maintain a certain level of air temperature in the room, to ensure uniformity of temperature horizontally and vertically and the heating sources must not affect the air quality.

Home heating can be achieved through systems like:

- local heating system (with local heating devices: stoves, resources, electric radiators);
- central heating system
- underfloor and wall heating (LTH) systems [11].

Comparing the two heating systems we can say that the central heating system has several advantages: it keeps the air temperature stable (regardless of the external temperature), it does not affect the air quality, it creates uniform temperature, and that is why it has now spread more and more worldwide [11].

At present, steam, water and air central heating systems are used to heat homes. In the case of the steam heating system, the heat agent is the steam, which, under pressure, through the pipes, reaches the rooms. The surface temperature of the heaters exceeds 100°C. The disadvantages of this system are: the impossibility of regulating the heat flow, the high temperature of the surface of the heaters, which contributes to the burning of dust [11].

The best hygienic and most common system is hot water heating. This system is used for heating homes, hospitals and public buildings. Within the hot water heating system, the permanent room temperature is maintained uniformly and allows the centralized regulation of the temperature in the rooms by adjusting the water temperature. The surface temperature of the heaters is usually lower than 80°C and there is no risk of burning.

For industrial enterprises is used a system that raises the pressure and temperature of boiling water. This closed system is called a pressure water heating system. In this system, the temperature of the heating surfaces of the heaters exceeds the permissible temperature, which can cause dust to burn on their surface and pollute the air in the room.

Another heating system successfully used today presented in international literature is that underfloor and wall heating (LTH) systems. It has many more advantages over traditional radiator-based heating (HTS) systems in terms of safety, health and comfort [11].

Using LTH system, we considerably improve a lot of aspects such as:

- more radiant heat transfer,
- less temperature gradients,
- more comfortable floor temperature,
- less draught and air turbulence [11].

It is well known that inhaling dust can cause allergic reactions. The sensitivity of humans for inhaled particles is not only dependent on the amount of particles inhaled, more important is the quantity of suspended matter. Conclusion, indications exist that the application of LTH systems not only results in less suspended particles, but also leads to less reactive (less irritating) particles due to the absence of dust singe [12].

In conclusion, the heating system must meet several conditions, namely:

- to create the optimal conditions of the thermal environment necessary for the activity that takes place in the respective room;
- not to produce harmful substances (smoke, dust, toxic gases, caloric radiation, etc.). To avoid caloric radiation, the maximum permissible temperature for the heating element must not exceed 80°C;
- to ensure a uniformity of the microclimate in the whole room both in time and in space;
- to have a high and sufficiently high caloric yield from an economic point of view.

## 4. What about humidity?

Air humidity refers to the amount of water vapor in the air and depends on the temperature differences between inside and outside [13]. For example, during the winter, when the outside temperature is much lower than inside, the air humidity increases. Water vapor can also occur as a result of activities such as cooking, showering, or ironing clothes. In the absence of proper ventilation, these water droplets from the air reach our lungs.

The weather undergoes sudden changes sometimes even from one day to the next, and the humidity of the house must be adapted according to the outside temperature. Although, in general, the humidity of the indoor air should be between 40 and 60% [14]. Experts recommend that we adjust the humidity of the home as follows:

- if the outside temperature is higher than 20°C, the humidity inside must not exceed 60%;
- if the outside temperature is between 10 - 20°C, the humidity of the house should be less than 50%;
- if the outside temperature is between 0 - 10°C, the humidity inside should be below 45%.

If you are not sure if the room is too dry or too humid, there are some signs that suggest that the humidity is not at an optimal level.

- dampness - too much moisture;
- itchy nose - a low level of humidity;
- the presence of mold and specific odor - too much moisture;
- dryness in the throat, lips, and nose, dry skin, thirst - a low level of moisture.

Although an invisible element, air humidity has, over time, a considerable effect on quality of life. Which is why we should pay close attention to the humidity level of the air we breathe.

Scientists have repeatedly concluded that exposure to mold greatly increases the risk of allergies and asthma (especially among children and people with a weakened immune system) and leads to fatigue, lack of concentration, and headaches. Moreover, large variations in humidity favor the development of flu virus, the humid environment maintain its life.

The air is too dry when the humidity has values between 0 and 30%. Too much dry air has adverse effects on both the human body and your home. Too dry air is felt especially during cool periods. Low humidity causes dryness of the nasal mucosa, skin irritations, allergies, and dry skin. Perhaps the most serious of these is the drying of the nasal mucosa. Lack of hydration at this level increases the chances of developing infections and diseases of the respiratory system [15].

It has effects on the house too, and they should not be neglected. Dry air can lead to cracks, wood cracks, and installation failure. The furniture may also suffer cracks in the painted areas and the dust will settle much faster. If you have plants in the house, a low level of humidity will lead to their drying and wilting.

Humid air (high humidity) or dry air (low humidity) can have adverse effects on our body [15]:

Signs of indoor air ***too humid***:

- feeling sweat or heat fatigue;
- wrinkled hair;
- sleep interruptions;
- respiratory problems;
- asthma and allergy problems.

Signs of indoor air ***too dry***:

- irritated skin and dry eyes;
- chapped lips;
- bleeding nose;
- itching in the throat;
- asthma and allergy problems.

In the winter, the indoor air is almost always dry. When the temperature drops significantly below zero, the relative humidity can drop to 20%. Adequate indoor relative humidity in winter is between 20% and 40%. The high indoor temperature increases the dryness of the air. The high room temperature is sometimes the reason behind the dry indoor air. It is recommended to keep the room temperature below 22°C. Humidity increases when the indoor temperature drops [15].

To prevent such effects, the literature recommends [16]:

- humidity measurement;
- regular ventilation of the rooms;
- placing a container of water on the radiator - helps to increase the humidity by a few percent;
- use of humidifiers.

## 5. Dust

Both in homes and in buildings where we carry out various activities, dust in the air is an important risk factor. Dust is a dispersion of solid material in the form of small particles that can float in the air for a certain period of time [17].

The diameter of the solid particles is generally less than 75μm [13]. The dust that floats in the air from a room, contains solid particles of different sizes and which deposit on different surfaces faster or later depending on their size and consistency. Compared with large particles, fine particles can remain suspended in the atmosphere for longer periods and be transported over longer distances. Some studies suggest that fine particles have stronger respiratory effects in children than large particles [2].

Most airborne particles are invisible to the naked eye, but their effect on health can be devastating. Dust particles from the air can come from the room but also from the outside (entered with the ventilation of the rooms, or through clothes or outside objects). The urban environment is a perfect place for all particles that are formed from pollution, pollen, traffic and other materials, that float easily through the atmosphere. It can be created from any type of material. For example, the dust that accumulates in our homes is made up of 90% of particles with impurities floating in the atmosphere, dead skin cells, lint, barely visible fabric fiber that are transported on the weakest stream of air, yarn, animal hair, our hair, soot and more.

Indoor dust is described as the dispersed distribution of solid material in indoor air. It may consist of both inorganic and organic particles as well as fibre of different sizes and can be very heterogeneous with regard to quantity and composition. Respirable particles and gases affect different parts of the respiratory process, depending upon their inherent characteristics [2,18].

- for gases, relative solubility is important;
- for particles, size is important [18].

Indoor sources include smoking and combustion processes, fibres from paper, wood and textiles, the occupants themselves, their pets, and activities taking place in the residence [17].

The Institute of Technology from Estonian University of Life Sciences made a study [17] to analyse, synthesise and assess correlation between dust measurement results and different heating systems in order to improve indoor climatology. To achieve the aim, the following tasks have been solved: classification of heating systems, selection of measuring method and equipment for indoor climate, selection of meters, dust measuring, data processing with software Microsoft Office Excel, and making a summary [17]. The idea of the study was to apply different heating systems in a dusty room, inside which 10 students together with their teacher carried out their research activities. The surface area of the laboratory is 32.49 m$^2$, height 3.23 m and volume 104.94 m$^3$. Two of the six windows in the room can be opened; the windows are covered with lamella curtains and can be opened in two ways: by tilting and turning. Measurement of dust content in room occurred in four points of measurement at three heights: 0.1 m, 1.0 m, and 1.5 m. Twenty measuring operations were performed in the conditions of laboratory indoor climate, guaranteed by different heating systems. The two doors and windows were closed during the measuring. In between the measuring, the room was aerated and all windows that could be opened and all doors were kept open at least 10 minutes. Two heating systems were alternative. These were the ground source heat pump floor and radiator heating [17].

The results of the test where [17]:

- in the case of the heating system using traditional iron radiators
  - the highest measured dust values (16.77 mg/m$^3$) were recorded for the variant with 10 people in the room;
  - the lowest values (1.54 mg/m$^3$) were measured in case of unventilated room and without people inside;
- in the case of the heating system using underfloor heating
  - the measured values were generally higher than in the case of heating with radiators, for ventilated (13.70 mg/m$^3$) and unventilated (11.80 mg/m$^3$) rooms without persons.
  - the lowest values (7.67 mg/m$^3$) were measured in the case of a room with 8 people inside.

Studying the results from the Institute of Technology from Estonian University of Life Sciences, we can say that the heating system adopted for a building can influence the movements of the air in the rooms. The underfloor heating system heats the lower area of the rooms, the floor being warm, it draws hot air, thus rooting the dust deposited on it and on the furniture inside, at the height of a person, thus affecting the respiratory organs.

In the case of the heating system using radiators placed next to the windows, the hot air moves vertically in front of them, between the upper part of the radiator and the ceiling. The movement of air currents in the room, in this case, is easier and because of this the dust in the room is less entrained.

## 6. Discussions and conclusions

People generally spend about 90% of their time indoors. Whether we work, learn, eat, rest, relax or do sports, all these activities, in the times, now strongly affected by the pandemic restrictions, take place indoors, in most cases in our homes. All the activity are strongly influenced by the environment of the rooms, and therefore it is important to ensure a high-quality indoor climate.

The scientific evidence on the many links between buildings and health has grown substantially in recent decades and covers a vast array of risks to health from flaws in the heating, ventilation, and air conditioning systems over contaminants produced by some building materials, volatile organic compounds (VOC) and as well the use of industrial chemicals. The main purpose of buildings design should be to create a comfortable living condition for all occupants. As presented before, a large part of the population spends more than 20 hours a day in an artificial environment: at home, at work, in shops, in various places of recreation (restaurants, cafes, theaters, cinemas, galleries etc.).

This aspect has led to increased interest, both from builders and beneficiaries, in how the indoor environment can influence human health and mood. Over time, more research has been conducted in this area, national and international standards have also improved. These standards refer to indoor air quality (purity), air temperature, average radiant temperature, relative humidity, air speed, human activity, clothing.

Microclimatic conditions at home or at work must ensure high-performance work and the maintenance of health. In the current context of the SARSCOV-2 pandemic, we point out that for many employees the microclimatic conditions at work coincide with those at home. Globally, studying the specialized articles, there is a widespread concern that the indoor environment in which we carry out our daily activities can affect our health, comfort and performance.

For example, the indoor environment in educational institutions is of special interest, because educational buildings, compared to other types of buildings, are considered to be particularly sensitive to environmental deficiencies that could lead to poor quality of the indoor microclimate.

Lack of school facilities, old buildings or not renovated in time, contribute to improper operation and maintenance of air conditioning, heating or ventilation systems. In general, school children are much more active than adults, breathe larger volumes of air and may be more sensitive to indoor pollutants than adults. Indoor air quality in schools not only affects the health and comfort of students, but can affect their intellectual performance and have an influence on learning efficiency and attention during classes, with repercussions on their performance [19].

Consequently, the state of comfort is a feature of great technical complexity, and it can be negatively influenced by a multitude of factors. The performance of each person is affected by the work environment, by the ability to adapt to new environments, but also by personal motivation. The working environment refers to: indoor climate, access to certain services and infrastructure, which separately and together at the same time affect health [20].

The quality of the indoor environment is also influenced by other elements related to the architecture of the rooms such as the surface and interior design of the building, the characteristics and thermal environment of the indoor air and the noise level and light intensity.

A good indoor climate is essential for health and well-being. The most important aspects regarding the indoor climate are represented by the thermal conditions in the room, the ambient air temperature, but also the air quality (humidity, amount of dust particles), important to stay comfortable regardless of weather and season [19].

Light levels are also important. Exposure to daylight helps people feel energized and is considered one of the most desirable features of a job.

For example, a 2014 World Green Building Council (World GBC) study shows that [20]:

• access to exterior view windows:
- in homes and educational institutions - improves cognitive functions and memory by 10 - 25%;
- in hospitals - reduces hospitalization time by 8.5%;

• natural light:
- at work - increases worker productivity by 18%;
- at school - improves student performance by 5 - 14%;
- in stores - sales increase by 15 - 40%;

• building systems:
- proper artificial lighting - increases productivity by 23%;
- adequate ventilation - increases productivity by 5 - 14%;
- individual temperature control - increases productivity by 3%.

A good design can greatly enhance the qualities that buildings should have, making them not only more affordable, but also more pleasant and healthier - productive to work and pleasant to live in. At present, 80 million Europeans live in houses with unfavorable indoor climate, and strict compliance with the provisions of technical regulations is a mandatory condition for project developers, verifiers and experts, as well as for investors and executors.

As designers and executors, we must follow and respect the hygiene and human health requirements (Law 10/1995 indicative NC001-99), and this involves the design and construction of residential, administrative, agreement buildings, etc., but also of their component parts, not to affect the health of the occupants over time, aiming to ensure minimum technical performance conditions regarding: air and water hygiene, hygrothermal hygiene of the indoor environment, sunshine, lighting, acoustic hygiene of the indoor environment, quality finishes.

**References**

[1] Ţiganas S 2020 When a house is a Healthy Home? Construction Arena, https://www.arenaconstruct.ro/cand-este-o-casa-sanatoasa accessed on 2020
[2] Norm for the design, execution and operation of ventilation and air conditioning installations indicative I5-2010
[3] Gordeljevic V 2018 Healthy buildings, healthier people, HEAL (Health and Environment Alliance), https://www.env-health.org/wp-content/uploads/2018/05/Healthy-Buildings-Briefing.pdf
[4] Braubach M 2011 Environmental burden of disease associated with inadequate housing, World Health Organization Regional Office for Europe.
[5] World Health Organization 2011, World health statistics 2011, ISBN 978 92 4 156419 9
[6] Norm for the design, execution and operation of central heating installations (revision and merging of regulations I 13-2002 and I 13 / 1-2002)", indicative I 13-2015
[7] Standard SR EN ISO 7730: 2006 Moderate thermal environments. Analytical determination and interpretation of thermal comfort by calculating PMV and PPD indices and specifying local thermal comfort criteria
[8] Roodman D M and Lenssen N 1995 *A building revolution: how ecology and health concerns are transforming construction* vol 124
[9] Normative SR 1907-2:2014 Heating installations, computational heat requirements. Conventional indoor temperatures
[10] Pacurar C 2013 Study on the efficiency of teaching activities depending on the environmental climate, doctoral thesis summary, "POLITEHNICA" University of Timişoara
[11] Boerstra A, Veld P O and Eijdems H 2000 The health, safety and comfort advantages of low temperature heating systems: a literature review *Proc. 6th Int. Conf. Heal. Build.* **2** 629–34
[12] Szer I, Błazik-Borowa E and Szer J 2017 The Influence of Environmental Factors on Employee Comfort Based on an Example of Location Temperature *Arch. Civ. Eng.* **63** 163–74
[13] Directive 2008/50/EC of the European Parliament and of the Council of 21 May 2008 on ambient air quality and cleaner air for Europe – European Directive or regulation No. 1. (https://eur-lex.europa.eu/legal-content/en/ALL/?uri=CELEX%3A32008L0050)
[14] Wolkoff P 2018 Indoor air humidity, air quality, and health – An overview *Int. J. Hyg. Environ. Health* **221** 376–90
[15] Framework normative regarding the detailing of the content of the requirements established by law 10/1995 indicative NC001-99
[16] Stevens S 2021 Ten Healthy Building Materials for Your Next Project, GBD magazine, https://gbdmagazine.com/healthy-building-materials/ accessed on 2021
[17] Nõu T and Viljasoo V 2011 The effect of heating systems on dust, an indoor climate factor *Agron. Res.* **9** 165–74
[18] Health Sector World Health Organization, Indoor air pollution, Children's Health and the Environment WHO Training Package for the www.who.int/ceh
[19] Moroianu I and Popescu O 2008 Environment and health, Romanian Institute for Human Rights Bucharest, ISBN 978-973-9316-66-8

# Soil stabilisation with hydraulic binders

A G Chiorean[1*], M D Vasile[1], A C Nagy[1] and N M Ilieş[1]
[1] Technical University of Cluj-Napoca, Romania

* chioreanadina@gmail.com

**Abstract.** The lack of a good foundation ground layer in some areas that could be used for construction proposes is a common problem all over the world. Today, more and more researches are focused on finding new solutions to improve the soil parameters in order to obtain a stabilized foundation layer. Finding an optimal admixture between the local soils, needed for construction purposes, hydraulic binders and the precise amount of water, may be the future of a revolutionary discovery in this field. The experimental study based on the admixture of a sandy silty clay with hydraulic binders from the Viacalco© [1] range, produced by Carmeuse Romania, in a percentage of 2%, 4% and 6% led to a significant improvement of the geotechnical parameters of the soil. The aim of this experimental study is to determine the optimal binder mixture in order to improve local soils and to obtain low investment costs.

## 1. Introduction

An experimental study focused on a comparative analysis of the soil mixed with the hydraulic binders, Viacalco D© [1] and Viacalco S© [1], was carried out in the geotechnical laboratory of the Technical University of Cluj County in order to obtain an optimal mixture that would improve the geotechnical parameters and the cost of the investment.

### *1.1. Objective of the study*

The main purpose of this study was to highlight how different amounts of hydraulic binders can change the properties of a weak foundation layer.

As more and more constructions are made on difficult foundation soils that involve high costs, the aim is to find a mixture that could improve the foundation layer and his geotechnical parameters and lead to a decrease of the execution cost.

Similar mixtures could be used on the hydrotechnical defense structures, to improve their stability, and due to the large number of floods in Romania, there is a special interest in this matter. Therefore, in the study, 3 sections of dam were compared in terms of stability, each dam body having a different composition.

### *1.2. State of the art*

At the basis of this chapter are standing the results of various tests presented in published articles, experimental studies, Romanian regulations and specialized guides.

An experimental study carried out on the Căpuş River [2], where hydraulic binders were used in order to stabilize the filling material of a riverbank defense and a dam, shows us that by stabilizing the filler material by mixing the local soil with hydraulic binder and water, followed by the compaction of the mixture, compared to the soil in its natural state, a significant improvement of the geotechnical parameters (internal friction angle, cohesion and linear deformation modulus) was made.

Another article published "Applications of stabilized soil in hydraulic constructions" [3], presents the benefits that the hydraulic binders have on the hydrotechnical constructions and on the environment. By mixing the local material with different types of cement, an increase of the cohesion is obtained, while the internal friction angle remains approximately the same. The deformation modulus and permeability were not affected by the mixture, but this technology has a considerably smaller carbon footprint than the conventional solutions, as most transport processes are avoided in the absence of reinforcement.

"Freeze-thaw testing of stabilized soil samples used for riverbank consolidation" [4], also shows that even the freeze-thaw action of the soil improved with hydraulic binders improves the resistance of a flood defense structure. The aim of the practical experiment, carried out during the winter-summer period, was to show that one of the binders used has a significant contribute regarding the resistance of the structure. Meanwhile the structure made from the first binder remained intact, the structure for which the second type of binder was used showed degradation and traces of erosion.

All existing studies show a significant improvement in the bearing capacity of soils based on mixtures with an optimal dosage of hydraulic binder, but we can say that further research is needed to improve this area.

## 2. Materials and procedures

### *2.1. Materials used for the experimental study*

The research was performed on a sandy silty clay mixed with the hydraulic binders Viacalco D© and Viacalco S© of Carmeuse Romania.

The soil used for the mixtures was taken from a site near Cluj county and his composition was determinate using the granulometric curve.

The soils initial geotechnical parameters were: $\gamma_s$ = 26.68 [kN/m$^3$] (bulk unit weight); $\gamma_d$ = 17.17 [kN/m$^3$] (dry unit weight); n = 35.51 [%] (porosity); e = 0.55 (voids ratio); w = 17.48 [%] (natural water content). The values of the parameters were taken from a previous study conducted on the same type of soil [5].

There were made 56 mixtures containing 2000 g of sandy silty clay each (figure 1), different amounts of water and different amounts of hydraulic binders. Each mixture was submitted to the Proctor test in order to obtain the optimum moisture content for the soil in natural state and for the soil mixed with the hydraulic binders.

**Figure 1**. The homogenizing of the mixture.

In the presence of the water, the hydraulic binders hydrate, forming crystals that lead to a common phenomenon known as "hydraulic input" that gives the soil a permanent cohesion.

The aim of mixing the soil with hydraulic binders is to improve the technical quality conditions that the form layers and the support layer of the foundations must fulfill.

Following the Proctor test 7 mixtures were obtained:

1$^{st}$: 2000 g of sandy silty clay + 18% water (figure 2)

2nd: 2000 g of sandy silty clay + 2% Viacalco D© + 20% water (figure 3)
3rd: 2000 g of sandy silty clay + 4% Viacalco D© + 20% water (figure 4)
4th: 2000 g of sandy silty clay + 6% Viacalco D© + 20% water (figure 5)
5th: 2000 g of sandy silty clay + 2% Viacalco S© + 20% water (figure 6)
6th: 2000 g of sandy silty clay + 4% Viacalco S© + 20% water (figure 7)
7th: 2000 g of sandy silty clay + 6% Viacalco S© + 22% water (figure 8)

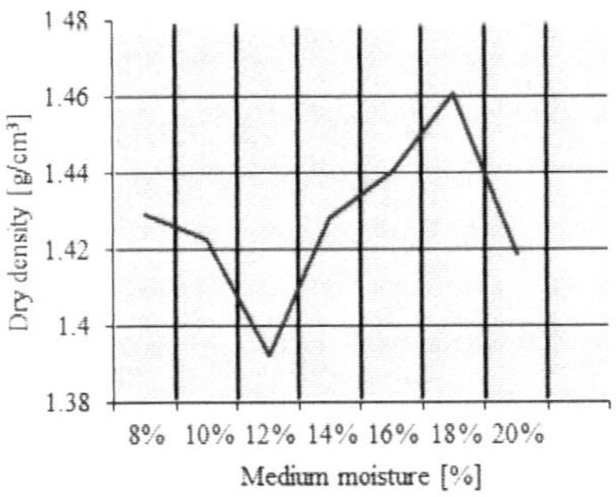


**Figure 2**. sandy silty clay + 18% water.

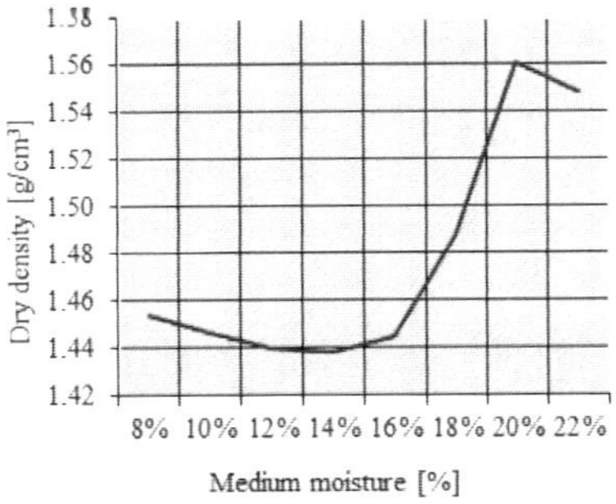


**Figure 3**. sandy silty clay + 2% Viacalco D© + 20% water.

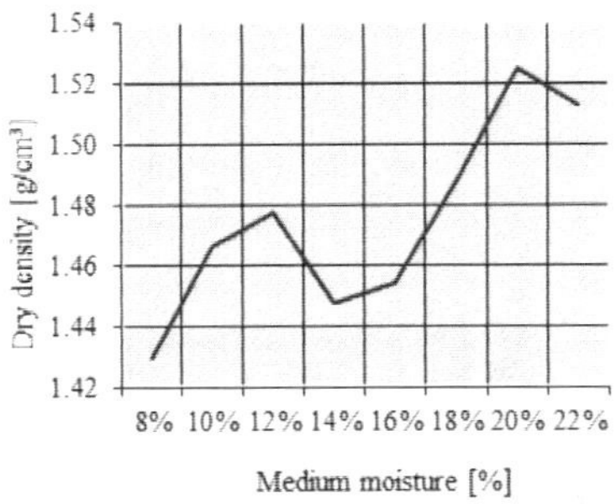


**Figure 4.** sandy silty clay + 4% Viacalco D© + 20% water.

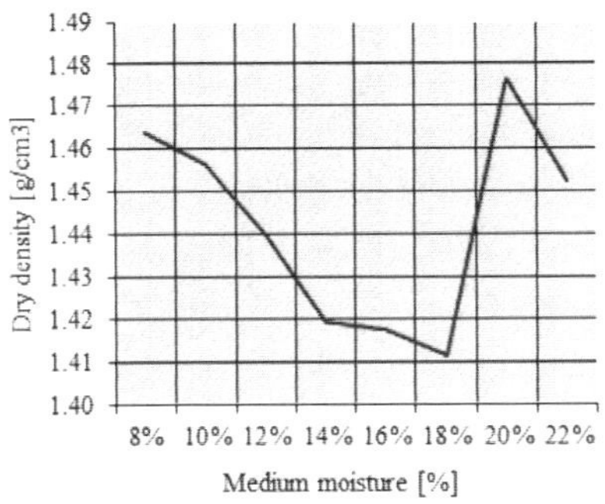


**Figure 5.** sandy silty clay + 6% Viacalco D© + 20% W

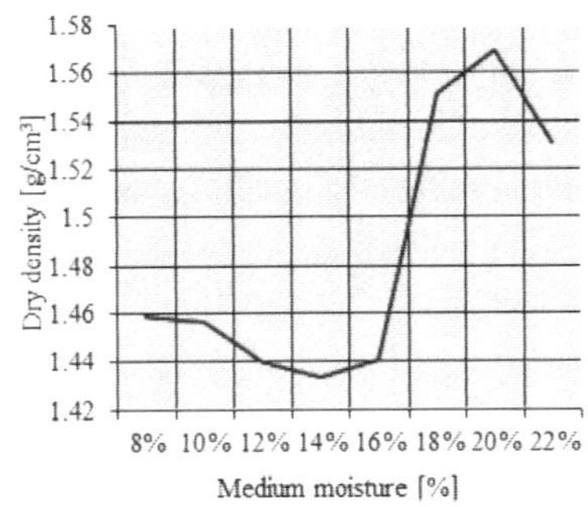


**Figure 6.** sandy silty clay + 2% Viacalco S© + 20% water.

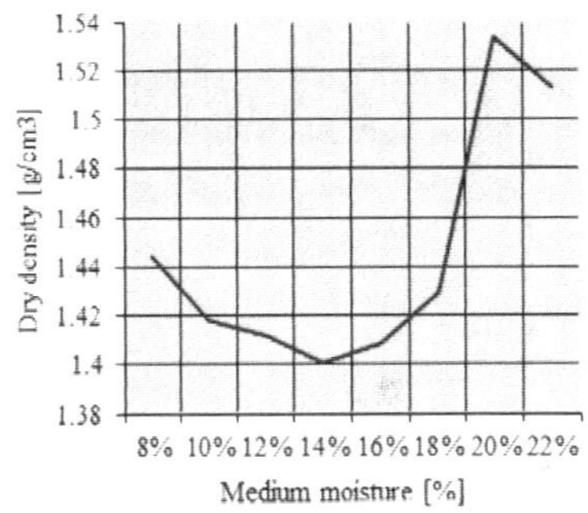


**Figure 7**. sandy silty clay + 4% Viacalco S© + 20% water.

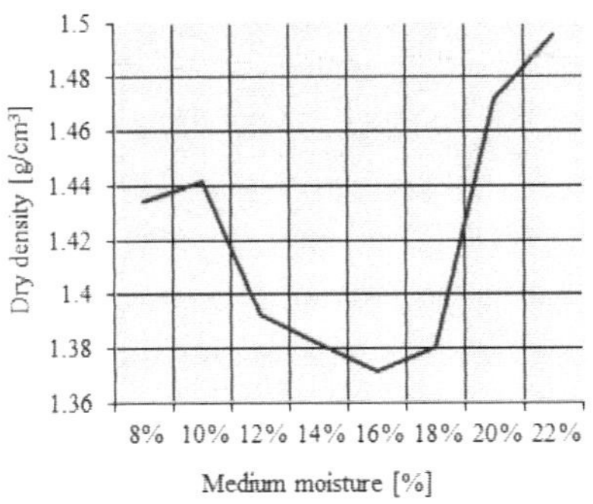


**Figure 8**. sandy silty clay + 6% Viacalco D© + 22% water.

*2.2. Benefits of the Viacalco© hydraulic binders*

Used in fine soils and cohesive soils, they optimize and neutralize the clay content, accelerating the development of the works. They help increase the load-bearing capacity, the mechanical strength of treated soils and the bearing capacity of embankments.

Thanks to them properties the binders can change the physico-chemical properties of soils and decrease rapidly the excess of moisture.

The installation of Viacalco D© and Viacalco S© must be carried out according to the technical data sheet that the suppliers provide.

*2.3. Testing procedures*

As presented before, the study started with the granulometric curve in order to determine the soil composition, followed by the Proctor test on the basis of which we obtained the optimum moisture content and we established the 7 mixtures we subjected to the direct shear test and to the oedometric test.

*2.3.1 Direct shear strength.* The principle of this method consists in subjecting the ground sample to a shear stress F, which deforms under the action of tangential unit forces τ in accordance with STAS 8942/2-82. After taking the 42 samples (figure 9), they were left to rest for 7 days, respectively 28 days to reach the initial resistance and the standard resistance and then, they were subjected to the direct shear test. Three samples were taken for each mixture, corresponding to the shear loads of 100 kPa, 200 kPa and 300 kPa.

**Figure 9**. Sampling and preparation of the samples for the direct shear test.

From a percentage point of view, there is a significant increase in the cohesion of the soil samples improved with Viacalco D© in a proportion of 2%, 4% and 6% and with Viacalco S© in a proportion of 4% and 6% compared to the soil sample in the natural state.

Regarding the residual values, it is observed that for the samples with binder in proportion of 2%, subjected to the test after 7 days of rest, the residual cohesion recorded is lower than the residual cohesion of the soil in its natural state.

The internal friction angle, in all cases, registers an increase with the increase of the hydraulic binder percentage, thus obtaining higher values than the internal friction angle for the soil sample in its natural state.

**Table 1.** The results of the 7-day direct shear test for the soil in his natural state.

| Sample type | Internal friction angle φ` | Cohesion c` | Residual Internal friction angle $\varphi_{res}$ | Residual Cohesion $c_{res}$ |
|---|---|---|---|---|
| Soil in the natural state + w 18% | 7.05° | 153.68 kPa | 7.00° | 45.65 kPa |

**Table 2.** The results of the 7-day direct shear test for the soil mixed with Viacalco D and Viacalco S.

| Sample type | Internal friction angle φ` | Cohesion c` | Residual Internal friction angle $\varphi_{res}$ | Residual Cohesion $c_{res}$ |
|---|---|---|---|---|
| Mixture with Viacalco D© 2% + w 20% | 17.85° | 160.16 kPa | 17.28° | 33.41 kPa |
| Mixture with Viacalco D© 4% + w 20% | 19.18° | 259.36 kPa | 18.53° | 85.19 kPa |
| Mixture with Viacalco D© 6% + w 20% | 20.88° | 293.47 kPa | 20.33° | 87.83 kPa |
| Mixture with Viacalco S© 2% + w 20% | 16.27° | 122.23 kPa | 11.45° | 42.72 kPa |
| Mixture with Viacalco S© 4% + w 20% | 24.27° | 168.00 kPa | 15.91° | 56.62 kPa |
| Mixture with Viacalco S© 6% + w 22% | 10.44° | 223.19 kPa | 15.56° | 61.15 kPa |

**Table 3**. The results of the 28-day direct shear test for the soil mixed with Viacalco D and Viacalco S.

| Sample type | Internal friction angle φ` | Cohesion c` | Residual Internal friction angle $\varphi_{res}$ | Residual Cohesion $c_{res}$ |
|---|---|---|---|---|
| Mixture with Viacalco D© 2% + w 20% | 26.49° | 175.04 kPa | 22.23° | 56.75 kPa |
| Mixture with Viacalco D© 4% + w 20% | 37.26° | 272.50 kPa | 15.39° | 95.77 kPa |
| Mixture with Viacalco D© 6% + w 20% | 40.21° | 300.69 kPa | 17.67° | 102.91 kPa |
| Mixture with Viacalco S© 2% + w 20% | 26.16° | 131.78 kPa | 9.62° | 62.34 kPa |
| Mixture with Viacalco S© 4% + w 20% | 42.08° | 225.24 kPa | 22.40° | 58.53 kPa |
| Mixture with Viacalco S© 6% + w 22% | 45.96° | 238.12 kPa | 19.31° | 63.11 kPa |

*2.3.2 Oedometric test*

The determination of the compressibility of the soil in the laboratory consists in following the settlement of the samples and their evolution in time under the effect of the vertical axial loads, applied in steps. The deformation measurement, under the effect of the applied loads, is made on cylindrical samples, with zero lateral deformations and with drainage on the lower and upper faces. The oedometric test was performed in accordance with STAS 8942/1-89.

**Figure 10**. Oedometric equipment.

The results are presented as follows:

**Table 4.** The results of the oedometric test.

| Sample type | Maximum point in the wet domain – w | Oedometric modulus $M_{2-3}$ |
|---|---|---|
| Soil in its natural state | 18 % | 7604.56 kPa |
| Mixture with Viacalco D© 2% | 20 % | 10638.30 kPa |
| Mixture with Viacalco D© 4% | 20 % | 12269.94 kPa |
| Mixture with Viacalco D© 6% | 20 % | 21739.13 kPa |
| Mixture with Viacalco S© 2% | 20 % | 14493 kPa |
| Mixture with Viacalco S© 4% | 20 % | 9662 kPa |

From a percentage point of view, there is a significant increase in the oedometric deformation modulus for all types of mixture compared to the natural soil sample.

## 3. Stability and cost simulation

A hypothetical study, which requires further investigations, was performed for three dam sections whose composition consists of soil mixtures with the two hydraulic binders, Viacalco D© and Viacalco S©, in a percentage of 4% and the soil without the hydraulic binder.

The study was performed to evaluate and compare the stability (figure 11, figure 12, figure 13) of the structures and the costs that are generated.

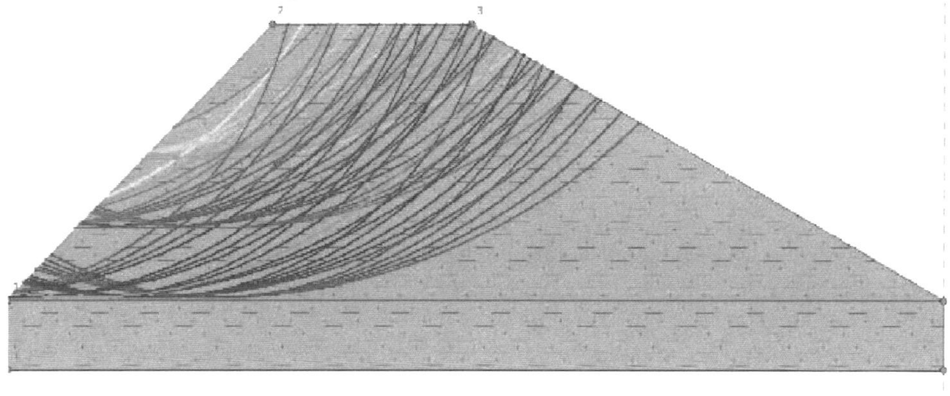

**Slope stability check (Fellenius / Petterson)**
The sum of the active forces: $F_a = 81.94$ kN/m
The sum of the passive forces: $F_p = 420.29$ kN/m
Sliding moment: $M_a = 81.94$ kN/m
Stability moment: $M_p = 81.94$ kN/m
Stability factor = 5.13 > 1.50
Slope stability ACCEPTABLE

**Figure 11**. Stability factor - soil in its natural state.

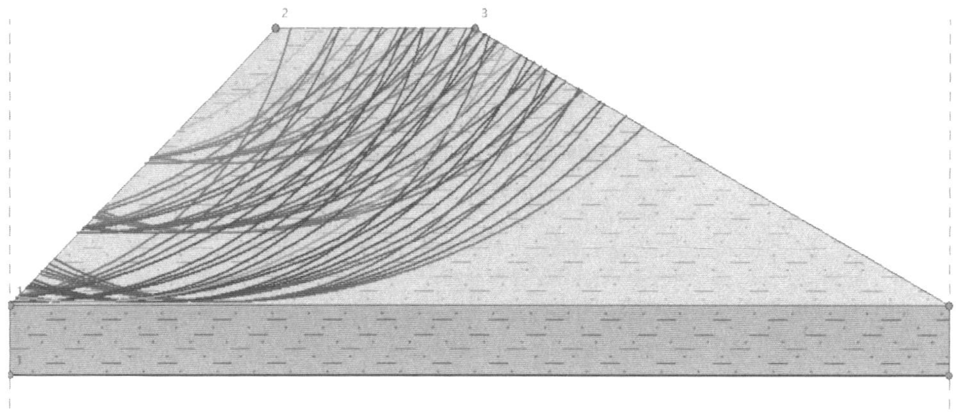

**Slope stability check (Fellenius / Petterson)**
The sum of the active forces: $F_a = 88.29$ kN/m
The sum of the passive forces: $F_p = 591.92$ kN/m
Sliding moment: $M_a = 640.08$ kN/m
Stability moment: $M_p = 4291.44$ kN/m
Stability factor = 6.70 > 1.50
Slope stability ACCEPTABLE

**Figure 12**. Stability factor - soil + Viacalco S© 4% + w = 20%.

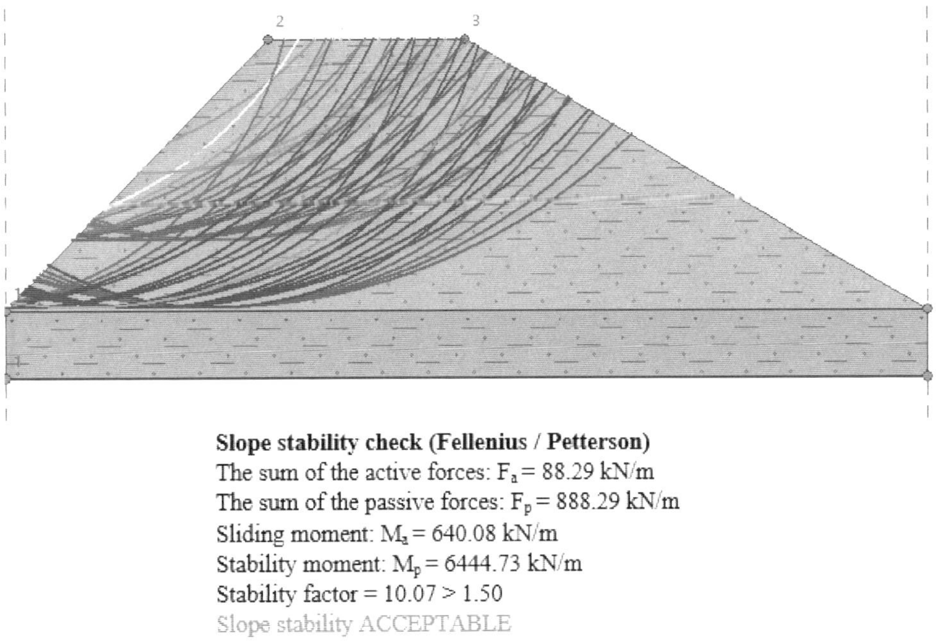


**Figure 13**. Stability factor - soil + Viacalco D© 4% + w = 20%.

The stability was verified using the Geo5 [6] program.

Both, economical evaluation and stability test were performed on a dam section having a height of 4 m, a length of 100 m, the upstream slope of 60 °, and the downstream slope of 45 °. The results were exagerated, they showed a great stability even without adding the hydraulic binder for the type of soil chosen. But, in any case, the results show a major improvement of the stability factor using specially Viacalco D© in the admixture with the soil and the optimum moisture content.

As for the economical evaluation, we cannot consider the cost as a decisional factor in choosing the type of binder that we want to put in place, since the price differences were insignificant.

## 4. Conclusion

After analyzing the studied soil (sandy silty clay) improved with the hydraulic binders Viacalco D© and Viacalco S©, compared to the soil in its natural state, the results were:

- a significant improvement of the geotechnical parameters, such as internal friction angle, cohesion and oedometric deformation modulus;
- also residual internal friction angle and residual cohesion had a significant improvement;
- the increase of the safety factor on dams;
- due to the high values of the safety factor, the dimensions of the sections of the defense structures, as dams, for example, can be reduced;
- by reducing the size of the sections and replacing the materials needed to be transported on the site, with local materials (local soil) improved with binders, a reduction in investment cost can be achieved.

The hydraulic binders are used more and more for soil stabilizations. Besides the improvement of the geotechnical parameters of the soil that we obtain, they reduce the carbon footprint and the environmental impact by being an eco-friendly product and they have an important role in the economic development of the construction field.

Not long ago, the mixture of local soils with hydraulic binders began to be used for embankments, dams and road works, but nowadays more studies are focused on developing solid base layers for other construction types.

The aim of this article was to emphasize the improvement that the hydraulic binders have on the local soil and to help other studies to develop based on the results obtained regarding the optimal amount of the hydraulic binder that needs to be used in an admixture.

**Acknowledgments**
The authors would like to thank Carmeuse Romania for the binders and the technical information provided and the Technical University of Cluj County, Faculty of Constructions as well for all the support and the access granted to the geotechnical laboratory and all the equipment needed to conduct this experimental study.

**References**

[1] Viacalco - http://www carmeuse-construction com/ro/viacalco - access on 2 June 2021
[2] Ciubotaru V S, Circu A P, Crăciunescu B M, Nagy A C, Topan A, ing Dârgâu A, Metode de apărare împotriva inundațiilor - sector experimental pe râul Căpuș
[3] Nagy A -Cs , Cîrcu A -P , Ciubotaru V -C , Crăciunescu B -M , Topan G , Chereches T , Fărcaş V S Applications of stabilized soil in hydraulic constructions
[4] A-Cs Nagy, A-P Cîrcu, V-C Ciubotaru, B M Crăciunescu, G Topan, D V Moldovan, L-E Moldovan, D M Pașca Freeze-thaw testing of stabilized soil samples used for riverbank consolidation
[5] Péter A and Gherman C, lucrare de disertație Studiu asupra stabilizării pământurilor argiloase cu lianți hidraulici de tip Dorosol
[6] Geo5 Software User's Guide

# Concrete consoles impact to a masonry walls building with a large hall at the top story and a lightweight timber roof

**S Constantinescu**[1*]

[1] Technical University of Bucharest, Romania

* sorina.constantinescu@yahoo.com

**Abstract**. The paper studies the behavior of a medium height masonry walls building, placed in Satu Mare (Romania). It contains large balcony consoles and a large hall at the top story, covered by a timber roof. The roof trusses sit at both ends on the external walls. The study's purpose is to determine the importance of large slab consoles and walls buttresses at the top story, for the structure. The paper compares the structure's behavior in 3 cases. The first contains both large consoles and wall spurs, the second only incorporates pier buttresses and the third holds just the balcony consoles. The structure's behavior for both the elastic and plastic stage is compared for the 3 cases. The plastic stage shows the bending moment's distribution in slabs and stresses in masonry walls during stages when they surpass the masonry strengths. The plastic hinges development at structure collapse and the pushover diagrams for each case along with the results interpretation are presented. A nonlinear analysis is done for the timber roof also, for both orthogonal directions. The study shows both large consoles and halls at the top floor are appropriate for masonry structures, if designed correctly.

## 1. Introduction

Timber trusses are used as roof supporting construction for objects with open inside space, especially halls or office buildings. In literature, there are studies about timber trusses behavior using the finite element method [1]. Long term behavior of timber structures is influenced by the environment, by exploitation conditions, by the structural elements state at the building time, by the connections and material degradation. Some of these parameters are taken into account in the design codes [2]. Joints connections will creep at a different rate to the structural timber and could cause stress redistribution [3]. It is important that if old timber elements are reused there are certain problems to be remembered. There may be cracks, screws and nail holes and biological damages, especially at the elements ends [4]. For bearing masonry walls both concrete blocks and clay bricks can be used. Their compressive strengths are similar. The concrete bricks have a more brittle failing mechanism and require more mortar [5]. Autoclaved aerated concrete blocks can also be used for bearing walls. The AAC seismic analyses show that for 1 and 2 story buildings and low to medium seismic areas, the damage limit may not be reached. For higher seismic areas and higher buildings, significant damage can be expected [6]. Masonry walls are thought to fail due to diagonal shear or sliding shear [7]. Confined masonry walls are a suited solution for buildings of low to medium rise. Earthquakes and laboratory tests show the effectiveness of confined masonry systems. The seismic performance is also influenced by the masonry properties, structure configuration and reinforcement in tie columns and tie beams [8]. The literature shows that slender masonry walls may experience buckling to axial force [9]. Thin walls also have a lower shear force bearing capacity [10]. Nonlinear analysis use, for masonry buildings, is seen in literature [11]. Pushover analysis gives information about the building's failure pattern and the building's safety. This is shown from the plastic hinges development as the structure turns into a mechanism [12]. This is a study about a medium height masonry building, placed in a low seismic area (Satu Mare, Romania). The aim is to establish the importance of large slab consoles and large halls with buttress walls, covered by a timber truss roof, to the structure's behavior.

## 2. Building description

The building in study is situated in Satu Mare, Romania. The codes in force used to design the building are: [13–22]. This is a low seismic area, the seismic acceleration is 0.15g (g is the gravity acceleration) [22]. The structure incorporates confined masonry walls, beams and reinforced concrete slabs. The building contains a basement and 5 stories above it. It will be used as a hotel. Story height is 3m for all 5 stories above the basement. There is a timber truss roof placed above the top story.

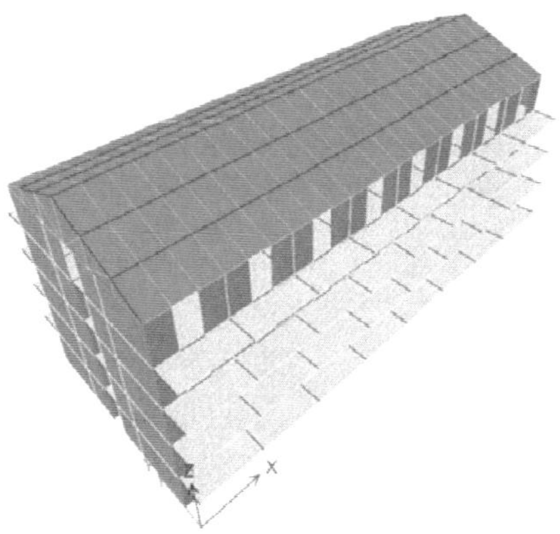

Figure 1. 3D image

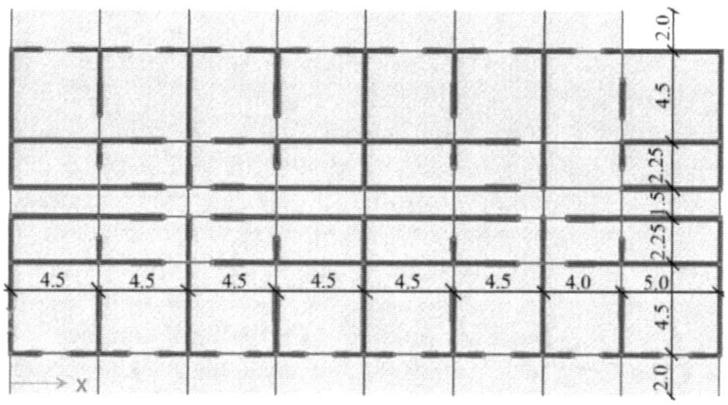


Figure 2. Current story layout

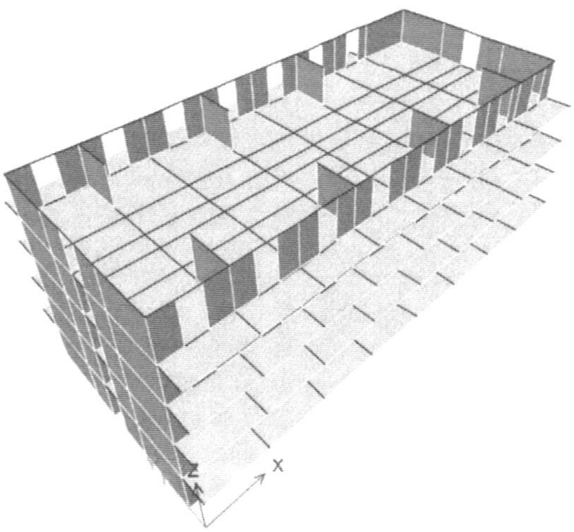

Figure 3. 3D image for case A

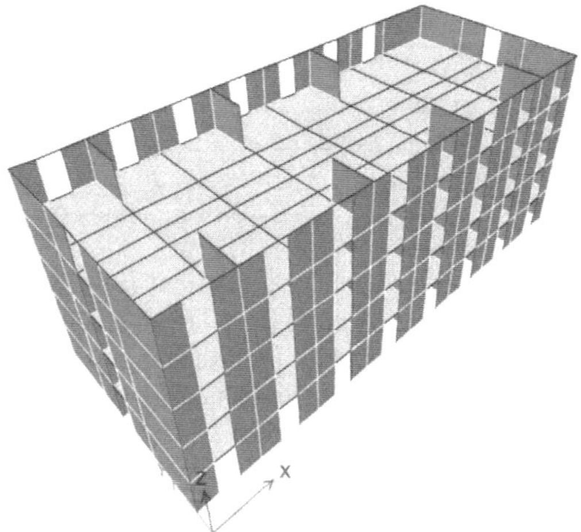

Figure 4. 3D image for case B

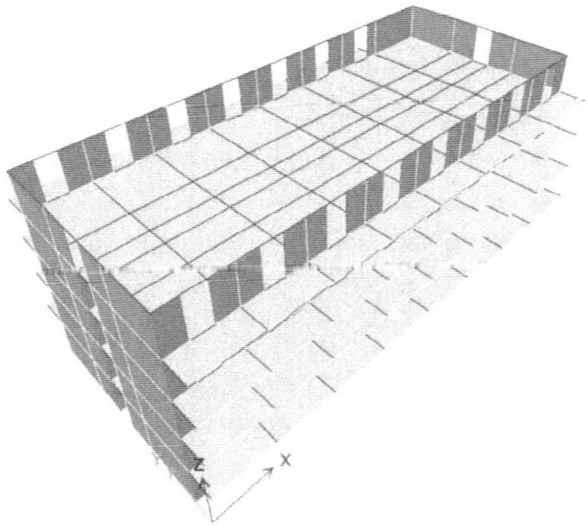


**Figure 5.** 3D image for case C

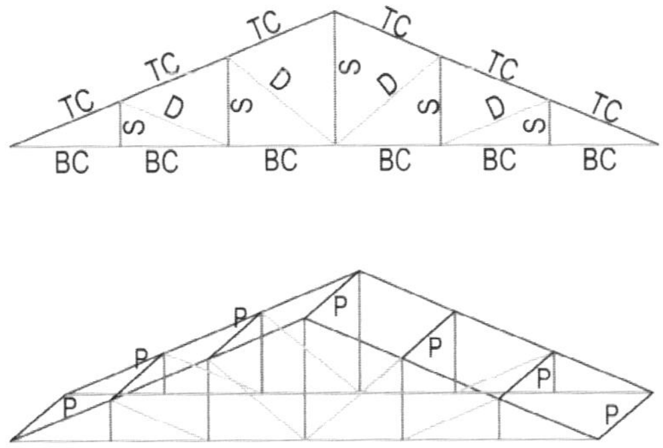


**Figure 6.** Timber elements

This roof is used because the top story is a big hall and there is a large space to be covered. The trusses sit on the edge masonry walls. They are placed at 1.5m between them and cover a 15m distance. This paper analyses 3 cases. In the first case (A) the slabs contain large consoles extended 2m beyond the building's edges and the longitudinal walls (on direction X) at the top story that bare the trusses, have spurs on direction Y. The second case (B) does not contain the large slab consoles. The third case (C) does not contain the walls buttresses at the top story. This is done to highlight the importance of both large consoles and walls spurs. The slabs bending moments change from the elastic to the plastic stage is also studied. The structure plans with bay dimensions and 3D images are seen in figures 1 to 5. The design program used is ETABS 18. Beams and braces are blue, columns are green, walls are red and slabs are grey. The truss elements are seen in figure 6. The roof contains 24 trusses. Only 2 are shown here, to highlight the purlins (P). The truss elements are: top chord (TC), diagonals (D), bottom chord (BC) and soldiers (S). Purlins are subjected to biaxial bending moment.

## 3. Theory elements used in paper

The permanent and variable loads and load combinations are determined using [15-19].

### *3.1 The seismic action*

The base force $F_b$ at the ground floor bottom is calculated using [19] and [22]. The coefficient $c_s$ is used to introduce the seismic load. The factor in the formula are: $\gamma_{I,e} = 1$, the building's importance-exposure coefficient, $\beta_0 = 2.5$ the maximum value in the elastic spectrum, $a_g = 0.15g$, q the structure's behavior factor, $q=2.25\cdot1.25$ [22], m = building's mass, $\eta = 0.88$ the reduction factor and $\lambda = 0.85$ [22]. The stresses analysed for the confined masonry walls are $\sigma_x$, $\sigma_z$, $\tau_{xz}$, $\tau_{xy}$ and $\tau_{yz}$. These stresses are compared to the masonry design strengths [13]: horizontal compression $f_{dh}$ (1.1N/mm$^2$), vertical compression $f_d$ (2.31 N/mm$^2$), shear strength for horizontal direction $f_{vd,0}$ (0.158 N/mm$^2$), horizontal and vertical strengths perpendicular to the wall $f_{xd1}$ (0.126 N/mm$^2$) and $f_{xd2}$ (0.252 N/mm$^2$). The design strengths for the full bricks used, are calculated using the characteristic values: $f_{kh}$, $f_k$, $f_{vk,0}$, $f_{xk1}$ and $f_{xk2}$. The shear diagonal masonry design strength is $f_{vd,i}=f_{vk,i}/\gamma_M$. The masonry strengths insurance factor is $\gamma_M=1.9$ [13]. The concrete compression design strength $f_{cd}$, is determined using the characteristic strength $f_{ck}$. For the steel reinforcement bars in tie beams and slender columns, $f_{yd}$ is the design strength and $f_{yk}$ is the characteristic value. The insurance factors are $\gamma_C=1.5$, for concrete and $\gamma_S=1.15$ for steel [20].

$$F_b = \gamma_{I,e}\cdot\beta_0\cdot a_g/q\cdot m\cdot\eta\cdot\lambda = c_s\cdot m\cdot g = 0.1\cdot m\cdot g \quad [kN] \tag{1}$$

$$f_d = f_k/\gamma_M \quad [N/mm^2] \tag{2}$$

$$f_{cd} = f_{ck}/\gamma_C = 16/1.5 = 10.67\ N/mm^2 \quad (3)$$

$$f_{yd} = f_{yk}/\gamma_S = 355/1.15 = 309\ N/mm^2 \quad (4)$$

*3.2. Confined masonry walls design rules and results*

$M_{Rd}$ is the wall's bearing bending moment [21]. $M_{Rd(M)}$ and $M_{Rd(As)}$ are the bearing bending moments from the masonry area and from the slender columns reinforcements respectively [21]. CA is the wall's compressed area [21]. As is the slender columns horizontal reinforcement area. $y_c$ is the distance between the compressed area weight center GC and the wall's weight center G [13]. ls is the distance between the slender columns at the wall's edges. S is the earthquake direction and tw=25cm is the wall's width. $V_{Rd}$ is the masonry wall bearing shear force and $V_{Ed}$ is the horizontal shear force from the seismic loads combination [13]. $V_{Rd,i}^* = A_w \cdot f_{vd,i} / b$, where b is a correction coefficient connected to the masonry walls dimensions $h_{pan}$ =h=story height- beam height and $l_{pan}$ are the height and length of the masonry area panel. $b=1.5$ if $h/l_w \geq 1.5$, $b=1.0$ if $h/l_w<1.0$ and $b=h/l_w$ if $1.0 \leq h/l_w<1.5$, $f_{vd,i}$ is the diagonal shear strength.

$$M_{Rd} = M_{Rd(M)} + M_{Rd(As)}\ [kNm] \quad (5)$$

$$CA = N_{Ed}/(0.85 \cdot f_d)\ [mm^2] \quad (6)$$

$$M_{Rd(M)} = N_{Ed} \cdot y_c\ [kNm] \quad (7)$$

$$M_{Rd(As)} = ls \cdot As \cdot f_{yd}\ [kNm] \quad (8)$$

$$V_{Rd} = \min(V_{Rd1}^*, V_{Rdi}^*) + V_{Rd2} + V_{RSC}\ [kN] \quad (9)$$

$$V_{Rd1}^* = 0.4 \cdot (N_{Ed} + 0.8 \cdot V_{Ed} \cdot h_{pan}/l_{pan})\ [kN] \quad (10)$$

$$V_{Ed} \leq l_{pan} \cdot t \cdot f_{vd,0} \quad (11)$$

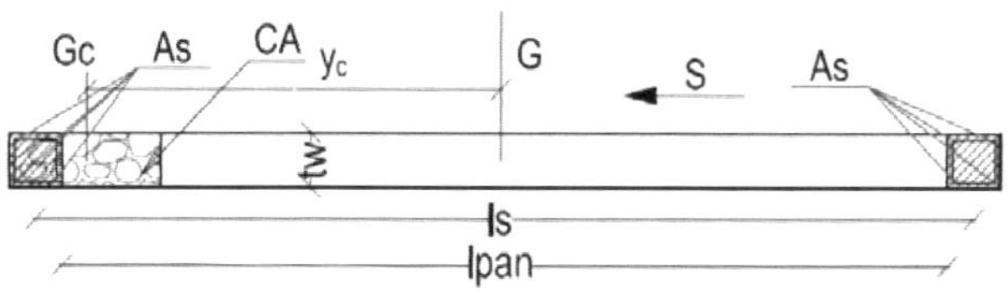


**Figure 7.** Confined masonry wall section

$h_{pan}$ =story height - tie beam height and $l_{pan}$ are the height and length of the masonry area panel. $V_{Rd2}$ is the bearing horizontal shear force from the reinforcement in the slender column at wall's compressed edge [13]. As is the reinforcement area in the slender column at that edge. $\lambda_c$ is the reinforcement participation factor. Here, $\lambda_c=0.30$, for longitudinal reinforcement Φ12.

$$V_{Rd2} = \lambda c \cdot As \cdot f_{yd}\ [kN] \quad (12)$$

$$V_{RSC} = Ac \cdot f_{cvd} \quad (13)$$

*3.3. Roof truss and purlin design rules and results*

TC and D are subjected to compression. BC and S are subjected to tension. The compression stress $\sigma_{c,0,d}$ and tension stress $\sigma_{t,0,d}$ are determined according to equations (14) and (15). $N_d$ is the compression force and A is the section area. The $N_d$ values for the truss elements are 105kN for BC, 86kN for TC, 51kN for D and 40kN for S.

$$\sigma_{c,0,d} = N_d/(A \cdot k_{c,y(z)}) \quad (14)$$

$$\sigma_{t,0,d} = N_d/A \tag{15}$$

Conditions in formulas (14) and (15), need to be respected [14]. $\beta_c$=0.2 for solid timber. $f_{c,0,d}$ and $f_{t,0,d}$ are the timber compression and tension strengths. They are determined from the characteristic values $f_{c,0,k}$ =23N/mm$^2$ and $f_{t,0,k}$ = 18N/mm$^2$ using the coefficients $k_{mod}$, $k_{sys}$, $k_h$ and $\gamma_M$. $k_{mod}$=0.8 here, is taken from the code according to the humidity class, the timber type and the load duration. $k_{sys}$=1 and $\gamma_M$=1.3 for solid timber. $k_h$ is calculated from formula (16), where h is the section height. A, I and i are the cross section area, moment of inertia and radius of gyration. $E_{0,05}$ is the timber 5% longitudinal elasticity modulus [14]. Le is the effective length for each compressed element. Here, both TC and D are hinged at both sides, so Le= the segment length. Le=2.7m for TC and 3.2m for the longest D. Timber class C30 is used here. If $\lambda_{rel,y(z)} \leq 0.3$, then $k_{c,y(z)}$ is ignored [14].

$$k_h=\min[(150/h)^{0.2}, 1.3] \tag{16}$$

$$f_{t,0,d}=k_{mod}\cdot k_{sys}\cdot k_h\cdot f_{t,0,k}/\gamma_M \tag{17}$$

$$f_{c,0,d}=k_{mod}\cdot k_{sys}\cdot k_h\cdot f_{c,0,k}/\gamma_M \tag{18}$$

$$f_{c,0,d}\cdot k_{c,y(z)} \geq \sigma_{c,0,d} \tag{19}$$

$$k_{c,y(z)}=1/[k_{y(z)}+(k^2_{y(z)}-\lambda^2_{rel,y(z)})^{0.5}] \tag{20}$$

$$k_{y(z)}=0.5[1+\beta_c(\lambda_{rel,y(z)}-0.3)+\lambda^2_{rel,y(z)}] \tag{21}$$

$$\lambda_{rel,y(z)}=\lambda_{y(z)} / \pi \cdot(f_{c,0,k}/E_{0,05})^{0.5} \tag{22}$$

$$\lambda_{y(z)} = Le/i_{y(z)}=Le(A/I)^{0.5} \tag{23}$$

The purlins (P) connecting the trusses are designed for bending about both y-y and z-z axes. The conditions that need to be followed are in formulas (24) and (25) because in this case the values are calculated for $\lambda_{rel,m,y/z} \leq 0.75$ [14]. $\sigma_{m,y,d}$ and $\sigma_{m,z,d}$ are the design bending stresses on axes y and z. $f_{m,y,d}$ and $f_{m,z,d}$ are the design bending strengths, calculated from the characteristic values $f_{m,y,k}$ and $f_{m,z,k}$. $k_m$=1 for this case, is a modification factor. $M_{y,d}$ (0.95kNm) and $M_{z,d}$ (2.6kNm) are the design bending moments. $W_y$ and $W_z$ are the associated section modulus. b is the section width and h is the height. $G_{0,05}$ is the timber 5% transversal elasticity modulus and $l_{ef}$ =0.9·l is the effective purlin length, calculated using the real length "l " [14].

$$\sigma_{m,y,d}/f_{m,y,d} + k_m\cdot\sigma_{m,z,d}/f_{m,z,d} \leq 1 \tag{24}$$

$$k_m\cdot\sigma_{m,y,d}/f_{m,y,d} + \sigma_{m,z,d}/f_{m,z,d} \leq 1 \tag{25}$$

$$\sigma_{m,y(z),d}=M_{y(z),d}/W_{y(z)} \tag{26}$$

$$f_{m,y(z),d}= k_{mod}\cdot k_{sys}\cdot k_h\cdot f_{m,k}/\gamma_M \tag{27}$$

$$M_{y(z),crit}=\pi\cdot b^3\cdot h(E_{0,05}\cdot G_{0.05})^{0.5}/(6\cdot l_{ef}) \tag{28}$$

$$\sigma_{m,y(z),crit}=M_{y(z),crit}/W_{y(z)} \tag{29}$$

$$\lambda_{rel,m,y(z)} =(f_{m,k}/\sigma_{m,y(z),crit})^{0.5} \tag{30}$$

Table 1 shows the designed truss elements. The axes y and z are considered in formulas (14-30). The elements composing BC, TC and D sections are connected to each other by glue.

**Table 1.** Timber elements sections [cm]

| BC | TC | S | D | P |
|---|---|---|---|---|

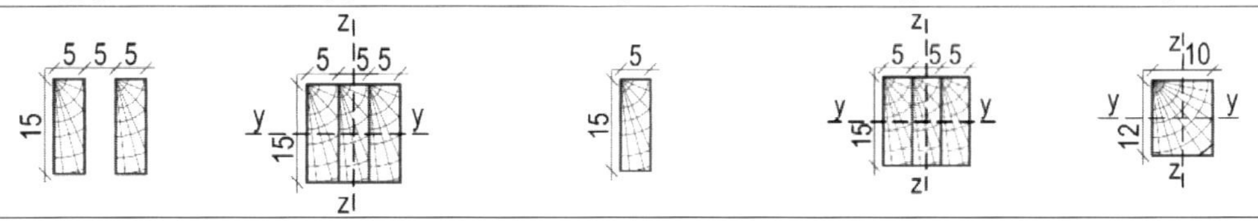


### 3.4. *Concrete slabs design rules and results*

The slabs reinforcement is designed for the elastic stage. As is the slabs reinforcement area, $M_{Ed}$ is the effective bending moment and $d \approx 0.9 \cdot h_{slab}$ (slab thickness). As=5Φ10/m (10 bars with 10mm diameter for 1 meter of slab) on direction X and 10Φ8/m on Y. The reinforcement is designed for cases A and C so it can also be used for cases B, where there are no slab consoles.

$$As = M_{Ed}/(0.9 \cdot d \cdot f_{yd}) \tag{31}$$

## 4. Elastic stage detailed results

### 4.1. *Efforts in confined masonry walls*

Efforts in confined masonry walls are seen in table 3. Piers are named according to figures 8 and 9. The walls with similar stresses and dimensions have the same labels. The design and bearing efforts are analysed only for cases A and B. It is important to see the way the big concrete consoles affect the masonry walls efforts. For case C piers stresses are clearly lower because there are some walls missing at the top story.

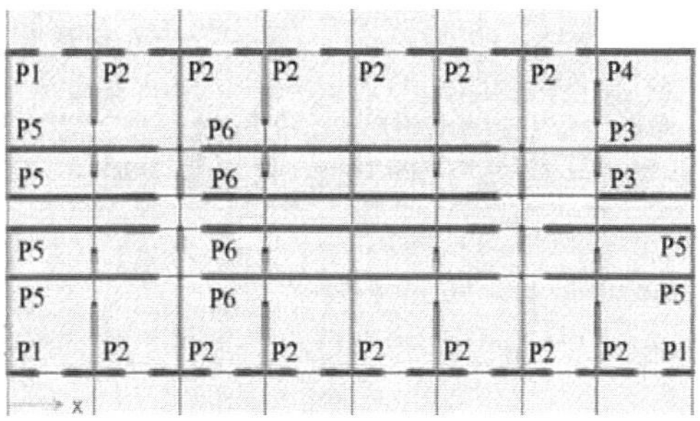


**Figure 8**. Piers on direction X

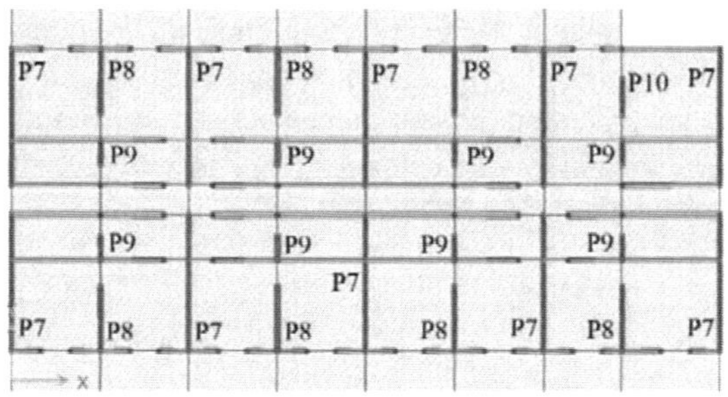


**Figure 9.** Piers on direction Y

In table 3, the piers axial forces generally decrease from case A to case B. They do increase in some walls top stories and in P10 on direction Y. The increase is caused by the seismic action and the lack of consoles in case B. Bending moments and shears decrease in all walls on direction Y situated close to the X direction edge. Table 2 shows the design and bearing efforts in masonry walls: $M_{Ed}$ [kNm] (bending moment), $N_{Ed}$ [kN] (axial force) and $V_{Ed}$ [kN] (shear force) and the bearing efforts respectively $M_{Rd}$ [kNm], $N_{Rd}$ [kN] and $V_{Rd}$ [kN].

**Table 2.** Piers design and bearing efforts

| wall | $N_{Ed}$(A) | $N_{Ed}$(B) | $N_{Rd}$ | $M_{Ed}$(A) | $M_{Ed}$(B) | $M_{Rd}$ | $V_{Ed}$(A) | $V_{Ed}$(B) | $V_{Rd}$ |
|---|---|---|---|---|---|---|---|---|---|
| P1 St1 | 711 | 539 | 769 | 90 | 76 | 404 | 18 | 19 | 258 |
| P1 St2 | 548 | 431 | 717 | 30 | 34 | 398 | 6 | 13 | 193 |
| P1 St3 | 391 | 319 | 717 | 10 | 9 | 368 | 5 | 5 | 150 |
| P1 St4 | 231 | 199 | 564 | 24 | 6 | 317 | 15 | 3 | 117 |
| P1 St5 | 64 | 78 | 709 | 23 | 7 | 243 | 12 | 3 | 73 |

| | | | | | | | | | |
|---|---|---|---|---|---|---|---|---|---|
| P2 St1 | 1375 | 960 | 1429 | 305 | 276 | 1056 | 83 | 71 | 508 |
| P2 St2 | 1061 | 761 | 1429 | 137 | 124 | 1092 | 61 | 52 | 378 |
| P2 St3 | 763 | 560 | 1334 | 82 | 65 | 1028 | 44 | 37 | 286 |
| P2 St4 | 459 | 345 | 934 | 74 | 61 | 862 | 28 | 23 | 192 |
| P2 St5 | 161 | 161 | 934 | 45 | 46 | 605 | 18 | 18 | 103 |
| P3 St1 | 1189 | 1162 | 2425 | 635 | 554 | 2203 | 162 | 146 | 424 |
| P3 St2 | 912 | 891 | 2425 | 350 | 312 | 2001 | 117 | 109 | 340 |
| P3 St3 | 625 | 610 | 2425 | 165 | 155 | 1700 | 78 | 75 | 251 |
| P3 St4 | 325 | 318 | 2274 | 58 | 64 | 1276 | 30 | 36 | 153 |
| P4 St1 | 1631 | 1426 | 2887 | 541 | 643 | 3124 | 206 | 161 | 563 |
| P4 St2 | 1279 | 1138 | 2707 | 261 | 304 | 2901 | 174 | 109 | 461 |
| P4 St3 | 933 | 842 | 2887 | 318 | 126 | 2548 | 138 | 75 | 356 |
| P4 St4 | 594 | 547 | 2093 | 90 | 103 | 2058 | 206 | 32 | 265 |
| P4 St5 | 236 | 242 | 480 | 95 | 73 | 1376 | 36 | 20 | 129 |
| P5 St1 | 1641 | 1613 | 3696 | 1338 | 1205 | 4452 | 185 | 160 | 561 |
| P5 St2 | 1258 | 1234 | 3696 | 880 | 798 | 3917 | 189 | 165 | 443 |
| P5 St3 | 863 | 844 | 3696 | 447 | 414 | 3198 | 152 | 134 | 320 |
| P5 St4 | 447 | 435 | 3465 | 170 | 134 | 2281 | 93 | 80 | 192 |
| P6 St1 | 3184 | 3134 | 7276 | 3785 | 3360 | 14977 | 496 | 438 | 1009 |
| P6 St2 | 2421 | 2383 | 7276 | 2554 | 2282 | 12908 | 484 | 431 | 785 |
| P6 St3 | 1776 | 1608 | 7276 | 1408 | 1269 | 10099 | 407 | 366 | 550 |
| P6 St4 | 798 | 787 | 6367 | 491 | 449 | 6410 | 247 | 227 | 297 |
| P7 St1 | 1588 | 1423 | 3032 | 1362 | 990 | 3749 | 306 | 275 | 595 |
| P7 St2 | 1225 | 1099 | 3032 | 819 | 602 | 3393 | 262 | 246 | 476 |
| P7 St3 | 873 | 786 | 3032 | 396 | 315 | 2877 | 189 | 194 | 352 |
| P7 St4 | 505 | 460 | 2959 | 202 | 246 | 1895 | 88 | 111 | 181 |
| P7 St5 | 261 | 182 | 2830 | 173 | 159 | 1781 | 29 | 4 | 159 |
| P8 St1 | 1126 | 900 | 1617 | 504 | 389 | 1200 | 127 | 132 | 399 |
| P8 St2 | 879 | 701 | 1617 | 249 | 235 | 1159 | 84 | 112 | 334 |
| P8 St3 | 626 | 509 | 1617 | 91 | 129 | 1019 | 39 | 85 | 234 |
| P8 St4 | 354 | 300 | 1617 | 40 | 70 | 866 | 9 | 45 | 166 |
| P8 St5 | 104 | 96 | 1617 | 49 | 79 | 584 | 26 | 37 | 89 |
| P9 St1 | 591 | 572 | 681 | 113 | 96 | 304 | 41 | 35 | 231 |
| P9 St2 | 457 | 443 | 681 | 42 | 37 | 301 | 22 | 19 | 191 |
| P9 St3 | 311 | 301 | 681 | 19 | 19 | 283 | 11 | 11 | 147 |
| P9 St4 | 145 | 143 | 639 | 8 | 6 | 237 | 3 | 1 | 95 |
| P10 St1 | 808 | 838 | 970 | 211 | 59 | 515 | 68 | 191 | 355 |
| P10 St2 | 606 | 621 | 970 | 79 | 35 | 527 | 41 | 69 | 261 |
| P10 St3 | 403 | 412 | 970 | 54 | 23 | 486 | 26 | 47 | 188 |
| P10 St4 | 196 | 202 | 970 | 43 | 11 | 393 | 13 | 36 | 118 |

*4.2. Concrete slabs design bending moments diagrams*

The bending moments values are generated by the load combination without seism. M11 [kNm] is the bending moment on direction X, horizontal in figures 10 to 13, M22 [kNm] is the effort on Y, vertical. The slab consoles generate high bending moment values at the restraints. This happens although there are beams supporting these slabs.

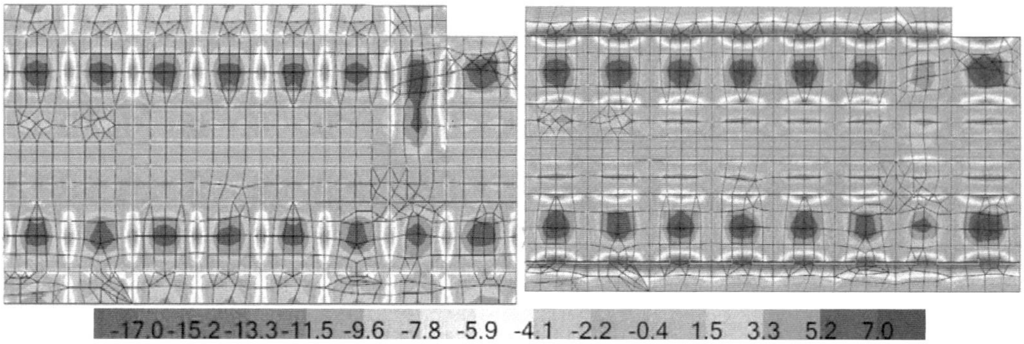


**Figure 10.** M11 for case A **Figure 11.** M22 for case A

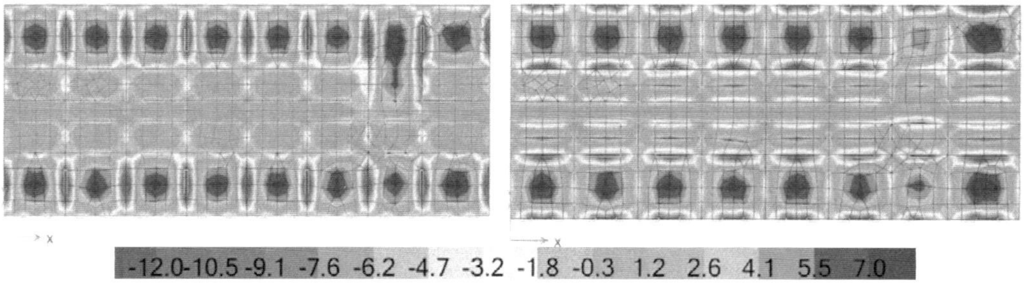


**Figure 12.** M11 for case B **Figure 13.** M22 for case B

### *4.3. Natural periods of vibration*

**Table 3.** Natural periods of vibration

| Case | T1 | T2 | T3 |
|---|---|---|---|
| A | 0.565s | 0.289s | 0.233s |
| B | 0.566s | 0.290s | 0.230s |
| C | 0.566s | 0.332s | 0.292s |

The natural periods of vibration are similar for cases A and B. They increase for case C, because there are no wall spurs at the top story. It is noticed that T3 increases if slab consoles are present and the buttresses are absent.

## 5. Plastic stage results

For each of the cases A, B and C, the nonlinear analysis is done using 2 pushover cases PX and PY. These cases develop on directions X and Y respectively.

### *5.1. Slabs bending moments distributions and values*

The bending moment values in figures 14 to 17, in kNm, are from case A. There are 179 steps in the plastic analysis for PX and 128 for PY. M11 is on direction X and M22 is on Y. M11 is seen for stages 8 and 12. M22 is for stages 10 and 16. Areas with high bending moments are larger in the plastic stage. On the X alignment (horizontal in figures 14 to 17), the highest values are reached near the walls edges on this direction. The bending moments values change signs close to the walls on direction Y (vertical in figures 14 to 17). This means the slabs are bent and may be cracked in these areas. The same phenomenon happens on the Y orientation. Also, on Y there is an evident crack longitudinal line in the centre. This is because there are no walls anywhere in that 1.5m bay, as it is seen in figure 2, and the slabs can easily bent. Walls have been drawn with grey to make the figures easier to read.

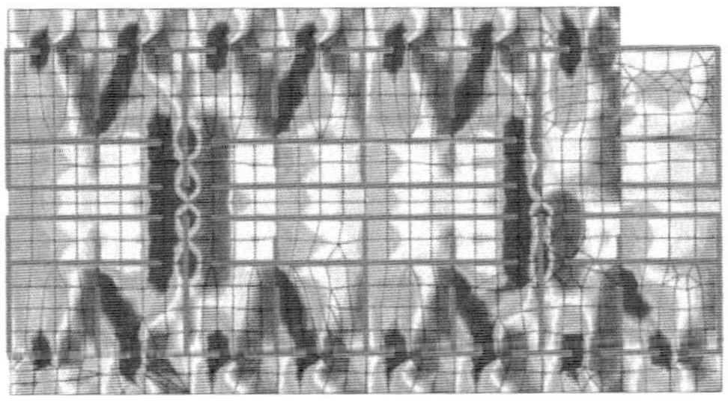

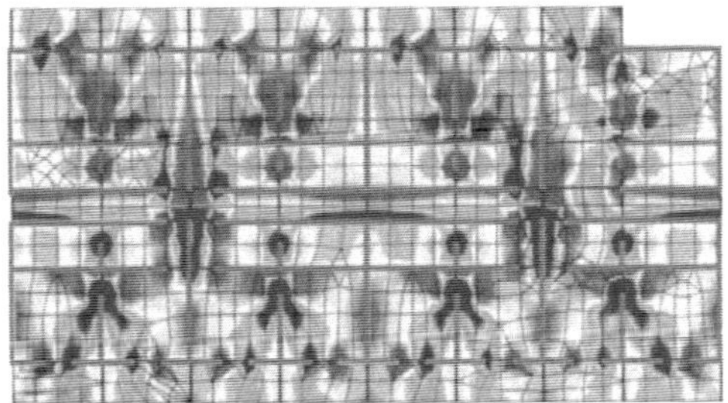

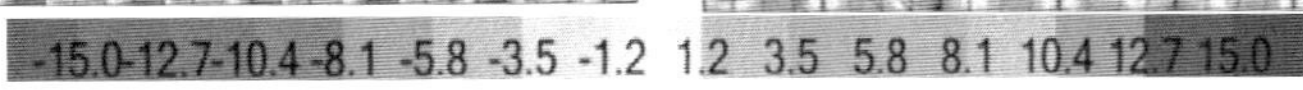


**Figure 14.** M11 PX 8/179

**Figure 15.** M11 PY 12/128

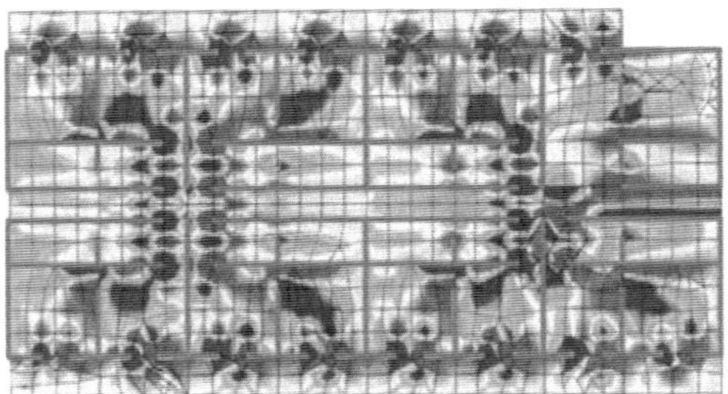

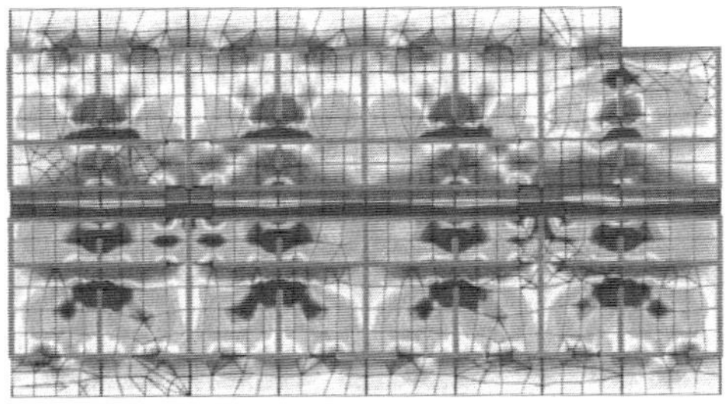

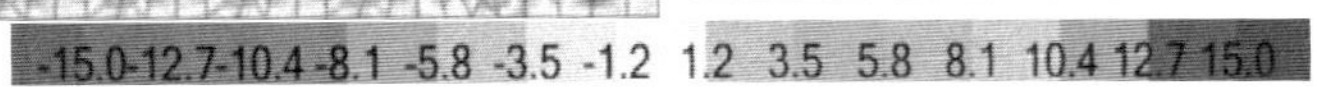


**Figure 16.** M22 PX 16/179

**Figure 17.** M22 PY 10/128

### *5.2. Stresses in confined masonry walls in the plastic stage*

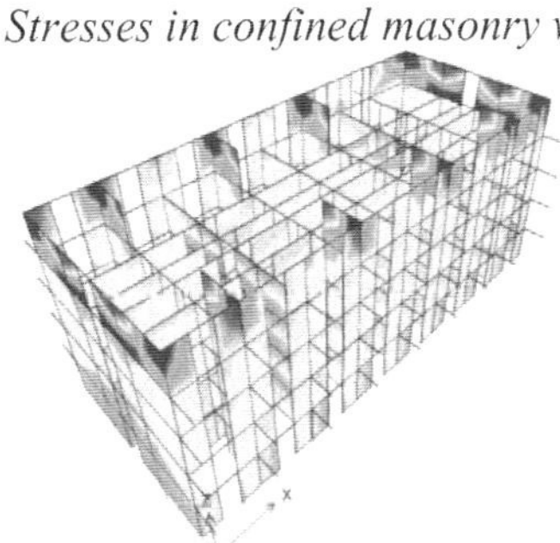

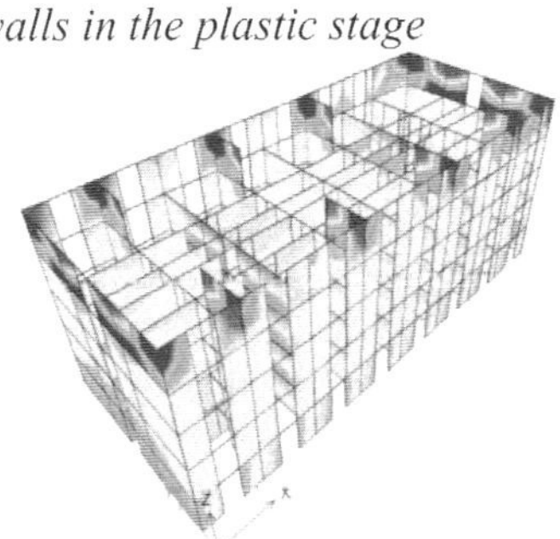

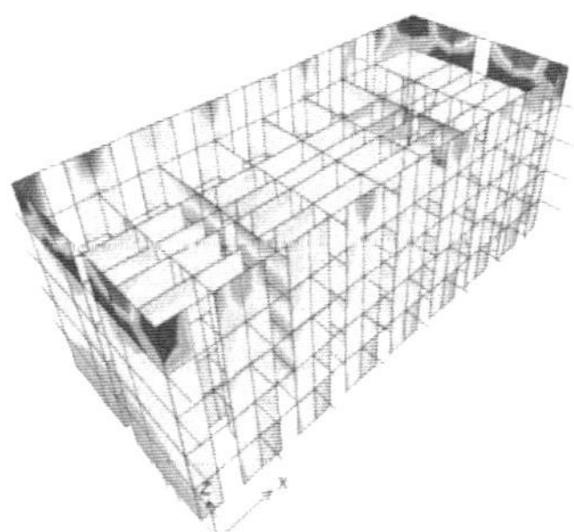

**Figure 18.** $\sigma_x$ X (A) 3/179

**Figure 19.** $\sigma_x$ X (B) 3/137

**Figure 20.** $\sigma_x$ X (C) 3/158

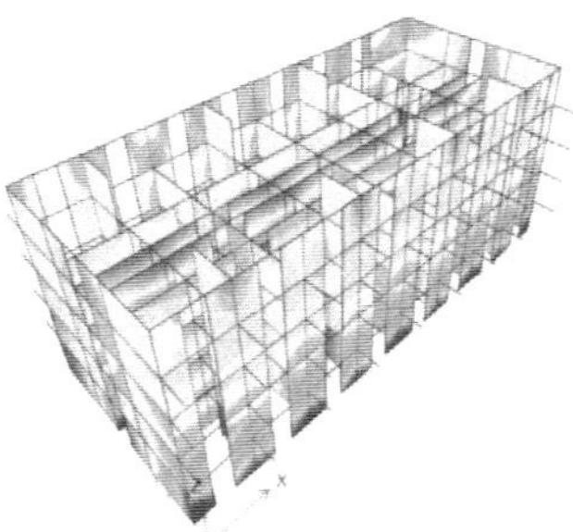

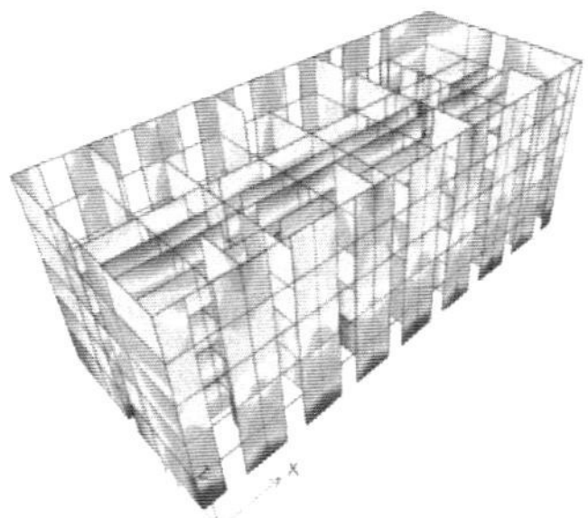

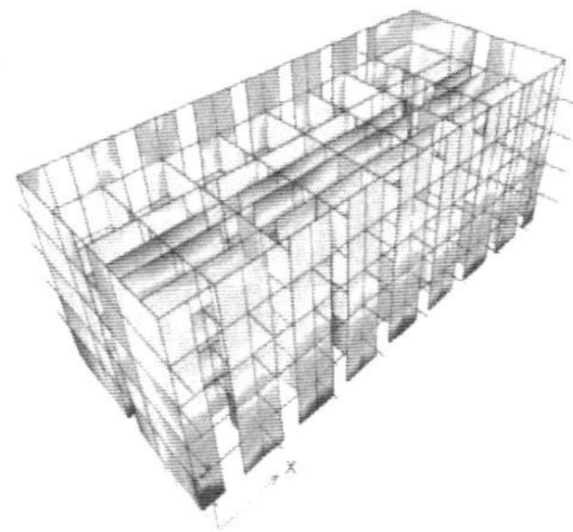

**Figure 21.** $\sigma_x$ Y (A) 4/128

**Figure 22.** $\sigma_x$ Y (B) 4/128

**Figure 23.** $\sigma_x$ Y (C) 4/45

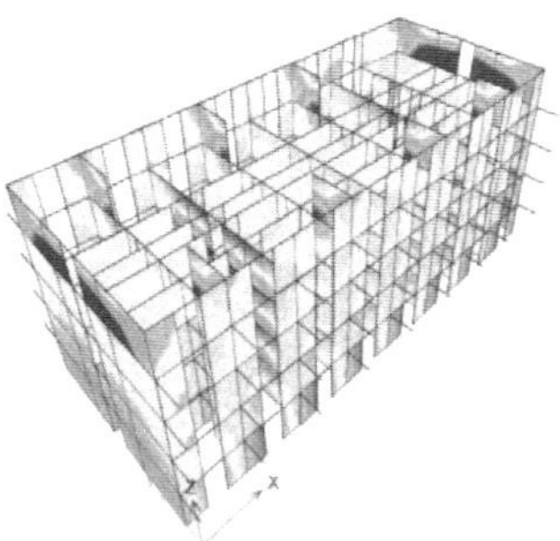
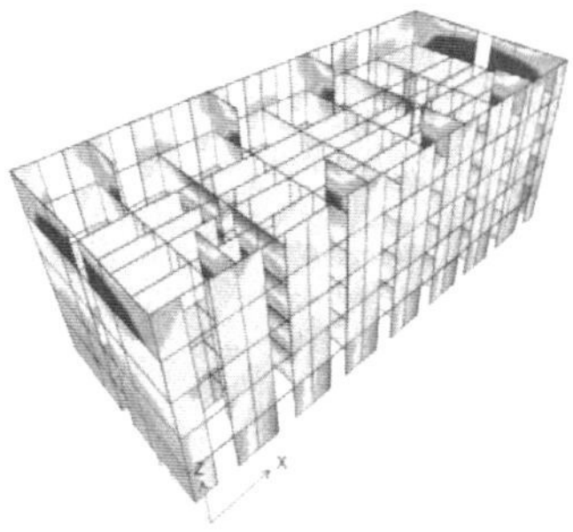
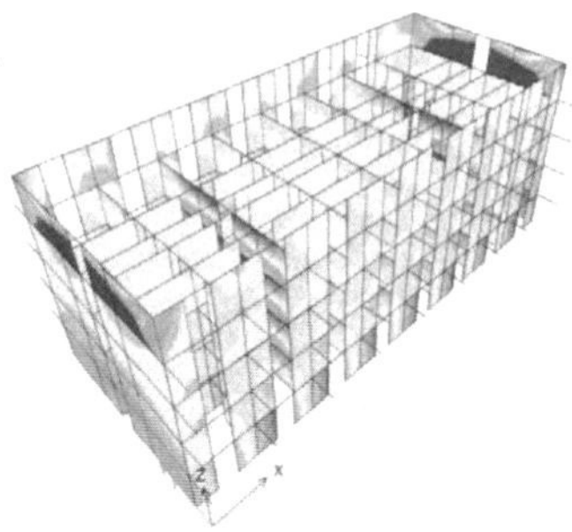

**Figure 24.** $\sigma_z$ X (A) 3/179 **Figure 25.** $\sigma_z$ X (B) 3/137 **Figure 26.** $\sigma_z$ X (C) 3/158

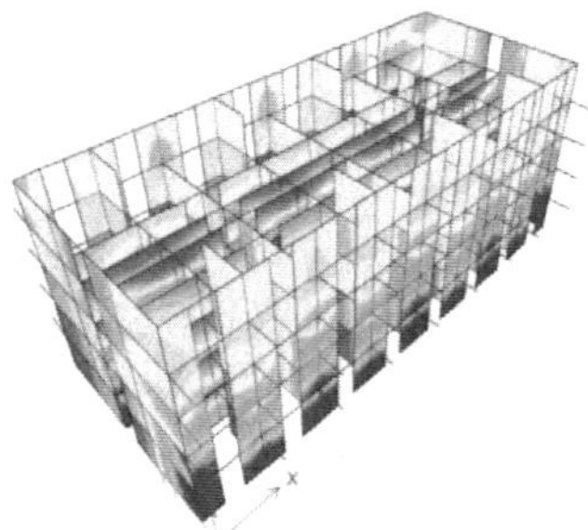
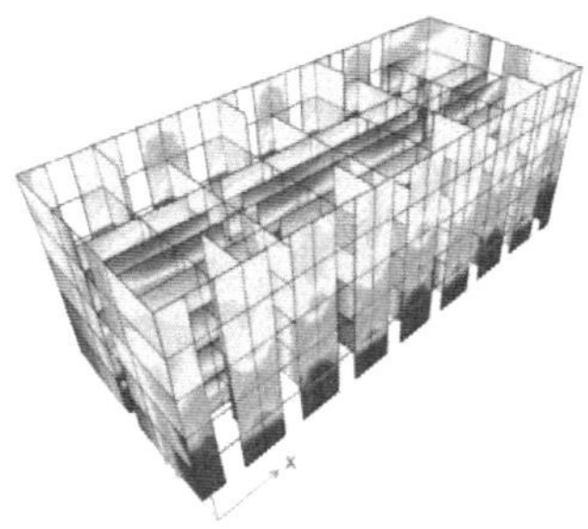
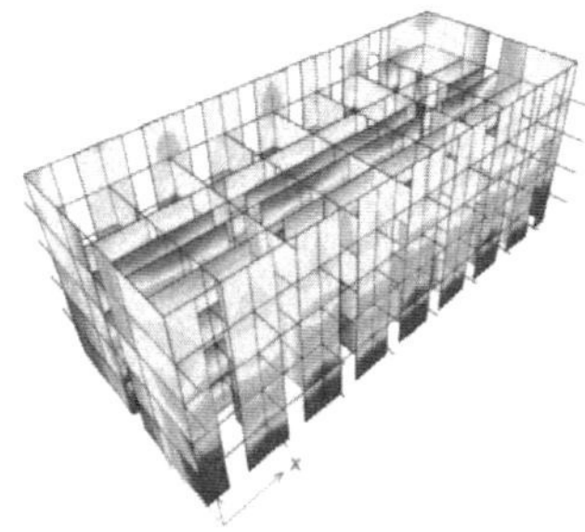

**Figure 27.** $\sigma_z$ Y(A) 4/128 **Figure 28.** $\sigma_z$ Y (B) 4/128 **Figure 29.** $\sigma_z$ Y (C) 4/45

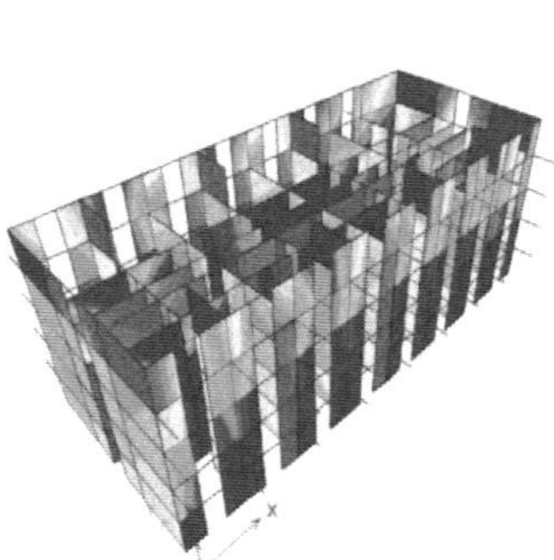
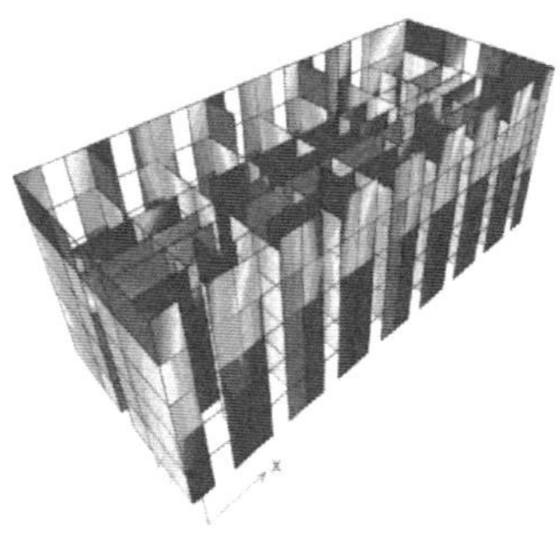
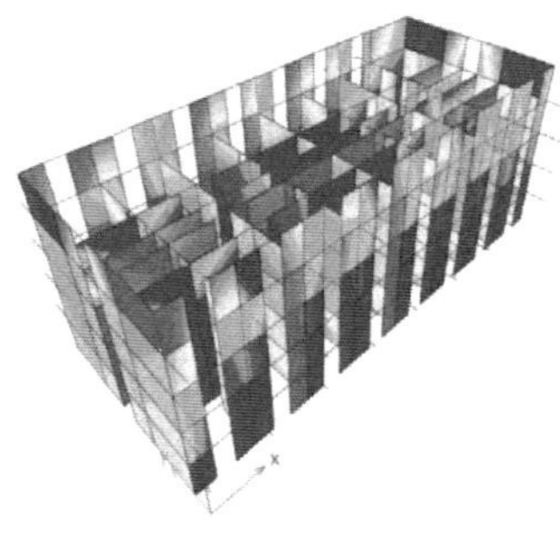

**Figure 30.** $\tau_{xz}$ X (A) 1/179 **Figure 31.** $\tau_{xz}$ X (B) 1/137 **Figure 32.** $\tau_{xz}$ X (C) 1/158

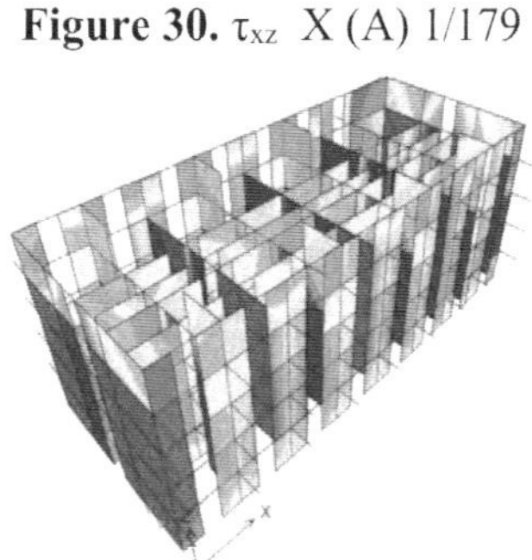
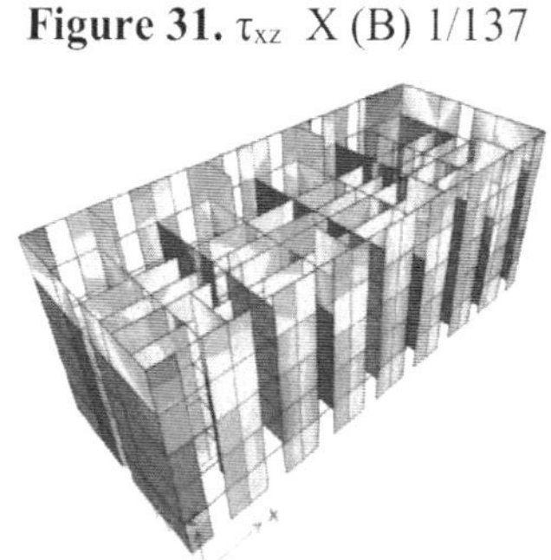
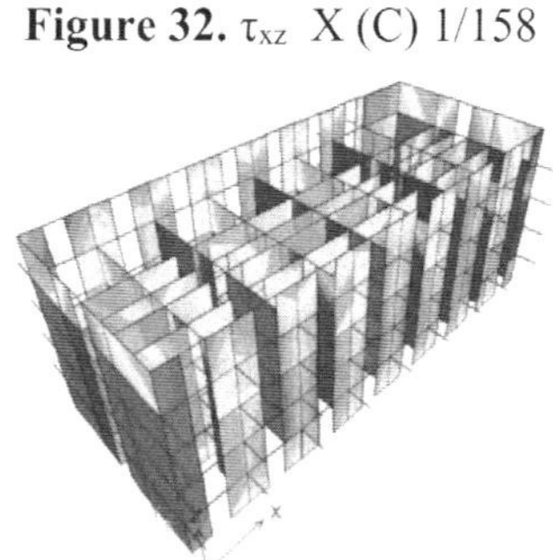
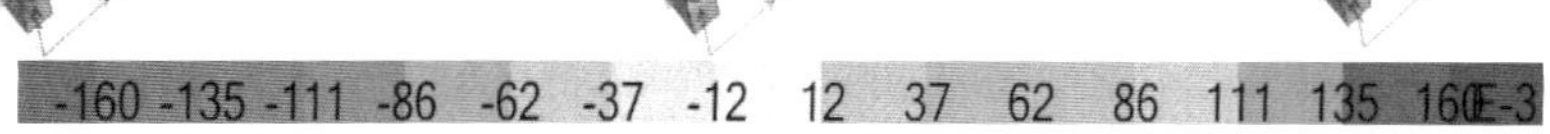


**Figure 33.** $\tau_{xz}$ Y (A) 1/128 **Figure 34.** $\tau_{xz}$ Y (B) 1/128 **Figure 35.** $\tau_{xz}$ Y (C) 1/45

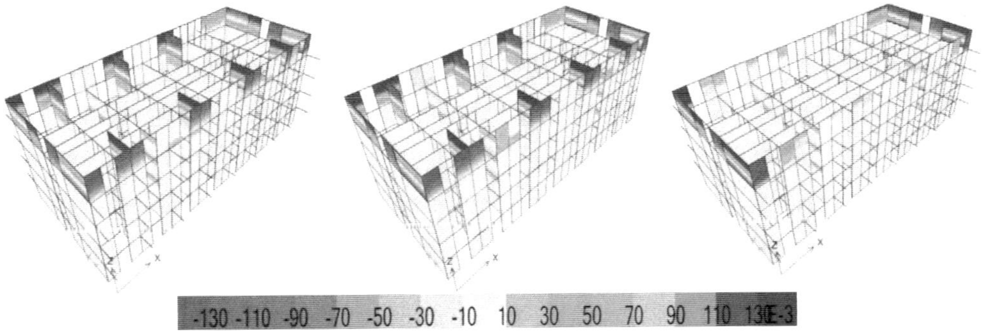


**Figure 36.** $\tau_{xy}$ X (A) 10/179 **Figure 37.** $\tau_{xy}$ X (B) 9/137 **Figure 38.** $\tau_{xy}$ X (C) 9/158

Figures 18 to 38 names are followed by the type of stress, the stress direction, the case analysed (A, B or C) and the step number/the maximum number of steps for that case, in that specific direction. Stresses $\sigma_x$ and $\sigma_z$ reach greater values in walls oriented perpendicular to the seismic action. For direction X, there are higher stress values at the top story walls. For all 3 cases, the stress values are increased in the edge walls. This is because they are long and stiff. On Y, the stress values are evenly distributed among the piers. They reach greater values in stiffer, longer ones. The walls on the X alignment at the top story don't get such high stresses, as they are short and bend over easily. Stresses $\tau_{xz}$ reach higher values in piers aligned with its direction. The only walls perpendicular to the stress orientation where high values are seen, are at the top story. This is because they are stiff and connected to walls on X only at the edges. They can't transmit so much effort to the walls on X. On direction Y, the stress values are high only in piers on this alignment and not at the top story. These are not as stiff in their own plane at the top story as they are at the others.

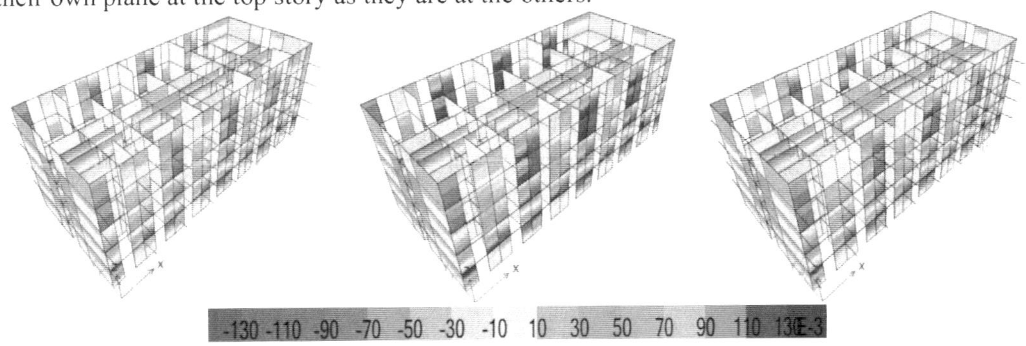


**Figure 39.** $\tau_{xy}$ Y (A) 40/128 **Figure 40.** $\tau_{xy}$ Y (B) 40/128 **Figure 41.** $\tau_{xy}$ Y (C) 40/45

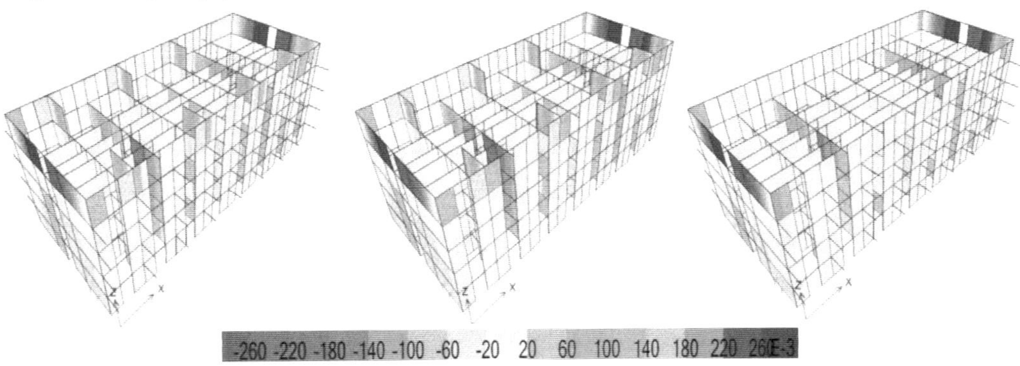


**Figure 42.** $\tau_{yz}$ X (A) 7/179 **Figure 43.** $\tau_{yz}$ X (B) 7/137 **Figure 44.** $\tau_{yz}$ X (C) 6/158

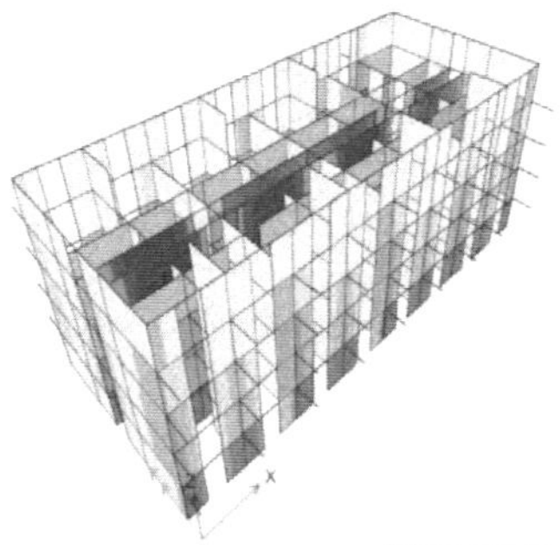
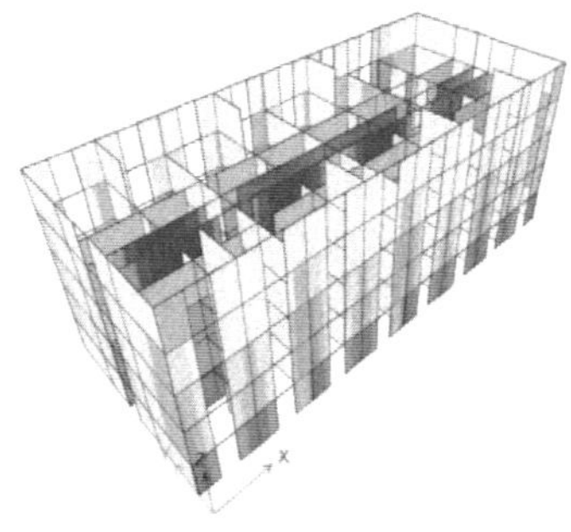
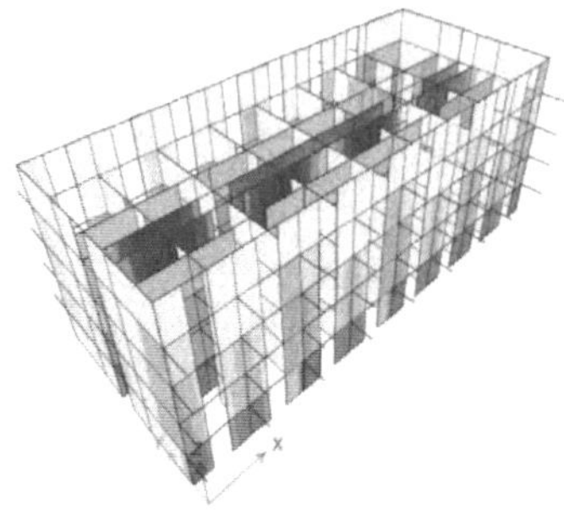

**Figure 45.** $\tau_{yz}$ Y (A) 12/128 **Figure 46.** $\tau_{yz}$ Y (B) 12/128 **Figure 47.** $\tau_{yz}$ Y (C) 12/45

Stress $\tau_{xy}$ folds the walls horizontally. For direction X, high stress values are seen in most walls at the top story. These piers are stiffened by walls on the other direction. For the Y orientation, stress $\tau_{xy}$ reaches higher values in walls on direction X, connected to long piers on alignment Y. These long walls stiffen them. Stress $\tau_{yz}$ folds the walls vertically. The highest values are in long piers perpendicular to the stress direction. For the X alignment, the highest stress values are in the top story walls on direction Y, placed at the edges. They are more vulnerable to vertical folding because there are only two short perpendicular piers, connected to their edges.

### *5.3. Plastic hinges development in the masonry walls*

Figures 48 to 53 show the last plastic analysis steps before the structure turns into a mechanism. The hinges colour code is: green for the plastic hinges being formed (stage B), blue for hinges that have reached their limit capacity (C), pink means the load has been redistributed (D) and red is for collapse (E). For direction X, the hinges reach the same stages at collapse for all 3 cases (A, B and C). It is noticed that collapse is reached at the top story in beams on the X orientation. The walls lack stiffness at that story and so they are more likely to be bent, causing high efforts in the beams connecting them. This phenomenon causes the structure collapse regardless of the presence of spurs for these walls. The console slabs also don't have an important role here. For direction Y, in case A, the plastic hinges reach stage E only in beams connecting the edge walls. This may be because these walls are likely to be bent, being at the edges, so the beams connecting them are subjected to higher efforts. There are also hinges developed up to stage D in the other short beams on the Y orientation.

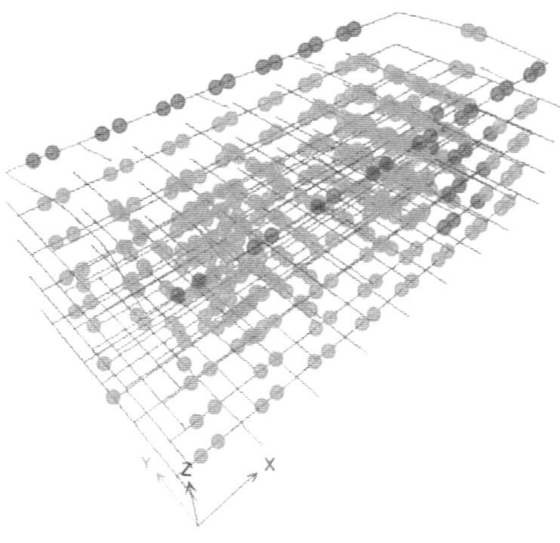

**Figure 48.** PX case A 179

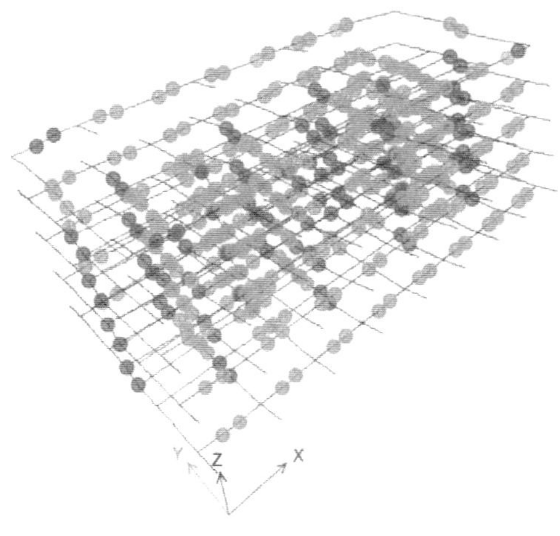

**Figure 49.** PY case A 128

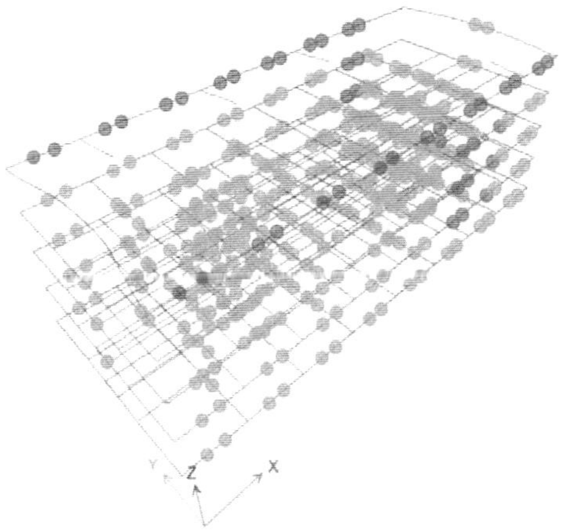


**Figure 50.** PX case B 137

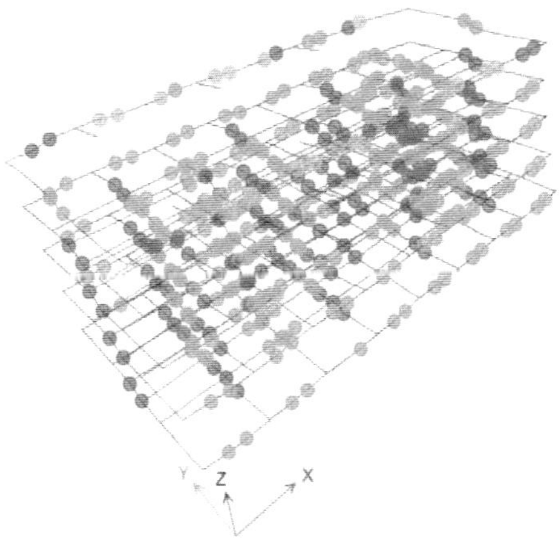


**Figure 51.** PY case B 142

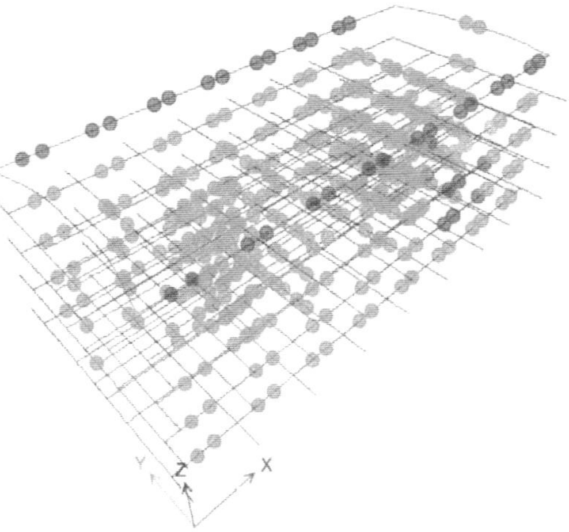


**Figure 52.** PX case C 158

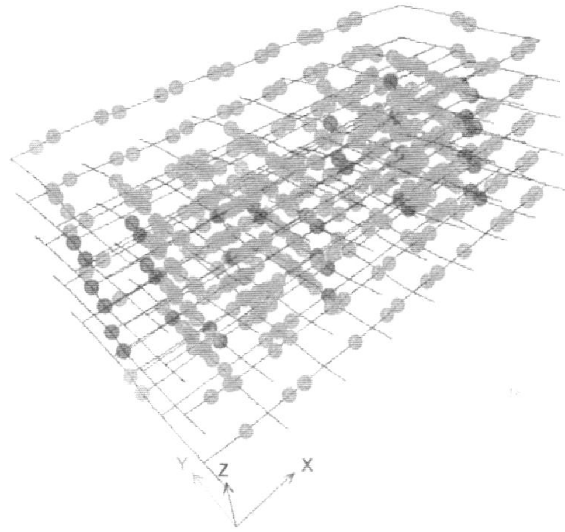


**Figure 53.** PY case C 45

In case B, the lack of slab consoles generates stage E plastic hinges in the interior beams as well. The interior beams are not as held down by the slabs weight, so they are free to rotate. The same conclusions are underlined by case C, where the consoles are present. Plastic hinges are more developed towards the building edges. Stage E is not reached here.

*5.4. Plastic hinges development in the truss roof*

The roof structure is analysed in the plastic stage. The pushover cases used are TX and TY, for the rectangular directions, in figures 54 and 55. The X axis is the long alignment, perpendicular to the trusses and Y is the short one. Collapse here means the bearing axial force is reached in the truss elements. There is a noticeable difference between the results on X and Y. For case TX, collapse is reached when stage E appears only in hinges from trusses at the roof ends. The timber structure is visibly deformed in this case. For TY, collapse is reached when stage E is seen in most hinges but there is no such deformation for the roof as it is in case TX. This kind of behavior is expected, because the trusses can only hold the roof stiffness in their own plane. The timber roof behaves the same in all 3 cases. There are 10 steps for the nonlinear analysis on direction X and 64 for Y.

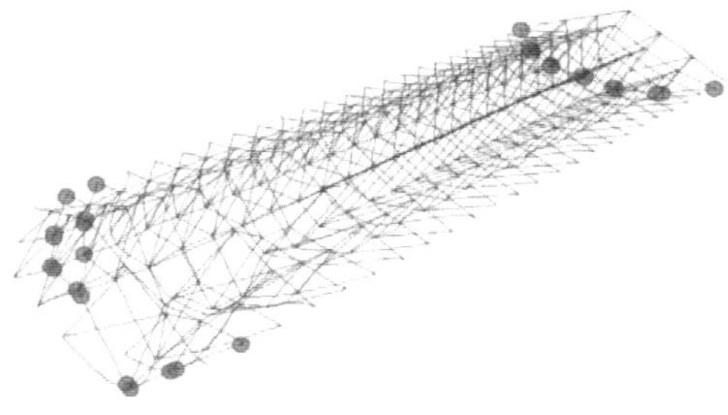

**Figure 54.** TX 10

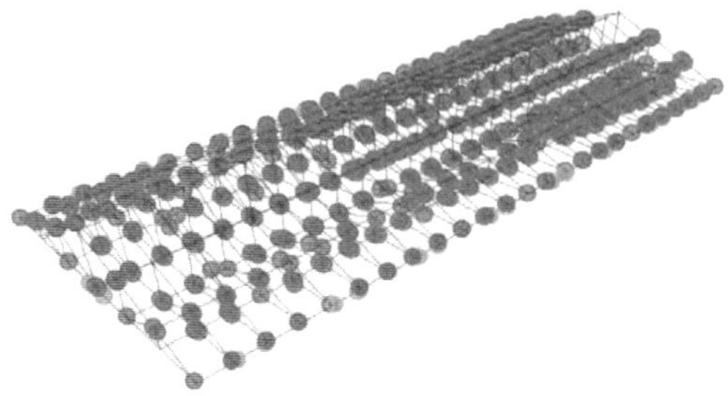

**Figure 55.** TY 64

*5.5. Pushover diagrams*

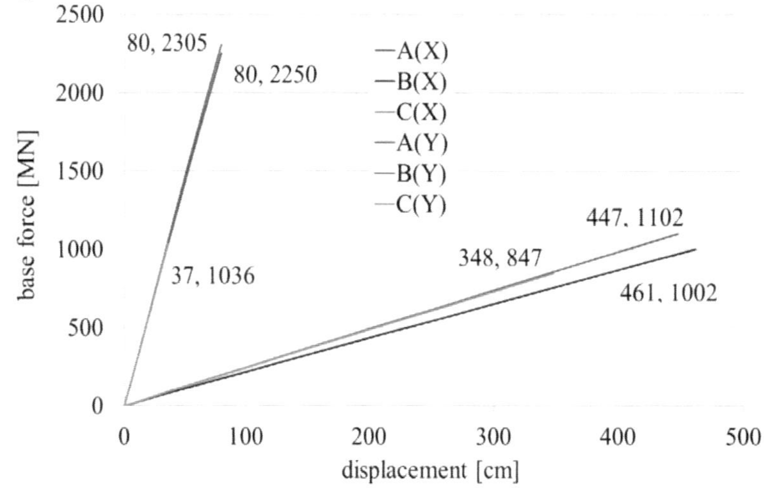


**Figure 56.** Pushover diagrams for directions X and Y for cases A, B and C

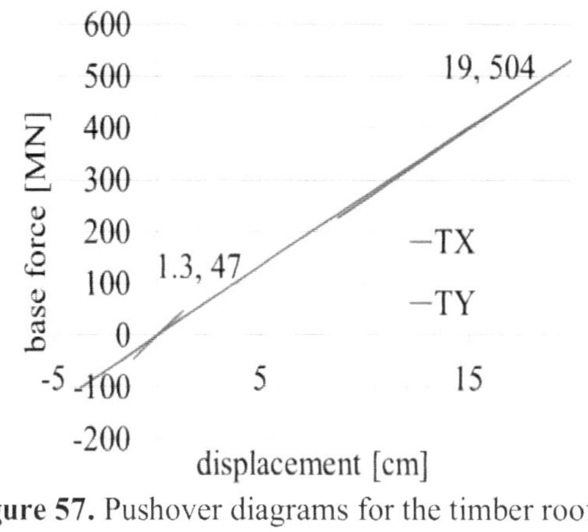


**Figure 57.** Pushover diagrams for the timber roof

The plastic behavior is further studied using the pushover diagrams. For the confined masonry structure, there are 2 pushover diagrams for each case, seen in figure 56, according to the X and Y alignments. For direction X, the maximum base force values are half those on Y, and the capable top displacements are 5.6 times greater. The values reached for cases A and B are similar. The slab consoles in cases A and C cause a stiffer behavior on X. They don't have an important influence to the diagrams for direction Y. This is because the consoles are attached to the building's sides on direction X. For case C, the values are visibly smaller.

The structure gives in much earlier if the stiffening walls at the top story are not present. The pushover diagrams in figure 57, for the roof, show 10 times greater maximum base force and displacement for direction Y. It is also seen that the values alternate from one sense to another for each direction.

## 6. Results and discussions

The elastic analysis for bearing masonry walls shows the piers axial forces decrease from case A to case B. They only increase in some walls top stories and in P10 on direction Y. This increase is caused by the seismic action and the lack of consoles in case B. Bending moments and shear forces decrease in all walls on direction Y situated close to the X direction edge.

The elastic stage bending moments in concrete slabs are greater in case A, near the consoles. This is because the consoles are fixed to the structure. In the plastic stage, the highest values are reached near the walls edges on direction X. On direction Y, the bending moments change values signs close to the walls. The slabs are bent and may be cracked in these areas. On Y there is an evident longitudinal crack line across the structure. It appeared because there are no walls anywhere in that 1.5m bay, so the slabs can easily bend.

The natural periods of vibration show similar values for cases A and B. They increase for case C, as there are no wall spurs at the top story. T3 increases if slab consoles are present and buttresses are absent.

Stresses $\sigma_x$ and $\sigma_z$ reach high values in walls oriented perpendicular to the seismic action and situated at the top story, for direction X in all 3 cases. They reach greater values in stiffer, longer ones. Stresses $\tau_{xz}$ reach higher values in piers aligned with its direction. On direction X, high $\tau_{xy}$ stress values are seen in most walls at the top story. On the Y orientation, stress $\tau_{xy}$ reaches higher values in walls on direction X, connected to long piers on alignment Y. Stress $\tau_{yz}$ reaches the highest values in long piers perpendicular to the stress direction. For the X alignment, the highest stress values are in the top story walls on direction Y, placed at the edges. The structure shows similar stress values and distributions for all 3 cases.

The plastic hinges reach the same stages at collapse for all 3 cases (A, B and C) for direction X. Collapse is first reached at the top story in beams on the X orientation. The walls there lack stiffness and so they are more likely to be bent, causing high efforts in the connecting beams. This causes the structure collapse regardless of the presence of spurs. In case A, on direction Y, stage E is reached only in beams connecting the edge walls. Hinges develop up to stage D in the other short beams on the Y orientation. In case B, there are stage E plastic hinges on direction X beams as well, as they are not as held down by the slabs weight, so they are free to rotate. For case C, plastic hinges are developed towards the building edges, but stage E is not reached.

For the confined masonry structure, there are 2 pushover diagrams for each case, according to the X and Y alignments. For direction X, the maximum base force values are half those on Y, and the capable top displacements are 5.6 times greater. The values reached for cases A and B are similar. The slab consoles in cases A and C cause a stiffer behavior on X. They don't influence direction Y, because the consoles are attached to the building's sides on direction X. For case C, the values are visibly smaller, as the structure gives in much earlier because the stiffening walls at the top story are not present.

The pushover diagrams for the timber roof show 10 times greater maximum base force and displacement for direction Y, parallel to the trusses.

## 7. Conclusions

The structure in case A will be used, as it can bare the efforts it is subjected to and it behaves well in the plastic stage too. Cases B and C do not show any major improvement in the structure's behavior.

## References

[1] Romlova J and Fojtik R 2015 The timber truss: the studying of the behavior of the spatial framework joint *Elsevier Perspectives in Science* **7** pp 299-303 doi.org/10.1016/j.pisc.2015.11.046

[2] Lanata F 2015 The analysis of the long term behavior of a timber truss structure: the case of the ESB building *3rd International Conference on structural Health Assessment of timber Structures Poland*, 2015

[3] Morris H, Worth M and Omenzetter P 2015 Monitoring modern timber structures and connections *3rd International Conference on structural Health Assessment of timber Structures Poland*,2015

[4] Brol J, Dawczynski S and Adamczyk K 2015 Possibilities of timber structural members reuse *3rd International Conference on structural Health Assessment of timber Structures Poland* ,2015

[5] San Bartolome A, Angles P and Quinn D 2012 Seismic behavior comparison of confined masonry walls of clay and concrete bricks *15th Interantional Brick and Block Masonry Conference Florianopolis- Brazil 2012*

[6] Costa A A, Penna A, Magenes G and Galasco A 2008 Seismic performance assessment of autoclaved aerated concrete (AAC) masonry buildings in Proc. *The 14th World Conference on Earthquake Engineering 2008*, pp 12-17 Beijing, China http://www.iitk.ac.in/nicee/wcee/

[7] Dhanasekar M 2011 Shear in reinforced and unreinforced masonry: response, design and construction *The Twelfth East Asia Pacific Conference on Structural Engineering and Construction, Elsevier Procedia Engineering* Vol **14**. p 2069- 2076, 2011. doi:10.1016/j.proceng.2011.07. 260

[8] Chourasia A, Bhattacharyya S K, Bharagava P K and Bhandari N M 2013 Influential aspects on seismic performance of confined masonry conctruction *Natual Science* Vol.5, No.8A1,56-62 (2013) doi 10.4236/ns.2013.58A1007

[9] Cornado C, Rosell J R, Leiva J and Diaz C 2016 Experimental study of brick masonry walls subjected to eccentric and axial load *RILEM International Conference on Materials, Systems and Structures in Civil Engineering, Technical University of Denmark,* p 33- 40, 2016. www.rilem.net/publications/proceedings-500218

[10] Perez-Gavilan J J, Flores L E and Manzano A 2014 A new shear strength design formula for confined masonry walls: proposal to the Mexican code *Tenth U.S. National Conference of Earthquake Engineering Frontieres Anchorage*, *Alaska*, 2014. July 21-25

[11] Salvatore M, Cattari S and Lagomarsino S 2019 Are the nonlinear static procedures feasible for the seismic assessment of irregular existing masonry buildings? *Engineering Structures Elsevier* https://doi.org/10.1016/j.engstruct.2019.109700

[12] Naik P and Annigeri S 2017 Performance evaluation of 9 story RC building located in North Goa. *11th International Symposium on plasticity and Impact Mechanics 2016 Elsevier Procedia Engineering* **173** (2017) 1841 -1846 doi:10.1016/j.proceng.2016.12.231

[13] CEN EN 1996-1-1-2006 Eurocode 6: Design of masonry structures - Part 1-1: General rules for reinforced and unreinforced masonry structures, 2006

[14] CEN EN 1995-1-1 Eurocode 5: Design of timber structures – Part 1-1: General rules for buildings

[15] CEN EN 1991-1-1-2004 Eurocode 1: Actions on structures - Part 1-1: General actions- Densities, self-weight, imposed loads for buildings, 2004

[16] CEN EN 1990-2004 Eurocode 0: Basics of structural design, 2004

[17] CEN EN 1991-1-3-2005 Eurocode 1: Actions on structures - Part 1-3: General actions- Snow loads, 2005

[18] CEN EN 1994-1-4-2005 Eurocode 1: Actions on structures- Part 1-4: General actions- Wind actions, 2005

[19] CEN EN 1998-1-2004 Eurocode 8: Design of structures for earthquake resistance. Part 1:

General rules, seismic actions and rules for buildings, 2004.
[20] CEN EN 1992-1-1-2004 Eurocode 2: Design of concrete structures - Part 1-1: General rules and rules for buildings, 2004
[21] CR-6-2013 Design code for masonry structures
[22] P100-1/2013 Seismic design code – Part 1- General rules for buildings, 2013.

# Plane shape impact to non-bearing masonry walls in a high rise reinforced concrete walls building

**S Constantinescu**[1*]

[1] Technical University of Bucharest, Romania

* sorina.constantinescu@yahoo.com

**Abstract**. Nowadays, due to architectural necessities, structures have emerged in many different plane shapes, such as U, V, L T and others. It is important to establish if a nonrectangular shape causes important changes both for the bearing and non-bearing elements, compared to a rectangular one. The paper shows the impact a buildings shapes on the partitioning walls. For this paper the structures in study are two high rise reinforced concrete walls buildings placed in a high seismic area (Bucharest, Romania). The partitioning walls are made of masonry. Both bearing and non-bearing walls are the same for both buildings. They are just placed in different locations, so that the structures have different shapes in plan. One has the shape of an I and the other of a V. The paper studies the buildings behavior in both cases. The masonry walls stresses are analysed for both the elastic and plastic stage. The conclusion is that in this case, for these shapes, the non-bearing walls are not affected very much by the building's plane shape.

## 1. Introduction

Nowadays high-rise buildings are very much used. Lately, next to economics, aesthetics have an important role in planning and design of these structures, creating interesting shapes [1]. A building's shape in plane has an important influence over the bearing elements efforts and the structure's behavior in general. This happens especially for high rise structures in high seismic areas [2]. According to previous studies, disadvantages of structural irregularity in plan of building under seismic loads are clear in T-plan floor shape in comparison with L and U-shape [3]. The behavior of buildings during earthquakes depends on many conditions like stiffness, strength, ductility and most probably on structure's configuration [4]. L-shaped floor buildings are significantly influenced by the lateral forces at its level because of the different lateral deformations [5]. Literature does contain nonlinear studies of 3D plan irregular buildings [6] and buildings with structural irregularities may show different drifts of adjacent stories or excessive torsion [7]. Masonry infill walls cause important changes to the structure's behavior [8]. Strong infill walls may prevent slender structures collapse if they are placed regularly throughout the structure [9]. The addition of infill masonry walls increases the structure's lateral strength and decreases the lateral displacement [10]. The building in study is a high rise reinforced concrete walls structure with masonry nonbearing walls. It is designed as a rectangle in the first case is and as a V in the second. All the bearing and nonbearing walls have the same lengths in both cases. Therefore they have the same stiffness. The building plane area is the same. The paper highlights the structure's behavior both in the elastic and plastic stage. The efforts in the masonry walls will be analyzed. The results from the 2 cases will be compared, so the building's shape effect to its behavior will be established. The codes in force used to design both structures are [11-18].

## 2. Building description

The structure is composed of a basement, a ground floor and 8 floors above it.

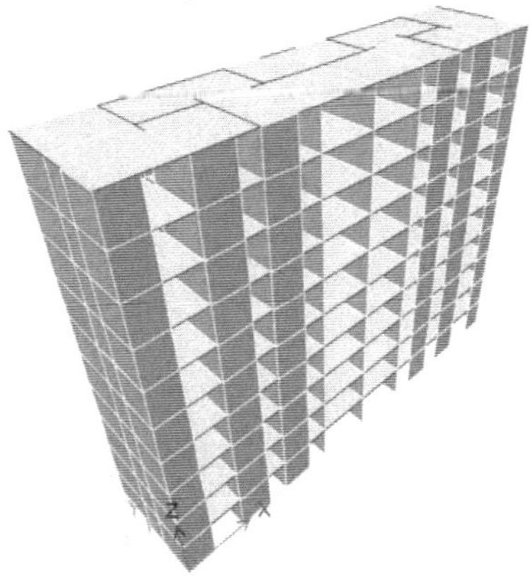

**Figure 1.** 3D for building I.

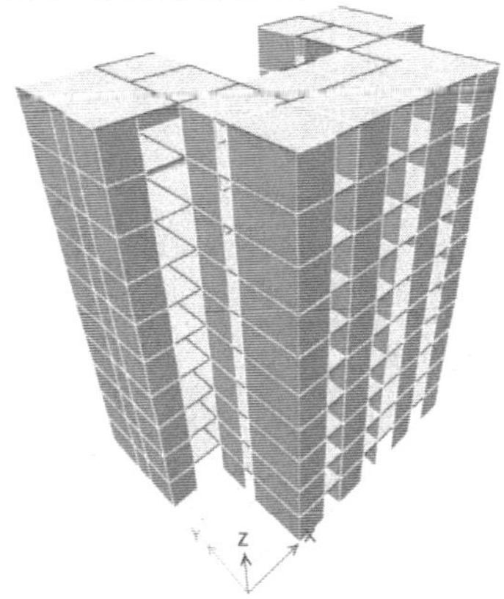

**Figure 2.** 3D for building V.

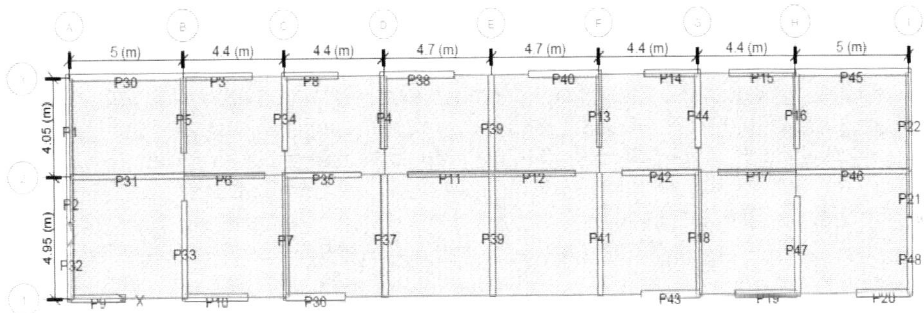


**Figure 3**. Building I plan.

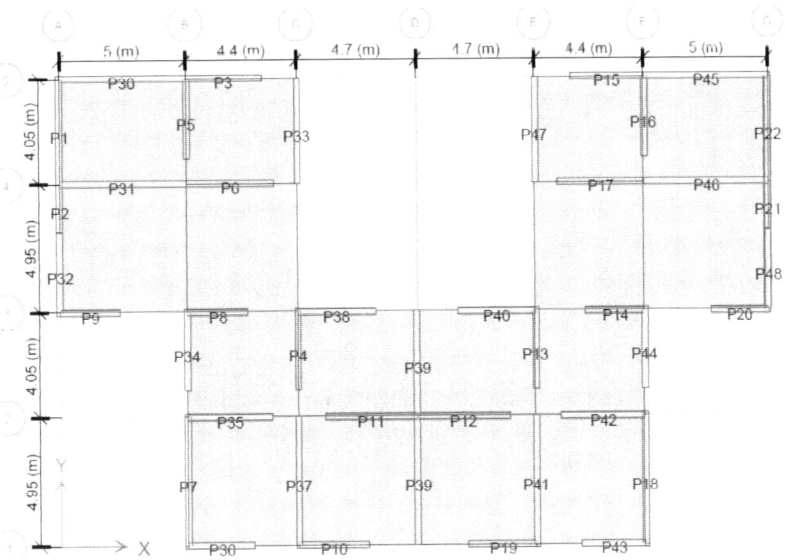


**Figure 4.** Building V plan.

**Table 1.** Identical walls in both I and V buildings.

| Masonry walls | Axis | Length [m] | Reinforced concrete walls | Axis | Length [m] |
|---|---|---|---|---|---|
| P1, P22 | Y | 4.05 | P30, P45 | X | 5.00 |
| P2, P21 | Y | 1.85 | P31, P46 | X | 5.00 |
| P3, P15 | X | 3.00 | P32, P48 | Y | 3.10 |
| P4, P13 | Y | 3.00 | P33, P47 | Y | 4.05 |
| P5, P16 | Y | 3.10 | P34, P44 | Y | 3.02 |
| P6, P17 | X | 3.50 | P35, P42 | X | 3.37 |
| P7, P18 | Y | 4.95 | P36, P43 | X | 2.63 |
| P8, P14 | X | 2.40 | P37, P41 | Y | 4.95 |
| P9, P20 | X | 2.35 | P38, P40 | X | 3.07 |
| P10, P19 | X | 2.75 | P39 | Y | 9.00 |
| P11, P12 | X | 3.65 | | | |

### *2.1. Materials used*

Concrete is class C25/30, reinforcement bars are S500 [15] and masonry for the nonbearing walls is vertical hollow bricks with standard strength $f_b$=10N/mm$^2$ and M10 mortar [11, 18]. The stresses analyzed for the confined masonry walls are $\sigma_x$, $\sigma_z$, $\tau_{xz}$, $\tau_{xy}$ and $\tau_{yz}$. These stresses are compared to the masonry design strengths [18]: horizontal compression $f_{dh}$, vertical compression $f_d$, shear strength for horizontal direction $f_{vd,0}$, shear for diagonal direction $f_{vd,i}$, horizontal and vertical strengths perpendicular to the wall $f_{xd1}$ and $f_{xd2}$. The design strengths are calculated using the characteristic values: $f_{kh}$, $f_k$, $f_{vk,0}$, $f_{vk,i}$, $f_{xd1}$ and $f_{xd2}$, in formulas (1) to (5) The masonry strengths insurance factor is $\gamma_M$=1.9 [18]. The concrete compression design strength $f_{cd}$, is determined using the characteristic strength $f_{ck}$. For the steel reinforcement bars, $f_{yd}$ is the design strength and $f_{yk}$ is the characteristic value. The insurance factors are $\gamma_C$=1.5, for concrete and $\gamma_S$=1.15 for steel [17].

$$f_{dh} = f_{kh}/\gamma_M = 2.05/1.9 = 1.0 \text{ N/mm}^2 \quad (1)$$

$$f_d = f_k/\gamma_M = 4.5/1.9 = 2.36 \text{ N/mm}^2 \quad (2)$$

$$f_{vd,0} = f_{vk,0}/\gamma_M = 0.3/1.9 = 0.158 \text{ N/mm}^2 \quad (3)$$

$$f_{xd1} = f_{xk1}/\gamma_M = 0.24/1.9 = 0.126 \text{ N/mm}^2 \quad (4)$$

$$f_{xd2} = f_{xk2}/\gamma_M = 0.48/1.9 = 0.252 \text{ N/mm}^2 \quad (5)$$

$$f_{cd} = f_{ck}/\gamma_C = 25/1.5 = 16.67 \text{ N/mm}^2 \quad (6)$$

$$f_{yd} = f_{yk}/\gamma_S = 500/1.15 = 435 \text{ N/mm}^2 \quad (7)$$

## 3. Theory elements

### *3.1. The seismic action*

The base force $F_b$ in equation (8) is calculated using [16, 17]. The seismic load is introduced using coefficient $c_s$. The factors in the base force formula are: $\gamma_{I,e}$ = 1.2, the building's importance-exposure coefficient, $\beta_0$ = 2.5 the maximum value in the elastic spectrum, $a_g$ = 0.30g for Bucharest, q the structure's behavior factor, q=3.5·1.35 for medium ductility DCM structures [16, 17], m = building's mass, $\eta$ = 1 the reduction factor and $\lambda$ = 0.85 [16, 17].

$$F_b = \gamma_{I,e} \cdot \beta_0 \cdot a_g/q \cdot m \cdot \eta \cdot \lambda = c_s \cdot m \cdot g = 0.16 \cdot m \cdot g \quad [\text{kN}] \quad (8)$$

### *3.2. Masonry non-bearing walls design subjected to parallel seismic action*

The load combinations used for the elastic stage are CSX and CSY with permanent load+ 0.4·variable load+ seismic load on direction X and Y respectively. $M_{Rd}$ is the wall's bearing bending moment [11, 18]. $M_{Rd(M)}$ is the bearing bending moments from the masonry area [18]. CA is the wall's compressed area, in figure 5 [18]. In formula (10), yzc is the distance between the compressed area weight center G1 and the wall's weight center G [18]. S is the earthquake direction and t=25cm is the walls width. $V_{Ed}$ is the horizontal shear force from the seismic loads combination and $V_{Rd}$ is the masonry walls bearing shear force, seen in equation (15) [11, 18].

$$C_A=N_{Ed}/(0.85 \cdot f_d)\ [mm^2] \tag{9}$$

$$M_{Rd}= N_{Ed} \cdot yzc\ [kNm] \tag{10}$$

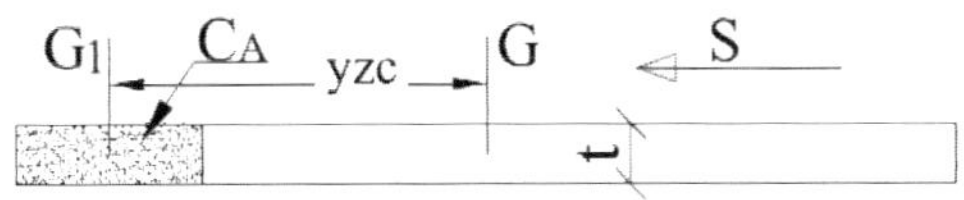


**Figure 5.** Confined masonry wall section.

In equation (18), b is a correction coefficient connected to the masonry walls dimensions $h_{pan}$ =h=story height- beam height and $l_{pan}$ =$l_w$ are the height and length of the masonry area panel. b=1.5 if $h/l_w \geq 1.5$, b=1.0 if $h/l_w<1.0$ and b=$h/l_w$ if $1.0 \leq h/l_w < 1.5$, $f_{vd,i}$ is the diagonal shear strength.

$\sigma_{0d}=N_{Ed}/(t \cdot l_w)$, $N_{Ed}$ and $A_w$ are masonry walls axial force and horizontal area [18]. $V_{Rd,1}^*$ and $V_{Rd,i}^*$ are the bearing shear forces on the horizontal and diagonal directions, calculated using equations (12) and (14) [18].

$$f_{vd,i}=f_{vk,i}/1.9=0.22 \cdot f_{bt} \cdot(1+5 \cdot \sigma_{0d}/f_{bt}) \cdot 0.5/1.9\ [N/mm^2] \tag{11}$$

$$V_{Rd,1}^*= 0.4 \cdot (N_{Ed}+0.8 \cdot V_{Ed} \cdot h_{pan}/l_{pan})\ [kN] \tag{12}$$

$$V_{Ed} \leq l_{pan} \cdot t \cdot f_{vd,0} \tag{13}$$

$$V_{Rd,i}=A_w \cdot f_{vd,i}/b\ \ [kN] \tag{14}$$

$$V_{Rd}=\min(V_{Rd,1}^*, V_{Rd,i}^*) \tag{15}$$

**Table 2.** Effective and bearing efforts in masonry walls

| | | efforts for building I | | | efforts for building V | | | bearing efforts | |
|---|---|---|---|---|---|---|---|---|---|
| | l (m) | $N_{Ed}$ [kN] | $M_{Ed}$ [kNm] | $V_{Ed}$ [kNm] | $N_{Ed}$ [kNm] | $M_{Ed}$ [kNm] | $V_{Ed}$ [kNm] | $M_{Rd}$ [kNm] | $V_{Rd}$ [kNm] |
| P1, P22 | 4.05 | 73 | 31 | 11 | 73 | 32 | 11 | 143 | 31 |
| P2, P21 | 1.85 | 34 | 14 | 5 | 34 | 15 | 5 | 30 | 15.7 |
| P3, P15 | 3.00 | 54 | 23 | 8 | 54 | 24 | 8 | 78 | 24 |
| P4, P13 | 3.00 | 54 | 23 | 8 | 54 | 24 | 8 | 78 | 24 |
| P5, P16 | 3.10 | 56 | 24 | 8 | 56 | 24 | 8 | 84 | 24 |
| P6, P17 | 3.50 | 63 | 27 | 9 | 63 | 27 | 9 | 107 | 27 |
| P7, P18 | 4.95 | 90 | 38 | 13 | 90 | 39 | 13 | 216 | 38 |
| P8, P14 | 2.40 | 44 | 18 | 6 | 43 | 19 | 7 | 51 | 19.6 |
| P9, P20 | 2.35 | 43 | 18 | 6 | 43 | 18 | 6 | 49 | 19 |
| P10, P19 | 2.75 | 50 | 21 | 7 | 50 | 22 | 7 | 66 | 22 |
| P11, P12 | 3.65 | 66 | 28 | 10 | 66 | 29 | 10 | 117 | 28.6 |

### *3.3. Masonry non-bearing walls design subjected to perpendicular seismic action*

The force $F_{NBW}(z)$ calculated in equation (16) is considered uniformly distributed, perpendicular to the non-bearing walls [18]. It reaches the highest value at the top story. This is because the nonbearing wall height placement in the structure increases the force value. $\beta_{NBW}$ =2.5 is the non-bearing walls amplification factor, $k_z$ is a coefficient according to the non-bearing wall's level height (the distance to the building's base), z is the non-bearing wall's top and bottom height and H is the building's total height [18].

$$F_{NBW}(z)=\gamma_{I,e} \cdot ag \cdot \beta_{NBW} \cdot k_z \cdot m_{NBW}/q_{NBW}=1.79kN/m^2 \quad (16)$$

$$k_z=(k_{z1}+k_{z2})/2=(1+2\cdot z1/H+1+2\cdot z2/H)/2=$$

$$=(1+2\cdot 27/30+1+2\cdot 30/30)/2=2.09 \quad (17)$$

$k_{z1}$ and $k_{z2}$ are coefficients connected to the highest and lowest points of the wall. The greatest value for $k_z$ is calculated at the top story. $q_{NBW}$ =4.725 is the behavior factor for non-bearing walls. The wall's mass/m$^2$ is $m_{NBW}$ in equation (18). Seismic force value $F_{NBW}$ is limited as (19) shows [18].

$$m_{NBW} =\gamma_{mas} \cdot t =18 \cdot 0.25= 4.5kN/m^2 \quad (18)$$

$$0.75\cdot \gamma_{I,e} \cdot ag \cdot m_{NBW} \leq F_{NBW} \leq 4\cdot\gamma_{I,e} \cdot ag \cdot m_{NBW} \quad (19)$$

$$0.75\cdot 1.2\cdot 0.30\cdot 4.5kN/m^2 \leq 1.79\ kN/m^2 \leq 4\cdot 1.2\cdot 0.30\cdot 4.5\ kN/m^2 \quad (20)$$

$$1.215\ kN/m^2 \leq 1.79\ kN/m^2 \leq 6.48\ kN/m^2 \quad (21)$$

Bearing bending moments perpendicular to the wall $M_{Rxd1}$ (horizontal) and $M_{Rxd2}$ (vertical) are calculated according to the design strengths perpendicular to the wall [11, 18]. These values will be compared to the design bending moments values, $M_{Exd1}$ (horizontal) and $M_{Exd2}$ (vertical) [11, 18] calculated in the elastic stage. $H_w$ =2.5m is masonry walls height, t is the wall thickness, $W_w$= 1000 · t$^2$/6 is the wall resistance modulus (in mm$^3$/m), $\sigma_{dw}$ is the compression stress at the wall's middle height section [11, 18] and $\gamma_{mas}$ =18kN/m$^3$ is the masonry weight.

$$\sigma_{dw} =\gamma_{mas} \cdot H_w/2=18 \cdot 2.5/2=0.0225\ N/mm^2 \quad (22)$$

$$W_w=1000 \cdot t^2/6=1000 \cdot 2502/6=10416666\ mm^3/m \quad (23)$$

$$M_{Rxd1}= W_w \cdot(f_{xd1} +\sigma_{dw})=10416666.67\cdot(0,126+0.0225)=1.546\ kNm/m \quad (24)$$

$$M_{Rxd2}= W_w \cdot f_{xd2} =10416666.67\cdot 0,25= 2.604\ kNm/m \quad (25)$$

## 4. Masonry walls elastic efforts

### *4.1. Bending moments perpendicular to the wall plan*

Figures 6 to 13 show higher moment values for the seismic combination for direction Y. Partitioning walls for buildings I and V show similar bending moment values for the same walls. $M_{Ed1}$ and $M_{Ed2}$ values increase from the ground floor up, just like the seismic force perpendicular to the walls planes. These walls are perpendicular to the seismic action. $M_{Ed1}$ and $M_{Ed2}$ are both perpendicular to the wall plane. Bending moment's diagrams are similar for $M_{Ed1}$ and $M_{Ed2}$, but $M_{Ed2}$ values are higher. It is noticed that some walls corners reach greater values than the bearing bending moments. This means the walls get cracked at the corners. Walls situated near a concrete wall reach high stress values near the contact area. They are influenced by the bearing walls stiffness.

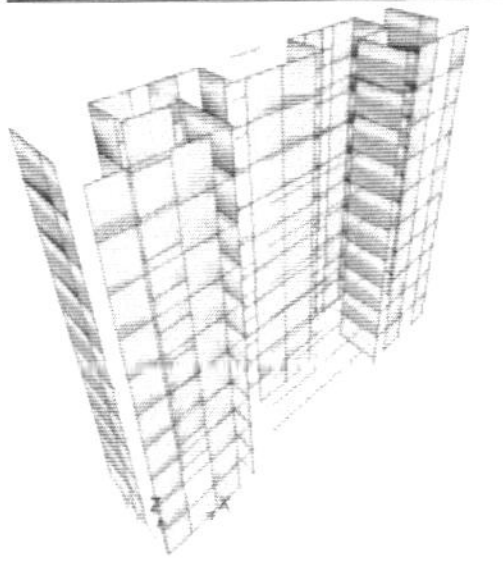

**Figure 6.** CSX $M_{Exd1}$ building I

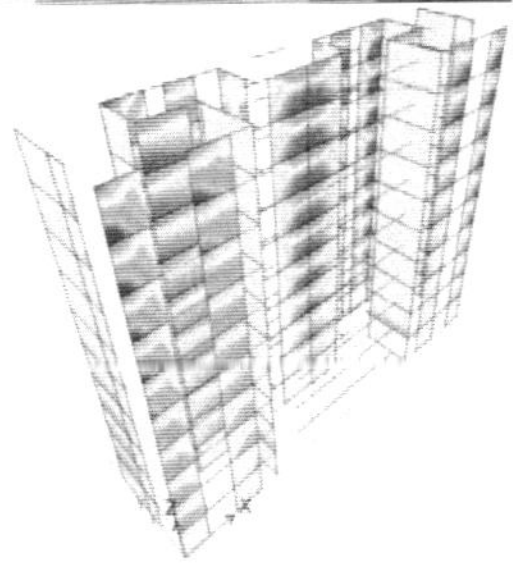

**Figure 7.** CSY $M_{Exd1}$ building I

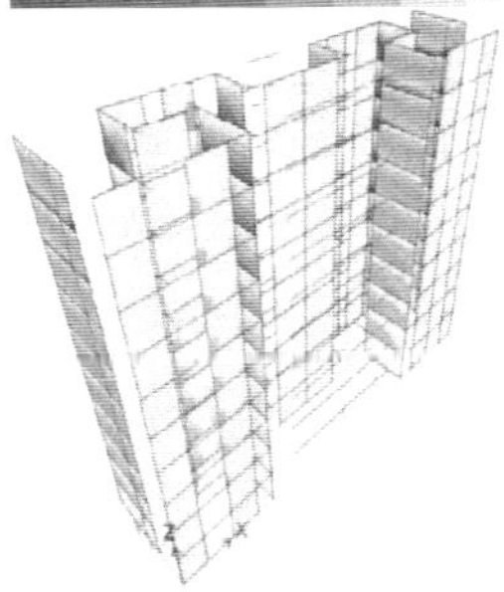

**Figure 8.** CSX $M_{Exd2}$ building I

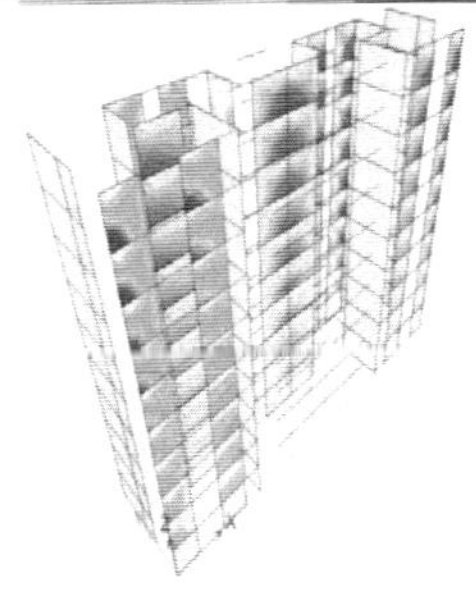

**Figure 9.** CSY $M_{Exd2}$ building I

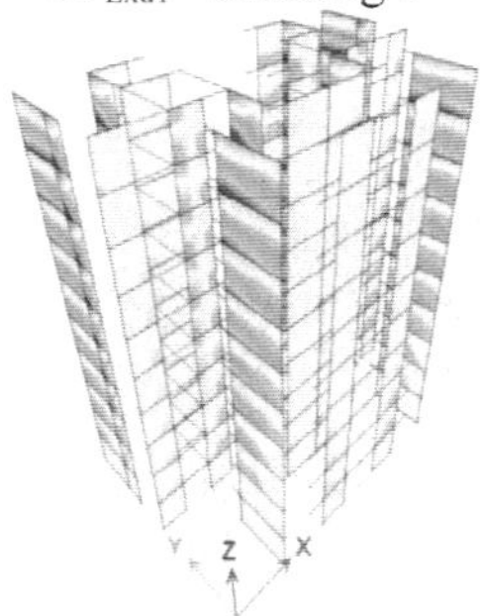

**Figure 10.** CSX $M_{Exd1}$ building V

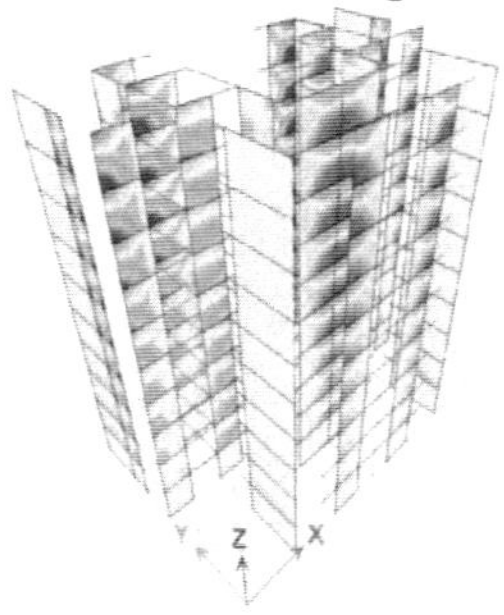

**Figure 11.** CSY $M_{Exd1}$ building V

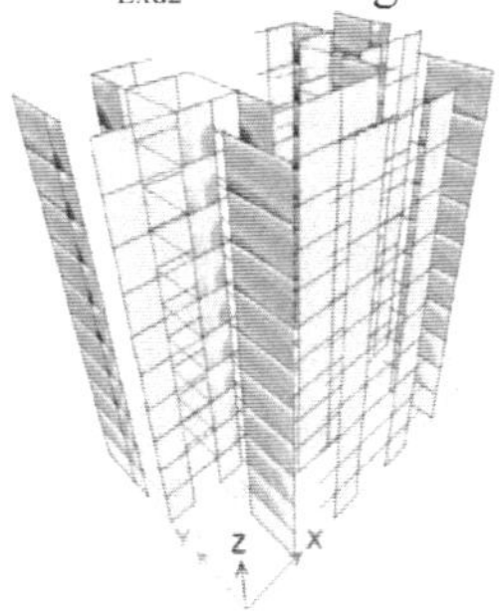

**Figure 12.** CSX $M_{Exd2}$ building V

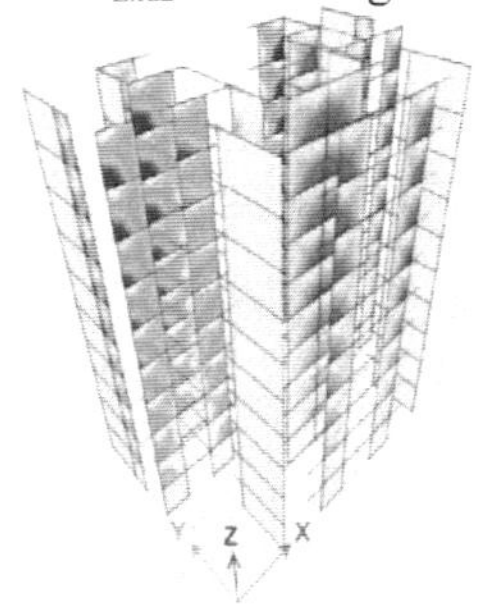

**Figure 13.** CSY $M_{Exd2}$ building V

### *4.2. Efforts in reinforced concrete walls*

To get a better view on the buildings behavior, the axial and shear forces (N and V), the bending moments (M), in the reinforced concrete walls in buildings I and V, are presented in table 3, although these walls are not designed in this paper. However, table 3 only contains the efforts values at the bottom story because they are higher there.

**Table 3.** Highest effective efforts in reinforced concrete walls.

| Pier | direction | efforts for building I | | | direction | efforts for building V | | |
|---|---|---|---|---|---|---|---|---|
| | X | $N_{Ed}$ [kN] | $M_{Ed}$ [kNm] | $V_{Ed}$ [kN] | X | $N_{Ed}$ [kN] | $M_{Ed}$ [kNm] | $V_{Ed}$ [kN] |
| P30 | | 1228 | 7486 | 931 | | 1093 | 8174 | 909 |
| P31 | | 1992 | 7806 | 1378 | | 3305 | 9630 | 1282 |
| P35 | | 3769 | 3210 | 833 | | 1528 | 4759 | 945 |
| P36 | | 3769 | 1769 | 404 | | 626 | 2646 | 456 |
| P38 | | 1297 | 2211 | 414 | | 1561 | 3276 | 584 |
| P40 | | 2711 | 2309 | 410 | | 3250 | 3450 | 595 |
| P42 | | 1255 | 3205 | 734 | | 3234 | 4897 | 971 |
| P43 | | 2153 | 1750 | 341 | | 2539 | 2643 | 427 |
| P45 | | 5547 | 7947 | 1203 | | 5799 | 8362 | 1089 |
| P46 | | 5209 | 8231 | 1698 | | 4349 | 9906 | 1581 |

| | Y | $N_{Ed}$ [kN] | $M_{Ed}$ [kNm] | $V_{Ed}$ [kN] | Y | $N_{Ed}$ [kN] | $M_{Ed}$ [kNm] | $V_{Ed}$ [kN] |
|---|---|---|---|---|---|---|---|---|
| P32 | | 532 | 2715 | 646 | | 1058 | 3204 | 849 |
| P33 | | 699 | 4644 | 704 | | 3138 | 5228 | 644 |
| P34 | | 3807 | 2408 | 437 | | 4114 | 2498 | 390 |
| P37 | | 1029 | 8298 | 982 | | 985 | 8902 | 1077 |
| P39 | | 4438 | 37775 | 2816 | | 4552 | 40936 | 3059 |
| P41 | | 1035 | 8299 | 981 | | 1164 | 8909 | 1081 |
| P44 | | 3809 | 2409 | 437 | | 4132 | 2506 | 393 |
| P47 | | 694 | 4647 | 704 | | 3147 | 5238 | 646 |
| P48 | | 605 | 2717 | 646 | | 1124 | 3215 | 853 |

Axial forces are more evenly distributed for walls on direction X, in building I. They also decrease more gradually. High shear forces are seen in the same walls. Bending moments are greater for building V. Both shear and bending moment's values show similar distributions for both buildings. Torsion moments are increased for the top story compared to the previous one.
All diagrams show similar patterns for both buildings, for the same effort, on the same direction. Slightly higher efforts values are seen for building V. The longest wall, P39 has the highest efforts. P39 on direction Y clearly shows the highest effort values, while the other walls are clearly less loaded. On direction X walls are loaded more evenly and subjected to lower efforts.

### *4.3. Natural vibration periods*
Natural vibration periods in the elastic stage are seen in table 4 for both buildings I and V with and without the partitioning walls (cases A and B). The natural periods are increased by up to10% for building V. It is interesting to see the natural vibration periods in case A, are higher for building I and lower for V, compared to case B. This means partitioning walls stiffen building I, while they make structure V heavier, so its vibration periods increase.

**Table 4.** Natural vibration periods.

| Building | T1 (s) | T2 (s) | T3 (s) | T1 (s) | T2 (s) | T3 (s) |
|---|---|---|---|---|---|---|
| | | Case A | | | Case B | |
| I | 0.504 | 0.472 | 0.404 | 0.489 | 0.421 | 0.387 |
| V | 0.558 | 0.485 | 0.409 | 0.600 | 0.560 | 0.498 |

## 5. Building plastic results

### *5.1. Plastic hinges development*
The nonlinear analysis contains the plastic hinges development and the nonbearing walls stresses diagrams, for both buildings. The pushover cases used here are PX and PY for directions X and Y respectively. The maximum top displacement here is 1.2m. It is chosen because any building becomes unusable for such a value and even for values below it. Figures 14 to 21 show the plastic hinges development for the mentioned top displacement for both buildings I and V. The plastic hinges colors mean: green for the formed hinge (B), blue for the hinge having reached its limit (C), pink for the load being redistributed (D) and red for collapse (E). When the set displacement is reached, building I has stage E hinges in short beams placed in the center of the building at most stories. Building V contains E stage hinges placed towards the bottom of the V. It is noticed that there are more red hinges for case PX for both buildings. Another important observation is the rotation around the weight center of both buildings, especially for V, for case PX. There is no rotation for case PY. Both buildings show displacements on direction Y, but the X coordinates are not changed.

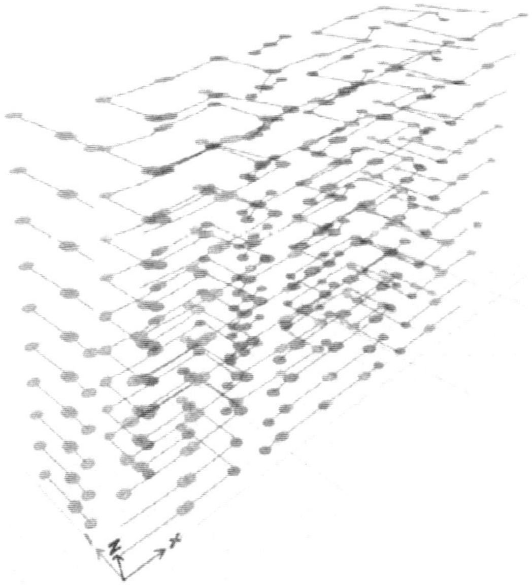

**Figure 14.** 3D plastic hinges PX (I).

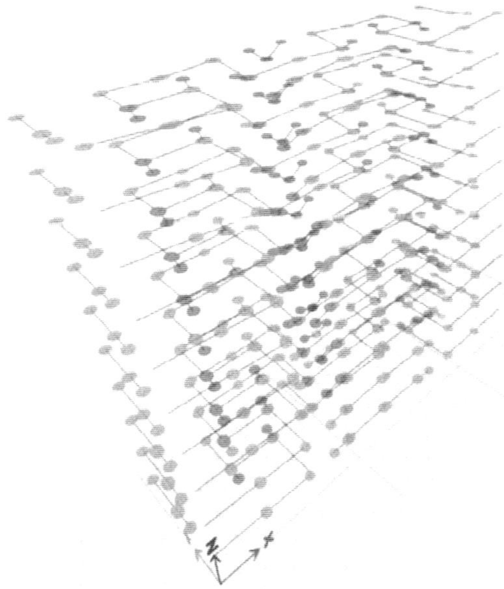

**Figure 15.** 3D plastic hinges PY (I).

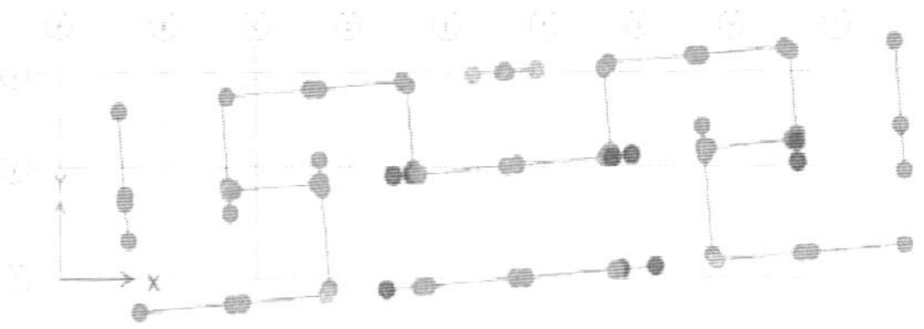

**Figure 16.** Plan plastic hinges PX (I).

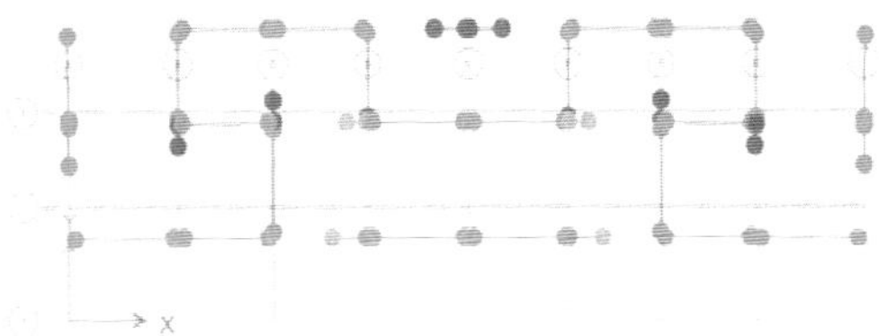

**Figure 17.** Plan plastic hinges PY (I).

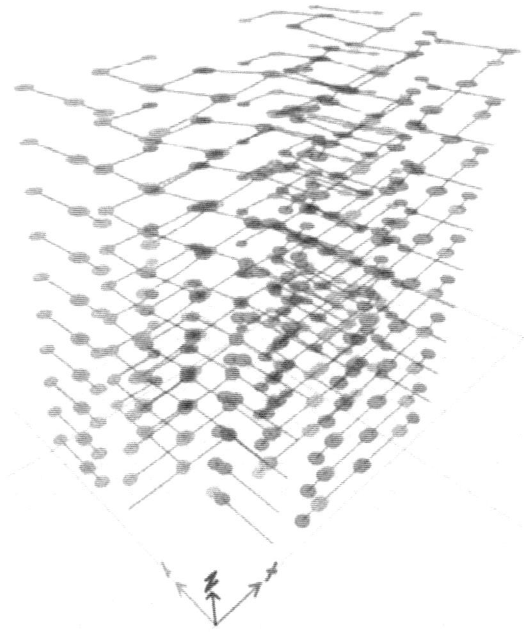

**Figure 18.** 3D plastic hinges PX (V).

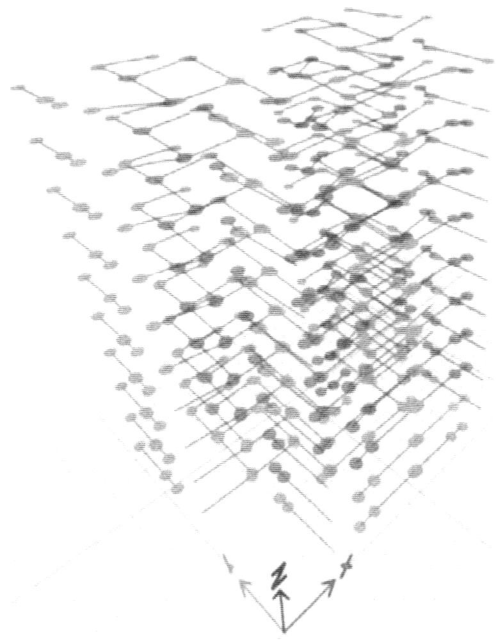

**Figure 19.** 3D plastic hinges PY (V).

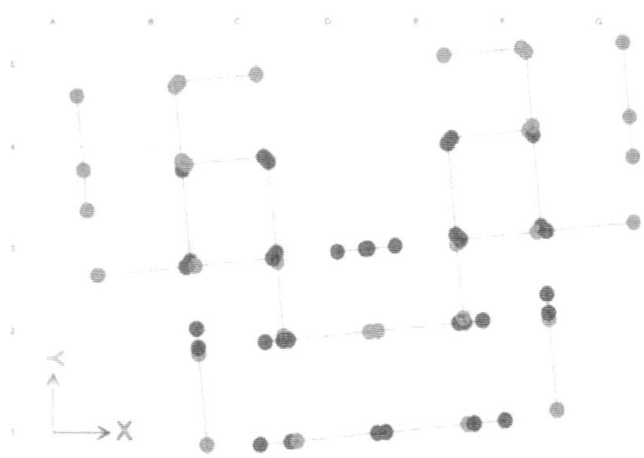


Figure 20. Plan plastic hinges PX (V).

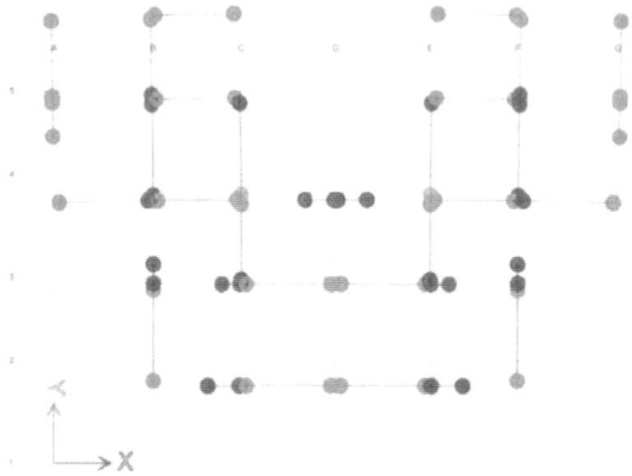


Figure 21. Plan plastic hinges PY (V).

*5.2. Stresses in masonry non-bearing walls*

Stresses seen in figures 22 to 41 are $\sigma_z$, $\sigma_z$, $\tau_{xz}$, $\tau_{xy}$ and $\tau_{yz}$ in N/mm$^2$. The highest stresses are reached for case PX, for building V. There, walls on both directions reach the bearing strength. Walls in building I reach higher stresses for PX compared to PY, but the highest values are seen on direction X. Case PY generates high stresses in walls at the edges and walls on X. The masonry walls are crushed first at the corners, as the $\sigma_x$ values are greater there. For building I, stress $\sigma_z$ surpasses the strengths at the bottom part of the masonry walls perpendicular to the pushover case direction. Also, for this building, the highest stress values are seen in the top walls. For building V, the highest stress values are in the bottom non-bearing walls. The stresses reach greater values at the walls bottoms and tops in building V, especially if they are perpendicular to the pushover case direction.

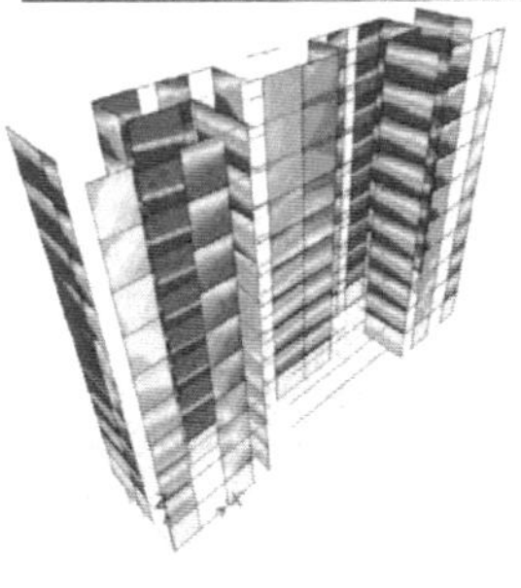

Figure 22. PX step 2 building I $\sigma_x$

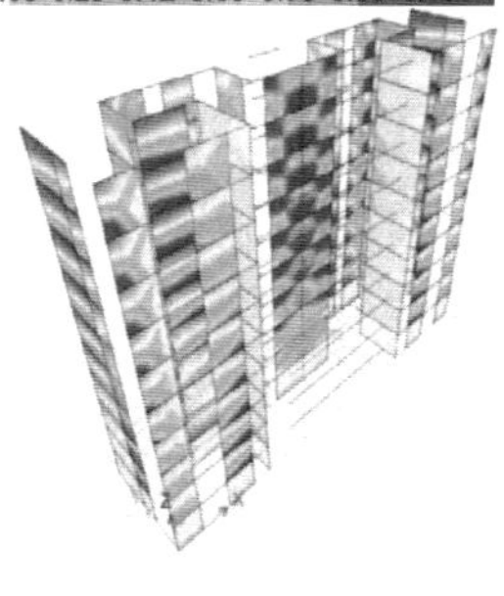

Figure 23. PY step 2 building I $\sigma_x$

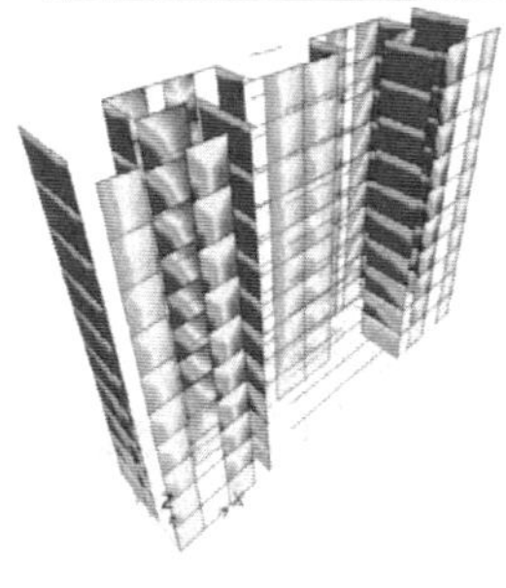

Figure 24. PX step 2 building I $\sigma_z$

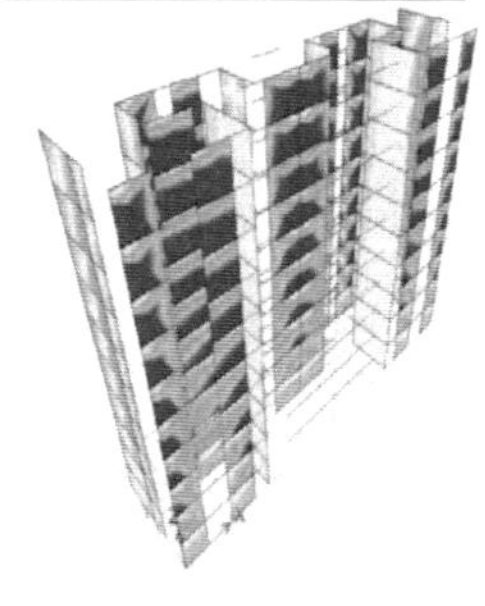

Figure 25. PY step 2 building I $\sigma_z$

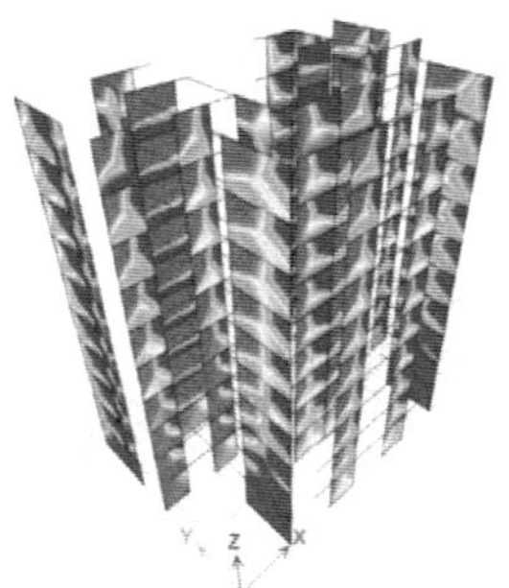

Figure 26. PX step 2 $\sigma_x$ building V

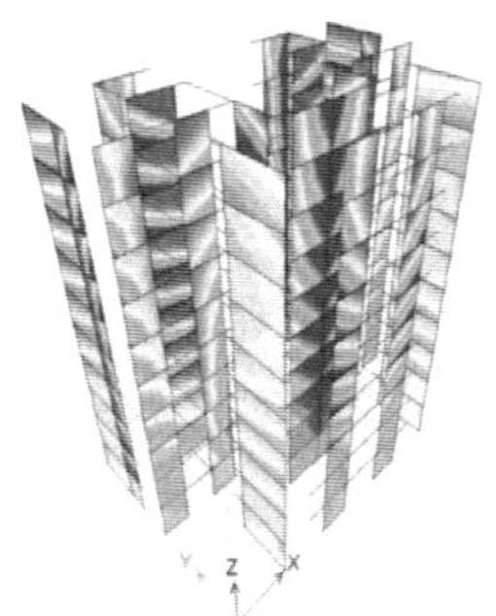

Figure 27. PY step 2 $\sigma_x$ building V

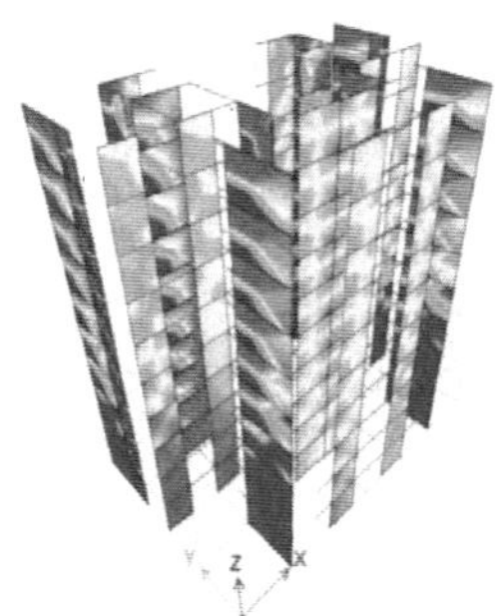

Figure 28. PX step 2 building V $\sigma_z$

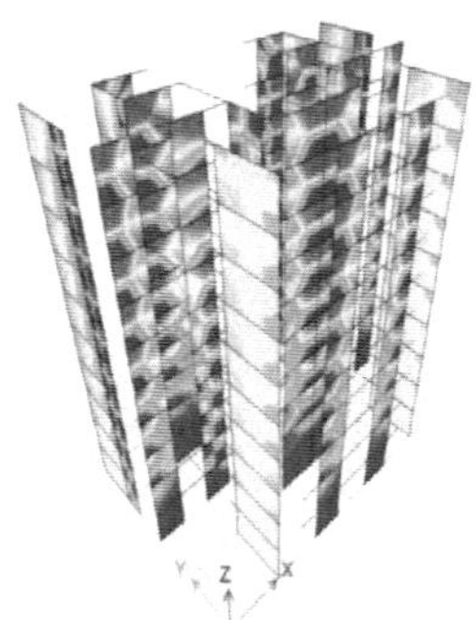

Figure 29. PY step 2 building V $\sigma_z$

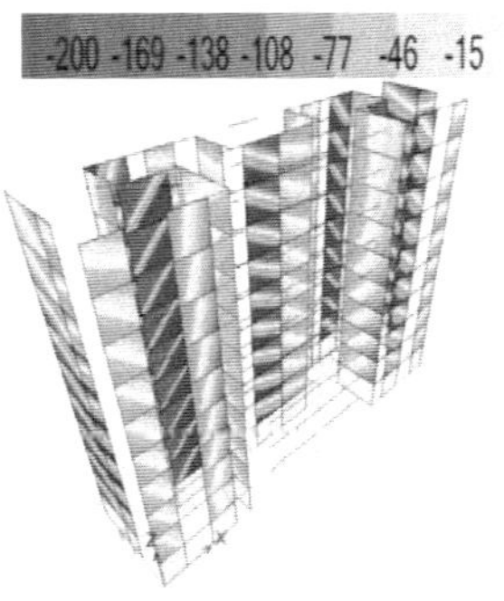


**Figure 30.** PX step 1 building I $\tau_{xz}$

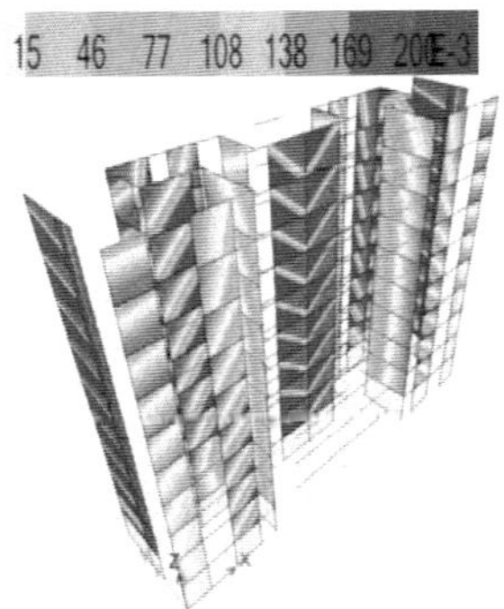


**Figure 31.** PY step 1 building I $\tau_{xz}$

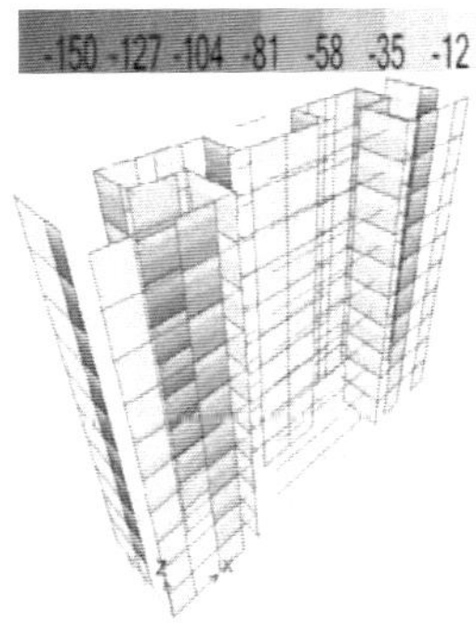


**Figure 32.** PX step 5 building I $\tau_{xy}$

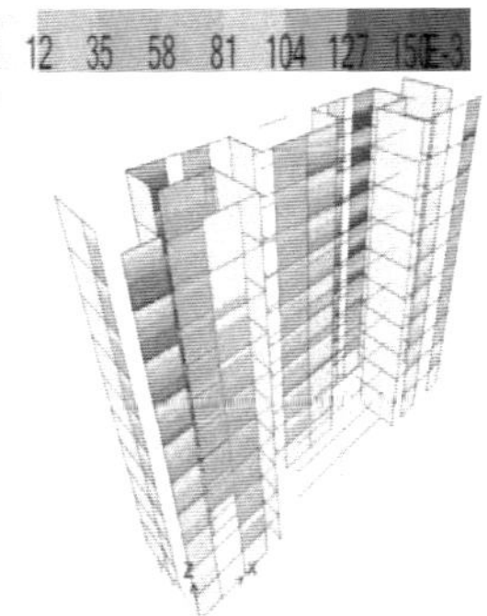


**Figure 33.** PY step 5 building I $\tau_{xy}$

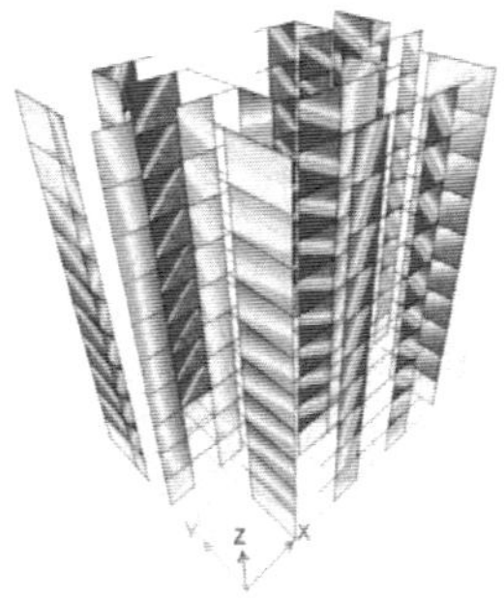

**Figure 34.** PX step 1 building I $\tau_{xz}$

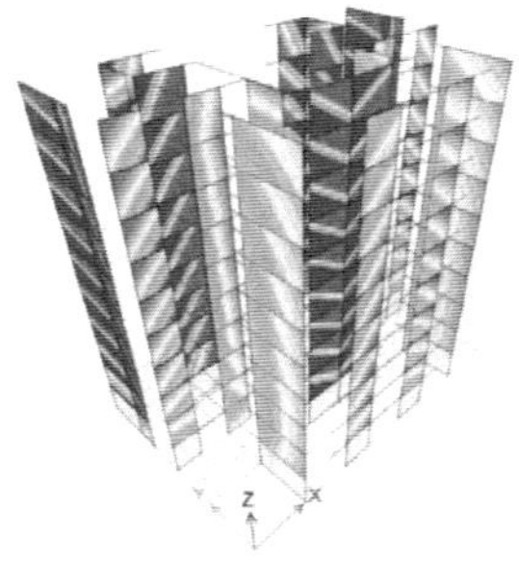

**Figure 35.** PY step 1 building I $\tau_{xz}$

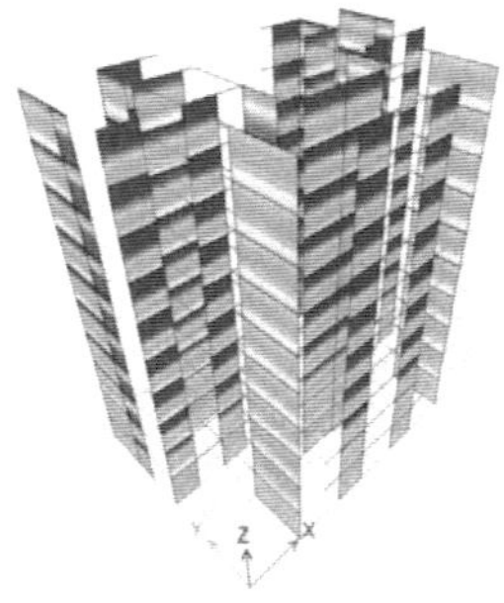

**Figure 36.** PX step 3 building V $\tau_{xy}$

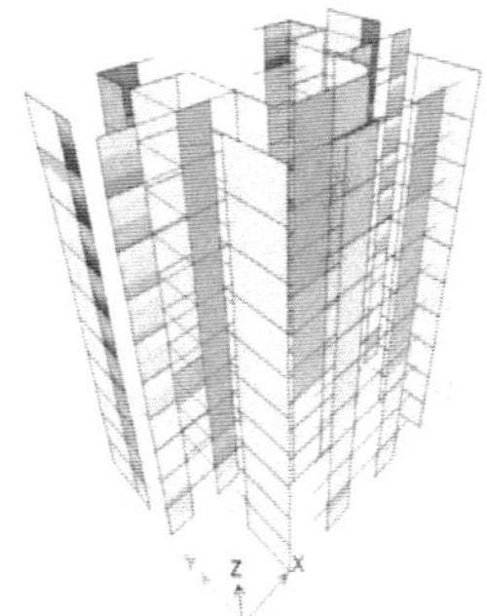

**Figure 37.** PY step 4 building V $\tau_{xy}$

Stresses $\tau_{xz}$ reach high values at the walls corners, particularly these on the same direction as the pushover case or walls connected to them. The lowest values are seen at the bottom story. Stresses reach similar values for buildings I and V. Stresses $\tau_{xy}$ fold the walls horizontally. They affect walls on the same direction as the pushover case, but also some slender walls on the other direction. This can happen because they are easier to bend. The stress values increase towards the higher stories. Masonry walls in building V look more affected. Stress $\tau_{yz}$ folds the masonry walls diagonally and vertically. In this case they create diagonal patterns on the masonry walls. This can be explained by the torsion effect, present especially for building V. Slender walls, and walls perpendicular to the pushover case direction reach higher stress values. Also, stresses increase toward the building top.

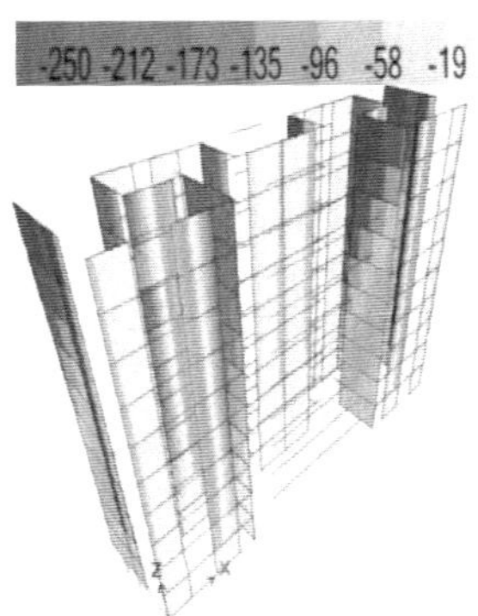


**Figure 38.** PX step 4 building I $\tau_{yz}$

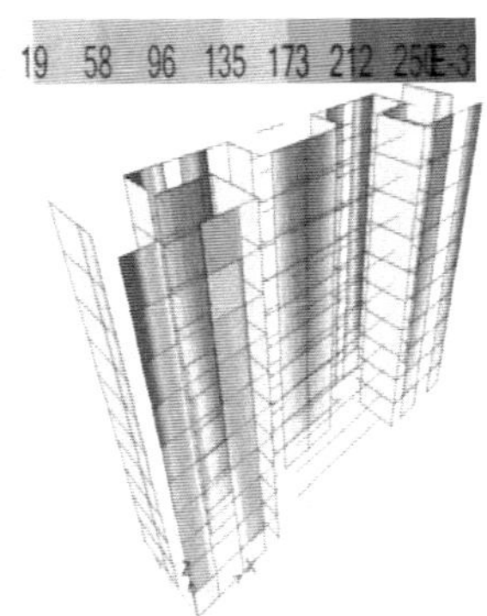


**Figure 39.** PY step 5 building I $\tau_{yz}$

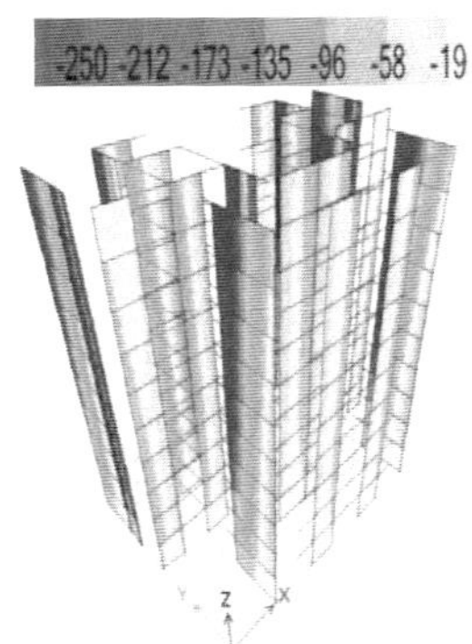


**Figure 40.** PX step 3 building V $\tau_{yz}$

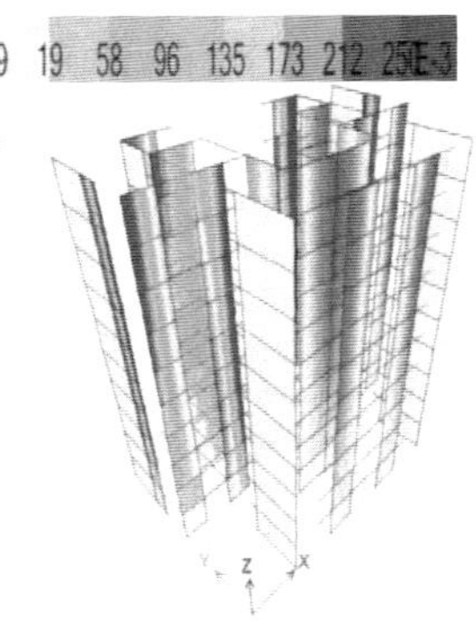


**Figure 41.** PY step 4 building V $\tau_{yz}$

### 5.3. *Pushover diagrams*

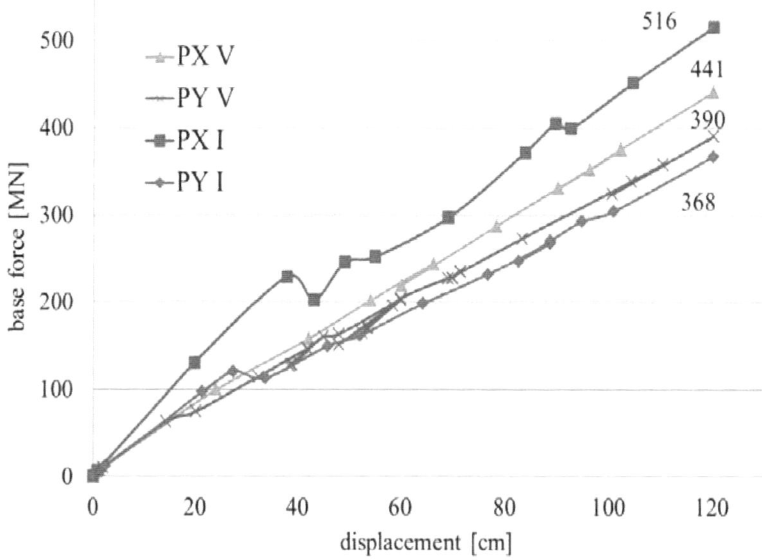


**Figure 42.** Pushover diagrams.

The pushover curves in figure 42, are influenced both by the bearing reinforced concrete walls and by the partitioning masonry walls. The same walls are used for both buildings I and V, but in different location. The highest base forces are reached for case PX. Diagrams for building I show some rigidity drops and load redistributions.

Diagrams for building V, show rigidity drops and fast stiffness recoveries, especially for case PX. Unlike the other diagrams, PXV maintains overall the same rigidity until the established top displacement is reached. PYI shows a gradual rigidity drop.

## 6. Conclusions

Masonry walls are subjected to almost the same efforts in the elastic stage in both buildings even if they are placed in different areas.

In the plastic stage, masonry walls are subjected to similar stress patterns for both buildings.

Case PX generates higher stresses in masonry walls on both directions in both buildings I and V, compared to PY. Case PY creates lower stresses, but for both PX and PY, the highest stress values are seen on direction X. The masonry walls reach stress $\sigma_x$ values in the corners. These stresses surpass the bearing values, so the walls are crushed there. Stresses $\sigma_z$ surpass the strengths at the bottom part of the masonry walls perpendicular to the pushover case direction in building I. Building I shows higher stress $\sigma_x$ and $\sigma_z$ values in the upper stories walls, while building V shows higher stresses in the lower story masonry walls, mostly those perpendicular to the case direction. Shear stresses $\tau_{xz}$ reach high values at the walls corners, mostly on the same direction as the pushover case and at lower stories, with similar values for structures I and V. Masonry walls are folded horizontally by stress $\tau_{xy}$ and higher values are seen at the higher stories. This happens particularly to the walls on the same direction, but also to some slender walls on the other direction. Slender walls are easier to bend. This stress affects the walls in structure V more. Stress $\tau_{yz}$ folds the masonry walls diagonally and vertically, so diagonal patterns are created. It mostly affects the slender walls and all walls placed perpendicular to the pushover case. Stress values increase towards the top story masonry walls.

The pushover diagrams show important stiffness drops followed by some increases. They happen much faster for building V.

Both bearing concrete walls and masonry walls influence the pushover diagrams. The same walls are used, but they are placed in different locations for buildings I and V. Case PX shows the highest base force for both structures. There are fast rigidity drops and fast regains in diagrams for building V, especially for case PYV. Case PXV maintains overall the same rigidity until the end of the analysis. There are gradual rigidity drops for diagram PYI.

Bearing walls are subjected to slightly higher efforts in structure V. Diagrams show similar patterns for the same effort and seismic direction.

The nonbearing walls aren't much affected by the building's shape, if it is like a V or an I. The impact another shape, such as L or T, may have on the partitioning walls, may be extended in future research.

**References**

[1] Carpinteri A, Lacidogna G and Cammarano S 2016 Conceptual Design of Tall and Unconventionally Shaped Structures: A Handy Analytical Method *Adavnces in Structural Engineering* https://doi.org/10.1260/1369-4332.17.5.767

[2] Lacidogna G, Scaramozzino D and Carpinteri A 2020 Influence of the geometrical shape on the structural behavior of diagrid tall buildings under lateral and torque actions. Development in the built environment Elsevier https://doi.org/10.1016/j.dibe.2020.100009

[3] Abd-El-Rahim H and Abd-El-Raheem Farghaly A 2010 Influence of structural irregularity in plan floor shape on seismic response of buildings *Journal of Engineering Sciences, Assiut University*, Vol. **38**, No. 4, pp. 911-928.

[4] Raagavi M T and Sidhardhan S 2021 A Study on Seismic Performance of Various Irregular Structure *International Journal of Research in Engineering and Science (IJRES)* ISSN: 2320-9364, ISSN: 2320-9356 www.ijres.org Vol **9** Issue 5 2021 pp. 12-19

[5] Ahmed M, Shehata E. Abdel Raheem S, Ahmed M and Abdel-Shafy A 2016 Irregularity effects on the seismic performance of L-shaped multi-story buildings *Journal of Engineering Sciences Assiut University Faculty of Engineering* Vol. **44** No. 5 2016 PP. 513 – 536

[6] Ferraioli M 2015 Case study of seismic performance assessment of irregular RC buildings: hospital structure of Avezzano (L'Aquila, Italy) *Earthquake Engineering and Engineering Vibration* volume 14, pp141–156 2015 https://doi.org/10.1007/s11803-015-0012-7

[7] Gokdemir H, Ozbasaran H, Dogan M, Unluoglu E and Albayrak U 2013 Effects of Torsional Irregularity to Structures during Earthquake *Engineering Failure Analysis Elsevier*, Vol. 35, pp: 713–717. https://doi.org/10.1016/j.engfailanal.2013.06.028

[8] Abd-Elhamed A and Mahmoud S 2017 Linear and Nonlinear Dynamic Analysis of Masonry Infill RC Framed Buildings *C.E.J.* Vol. **3** Nr. 10, 2017. http://dx.doi.org/10.28991/cej-030922

[9] Teguh M 2016 Experimental evaluation of masonry infill walls of RC frame buildings subjected to cyclic loads *Elsevier Procedia Engineering 171 Sustainable Civil Engineering Structures and Construction Materials 2016* pp 191-200, 2017. doi:10.1016/j.proeng.2017.01.326

[10] Jiang H, Liu X and Mao J 2015 Full scale experimental study on masonry infilled RC moment-resisting frames under cyclic loads *Elsevier Engineering Structures* 91, p70-84, 2015. http://dx.doi.org/10.1016/j.engstruct.2015.02.008

[11] CEN EN 1996-1-1-2006 Eurocode 6: Design of masonry structures - Part 1-1: General rules for reinforced and unreinforced masonry structures, 2006.

[12] CEN EN 1991-1-1-2004 Eurocode 1: Actions on structures - Part 1-1: General actions-Densities, self-weight, imposed loads for buildings, 2004.

[13] CEN EN 1990-2004 Eurocode 0: Basics of structural design, 2004.

[14] CEN EN 1991-1-3-2005 Eurocode 1: Actions on structures - Part 1-3: General actions- Snow loads, 2005.

[15] CEN EN 1992-1-1-2004 Eurocode 2: Design of concrete structures - Part 1-1: General rules and rules for buildings, 2004.

[16] CEN EN 1998-1-2004 Eurocode 8: Design of structures for earthquake resistance. Part 1: General rules, seismic actions and rules for buildings, 2004.
[17] P100-1/2013 Seismic design code–Part 1 - General rules for buildings, 2013 updated P100-1/2019.
[18] CR 6/2013 Masonry design code, 2013.

# Assessing the vulnerability of buildings to floods in the lower sector of Slănic River. Case Study of Cernăteşti village, Buzău County, Romania

**R Costache**[1], **C Popescu**[2,*] **and A Barbulescu** [1,2]

[1] Transilvania University of Braşov, Braşov, Romania
[2] Technical University of Civil Engineering, Bucharest, Romania

* cristiannicolae.popescu@gmail.com

**Abstract.** Assessing the vulnerability of buildings is very important to determine the settlements affected by floods since authorities can take measures before actual flooding takes place. The present research aims to identify areas affected by flooding in Cernăteşti. The village is located in an area where floods are favored by the elongated shape of the catchment, the horizontal slope, relatively small distance between the hills bordering the valley, water accumulation resulting from flash floods formed upstream, and heavy local rainfall. Field observations, GIS techniques, and mathematical calculations have been used in this research. Flood simulations have been made to determine the degree (number of houses and roads affected, estimated losses in lei) of impact at flows with different probabilities of exceedance (10%, 5%, 2%, and 1%). Depending on the exposure, calculated using factors like distance from the river or construction material, the buildings on the left bank of Slănic are more affected and the flooded area is more extensive. Buildings in the south-eastern, north-eastern, and north-western parts of the flooded area are highly vulnerable, but most of the other settlements are less vulnerable.

## 1. Introduction

Extreme hydrological phenomena are unpredictable and negatively impact human settlements. Due to the continuous climate dynamics, they became more frequent than before [1-5]. Other contributors to hydrological hazards are relief, Earth rotation, exposure to the Sun, and anthropic activities. In Europe, hydrological phenomena take place everywhere. In the coastal zones of western areas, there are thunderstorms, while in the east, high thermal differences favor the appearance of flooding in spring when the snow melts fast.

Last period, a significant increment in the floods occurrence all over the world has been reported, significantly affecting the human life and economy [6,7]. It has been noticed that that most of the rural area lack the flood defense and control [8]. Evaluation of the hazard probability and building an early warning system is a must for saving life and minimize the material loss of such events.

Deepak et al [6] analyzed the factors responsible of hazards by GIS, machine learning techniques and random forest and took a case study from Aluva municipality, Kerala, India. Kaoje et al [9] proposed the use of GIS and Participatory Analytical Hierarchy Process for assessing the Physical Flood Vulnerability in Kota Bharu, Malaysia, whereas Len et al [8] reviewed the vulnerability indices employed by different authors.

Romania is a country highly exposed to flash floods [10-12], especially in areas with steep slopes, and to floods in the extra-Carpathian areas where water accumulates at heavy rainfalls. Valleys with high population density are affected from time to time. Exposure of settlements can vary depending on the distance from the river, materials used for construction, their position on the relief (if they are built at a high altitude or if a dam defends a village from floods).

Cernăteşti village, the study area from Buzău county, makes no exception when it comes to hazards. The village is located in a hilly region, favorable to the production of flash floods and flooding. Settlements and public buildings are in the proximity of the river.

Studies regarding the exposure and vulnerability to flooding and the flooding simulation can help the authorities to have a precise overview of the area of interest. Measures, like evacuation and saving as much as possible from the households can be taken before the flooding occurs. Therefore, the scope of the present research is to estimate the number of settlements affected in Cernăteşti at different flow rates, taking into account different factors.

## 2. Study area

Cernăteşti, the village chosen for the study, is located in the central part of Buzău County, Romania (figure 1) at an altitude of approximately 150 meters, at the contact between the hilly Sub-Carpathians and Romanian plain [13]. Slănic, the river that crosses the village, flows into Buzău River near Săpoca, not far from the village (5 km from the confluence), being one of the main tributaries of this river.

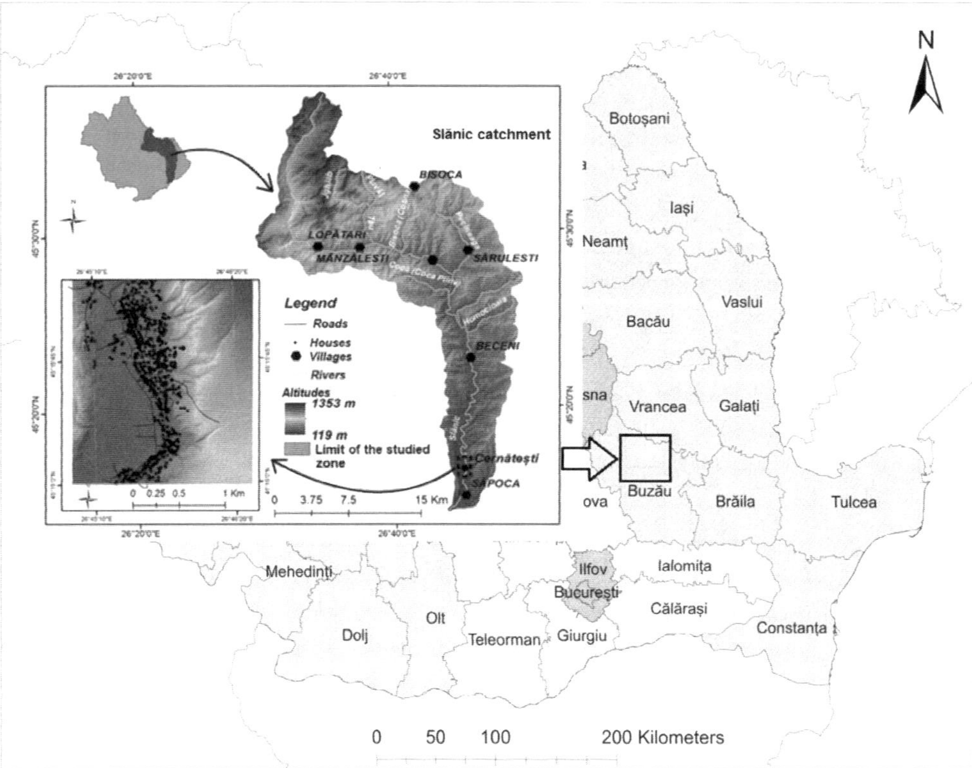


**Figure 1.** Location of Cernăteşti.

Cernăteşti is situated in the lower part of Slănic catchment, where the basin is elongated, the valley is quite narrow, a characteristic that makes it susceptible to flooding. In upper areas, the slope is steeper, the shape of the subbasins is circular and precipitations are higher [14]. These properties can favor the appearance of flash floods [6] that can lead further to the appearance of floods downstream

[15,16], and subsequently to pollution with carried materials [17,18], especially in an area like Cernăteşti, where the distance between the hills bordering the valley is under 1 km.

The vegetation is composed of deciduous forests with species of oak, lilac, shrub in the upper part of Cernăteşti, while in the lower parts, one can find meadow vegetation (willow), steppe (most of it was replaced with crops, settlements, and economic zones). Important areas, especially in the lower parts, are not forested.

The commune is formed of eight small villages, some of them located very close to the river.

Soil patterns influence the infiltration and further the percolation in aquifers [15]. If soil is saturated, then there is a high risk of flooding because water cannot drain into the ground, especially in areas with heavy rains and steep slopes. This also applies to clay soils that have a specific structure that reduces the infiltration. Most of the studied area has meadow soils, chernozem, and clay-alluvial.

Climate is fundamental in influencing the appearance of hydrological phenomena.

The precipitation is 500 - 600 mm/m$^2$/year in this region, with higher values in May and June. In the upper areas, they exceed 800 mm/m$^2$/year. The average temperature is 9$^o$C, with a high annual amplitude. Rapid melting of snow in spring, heavy rainfalls, accumulation of water downstream from the flash floods upstream can cause the appearance of flooding in Cernăteşti. The dominating wind is “Crivăţul” (chill wind), which is very cold and blows in winter at high speeds from the east.

## 3. Methodology

Field exploration consisted of GPS measurements made in Cernăteşti, with a high precision device, observing the riverbed slope (river bank line, river centerline and cross section), the surrounding areas (if the hills favor a rapid runoff), valley configuration, land use, and the field location of the buildings (if the settlements are close to the river) [19].

Digital Terrain Model (DTM) was used to extract the geometric elements of the riverbed (river bank, river centerline, and cross section profiles). DTM was obtained by interpolating the level curves vectored from a Topographic Plan 1:5000 and some 120 altimetric points taken with a GPS device (TRIMBLE GeoXH 2008 Series) from Slănic riverbed [20]. Thus, 23 cross section profiles were established, located 100 meters from each other, with a width between 780 and 1030 meters (figure 2) [21].

Riverbed geometry elements have been exported in the HEC-RAS 4.1 (CEIWR-HEC, USA) [22] format to obtain flood simulations based on river flow values.

The vulnerability is a value determined by combining the exposure with the susceptibility [23]. The vulnerability to floods includes the presence of hazards and potentially affected settlements.

To compose the vulnerability index of settlements to floods, specific steps must be followed.

The identification of the settlements was possible using aerial images (orthophotomap). Regarding the building properties, the following have been taken into account: function, height, surface, construction material, construction condition, number of floors, and Euclidian distance from the river [24]. A point containing the above properties corresponding to the position of each building was obtained using the GPS. GIS techniques have been used based on these points [25].

After computing the degree of exposure (value calculated by summing up house characteristics like building material, height or building condition), the water depth for each building was extracted. The constructions are represented as points. The depth corresponding to the flow with the probability of exceedance of 1% from Cernăteşti was estimated using hydraulic modeling techniques implemented in HEC-RAS 4.1 software [22] based on the flow value, digital terrain model (DTM), and the geometry elements of the riverbed extracted based on the digital terrain model. Since the water depth is represented spatially in raster format, the assignment of the values of each building was done by using the Extract Value to Points tool from ArcGIS 10.2 software [22].

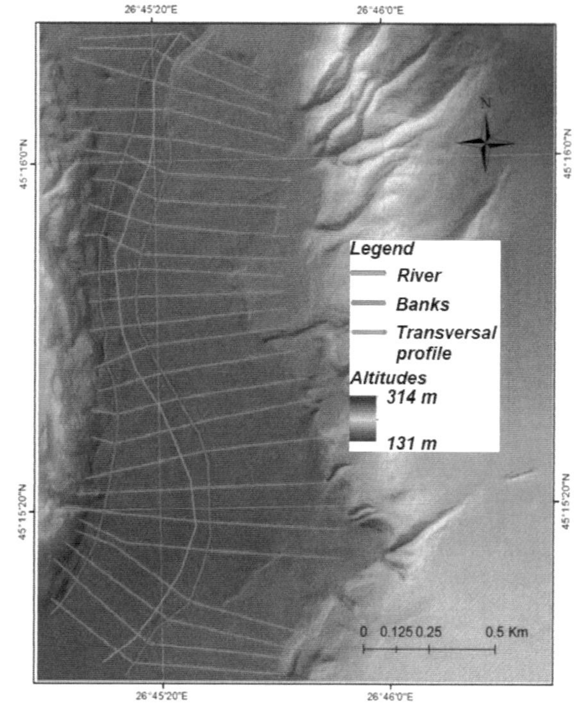


**Figure 2.** The riverbed geometry of Slănic in Cernăteşti.

Thus, the level of vulnerability of each building was calculated by applying the following equation [26]:

$$V = E \times H, \qquad (1)$$

where $V$ - vulnerability, $E$ – exposure, and $H$ – codified values of the water depth (hazard).

For the computation and spatialization of the Buildings Vulnerability Index (*IVC*) values, a first phase was used to create a grid with a cell size of 30 m. The sum of all vulnerability values for each building located in a specific cell was obtained automatically using the Spatial Join tool from ArcGIS 10.2 (Esri, USA) [22]. Thus, the *IVC* was obtained by summing up the individual indexes [18].

To facilitate the analysis, the *IVC* values obtained on each cell of the grid were standardized using the following formula [27]:

$$IVC_{t,i,j} = \frac{IVC'_{i,j} - IVC'_{min,j}}{IVC'_{max,j} - IVC'_{min,j}}, \qquad (2)$$

where $IVC_{t,i,j}$ – standardized value for each cell, $IVC'_{i,j}$– initial value of IVC on each cell, $IVC'_{min,j}$ - minimum value of IVC from the whole range of values; $IVC'_{max,j}$ – maximum value of IVC from the whole range of values.

The standardized *IVC* values are in the interval [0,1]. These values were grouped into five classes using the automatic Natural Breaks classification method [28].

## 4. Results and discussion

Using the methodology explained in Section 3, the values that show the exposure and vulnerability of settlements in Cernăteşti have been obtained. According to table 1, an important number of houses, roads, and lands are flooded when high flow values occur.

For probability of exceedance of 1% (620 $m^3/s$), 179 houses are affected, while for 10%, the number of affected houses is 78. Thus, there is a high risk for material losses. Regarding the roads, significant distances are flooded, resulting in isolation for the inhabitants. When lands are flooded, crops can be compromised. For example, for a probability of exceedance of 1%, more than 1 $km^2$ is flooded.

For a probability of exceedance of 1%, one may notice (figure 3) an important expansion in the south-eastern part of the river (where a small number of houses are flooded) and in the north-eastern area (where an important number of houses are flooded). For the other probabilities of exceedance, one may observe that houses and roads are flooded even in the case of flows with lower values.

**Table 1.** Potential effects of flood occurrence for the probabilities of exceedance of 1%, 2%, 5%, and 10% at Cernăteşti hydrometric point.

| Flood values with different probabilities of exceedance | 620 $m^3/s$ ($P_{1\%}$) | 495 $m^3/s$ ($P_{2\%}$) | 367 $m^3/s$ ($P_{5\%}$) | 236 $m^3/s$ ($P_{10\%}$) |
|---|---|---|---|---|
| Flooded area (Ha) | 120 | 108 | 91 | 78 |
| Number of affected houses | 179 | 94 | 34 | 26 |
| Number of kilometers of affected roads | 3.3 | 2.6 | 1.8 | 1.35 |

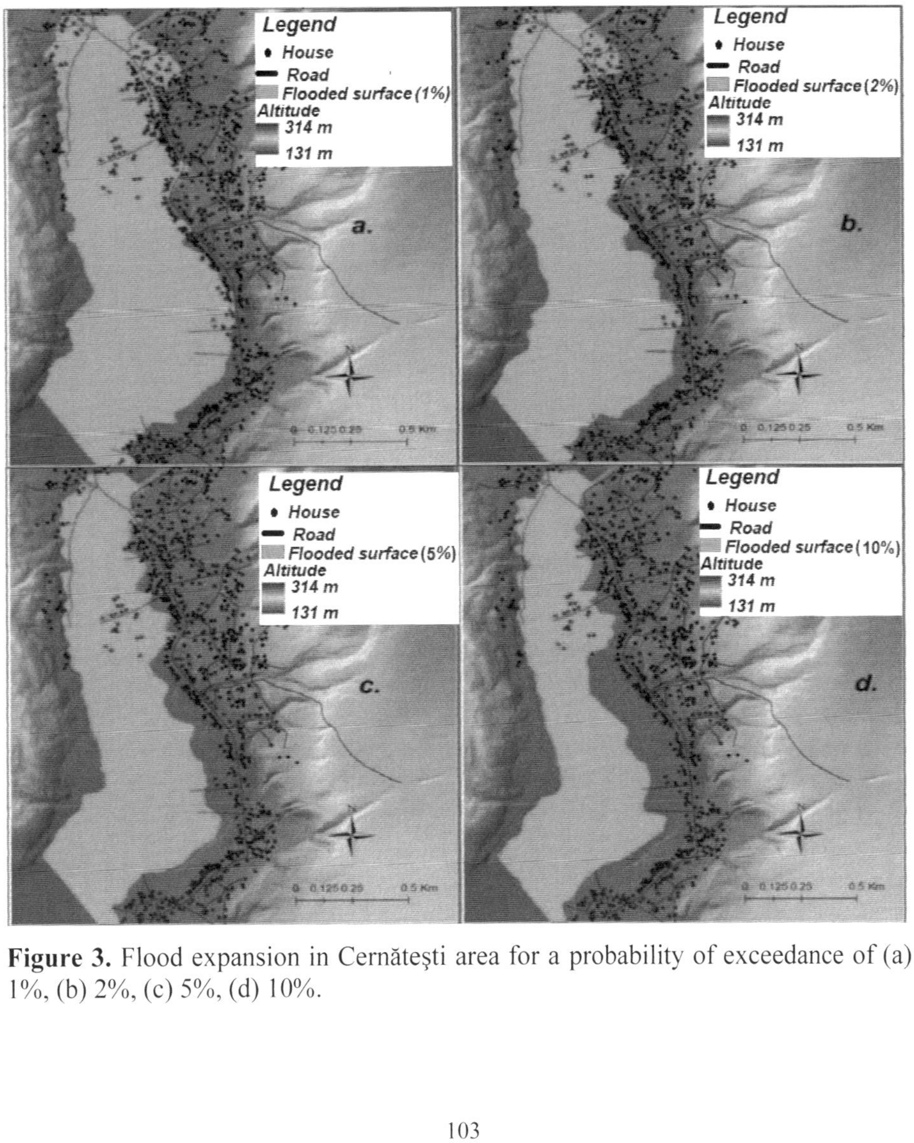


**Figure 3.** Flood expansion in Cernăteşti area for a probability of exceedance of (a) 1%, (b) 2%, (c) 5%, (d) 10%.

An estimation of the damage caused to the structural elements in case of a flood occurrence with different probabilities of exceedance (1%, 2%, 5%, and 10%) is presented in table 2. For example, for 1%, 0.8 km of the county and village roads (modernized roads) are flooded. These roads are the ones that connect the settlement with the rest of the network. When the value of 620 $m^3/s$ occurs, the village can be isolated, and the intervention of the authorities can be difficult. The total estimated losses for houses and roads are approximately 1500 thousand lei. For the probability of exceedance of 2%, 2.6 km of roads and 94 houses and annexes are affected. The total estimated losses are 848 thousand lei. For the probability of exceedance of 10%, the total losses recorded are 282 thousand lei.

The above results show that the Cernăteşti area presents a high risk of water accumulation.

**Table 2.** Potential damage for housing and transport infrastructure in case of discharge values with different probabilities of exceedence: 1%, 2%, 5%, and 10%.

| Probability of exceedance | Infrastructure element | Number | Value (USD) | Estimated losses (USD) |
|---|---|---|---|---|
| 1% | Houses and household annexes (no) | 179 | 1572 | 281353 |
| | County and village roads (km) | 0.8 | 28570 | 22851 |
| | Residential roads (km) | 2.5 | 14565 | 36412 |
| 2% | Houses and household annexes (no) | 94 | 1572 | 147763 |
| | County and village roads (km) | 0.5 | 28564 | 14292 |
| | Residential roads (km) | 2.1 | 14565 | 30584 |
| 5% | Houses and household annexes (no) | 34 | 1572 | 53422 |
| | County and village roads (km) | 0.33 | 28569 | 9429 |
| | Residential roads (km) | 1.45 | 14565 | 21108 |
| 10% | Houses and household annexes (no) | 26 | 1572 | 40881 |
| | County and village roads (km) | 0.25 | 28569 | 7134 |
| | Residential roads (km) | 1.1 | 14565 | 16019 |

Flooding results from heavy rainfalls, flash floods that feed the river, the riverbed's almost horizontal slope, elongated catchment shape, and hills that border the river and act as dams [11].

The exposure analysis was established based on the building properties mentioned in the methodology. A point corresponds to each building in Cernăteşti. Therefore, the type of buildings that are affected can be identified. According to their function, the library and the kindergarten located in the river proximity are the most exposed buildings. Thus, these buildings have been marked with 3 because of the risk. The least exposed are the warehouses that received a grade of 1. The higher the rate, the more exposed a building is considered Our findings show that approximately 94% of the structures are residential (figure 4(a)).

The height of the buildings was identified based on the field observations. Tall settlements are more exposed to floods than low-rise buildings. Buildings have been grouped as in figure 4(b). The class with a height of 2-3 m received a rate of 3, while the other two received 2 and 1.

An estimation of the condition of each building was made. They have been included in categories four categories: *good* (39.9%), *very good* (34.1%), *medium* (14.7%), and *poor* (11.3%). Each category received a rating score from 1 to 4, depending on the exposure to floods and flash floods.

The construction surface is another variable that influences the degree of its exposure to floods and flash floods. The larger the surface area is, the greater the destructive potential of the building is. To determine the building area, GIS techniques have been applied (constructions have been digitized in polygon vector format).

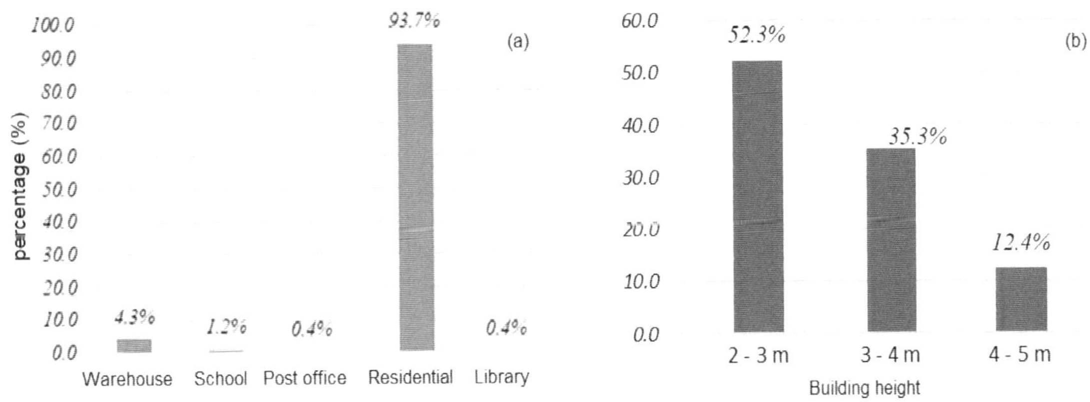


**Figure 4.** Distribution depending on (a) the function of the buildings and (b) their height.

The buildings received rates depending on their size (from 1 to 4). 47.7% of the buildings have a surface less than 100 $m^2$, 30.2% between 100 and 200 $m^2$, 15.1% - between 200 and 300 $m^2$, and 7% - greater than 300 $m^2$.

The number of floors was the fifth variable taken into account. Based on field observations, constructions with only two levels have been identified. Those with a single floor (86.8%) have been considered more exposed and received the rating of 2, while the buildings with two levels (13.2%) received 1.

For the variable represented by the distance of the houses from the river, the Euclidian distance was used (ArcGIS 10.2). Lower ratings were given to buildings located at a greater distance from the river, unlike houses near the river, which received a higher rating due to flood exposure. More than 36% of the houses are located under 200 meters away from the river.

Construction material is a very important element to determine the degree of exposure to future floods. The majority of buildings are built of three main materials BCA (lightweight aerated autoclaved concrete blocks masonry) (40.7%), brick (43%), and adobe (16.3%). Constructions built of BCA are considered to be the least exposed to damage caused by floods.

Once the ratings have been assigned, the exposure degree ($E$) of the houses in case of floods was calculated by,

$$E = F + H + Sf + S + M + DR + N, \tag{3}$$

where: $E$ – exposure, $F$ – construction function, $H$ – construction height, $Sf$ – construction condition, $S$ – building surface, $Mc$– construction material, $DR$ – distance from the river, $N$ – number of floors.

Figure 5 shows that for a probability of exceedance of 1%, more than 34% of the buildings present a high and very high risk of exposure, while a third present a medium risk. Thus, Cernătești is highly exposed to flooding.

Figure 6 indicates that the buildings with very high and high exposure are located on the left bank, in the north-eastern part of the flooded area. Due to the valley configuration, there is a favourable condition for flooding expansion on the left bank, where many buildings are situated.

The level of vulnerability is an index proposed by the authors taking into account the depth of the water for each building and its exposure. The *IVC* classes and the percentage of buildings included in those classes are: 0 - 0.08 (44%), 0.09 - 0.20 (28%), 0.21 – 0.34 (16%), 0.35 – 0.60 (10%), and 0.61 - 1 (2%). So, 44% of buildings present a low vulnerability to flooding, while only 12% of buildings present a high and very high vulnerability. IVC high values, over 0.35, correspond to buildings located in the north-western, north-eastern, and south-eastern flooded area (figure 6a).

The Building Vulnerability Index (IVC) is spatialized in figure 6b. Remark that the depth exceeds 7 meters in the river cross section area when the probability of exceedance for flooding is 1% (a very high value for a small river).

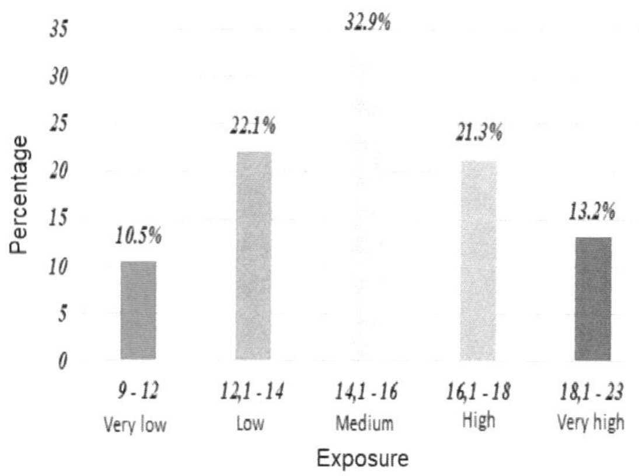


(a)

(b)

**Figure 5.** (a) Exposure to flooding risk at 1% probability of exceedance and (b) Flood band at 1% probability of exceedance.

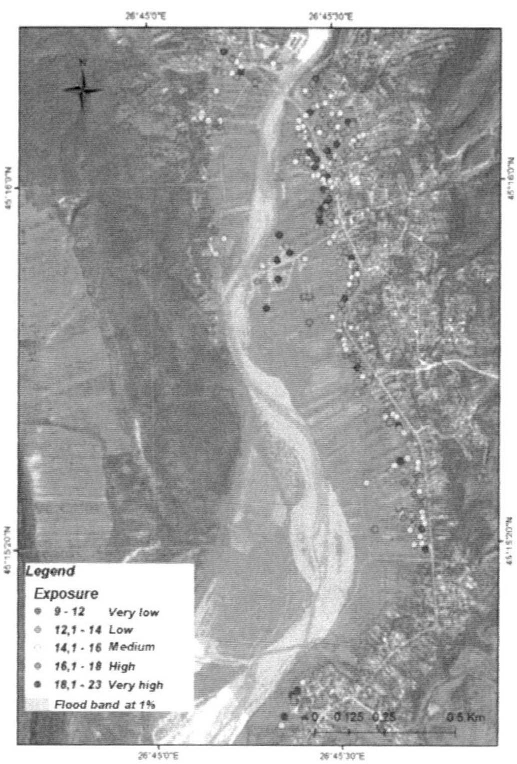


(a)

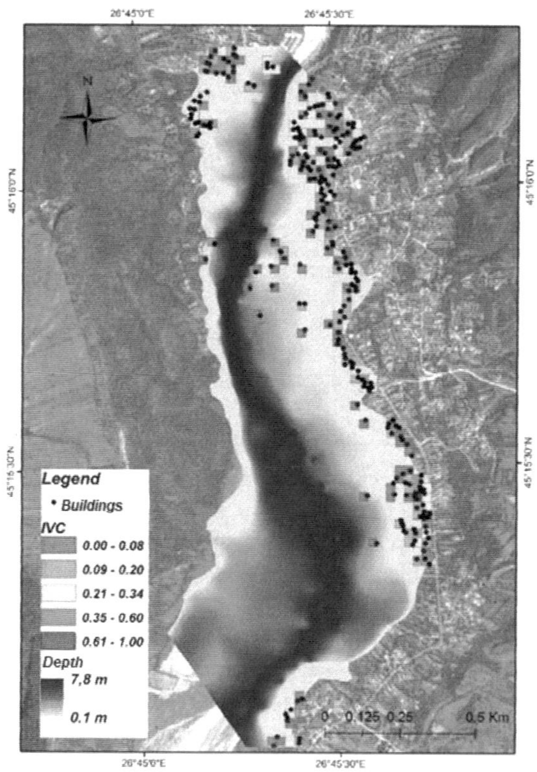


(b)

**Figure 6**. (a) Distribution of the exposure index for each building, corresponding to an exceedance probability of 1%; (b) IVC distribution at the exceedance probability of 1%.

## 5. Conclusion

Field observations and GIS techniques are essential to perform the present research. Using the high-precision GPS, databases, ArcMap, and HEC-RAS software, we could determine a precise configuration of Slănic Valley at Cernăteşti, flooding simulations that show areas exposed and buildings vulnerable to flooding. Different indexes have also been computed to assess the vulnerability to flooding at different exceedance probabilities.

The study should be extended in the future to assess the flood impact on the agricultural soils' pollution using GIS and statistical analysis and how to remove the potential flooding effects [29-31].

It was shown that Cernăteşti presents a risk to floods due to factors as the horizontal slope of the riverbed, the elongated shape of the catchment, the relatively small distance between the hills that border the valley (under 1 km), and water resulting from flash floods from upstream that favor the accumulation of high quantities. Local heavy rainfalls, melting slow, and runoff from the slope are added to these factors. The potential effects of flood occurrence for specific probabilities of exceedance have been computed. It was shown that in the cases of flood events, the roads are flooded, and the authorities' capacity for intervention is reduced because access is cut off. Tens of houses are vulnerable to floods, especially those located on the left bank.

Improvements should be made to avoid future floods and flash floods in the lower catchment of Slănic River. One of them is the construction of an artificial dam lake upstream that can control the river's flow when flash floods occur during heavy rains.

## Acknowledgments

This work was supported by a grant from the Romanian Ministry of Education and Research, CNCS – UEFISCDI, project number PN-III-P1-1.1-PD-2019-0424-P, within PNCDI III.

## References

[1] Bărbulescu A, Dumitriu C S and Maftei C 2022 *Rom J Phys* **59(5-6)** 590
[2] Bărbulescu A, Postolache F and Dumitriu C S 2021 *Hydrology* **8(3)** 125
[3] Bărbulescu A, Şerban C and Caramihai S 2021 *Rom J Phys* **66(3-4)** 806.
[4] Bărbulescu A, Dumitriu C S and Dragomir F L 2021 *Proc Int Conf Electrical, Computer, Commun Mechatronics Eng* (ICECCME 2021), 7-8 Oct 2021 Mauritius DOI: 10.1109/ ICECCME52200.2021.9591085
[5] Şerban C, Bărbulescu A and Dumitriu C S 2021 *IOP Conf. Ser.: Earth Environ. Sci* **958** 012006
[6] Deepak S, Rajan G and Jairaj P G 2020 *Geoenv Disasters* (2020) 35 https://doi.org/10.1186/s40677-020-00172-w
[7] Kourgialas N N and Karatzas G P 2011. *Hydrol Sci J* **55(2)** 212-255
[8] Len N L S et al 2018 *ASM Sc J* **11** Spec issue **3** for SANREM 134-146
[9] Usman Kaoje I et al 2021 *Water* **13**, 1786 https://doi.org/10.3390/ w13131786
[10] Costache R et al 2020 *Remote Sens* **12**, 1422
[11] Costache R, Hong H and Pham Q B 2020 *Sci Total Environ* **711** 134514
[12] Costache R, Bărbulescu A and Pham Q B 2021 *Water* **13** 758
[13] Ghinea D 2000 Romanian Geographical Encyclopaedia (Bucharest: Editura Enciclopedică) (in Romanian)
[14] Prăvălie R and Costache R 2014 *Forum Geogr* **13** 39–49
[15] Drobot R 2020 *Lectures of hydrology and hydrogeology* (Bucharest: Editura Didactica si Pedagogica) (in Romanian)
[16] Mătreaţă M and Mătreaţă S 2010 *Comunicări de Geografie* **XIV** (Bucharest: Editura Universităţii din Bucureşti) (in Romanian)
[17] Nazzal Y at al. 2021, *Toxics* **9(5)** 95, https://doi.org/10.3390/toxics9050095
[18] Bărbulescu A and Dumitriu C S 2021 *Water* **13(8)** 1026
[19] Schillaci C, Acutis M, Lombardo L, Lipani A, Fantappie M, Märker M and Saia S 2017 *Sci Total Environ* **601** 821–32

[20] Costache R 2019 *Stoch Environ Res Risk A* **33** 1375–1402
[21] Skentos A 2018 *Acta Geobalcanica* **4**, 7–15
[22] Costache R, Pravalie I, Mitof I and Popescu C 2015 *Carpat. J Earth Environ Sci* **10(1)** 161-169
[23] Khosravi K, Shahabi H, Pham BT, Adamowski J, Shirzadi A, Pradhan B, Dou J, Ly H-B, Gróf G and Ho H L 2019 *J Hydrol* **573**, 311–23
[24] Tehrany M S, Pradhan B and Jebur M N 2014 *J Hydrol* **512**, 332–43
[25] Wang Y et al 2019 *J Environ Manag* **247**, 712–29
[26] Sanyal J and Lu XX 2005 *Hydrol Proc* **19(18)** 3699-3716
[27] Salvati L and Zitti M 2009 *Ecol Ind* **9(2)** 357-363.
[28] Data classification metod. ArcGIS – Pro. https://pro.arcgis.com/en/pro-app/latest/help/mapping/layer-properties/data-classification-methods.htm
[29] Al-Taani et al 2021 *Toxics* **9(11)** 927
[30] Bărbulescu A, Barbes L and Dumitriu C S 2021 *Toxics* **9(11)** 927
[31] Mihăilescu M et al 2021 *Toxics* **9(5)** 111, https://doi.org/10.3390/toxics9050111

# Separating function of timber floor and wall assemblies

**R Dârmon**[1*]

[1] Technical University of Cluj-Napoca, Romania

* ruxandra.darmon@ccm.utcluj.ro

**Abstract.** The basic concern on the fire compartmentation is focused on to limit the spread of the fire and smoke outside of the room of the fire origin. Where the fire compartmentation is required, the elements forming the boundaries of the fire compartment should be designed and constructed in such a way that they maintain their separating function during the relevant fire exposure [1]. The general accepted method for assessing the fire resistance of a structural or non-structural member is the full-scale testing of the fire performance, however, several calculation methods for verification of the separation function of timber assemblies can also be used in order to prove compliance with the code regulations [2].

## 1. Introduction

Timber is a natural and renewable material, with very low embodied energy, which makes it a sustainable alternative for steel and concrete structures, especially for the residential occupancy. In the United States, the wood light-frame systems are currently widespread to the housing sector and light commercial construction [3], but their use is also expanding to the medium-rise multi-storey buildings [4]. Compared to steel, timber has a strength-to-weight ratio with 40 - 80% higher [5] and thus a better fire behaviour at the initial stages of a fire.

The design rules for the verification of the load-bearing function and/or separation function for construction elements are specified in the EN 1995-1-2 [1]. The compliance with the code requirements, for the passive protection design of the timber structures, can be demonstrated by fire tests certification or, by calculation methods.

The main objective of the passive fire protection is to mitigate the fire risk with respect to individual, society and neighboring property in case of fire. Applying the principles and the calculation approach is a sustainable method for the timber structural design, significantly reducing the cost and the environmental impact [6].

The load carrying capacity of the structural elements is required for certain periods of time through the Mechanical Resistance criterion R [7]. The basic concern on the fire compartmentation is focused on to limit the spread of the fire and the smoke and the hot gas outside the fire origin room. The requirement, listed in the EN 1995-1-2 [1], states that the bounding elements of a fire compartment should be designed and executed in such manner that they maintain their separating function over a specified time period. This includes, when relevant, that integrity failure does not occur - criterion E, insulation failure is prevented – criterion I and the thermal radiation from the unexposed side of the construction element is limited.

The integrity criterion, E assumes that the separation function is lost when there is sustained flaming or hot gases emerging through fissures and cracks, to ignite a cotton pad, placed on that side of the element which is not exposed to fire.

The Insulation criterion, I, is fulfilled if the average temperature rise over the whole non-exposed side of the element remains below 140K and the maximum temperature rise at any point on that surface does not reach 180 K when the opposite face of the element is exposed to a standard fire.

The thermal radiation criterion would be satisfied when the insulation criterion, (I), is not exceeded.

The general accepted method for assessing the fire resistance of a structural or non-structural member is the full-scale testing of the fire performance, however, several calculation methods [2] for verification of the separation function of timber assemblies can also be used in order to prove compliance with the code regulations.

The simplified calculation procedure from Annex E (informative): Analysis of the separating function of wall and floor assemblies of EN 1995-1-2 [1], requires that the time, $t_{ins}$, necessary for the temperature to rise by 140K/180K on the side of assembly that is not exposed to fire is equal or greater than the required fire resistance period, $t_{req}$, for the separating function.

$$t_{ins} \geq t_{req} \tag{1}$$

The insulation performance criteria will be assessed through a simplified method based on the sum of the contribution of the individual layers in the construction according to the following equation:

$$t_{ins} = \sum_{i=1}^{n} t_{ins,0,i} \cdot k_{pos,i} \cdot k_j \tag{2}$$

The insulation time, $t_{ins}$ is a function of the fire behavior of each layer of the assembly and the position and joint configuration of the layers. The contribution of each layer $t_{ins,i}$ is assessed using the basic insulation value, $t_{ins,0,i}$, the position coefficient, $k_{pos,i}$, and the joint coefficient, $k_j$. The basic insulation value, $t_{ins,0}$ represents the fire resistance of a single layer, without the influence of adjacent materials and it can be assessed by testing in line with EN 1364-1 [8], or finite element thermal analysis.

The joint coefficient, $k_j$ accounts for the influence of the joints on the protection and insulation time. According to EN 1995-1-2, the joints should be less than 2 mm. When the joints are backed by a layer, their influence on the protection time is negligible [9].

Based on BS EN 1995-1-2 it is considered that the requirements with respect to integrity are assumed to be satisfied where the requirements for the insulation criterion have been satisfied and the panels remain fixed to the timber frame unexposed side.

## 2. Verification of the separation function for timber joist floor assemblies

The timber joist floor assemblies are exposed to fire from below, on one side. The position coefficients, in this case, should be decreased with 20% [1].

### *2.1. Timber joint floor assembly $A_1$*

The timber joist floor assembly – has 145 x 45 mm joists laid at 600 mm inter axis distance. The layout of the timber floor – Type $A_1$, illustrated below, is comprised of the following layers:

- Layer 1: plasterboard, type A, - Soundshield – 15 mm
- Layer 2: Plywood, 12 mm
- Layer 3: Rocksilk RS45 Insulation, of density 45 kg/m$^3$, 100 mm

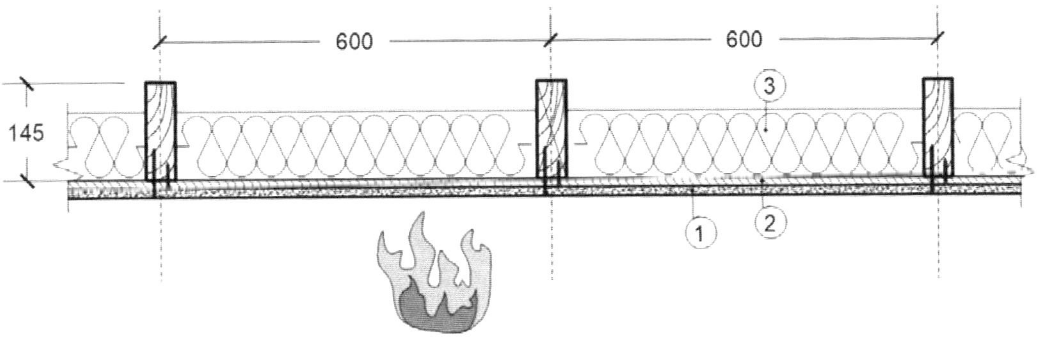


**Figure 1**. Detail of the joint floor assembly – $A_1$.

*2.1.1. The insulation time calculation for assembly $A_1$*

**Table 1.** Insulation time calculation for Type $A_1$ system.

| | $t_{ins,0,1}$ - the basic insulation value for Layer i [min] | $k_{pos,1}$ – the position coefficient for Layer i [-] | $k_j$ - joint coefficient [-] |
|---|---|---|---|
| Layer 1: plasterboard, type A, – 15 mm | $t_{ins,0,1}$ = 21 min | $k_{pos,1}$ = 0.84 | |
| Layer 2: Plywood, - 12 mm | $t_{ins,0,2}$ = 11.4 min | $k_{pos,1}$ = 0.78 | $k_j$ = 1.0 for filled joints |
| Layer 3: Rocksilk Insulation, of density 45 kg/m$^3$, 100 mm | $t_{ins,0,3}$ = 21.6 min | $k_{pos}$ = 1.0 for a void cavity filled with rock fibre insulation | |

$$t_{ins} = \sum_{i=1}^{3} t_{ins,0,i} \cdot k_{pos,i} \cdot k_j = [21 \cdot 1 + 11.4 \cdot 0.78 + 21.6 \cdot 1] \cdot 0.8 = 43\ min \qquad (3)$$

*2.2. Timber joint floor assembly $A_2$*

The timber joist floor assembly – has 95 x 35 mm joists laid at 450 mm inter axis distance. The layout of the timber floor – Type $A_2$, illustrated below, is comprised of the following layers:

- Layer 1: plasterboard, type A, - Soundshield – 15 mm
- Layer 2: Plywood, 12 mm
- Layer 3: Rocksilk Insulation, of density 45 kg/m$^3$, 100 mm
- Layer 4: Rocksilk Insulation, of density 45 kg/m$^3$, 60 mm
- Layer 5: Rocksilk Insulation, of density 45 kg/m$^3$, 60 mm

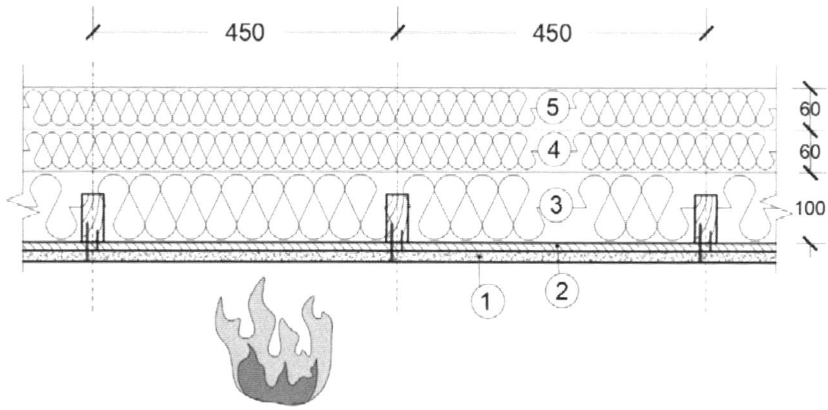


**Figure 2**. Detail of the joint floor assembly – $A_2$

*2.2.1. The insulation time calculation for assembly $A_2$*

**Table 2. Insulation time calculation for Type $A_2$ system**

| | $t_{ins,0,1}$ - the basic insulation value for Layer i [min] | $k_{pos,1}$ – the position coefficient for Layer i [-] | $k_j$ - joint coefficient [-] |
|---|---|---|---|
| Layer 1: plasterboard, type A, – 15 mm | $t_{ins,0,1}$ = 21 min | $k_{pos,1}$ = 0.84 | |
| Layer 2: Plywood, - 12 mm | $t_{ins,0,2}$ = 11.4 min | $k_{pos,1}$ = 0.78 | |
| Layer 3: Rocksilk Insulation, of density 45 kg/m³, 100 mm | $t_{ins,0,3}$ = 21.6 min<br>$k_{dens}$ = 1.08 | $k_{pos,3}$ = 1.0 for a void cavity filled with rock fibre insulation | $k_j$ = 1.0 for filled joints |
| Layer 4: Rocksilk Insulation, of density 45 kg/m³, 60 mm | $t_{ins,0,4}$ = 21.6 min<br>$k_{dens}$ = 1.08 | $k_{pos,4}$ = 1.0 for a void cavity filled with rock fibre insulation | |
| Layer 5: Rocksilk Insulation, of density 45 kg/m³, 60 mm | $t_{ins,0,5}$ = 21.6 min<br>$k_{dens}$ = 1.08 | $k_{pos,5}$ = 1.0 for a void cavity filled with rock fibre insulation | |

$$t_{ins} = \sum_{i=1}^{3} t_{ins,0,i} \cdot k_{pos,i} \cdot k_j = 68.75\ min \qquad (4)$$

**3. Verification of the timber stud wall assembly**

The timber wall assembly – has 89 x 38 mm timber stud at 600 mm inter axis distance. The layout of the timber floor – Type B, illustrated below, is comprised of the following layers:

- Layer 1: Plasterboard, type A, 15 mm
- Layer 2: Plywood, 12 mm
- Layer 3: Glasswool Insulation, of density 45 kg/m³, 50 mm
- Layer 4: Plywood, 12 mm
- Layer 5: Plasterboard, type A, 15 mm

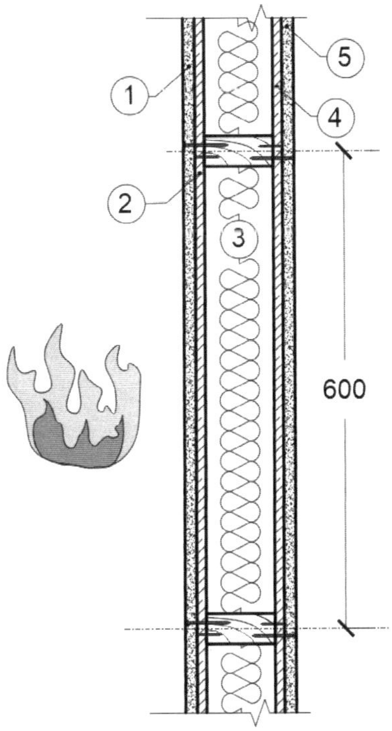


**Figure 3**. Detail of the timber stud wall.

The insulation time $t_{ins}$ of the timber wall should be assessed for different heat transfer paths, as through the field in between the studs and through the stud. Given that the contribution of the mineral wall to the total insulation time is 5 minutes, it is only analysed the first case of heat transfer through the panel field.

**Table 3. Insulation time calculation for timber stud wall assembly.**

| Wall' structure | $t_{ins,0,1}$ - the basic insulation value for Layer i [min] | $k_{pos,1}$ – the position coefficient for Layer i [-] | $k_j$ - joint coefficient [-] |
|---|---|---|---|
| Layer 1: plasterboard, type A, – 15 mm | $t_{ins,0,1}$ = 21 min | $k_{pos,1}$ = 1.0 | |
| Layer 2: Plywood, - 12 mm | $t_{ins,0,2}$ = 11.4 min | $k_{pos,2}$ = 0.8 | |
| Layer 3: Glasswool Insulation, of density 45 kg/m$^3$ – 50 mm | $t_{ins,0,3}$ = 5.0 min<br>$k_{dens}$ = 1.0 | $k_{pos,3}$ = 1.0 | $k_j$ = 1.0 for filled joints |
| Layer 4: Plywood, - 12 mm | $t_{ins,0,4}$ = 11.4 min | $k_{pos,4}$ = 1.0 | |
| Layer 5: plasterboard, type A, – 15 mm | $t_{ins,0,1}$ = 21 min | $k_{pos,5}$ = 1.2 | |

$$t_{ins} = \sum_{i=1}^{5} t_{ins,0,i} \cdot k_{pos,i} \cdot k_j = [21 \cdot 1 + 11.4 \cdot 0.8 + 5.0 \cdot 1 + 11.4 \cdot 1 + 21 \cdot 1.2] \cdot 1 = 71.72\ min \quad (5)$$

$$t_{ins} < t_{req} = 60\ min \quad (6)$$

## 4. Results

The calculation results for the two floor assemblies and for the timber stud wall, are summarised in the table below:

**Table 4**. Separation function summary for the floor and wall assemblies.

| System | Integrity (E) | Insulation (I) | Fire Rating |
|---|---|---|---|
| Timber joint floor A | 43 min | 43 min | EI 30 |
| Timber joint floor A | 43 min | 43 min | EI 30 |
| Timber stud wall | 71.72 min | 71.72 min | EI 60 |

The calculation procedure given in the Eurocode 5, does not consider the size of the element, nor the support conditions for the assessment of the fire separation function. Although it is a simplified method, it can be used to predict the time of failure for the fire integrity criterion, E and implicit, for the insulation criterion, I. The fire rating it is usually expressed in multiples of 30 minutes. In a standard fire test, the fire would cease as soon as the required rating time is attained, therefore, the calculation procedure could be used to optimise the element layer thicknesses in order to avoid the over dimensioning and wasting resources.

## 5. Conclusion

This study case is a practical calculation example, according to the Annex E of EC5 [1] for two joist floor assemblies and a timber stud wall for the assessment of the fire rating, as an alternative to the application of the fire testing, which is often done at prohibitive costs.

## References

[1] EN 1995-1-2:2004 (E) 2004 Eurocode 5 – Design of timber structures Part 1-2: General – Structural fire design *Eurocode 5 – Des. timber Struct.* **2004** 1–69
[2] König J, Oksanen T and Towler K 2000 A Review of Component Additive Methods Used for the Determination of Fire Resistance of Separating Light Timber Frame Construction *33th Proc. CIB-W18 Meet. Pap. Delft, Netherlands*
[3] Mehta M, Scarborough W and Armpriest D 2013 *Building Construction: Principles, Materials and Systems* (Prentice Hall)
[4] Buchanan A H 2017 Fire Resistance of Multistorey Timber Buildings *Fire Sci. Technol. 2015* 9–16
[5] Structural Timber Association, www.structuraltimber.co.uk, 11 11 2014. [Online]. Available: https://www.structuraltimber.co.uk/assets/InformationCentre/eb1.pdf. [Accessed 11 10 2021]
[6] Jones D and Brischke C 2017 *Performance of Bio-based Building Materials* (Woodhead Publishing Series in Civil and Structural Engineering)
[7] Frangi A 2012 *Fire resistance assessment of timber structures* Brussels
[8] EN 1364-1 2003 Fire resistance tests for non-loadbearing elements - Part 1: Classification using data from fire resistance tests, excluding ventilation services *Eur. Stand. Brussels CEN*
[9] Richardson L R and Batista M 2001 Fire resistance of timber decking for heavy timber construction *Fire Mater.* **25** 21–9

# Using fibers in construction

**O Deaconu[1*] and GC Chițonu[1]**

[1] Transilvania University of Brasov, Faculty of Civil Engineering, Romania

* ovideaconu@unitbv.ro

**Abstract**. This article is an overview about alternative solutions for reinforced concrete by using different types of fibers. The use of fiber reinforced concrete when is compared to the conventional reinforced concrete solutions. This study has taken in consideration structural performance and the total cost. The use of fibers or dispersed reinforcement also improves some of the characteristics of concrete, such as those related to: cracking, freezing, durability, erosion of ordinary or marine water, wind erosion, permeability, etc. In order to correct to a large extent, the unfavorable characteristics of the reinforced concrete, in the mass of the fresh concrete various types of fibers can be mixed and incorporated in the use of concrete with dispersed reinforcement. As materials often used as fibers, the most commonly are: hooked-end steel (steel fibers), straight polypropylene and straight polyolefin, glass fiber, carbon fiber, aramid fiber, natural hemp fibers, jute, hair, straw, etc.

## 1. Introduction

This article looks at the use of organic and inorganic materials for reinforcing the most common building materials like concrete for reductions of ecological footprint, while preserving structural and architectural options.

The use of fibers to reinforce concrete materials has been used since ancient times. While the Göbekli Tepe temple in modern-day Turkey used pillars of carved limestone around 10000 BC, early concrete is known to have been used from around 6000 BC. The concrete in the Pantheon in Rome has bricks stamped with the years 114, 115 and 116. As Robert Courland's book "Concrete Planet" spells out, "Unlike the Pantheon, virtually all the concrete structures one sees today will eventually need to be replaced" costing us trillions of dollars in the process [1].

The scale of use of modern concrete is almost unfathomable. Professor emeritus Vaclav Smil at the University in Winnipeg, Canada found the statistics which show that between the years 2011 and 2013, China used 6,6 gigatons of concrete, which is more than the 4,5 gigatons of concrete that the United States used between years 1901 - 2000 [2].

With such enormous quantities of one of the most energy intensive and polluting products in large scale use, there is a desperate need for reductions in total pollution output if the world is to reach the Paris Climate Agreement Goals.

**Figure 1.** Ancient Roman concrete consists of coarse chunks of volcanic tuff and brick bound together by a volcanic ash-lime mortar that resists microcracking, a key to its longevity and endurance. Foto: Roy Kaltschmidt/Berkeley Lab [3].

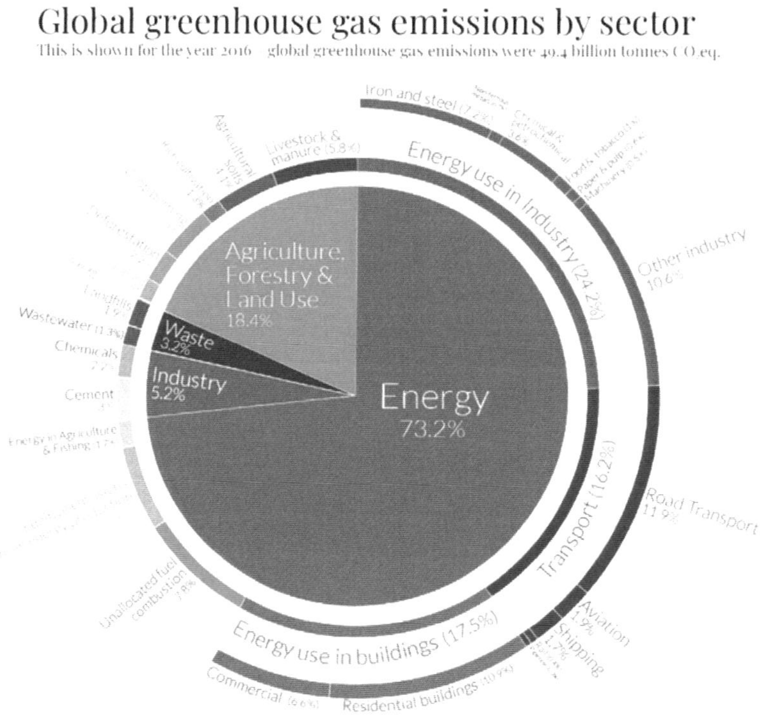


**Figure 2.** Global gas emissions by sector [4].

For buildings, the use or operational life cycle stage impacts are significantly greater than those in the other life cycle stages. A building usually operates for decades consuming energy and raw materials with associated environmental releases. These operational stage impacts typically dwarf the environmental impacts from material extraction, manufacturing and end-of-life stages for commercial buildings.

Although it depends on the type of building and the impacts being measured, the operational stage impacts are typically 5 to 20 times larger than stages associated with building product manufacturing and demolition. In fact, operating buildings in the U.S. consumes 19% of the nation's energy and 37%

of the nation's electricity. In total, commercial buildings account for 19% of the CO2 emissions in the U.S.4 Therefore, when conducting an LCA for buildings, it is extremely important to include the operational stage [5].

There is a large scope for improving the use of fibers and reinforcing materials in buildings, and to choose where in buildings it could be possible to use more environmentally friendly fibers, without damaging the structural or architectural properties.

Mixing materials and reinforcing concrete, mortars and other materials are known to have been done for at least ten thousand years.

Straw has been mixed into mud bricks, much like adobe clay is still made today, animal hair has been used in mortar, and fibers, wood types and other materials have been extensively mixed throughout history. Adding wood fibers to gypsum boards increases the strength of the board while providing acoustic benefits and reducing the ecological footprint.

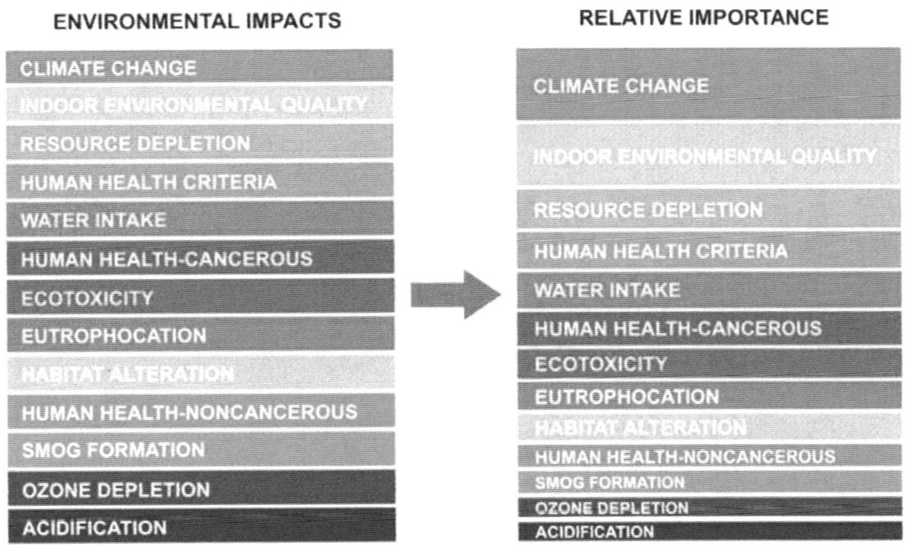


**Figure 3.** Environmental impacts and relative importance [5].

## 2. Applications of fiber reinforced concrete (FRC)

Reinforced concrete can be considered as a composite material with very good physical and mechanical properties, due to the very good cooperation between concrete and reinforcement, but also of other essential factors such as the coefficient of thermal expansion and the compatibility of deformations. For this reason, reinforced concrete is currently the main material used to make the strength structures of buildings.

The use of fibers or dispersed reinforcement also improves some of the characteristics of concrete, such as those related to: cracking, freezing, durability, erosion of ordinary or marine water, wind erosion, permeability, etc. [6-11]

However, reinforced concrete also has some disadvantages such as inhomogeneity, porosity and/or the existence of micro-cracks in the concrete mass, the gelation phenomenon specific to hardened concrete, especially under the stress of cracking concrete.

Under unfavorable physical and chemical conditions that persist over time, the disadvantages may increase or new ones may appear. In order to correct to a large extent, the unfavorable characteristics of the reinforced concrete, it is possible to incorporate in the mass of the fresh concrete various types of fibers for dispersed reinforcement. [12-16] The most commonly used fiber materials are: hooked-end steel (steel fibers), straight polypropylene and straight polyolefin, glass fiber, carbon fiber, aramid fiber, natural hemp fibers, jute, hair, straw, etc.

### *2.1. Applications of steel fibers*

Bellow we can see the constructive method of fiber reinforced concrete. The use of fiber reinforced concrete in applications with major contribution in structural performance is still relatively limited when compared to its wider use in applications as non-structural elements such as pavements, flooring and tunnel lining segments [17]. The disperse reinforced concrete is a particular type of concrete made of reinforced nets on continuous rows with small diameter wires, arranged at short distances between them as shown in the next figure. The mesh can be made of steel wires, glass fibers fabrics, polypropylene fabrics or natural fabrics like bamboo, hemp or jute.

**Figure 4.** Constructive method of fiber reinforced concrete [authors].

The most prevalent fiber types currently utilised in industry are steel (crimped or hooked end, figure 5) and macrosynthetic fibers (figure 6).

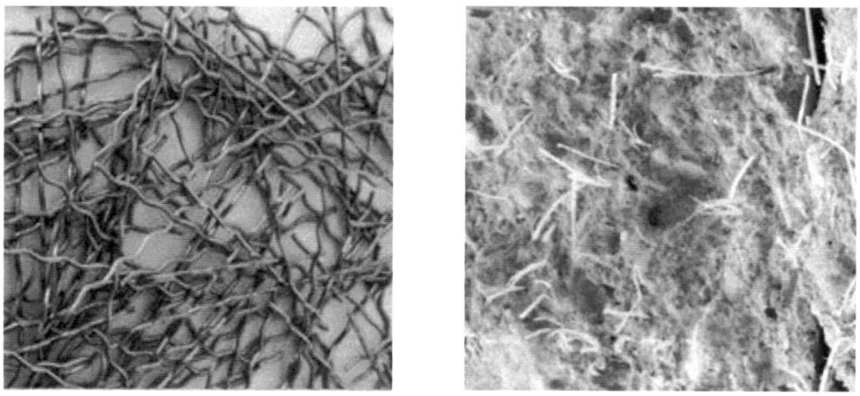

**Figure 5.** Steel fibers reinforced concrete [authors].

For modern concrete there are numerous alternatives to reduce both the carbon and the ecological footprint of the production, and to reduce the negative impact of the concrete produced throughout its lifecycle [18]. These include replacing sand with crusher dust (a byproduct of stone quarries), sand manufactured by crusher, washed bottom ash, granulated blast furnace slag, glass powder remains. The amount of cement needed in a batch of concrete can be reduced by replacing it with mining, construction, demolition and production waste.

### *2.2. Applications of glass fibers, aramid fibers, carbon fibers*

Using waste materials from other industrial operations both provide a waste management benefit, and very often an economic benefit to industries who can sell their waste as a resource rather than to pay for depositing it. Most significantly it reduces the enormous amounts of energy needed in extracting the materials from their natural state in the rocks, mountains, soil, water etc.

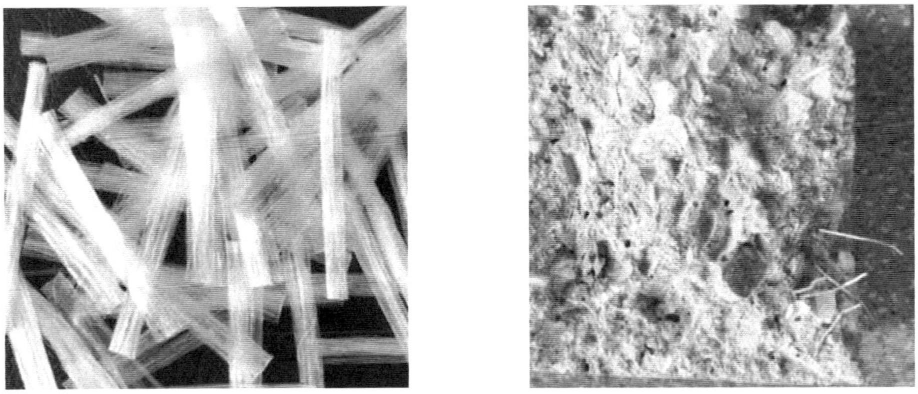

**Figure 6.** Macrosynthetic fibers reinforced concrete [authors].

Fiber reinforced concrete (FRC) with modern synthetic based fibers like polypropylene, nylon, polythene, polyester, glass fibers have been used for the last 50 years [19, 20].

Fiber reinforced concrete has better strength, ductility and durability than reinforced concrete. [6, 7, 9, 10, 14, 15, 18]

Due to these good physics - mechanical characteristics, the disperse reinforced concrete has a wide range of applications, allowing for building elements with more complex geometry. It can be used in the following areas of the civil, industrial and agricultural sectors: roofs made of thin blades; as a substitute for heat insulation in roof framing and roof deck, structural elements of the framework, precast panels for walls and floors, the panels fences, single-family houses, silos, structural elements: columns, beams, walls, thin-walls, animal shelters, floorings, pools of water treatment, pools or water tanks, adduction pipe, drains, bracket panels to capture solar energy, maritime constructions etc.

Fiberglass was invented in 1893 and was first marketed in 1936 as an insulating material. It became popular in the 1950s, when some of the health risks associated with asbestos use became apparent. Due to the similarity in shape between fiberglass and asbestos fibers, fiberglass has been able to effectively replace asbestos in many applications, such as electrical, thermal, and acoustic insulation, and structural reinforcement.

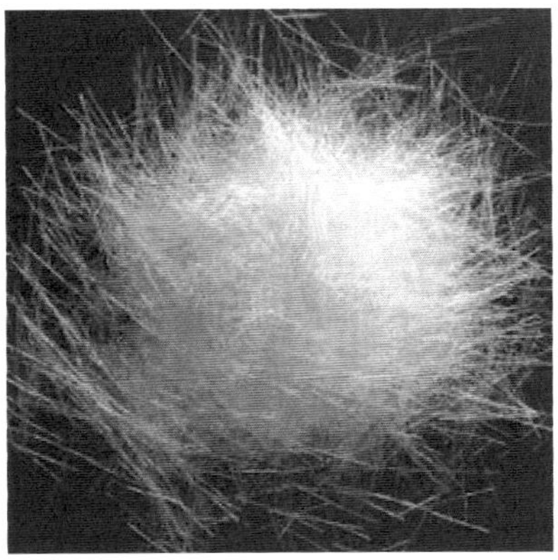

**Figure 7.** Glass fibers [authors].

Today glass fibers are the dominant reinforcing fibers in the construction of composites, representing over 90% of global consumption. This is due to the good weight to mechanical strength ratio, easy processing and low-cost price. Glass filaments are made relatively easily by extruding molten glass, obtaining fibers with a diameter in the range of 5 - 25 microns. The different types and qualities of fiberglass reinforcement have significantly different costs [17].

Aramid has a very good resistance in any field of stress such as stretching, compression or shearing. It has a very low density, a high modulus of elasticity, an excellent toughness, impact resistance, wear, heat and very high chemical attacks. These features make it ideal for use in military applications, armor, protection against high temperatures such as those in industrial chimneys. Aramid is generally used in structures of particular importance that may be subject to the risk of impact, abrasion or excess heat. Aramid is compatible with other fibers, such as fiberglass or carbon fiber.

**Figure 8.** Aramid fibers [authors].

Carbon fiber has substantially better mechanical strength and higher stiffness values than all other fibers. Resistant to high temperatures, it has a high electrical conductivity and low thermal conductivity. The tolerance to impact or damage of carbon fiber products is relatively low. When the weight of the structural elements is very important, then carbon fiber is the ideal reinforcement material. The most common applications of carbon fiber are in the field of consolidations.

**Figure 9.** Carbon fibers [authors].

## 3. Discussions

The architectural approach for sustainable buildings includes adapting to the newest technologies, for preparing for known, expected, worst case, and unforeseen climate change effects. The benefits of fibers in concrete provide ingenious solutions for contemporary architecture.

Fibers inhibit and control the formation of intrinsic cracking in concrete, that occur both in the plastic and the hardened stage of concrete. It reinforces the concrete against impact forces, improves the resistance to shattering forces caused by vibrations, earthquakes. Fibers reduce the permeability and water migration in concrete and enhances the hardness of surfaces against wear and tear of objects on the surfaces. Structures which are reduced to being just the right size, thickness and weight for the stress it will be exposed to during the lifecycle, are more environmentally friendly, often cheaper to build, and allows for unique architectural shapes and approaches.

**Figure 10.** Fiber cement facade in architecture: House in Norway, Sandefjord, arch. Knut Hjeltnes, foto: Inger Marie Grini [21].

**Figure 11.** Fiber cement facade in architecture Vila Do Bispo, Portugal, CORE Architects, foto: Alexander Bogorodskiy [22].

## 4. Conclusions

As the world community is working arduously to avoid the life changing planet changes that are coming with global warming and trying to get some control over the changing environment, reinforced materials including concrete will both need to play an important role in support structures in nature, and in heavily reducing the carbon and ecological footprint [23]. The combined emissions from the global cement industry are estimated to be between 5 - 7% of total $CO_2$ emissions [4, 24]. Making structures with relatively short lifecycles with large energy demands, from extraction, production and demolition is not financially or environmentally sustainable. Fibers in construction materials is ancient technology, but modern techniques, knowledge and equipment can significantly reduce the ecological footprint of concrete, while enhancing its abilities.

Fibers and dispersed reinforcement improve some of the characteristics of concrete, such as: cracking, freezing, durability, erosion of ordinary or marine water, wind erosion, permeability.

Reinforced concrete has disadvantages such as inhomogeneity, porosity and the existence of microcracks in the concrete mass, the gelation phenomenon specific to hardened concrete, especially under the stress of cracking concrete. Under unfavorable physical and chemical conditions that persist over time, the disadvantages may increase or new ones may appear. Various types of fibers can be incorporated into the mass of fresh concrete, which largely corrects the unfavorable characteristics of reinforced concrete such as steel fibers, glass fibers, carbon fibers, aramid fibers, etc.

The disadvantage of using fibers in reinforced concrete could be increase the cost per cubic meter of concrete used. The economic disadvantage is greater when using carbon or aramid fibers, because the production cost of these materials remains high.

**References**

[1] Courland R 2021 Concrete Planet: The Strange and Fascinating Story of the World s Most Common Man Made Material – United States
[2] Smil V 2019 Growth: From Microorganisms to Megacities- Cambridge
[3] https://www.universityofcalifornia.edu/news/what-makes-ancient-roman-concrete-so-tough accesed in September 2021
[4] OurWorldinData.org 2020 Climate Watch, the World Resources Institute accesed in September 2021
[5] Lemay L 2011 Sustainable Development, NRMCA - Concrete Sustaninalbility report – CSR04
[6] CEN 2005 Test method for metallic fibre concrete-measuring the flexural tensile strength (limit of proportionality (LOP), residual) BSI: London UK
[7] ASTM, ASTM C1609 / C1609M 2019 19a Standard Test Method for Flexural Performance of Fiber-Reinforced Concrete (Using Beam with Third-Point Loading). ASTM International: West Conshohocken, PA, USA
[8] Mesbah H A, Buyle-Bodin F 1999 Efficiency of polypropylene and metallic fibres on control of shrinkage and cracking of recycled aggregate mortars *Construction and Building Materials* **13** pp 439-447
[9] Banthia N and Gupta R 2006 Influence of Polypropylene Fibre Geometry on Plastic Shrinkage Cracking in Concrete *Cement and Concrete Research* **36** pp 1263-1267
[10] Gong J, Wei Z and Zhang W 2017 Influence of shrinkage-reducing agent and polypropylene fiber on shrinkage of ceramsite concrete *Construction and Building Materials* pp 155-163
[11] Stang H and Krenchel H 1993 Micromechanics of crack bridging in fibre-reinforced concrete, *Materials and Structures* **26** pp 486-494
[12] ACI Committee 544. ACI 544.4R-18: Guide to Design with Fiber-Reinforced Concrete. American Concrete Institute: Michigan, USA, 2018
[13] ACI Committee 544. ACI PRC-544.7-16 2016 Report on Design and Construction of Fiber-Reinforced Precast Concrete Tunnel Segments *American Concrete Institute: Farmington Hills* USA
[14] McLeod C H 2013 Investigation into Cracking in Reinforced Concrete Water-Retaining Structures *MScEng Stellenbosch University* Stellenbosch South Africa
[15] Facconi L and Minelli F 2016 Verification of structural elements made of FRC only: A critical discussion and proposal of a novel analytical method *Engineering Structures* **131** pp 530-541
[16] Qian C X and Stroeven P 2000 Development of hybrid polypropylene-steel fibre-reinforced concrete *Cement and Concrete Research* **30** pp. 63-69. https://doi.org/10.1016/S0008-8846(99)00202-1
[17] Ryan C and Garcia-Taengua E 2021 Fibre-Reinforced Concrete Is Sustainable and Cost-Effective for Water-Retaining Structures- *MDPI Sustanability* **13(20)** 11479
[18] Zollo R F 1997 Fiber-reinforced concrete: an overview after 30 years of development - Cement and Concrete Composites
[19] Muntean R, Grădinaru C M, Şerbănoiu A A and Sârbu G C 2020 Improved cement based construction elements with dispersed fibers as contribution for environmental protection, *Environmental Engineering and Management Journal* pp 85 – 94
[20] https://fiberreinforcedconcrete.org/fiber-reinforced-concrete/fiber-types accesed in September 2021

[21] https://www.archdaily.com/search/projects/categories/houses accesed in September 2021
[22] Naaman A E 2007 Deflection-Softening and Deflection-Hardening FRC Composites: Characterisation and Modeling *ACI Special Publications* **248** pp 53-66
[23] https://apnews.com/article/climate-science-europe-united-nations-weather-8d5e277660f7125ffdab7a833d9856a3 accesed in September 2021
[24] https://www.norcem.no/en/Cement_and_CCS accesed in September 2021

# Numerical model validation of non-composite slim floor beam subjected to fire

**D Duma[1*], R Zaharia[1], D Pintea[1] and I Both[1]**

[1] Politehnica University Timişoara, Romania

* diana.duma@upt.ro

**Abstract.** Compared to traditional composite steel-concrete beams, where the concrete slab is supported by the top flange of a steel profile, in slim floor systems, the beam is integrated into the height of the concrete slab. These configurations are popular due to their cost-effectiveness and fire performance. However, fire design requires advanced calculation models, as Eurocodes do not provide a simplified method for the analysis of these types of beams when exposed to fire. The validation of the numerical models is an important issue for researchers, designers, software developers, and authorities. The European Standard for the fire design of composite steel and concrete structures, EN 1994-1-2, states that advanced calculation models must be validated against experimental test results. The paper presents the validation of a numerical model, for the thermal and mechanical analysis of slim floor beams, using SAFIR software. SAFIR is a special purpose computer program, based on the Finite Element Method, developed for the analysis of structures under elevated temperature conditions. The model is validated against an existing experimental fire test, available in the literature. The temperatures on the cross-section and the vertical beam displacements obtained numerically are reproducing very well the experimental results.

## 1. Introduction

Slim floor systems consist of steel beams encased in the concrete slab, which is supported on the bottom flange of the steel profile. These types of floor became very popular in the last decades, considering that this solution leads to a reduced storey height and built-in fire resistance through its geometry. Compared to a traditional solution, in which the concrete slab rests on the top flange, in the case of slim floor systems, the steel profile may need fire protection only on the exposed part of the lower flange.

This configuration can behave as a composite beam through a classical solution with studs welded to the top flange (figure 1a). However, considering the reduced height of the concrete above the upper flange, the usual solution to ensure the composite behaviour is the transversal reinforcement passing through the web of the steel profile, as shown in figure 1b [1].

Arcelor Mittal [1] presents two geometries made of I-shaped European profiles, which are commonly used for slim floor beams:

- the Integrated Floor Beam (IFB), which has one of the flanges welded onto a cut HE or IPE European section, either as a wider bottom flange (figure 2a) or as a top flange (figure 2b)
- the Slim-Floor Beam (SFB), which is realized by welding a plate under an HE or an IPE European section (figure 2c)

a. b.

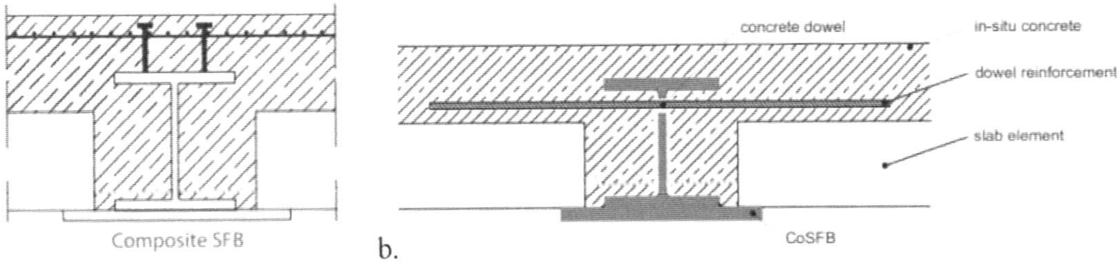


**Figure 1.** Slim floor beams with composite action [1].

a. b.

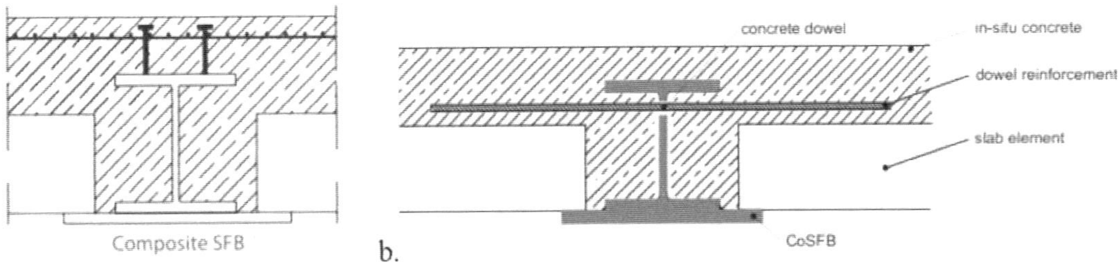


**Figure 2.** IFB type A section, IFB type B section, SFB section [1].

The slab can be constructed in different ways as: i) directly cast on site, ii) precast concrete elements resting on the lower flange of the steel profile with a concrete layer on top of it, or iii) composite slab with corrugated sheet as built-in form.

As slim floors became popular, engineers faced the problem of fire design, since Eurocodes do not provide a direct and simplified method for determining fire resistance. Therefore, the most common alternative remains numerical analysis using specialized software. A numerical model has to be validated before being used for further investigations or other configurations. The article presents a numerical model using the specialised computer program for structural analysis at elevated temperature SAFIR [2], for a slim floor beam made of an IFB steel beam and a concrete slab with a corrugated sheet. This model is validated against a laboratory experiment carried out in 1996 at the Warrington Fire Research Centre [3].

## 2. SAFIR software

As a special purpose computer program, based on the Finite Element Method (FEM), SAFIR [2] was developed for the analysis of structures under elevated temperature conditions.

In order to determine the load bearing capacity of a structure after a certain duration of fire, SAFIR tracks the behaviour of the structure step by step throughout the development of the fire. Therefore, by thermal analysis, the user obtains the temperature distribution in the structure during the fire. Thermal and mechanical analyzes are performed separately and subsequently, so that mechanical analysis will be influenced by the thermal response [2].

To describe the geometry of the cross section, the fibre model is used. The cross section of the beam is subdivided here into small triangle fibres. The material characteristics are calculated at the centre of each fiber and they are constant for the whole fibre. Each fibre has its own material, allowing for the building of composite sections made of different materials [2].

For the structural model, the stress-strain material laws are generally linear-elliptic for steel and non-linear for concrete. As a finite element program, the SAFIR model considered here is realized with 2-D BEAM finite elements. The beam element is straight in its non-deformed geometry. Its position in space is defined by the position of three nodes: the two end nodes (N1-N2), and a third node (N4) defining the position of the local y axis of the beam. Node N3 is used to support an additional degree of freedom (figure 3) [4]. A 2D cross-section with temperature distribution which varies in time is allocated. This means that no heat flux is considered along the longitudinal axis of the beam finite element. For beam finite elements, the Bernoulli hypothesis is considered, i.e., the cross section remains plane under bending moment.

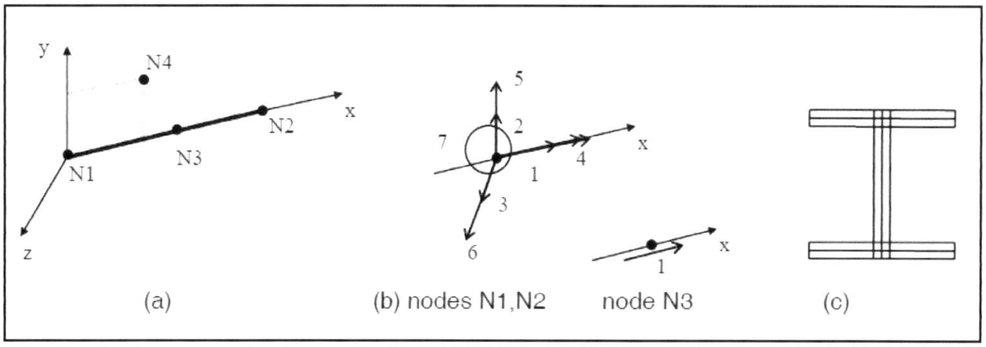


**Figure 3.** Beam element: (a) Local axes (b) Degrees of freedom at nodes (c) Cross section [4].

## 3. Experimental set-up: Test WFRC 66162

According to the Slimflor Compendium [5], in the Warrington Fire Research Centre, 21 fire tests were carried out on slim floor beams between 1985 and 1996. The test WFRC 66162 [3, 5] carried out in 1996 is used here for the validation of the numerical model. The test involved a non-composite slim floor beam assembly subjected to ISO fire. The steel profile was a hot rolled steel beam with 280 mm x 280 mm/180 mm asymmetric section (figure 4a), grade 355 JR steel. Although a tensile test was made on the sample taken from the steel beam. It showed a yield strength of 402 N/mm$^2$ [3]. The steel profile was supporting a 210 mm deep metal decking on the lower flange (figure 4b). On top of the metal decking was cast 1 m wide and 80 mm thick normal weight Grade 30 concrete slab, with A142 reinforcing mesh. The moisture content of the concrete was measured 4.2%

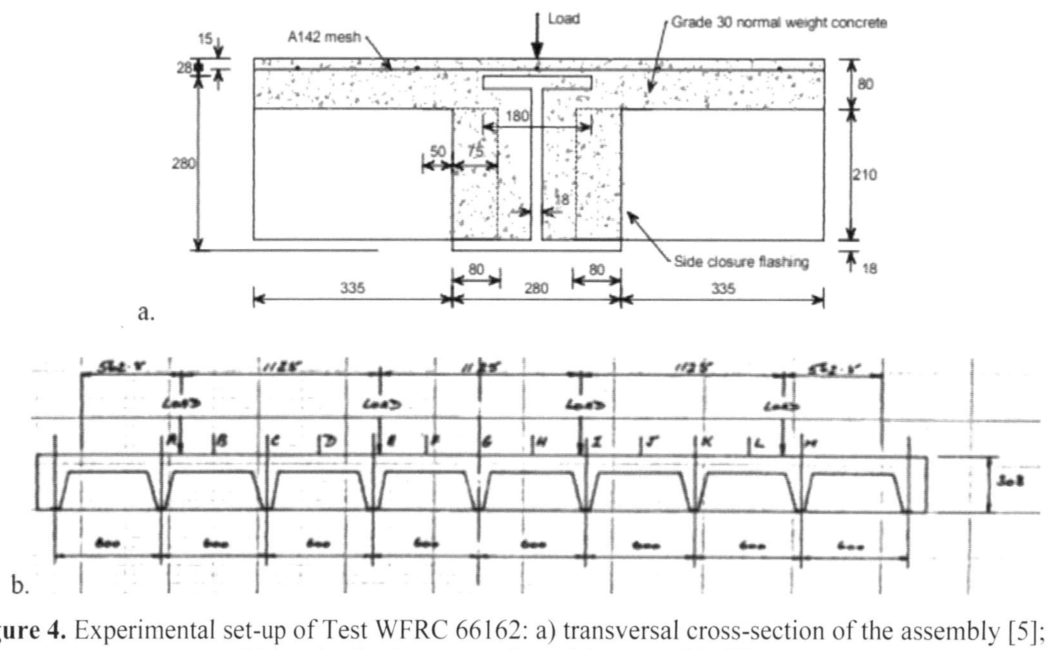


**Figure 4.** Experimental set-up of Test WFRC 66162: a) transversal cross-section of the assembly [5]; b) longitudinal cross-section of the assembly [3].

The beam was simply supported over 4500 mm and loaded by 4 hydraulic rams placed along the centre line of the web at 1/8, 3/8, 5/8 and 7/8 of the supported span, with 84.6 kN at each point (figure 4b). The specimen was placed on top of the oven, as a roof, so that the supports were placed outside of the oven walls, and only 4 m of beam length was heated [5].

The failure criteria were recorded at 91.75 minutes with a deflection of span/30 (i.e.150 mm) and 107.5 minutes with a deflection of span/20 (i.e. 225 mm).:

## 4. Numerical model - Thermal analysis

The thermal analysis is made for the cross-sections shown in figure 5. Three cross-sections were modelled and assigned along the beam axis in the structural model, in order to reproduce properly the concrete slab with corrugated sheet. Therefore, section F (figure 5a) represents the thinnest concrete depth of the slab, section G (figure 5b) represents the thickest concrete depth and the third section is a transition section (figure 5c). Only half of each cross-section was created in the SAFIR pre-processor, as shown in figure 5, but the symmetry function was applied in the mechanical analysis, in order to account for the entire cross-section. This design allows a finer mesh to be considered in a reduced running time of the analysis, without influencing the thermal results. The time step for the calculation was considered 12 s [6].

The material properties used in the three cross sections are in accordance with the Eurocodes for fire design [7, 8], considering the upper limit of the thermal conductivity for concrete. The moisture content of the concrete was taken 42 kg / m$^3$, according to the technical note of the experimental test [3] and the density of the concrete was considered 2300 kg/m$^3$, according to the Eurocode values [9]. The value of concrete density is not reported in the technical note following the experiment.

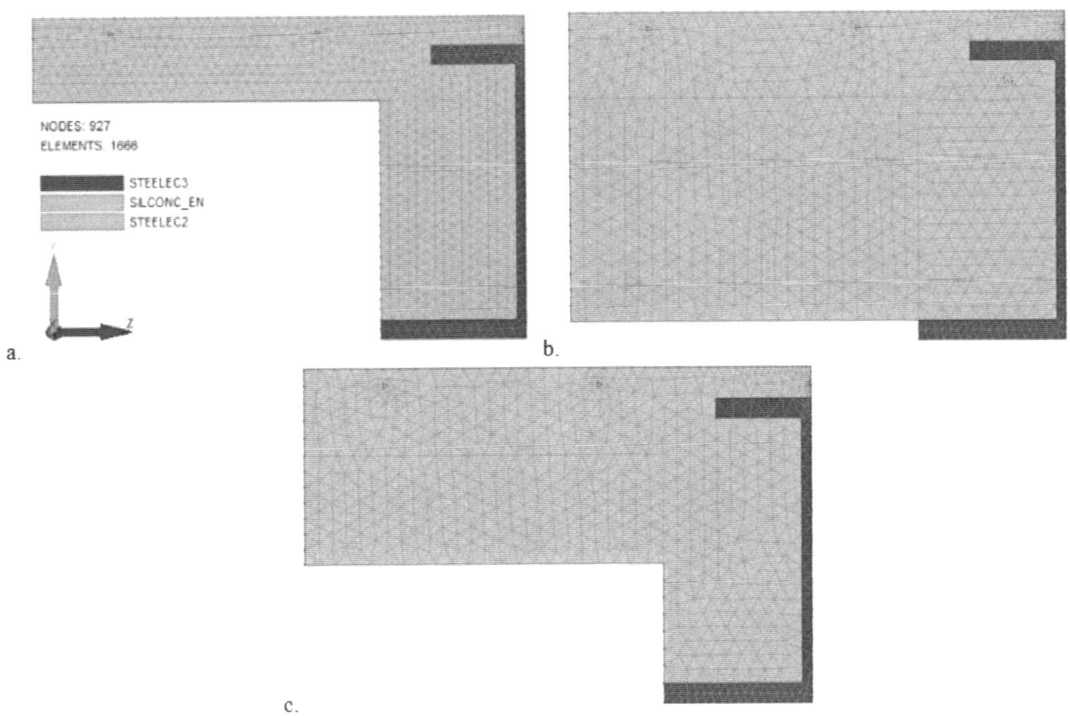


**Figure 5.** Modelled cross-sections: materials and mesh;
a) section F, b) section G, c) transition section.

Figure 6 shows the temperature distribution on the cross-section and the nodes considered for the comparison with the experimental data. The thermocouples were placed on the lower flange of steel profile at 80 mm from its edge, and in the steel web, at 10 mm above the bottom flange, 30 mm, and 50 mm, respectively.

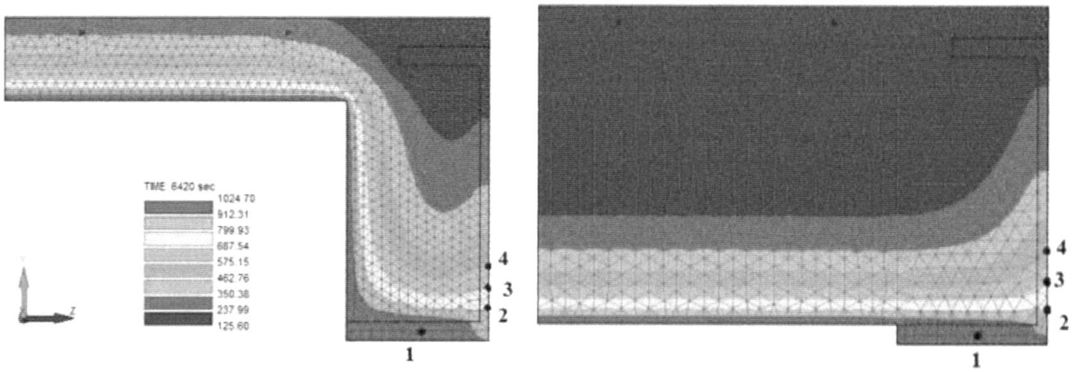


**Figure 6.** Temperature distribution over cross-sections and position of the thermocouples.

The temperature comparison between the numerical simulation and the values reported from the WFRC 66162 test are represented graphically in figures 7 and 8. Figure 7 represents a comparison of the temperature values recorded in Section F with those obtained with SAFIR at equivalent locations, and figure 8 represents a comparison of the temperature values from Section G, respectively.

The temperature variation in the steel profile obtained in the numerical simulation with SAFIR reproduces very well the temperature recorded during the WFRC 66162 laboratory test. In Section G, the numerical model offers slightly higher values than the experimental values, for the temperature in the lower flange, which placed the results on the safe side. Therefore, the thermal computation obtained with SAFIR was accepted to be used further in the mechanical analysis.

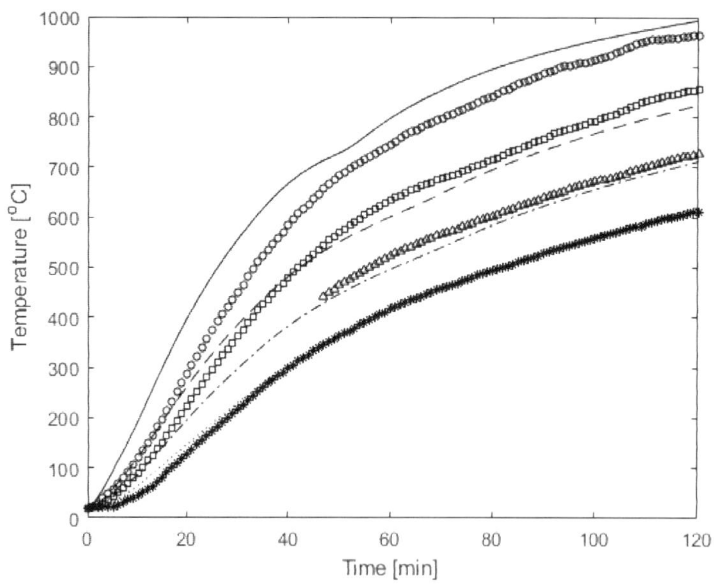


**Figure 7.** Comparison of the temperature variation obtained with SAFIR and the recorded values in section F of the WFRC 66162 test.

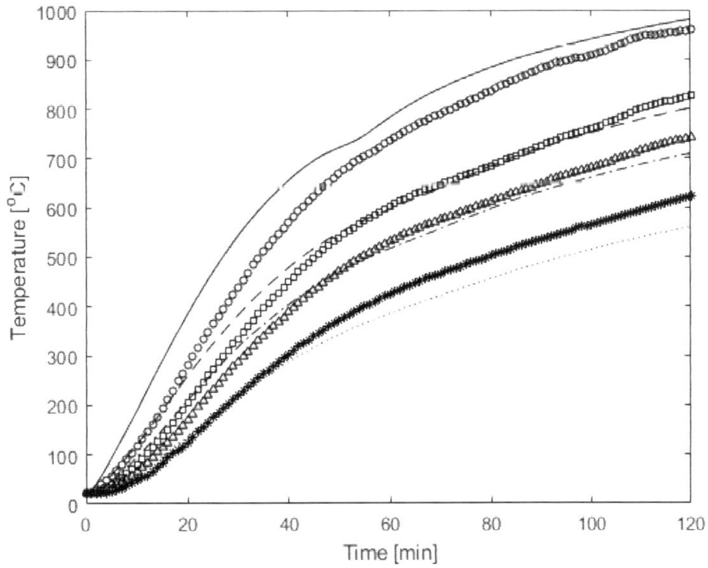


**Figure 8.** Comparison of the temperature variation obtained with SAFIR and the recorded values in section G of the WFRC 66162 test

## 5. Numerical model – Mechanical analysis

The mechanical model, here referred to as model 1, was designed as a simply supported beam with a span of 4.5 m and heated over 4 m (figure 9a). The model was realized with BEAM finite elements and loaded with 4 loads of 84.6 kN each at 0.5625 m, 1.6875 m, 2.8125 m and 3.9375 m, respectively.

The mechanical properties of the model were considered as those reported in the technical note [3]. The yield strength of the steel profile was taken 402 N/mm$^2$, according to the tensile test performed on the material sample. In order to simulate the non-composite behaviour, the concrete strength was not considered.

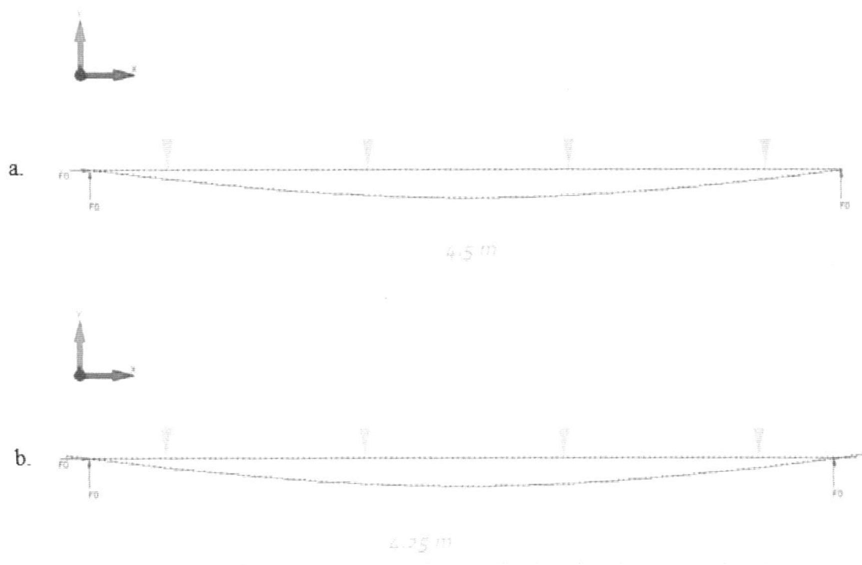


**Figure 9.** The mechanical model: a) model 1; b) model 2.

Figure 10 shows that the numerical displacement at mid-span is larger than that measured in the laboratory, leading to 94 minutes failure time, compared to 107 minutes obtained in the experiment. Therefore, the results are on the safe side, even if the numerical model presents a 12% reduction of beam fire resistance, compared with the experiment. Zanon et al [10], during a similar study, also noticed this phenomenon. The authors [10] used tests from the same experimental campaign [5] in order to validate a numerical model for SFB type slim floors.

The reduction in fire resistance of the analysed beam might be explained by the boundary conditions. In the numerical model, the beam was considered simply supported. But the support conditions within the experimental set-up were not clearly presented in the laboratory report [3]. As the assembly was part of the oven roof and the supports were outside of the oven, the oven walls might have interfered with the beam during the experiment. This would increase the support degree of restrain and reduce the free span of the tested beam. Therefore, a reduced span length of 4.25 m was considered in a second numerical analysis, referred here as model 2 (figure 9b). This numerical model offered better results, closer to the experimentally obtained ones. As shown in figure 10, a very good fit was obtained, in terms of both vertical displacement and failure time of the slim floor beam.

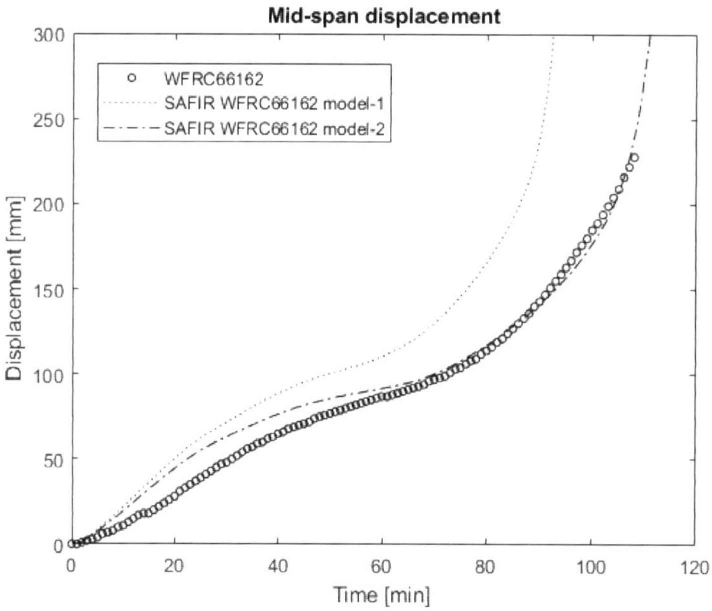


**Figure 10.** Vertical displacement at mid-span.

## 6. Conclusions

The paper presents the validation of a finite element numerical model realized for the thermal and mechanical analysis of the non-composite slim floor beam, IFB type, using SAFIR software.

For the validation of the model, an experimental test existing in the literature was used. The experimental test was carried out in 1996 at the Warrington Fire Research Centre. In this work, the temperature evolution on steel cross-section at different points along the beam length were analysed, as well as the beam vertical displacement at mid-span. These values, obtained numerically, are further compared with the experimental ones.

Very good results in terms of temperature evolution on the cross-sections were obtained, while the structural model offered slightly safer results. This phenomenon was noticed previously in the literature during similar studies, using tests from the same series of experimental tests. It might be explained through the support conditions, which are not clearly presented in the laboratory technical note. In order to simulate the support conditions, considering an interference between the tested specimen and the oven wall, here, a reduced span length was modelled. The results in terms of vertical displacement and fire resistance offered a very good fit with the experimental data.

The numerical model can represent a strong tool for the design of IFB beams, with non-composite behaviour.

**References**

[1] Arcelor Mittal Europe 2018, Long Products Sections and Merchant Bars: Slim Floor. An innovative concept for floors *Slim Floor Catalogue*

[2] Franssen J M 2005, Safir – A thermal/structural program modelling structures under fire, *Engineering Journal*, AISC, Vol **42**, No 3, 143-158

[3] Wainman D E 1996 Preliminary assessment of the data arising from a standard fire resistance test performed on a slimflor beam at the warrington fore research centre on 14th February,1996 Test number WFRC 66162, *Technical Note*, British Steel pic – Swinden Technology Centre

[4] Franssen J M 2019 User's manual for SAFIRa a computer program for analysis of structures subjected to fire, *User Manual*, University of Liege, Belgium

[5] The Steel Construction Institute (SCI): 2008 Slimflor Compendium SCI, Silwood Park, Ascot, Berkshire, UK

[6] Ferreira J, Franssen J M , Gernay T , Gamba A 2018 Validation of SAFIR through DIN EN 1992-1-2 NA, University of Liege, Belgium

[7] EN 1992-1-2, 2005 Eurocode 2 – Design of concrete structures Part 1-2 General rules – Structural Fire Design, CEN, Brussels

[8] EN 1993-1-2, 2005 Eurocode 3 – Design of steel structures Part 1-2 General rules – Structural Fire Design, CEN, Brussels

[9] EN 1994-1-2, 2005, Eurocode 4: Design of composite steel and concrete structures – Part 1-2: General rules – Structural fire Design Brussels

[10] Zanon R, Yildiz S , Obiala R , Braun, M , 2020, Shallow composite floor beams – Proposal of a simplified analytical method for standard fire rating, *Eurosteel European Conference*, Sheffield, UK, 2020

# The study on flexible manufacturing system in building decoration stone processing

**F Du[1*], M Zhao[2] and P Zhang[1]**

[1] Liaoning Communication University, Shenyang, China
[2] College of Arts and Information Engineering, Dalian Polytechnic University, China

* jinger1826@163.com

**Abstract.** Stone materials are widely used in interior and exterior decoration design and public facilities construction. They are one of the essential materials in the construction and decoration field. With the increase in demand of the personality and quality of the decoration materials, intelligent manufacturing of stone processing must be required. With the development of manufacturing technology, especially the rapid development of automatic control technology, CNC machining technology and industrial robot technology, flexible manufacturing technology (FMI) have been used. In the paper an intelligent flexible processing system of stone is designed, which is composed with CNC processing tool, material transmission and storage system, computer control system and stone processing process monitoring system. It can realize fully automatic and intelligent flexible production from building stone processing, stone handling, comprehensive waste treatment, tool optimization and computer control. The calculation method of optimal equipment layout is explained and the field layout of each unit of FMS is given. The purpose of this project is to improve the intelligent level of building decoration stone processing through flexible manufacturing technology and increase production efficiency and material utilization as well as to satisfy needs of diversified product.

## 1. Introduction

As a high-grade building decoration material, stone is widely used in indoor and outdoor decoration design, curtain wall decoration and public facilities construction. With the improvement of people's pursuit of architectural decoration quality, the demand for stone products has been continuously upgraded, and the individualized and diversified demand of stone products makes stone processing develop from traditional single-piece production to production line and flexible processing technology.

Flexible manufacturing system takes agile manufacturing as its core. It comprehensively applies advanced technologies such as group technology, just-in-time technology, lean production technology and computer integrated manufacturing technology [1,2].

It has the characteristics of high automation and strong adaptability, which can better adapt to the needs of small and medium batches, diversified products and short manufacturing cycle. It has important application value and research significance in the stone industry.

In this paper, the intelligent flexible processing mode is adopted to optimize the design of stone processing process. CNC machine tools, material transmission and storage system, computer control system and stone processing process monitoring system are integrated into a whole, and the whole process is controlled from stone raw material processing to plate grinding and polishing, and then to specification plate processing, so as to realize fully automatic and intelligent flexible processing production integrating building stone processing, stone handling, waste comprehensive treatment, tool optimization and computer control.

## 2. Principle and composition

The basic model and principal framework of the stone flexible manufacturing system are shown in figure 1 [3].

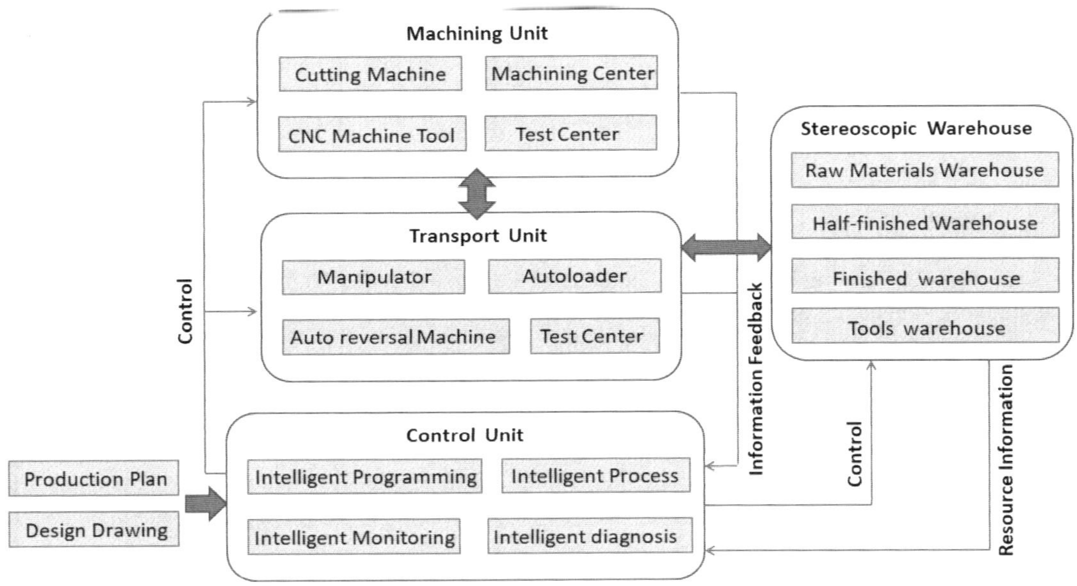


**Figure 1.** Basic model of stone flexible manufacturing system.

The composition of flexible manufacturing system of stone material includes system layer, hardware layer and application layer [4]. The system layer is mainly used for production scheduling, resource allocation, process scheduling, operation monitoring and abnormal state handling of the whole production line [5]. Hardware layer is the entity of flexible production line, which generally includes multiple machine tool units, clamping units, storage units and transfer units, among which the transfer units organically connect other units.

The application layer is mainly used for part processing control, including process scheme design, automatic alignment technology, processing process control technology, tool life control technology and automatic detection technology, as shown in figure 2.

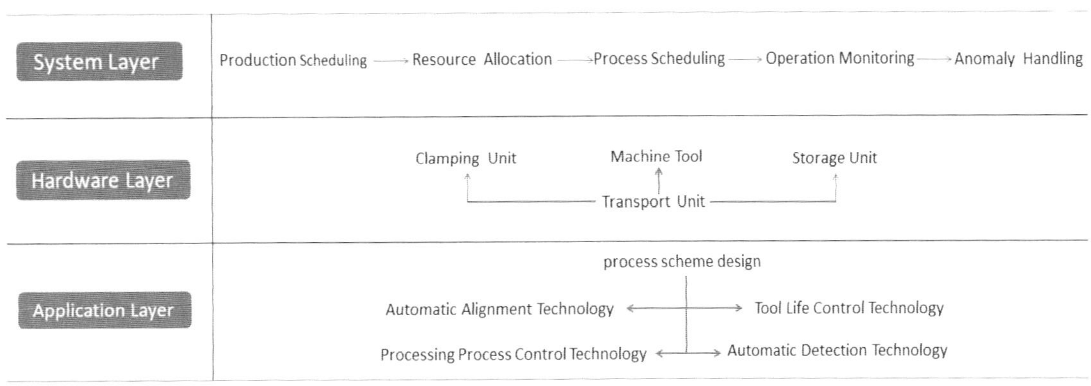


**Figure 2.** Framework of stone flexible manufacturing system.

## 3. Production line construction

The flexible stone production line is from block processing, plate grinding and throwing to specification plate processing, realizes a fully intelligent processing line integrating stone processing, stone handling, waste comprehensive treatment, tool optimization and computer control [6]. The main process is after cutting the raw material, the rough board is divided by an automatic divider, and is transported to a three-dimensional warehouse by a conveyor belt, and the rough board is transported to a stone production grinding and polishing line by a plate conveyor. After grinding and polishing, the plate is scanned, labeled automatically, and transported to the three-dimensional warehouse by the conveying roller table.

Figure 3 shows the intelligent flexible processing system for stone. The blocks are transported from the raw material warehouse 1 to the slabs processing center 15 through the automatic logistics system, then the block is transported to the semi-finished product warehouse 13 through the manipulator 14, and then the block is transported to the grinding and polishing processing center 11 by the tunnel stacker 2 and the manipulator 12.

The products are sent to the three-dimensional warehouse 10 through the chain rail machine, and then transported to the finished product processing center 5 by the bar conveyor, and the finished products are sent to the finished product three-dimensional warehouse 4 by the chain conveyor. The two tool libraries 6 provide tools to the three machining centers respectively, while the waste library 3 automatically recovers waste products. The whole machining system is completely controlled by the flexible machining center 8. The whole processing system includes information flow, tool flow and material transport flow.

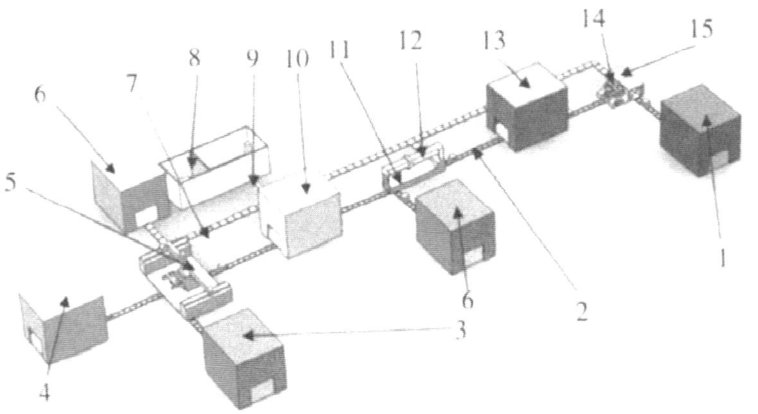


**Figure 3.** Simulation of flexible system for stone processing.

### *3.1. Blocks processing*

The cutting machine is equipped with a measuring device, which can measure the physical and mechanical properties of stone and evaluate and optimize the processing parameters. In the process of processing, water and waste residue flow completely in the closed space, which can realize self-recovery and cleaning [2].

At the same time, the whole equipment is equipped with noise reduction and sound insulation cabin, which completely controls the noise generated by the friction between tools and stone in the processing process in a closed sound insulation cabin.

### *3.2. Automatic segmentation*

After cutting the blocks, the rough board is divided by an automatic divider and transported to a three-dimensional warehouse by a conveyor belt, and the rough board is transported to a grinding and polishing line for stone production by a plate conveyor.

### *3.3. Stone grinding and polishing*
The stone grinding and polishing production line is equipped with an infrared sensing lens, which can automatically position the size of the plate. At the same time, it is equipped with a stone hardness tester, which can automatically detect the hardness of the plate, and select a grinding tool matching the hardness of the plate from the grinding tool library, so as to better determine the grinding and polishing process parameters, scan the ground and polished plate, automatically label it, and transport it to the three-dimensional warehouse via the conveying roller.

### *3.4. Stone cutting*
The cutting machine with scanner scans the large stone plate, and optimizes the surface of the large stone plate according to the different needs of customers. The stone cutting machine is equipped with a tool library, from which the matching diamond tools are selected according to the characteristics of the plate, and the processing path and parameters are optimized to complete various curves and straight lines of the plate. For stone products with complex shapes, a machining center which integrates milling, turning, grinding, engraving, chiseling and drilling, and with a tool magazine and a robot, makes the machining more intelligent. For some profiles, 3D volume graphics can be formed to engrave stone.

### *3.5. Stone processing auxiliary device*
The development of stone processing auxiliary devices is also very fast, and it is moving towards high precision. At present, there are automatic plate loading and unloading machines, automatic plate turning machines, robots and so on. Automation of these auxiliary equipment plays an important role in the intelligentization of stone processing, while greatly reducing labor intensity and improving labor productivity.

### *3.6. Virtual stone manufacturing technology*
The virtual design and processing of stone is based on CAD and CAM, and the CCD scanner is used to scan the large stone plate, and the surface features of the large stone plate, such as the size, texture and surface defects of the large plate, are collected. Then, the optimized cutting software is used to optimize the layout of the large plate, and the model is converted into the NC processing software, which simulates it [7].

Finally, the NC software gives the optimized processing parameters and corresponding processing trajectory. When the scheme is determined to be feasible, the NC code of the machining is delivered to the NC machining equipment to complete the virtual design to the real machining. Using virtual design technology, the specifications, varieties, colors, patterns and patterns of stone can be retrieved from the stone library to simulate the design and assembly of stone decoration space. Processing enterprises carry out simulation of machining according to the creative scheme, so as to realize design personalization and processing diversification.

## 4. Conclusions
In this paper, a flexible production line for stone processing is designed, and the basic model of flexible manufacturing system and the field layout of each unit are given. The flexible stone processing production line is composed of information control system, material storage and transportation system and digitally controlled manufacturing equipment, which can reasonably select processing equipment and processing technology according to factors such as batch size, shape, structure and texture of stone products, carry out production line layout, process diversified and personalized stone products in an optimized way, improve production efficiency and material utilization rate, enhance the intelligent level of stone processing enterprises, and meet the ever-changing market demand.

## References
[1] Fang L 2015 Design and Research on transmission lines of flexible manufacturing system [D] *Wuhan University of Technology*

[2] Tempelmeier H and Kuhn H 1993 Flexible manufacturing systems: decision support for design and operation vol 12 (John Wiley & Sons)
[3] Liu M and Chen R 2019 Construction of flexible manufacturing system with intelligent manufacturing characteristics [J] Electronic Technology & Software Engineering 116-117
[4] Liu M Q 202 Research on design and evaluation of warehouse automation reformation solution for Penterprise [D] *Jinan Shandong University*
[5] Shu H S, Li Q F, Yan S Y and Song B J 2005 Research on dynamic scheduling simulation model of FMS *Harbin Gongcheng Daxue Xuebao/Journal Harbin Eng. Univ.* **26** 192–6
[6] Zhao M 2015 Digital technology of stone processing [J] *Stone* **2**(8):39-48
[7] Wang F 2021 Research on NC machining process based on flexible production line [J] *Modular Machine Tool & Automatic Manufacturing Technique* **8**(8):109-117

# Tracking control a twin-rotor helicopter used for building measuring with PID-MPC controller

**F Guo**[1*]

[1] College of Arts and Information Engineering, Dalian Polytechnic University, Dalian City, China

* 17705497962@163.com

**Abstract**. In recent years, due to the development of UAV (Unmanned aerial vehicle) technology, UAVs have been used in an increasingly wide range of applications, including the architecture industry. In the field of architecture, the technologies of capturing images of building conditions with UAVs have a huge development, such as architectural planning, remote terrain detection, construction safety supervision and so forth. In general, the applications can be expanded widly, owing to the convenient operation of UAVs for the flyers and that does not require a high degree of precision for the control. However, when it comes to penetration between and within buildings. The lack of flight vision and signal interference often leads to collisions and crashes. Hence, this paper studies how to further optimise the control accuracy of construction measurement UAVs. The aim is to reduce the difficulty of operating the UAV when it faces to interference during architecture measurement. In this paper, a three DOFs two-rotor UAV is selected as the object of study. A mathematical model of the UAV is constructed, which is followed by the designing of an advanced controller based on a PID controller and an MPC (model predictive controller) controller, used for realising offset free reference tracking when changing elevation, pitch and travel angles. The usage of this controller can significantly reduce the cost of control during flight to reduce control bias in the presence of disturbing factors. The application of the controller to UAVs allows for automated and intelligent UAV control.

## 1. Introduction

With the development of unmanned aerial vehicles (UAV) and the advance of automatic control technologies, the application scenarios of unmanned helicopters have been widely extended, especially in the field of architecture such as power patrol, earthwork measurement, tower crane inspection architectural planning, construction safety supervision and so forth. In this study, a twin-rotor helicopter is chosen as the objective since it is lighter and has more endurance, which means it is ideal for construction applications and it is important to improve the performance of UAVs in architecture measurement.

As a sort of helicopter, the twin-rotor helicopter is designed with nonlinear multi-input-multi-output (MIMO) systems, which are underactuated systems with more degrees of freedoms than the executors [1]. In the representative twin-rotor helicopter, the aerodynamic thrusts generated by fans are controlled by varying the input voltages, as well as the angle of propellers blades [2]. Therefore, in this project, the twin-rotor helicopter with the fixed propellers blades angles can be controlled by inputting proper variable voltages in each control step, to adjust the angles of the degrees of freedoms.

However, in most cases, helicopter including twin-rotor helicopter systems own highly nonlinear and complicate dynamic models [3]. As a result, the modelling and control of architecture helicopters are commonly complex. This study concerned with the modelling, and control of a helicopter with two rotors.

## 2. Modelling and linearisation of the helicopter

The twin-rotor helicopter is a complicate nonlinear system, it has three degrees of freedom, with two control inputs and two measured outputs (elevation and travel angles) [4]. In this paper, the twin-rotor helicopter system is decomposed into some different modules, then it can be formulated with forces, voltages and physical parameters.

**Figure 1.** The composition of the helicopter system [5].

### *2.1 The composition of the twin-rotor helicopter system*

Twin-rotor helicopter system is an integrated system, it consists of four main modules: (1) NI myDAQ interface, (2) Signal conditioning and power electronics, (3) The helicopter mechanical structure, (4) Controllers PC [5]. Besides, the helicopter mechanical structure includes two fans and three axes. They are displayed in figure2.

**Figure 2.** The modules consisting of the system [5].

In practice, after inputting voltages through NI myDAQ interface to the system from Controllers PC, the signal will be transmitted through signal conditioning and power electronics, and the helicopter mechanical structure successively, then the three axes can finally react and output angle variations of elevation, pitch and travel. Therefore, for the modelling, apart from the interface and controllers, there are three parts that need to be considered: power electronics, fan dynamics and rigid body dynamics. The working process and signal converting can be described as figure 3.

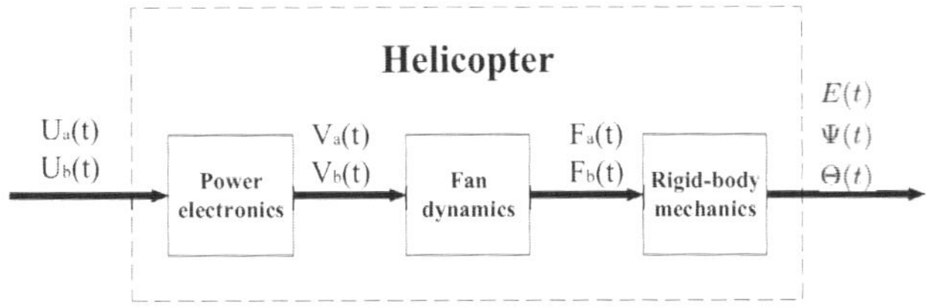


**Figure 3.** The process of helicopter responses to voltages.

*2.2 The modelling of helicopter system modules*

*2.2.1 Power Electronics*

In helicopter system, power electronics is important in matching the peculiarities of the dispersed generation units, and the requirements of the grid-connections containing voltage, frequency, harmonic minimization, control of active and reactive power, and so forth [6].

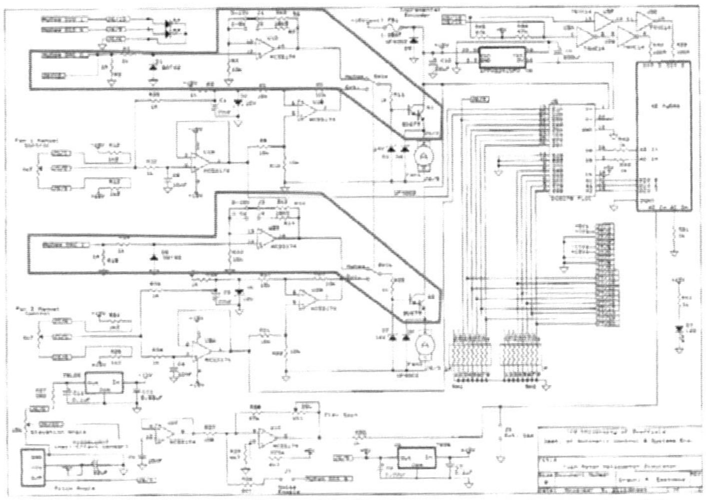

**Figure 4.** The overview of helicopter power electronics [5].

In figure 4, the drive circuitry is highlighted with red lines. This section is emphasized in the system not only because it can stabilise the voltage input by introducing a saturation limit, but also it can amplify signals to meet the instrument driving requirement. Therefore, as to the simulation of the power electronics module, the effect of drive circuitry should be considered.

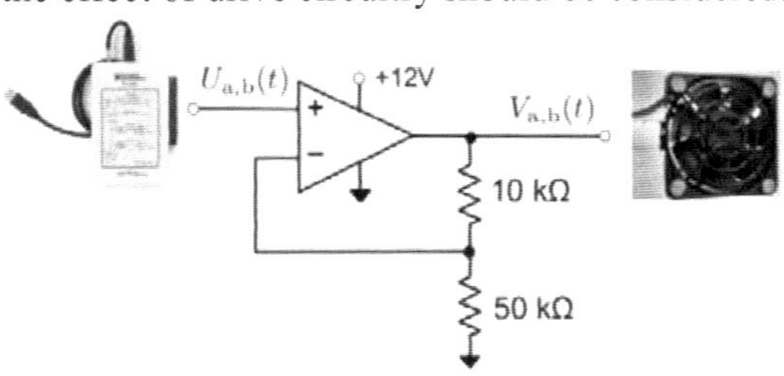


**Figure 5.** The simplified drive circuitry [5].

The drive circuitry can be simplified as an amplifying circuit shown as figure 5. It can be calculated that the relationship between the output voltage $V_{a,b}$ and the input voltage $U_{a,b}$ is

$$V_{a,b} = \frac{10+50}{50} U_{a,b} = 1.2\, U_{a,b} \tag{1}$$

which means the amplifier gain is 1.2, with the lower limit and the upper limit are 0V and 12V.

### *2.2.2 Fan dynamics*

For the twin-rotor helicopter system, two fans installed on the axis are the main power producer, it can also be regarded as a converter from electricity to force. The process of converting is shown below.

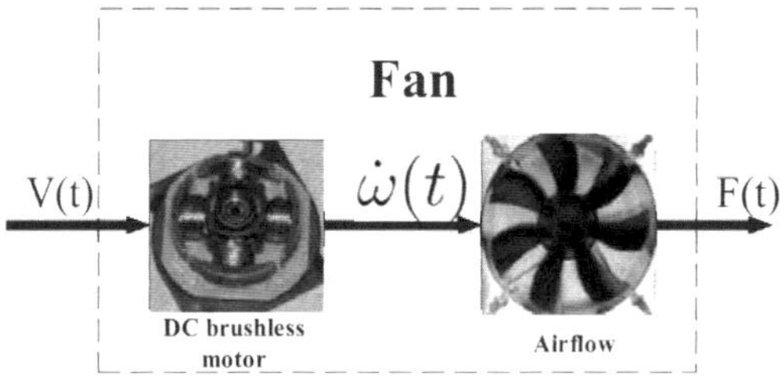


**Figure 6.** The working model of fan [5].

Generally speaking, fan modules are nonlinear systems with complicate dynamics. An accurate model can describe the fan system to some extent, but it is usually too complex to be inconvenient for the integration of the whole model, which increase the difficulty to control. Therefore, an alternative method based on experiments and system identification is recommended in this paper.

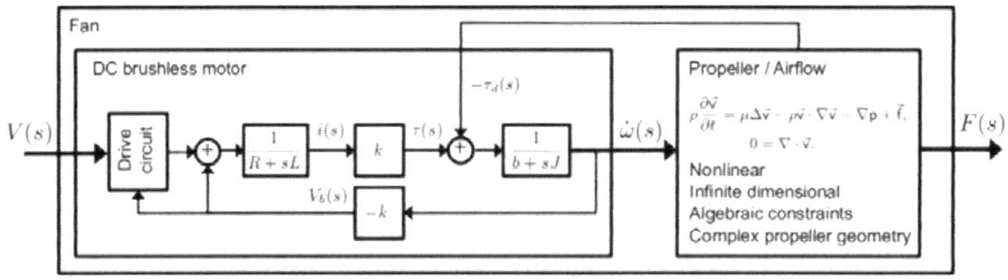


**Figure 7.** The complete general fan dynamics model [5].

The core idea of this method is to apply step signals to the fan system, then measure and identify the output response, to obtain the dynamics that can emulate the real response closely [4]. In practice, conduct experiments with adequate amplified voltages as first, then measure and record the output force by fans. The experimental data are shown in table 1.

**Table 1.** The input voltages and the output force [5].

| *V* (V) | $F_{ss}(V)$ (N) | *V* (V) | $F_{ss}(V)$ (N) |
|---|---|---|---|
| 0.0 | 0.0 * g/1000 | 6.5 | 1.2 * g/1000 |
| 0.5 | 0.0 * g/1000 | 7.0 | 1.4 * g/1000 |
| 1.0 | 0.0 * g/1000 | 7.5 | 1.6 * g/1000 |
| 1.5 | 0.0 * g/1000 | 8.0 | 1.8 * g/1000 |
| 2.0 | 0.0 * g/1000 | 8.5 | 1.9 * g/1000 |
| 2.5 | 0.2 * g/1000 | 9.0 | 2.1 * g/1000 |
| 3.0 | 0.3 * g/1000 | 9.5 | 2.3 * g/1000 |
| 3.5 | 0.4 * g/1000 | 10.0 | 2.5 * g/1000 |
| 4.0 | 0.5 * g/1000 | 10.5 | 2.6 * g/1000 |
| 4.5 | 0.6 * g/1000 | 11.0 | 2.8 * g/1000 |
| 5.0 | 0.7 * g/1000 | 11.5 | 3.0 * g/1000 |
| 5.5 | 0.9 * g/1000 | 12.0 | 3.2 * g/1000 |
| 6.0 | 1.0 * g/1000 | | |

The functional image of the experiment data that can reflect the relationship is shown on figure 8.

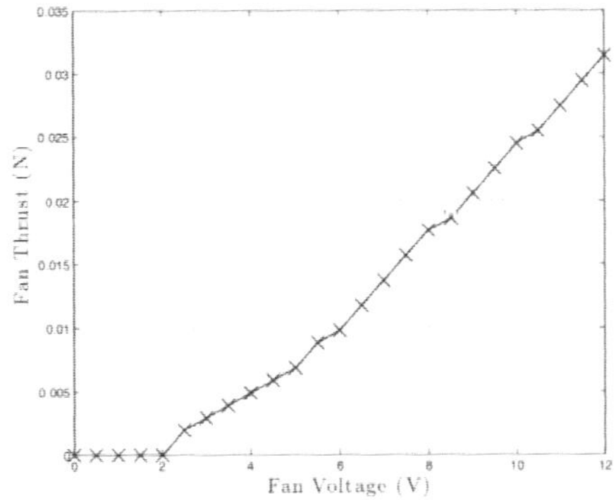


**Figure 8.** The functional image of the data [5].

Then, the step voltage input is applied with the data to emulate the real response. To build the system transfer function, the example response measured in the experiment with the input 12v step voltage is recorded as the left image of figure 9. The response can be regarded as a 1st order response with noise, so the transfer function can be built to emulate the real response. Based on the image, parameters of 1st order model from step response can be recognised using system identification, the identified response is shown as the right side.

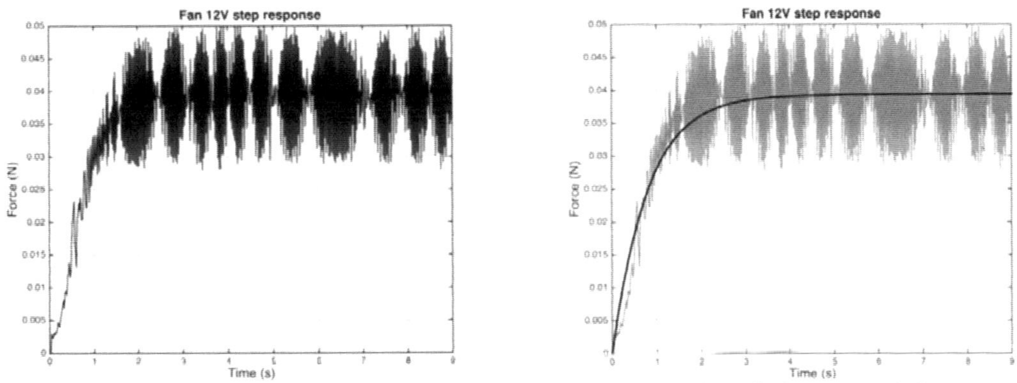


**Figure 9.** The experimental response with 12v and the identified response [5].

An additional experiment [5] is conducted to research the identification performance using different order model, the example orders are chosen as 1st, 3rd, 10th. The identified models are shown below.

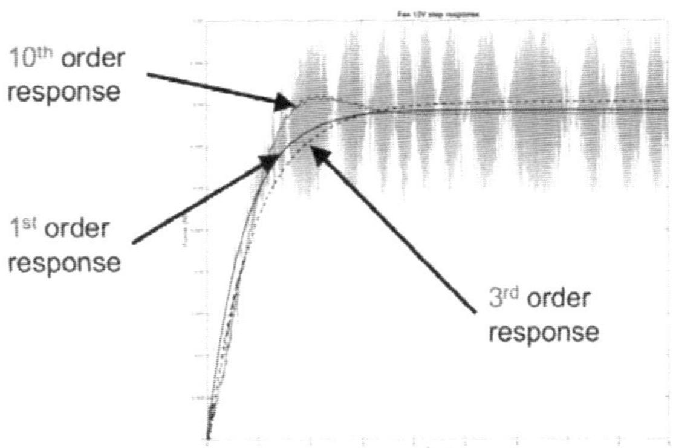


**Figure 10.** The identified response with1st, 3rd, 10th order [5].

From figure 10, it can be seen that the model with 1st order has the best correlation to the real response and the ideal output, trading off with accurate and complexity. Therefore, the 1st order model is determined in this part.

According to the above analysis, a first-order model with the step input can be built as below.

$$F(s) = G(s)V(s) = \frac{1}{\tau s+1}\cdot\frac{F_{ss}(V)}{s} \tag{2}$$

where $\mathbf{F_{ss}}$ is the steady-state thrust of the 12v step voltage input, $\boldsymbol{\tau}$ is time constant, it can be determined using the equation (3).

$$t_{10\%\text{-}90\%} = 2.2\tau \tag{3}$$

where $\mathbf{t_{10\%\text{-}90\%}}$ is the rising time.

After exploring the relationship between the time constant and rise time by experiments, the results show that the values which are inside the section [0.5, 1.3] are satisfied. So, set the time constant as 0.7, the rise time of the system is 1.5379s. Under this circumstance, the derivation rate is 0.14%, which means the response is very close to the real. Therefore, the first-order model becomes

$$F(s) = G(s)V(s) = \frac{1}{0.7s+1}\cdot\frac{F_{ss}(V)}{s} \tag{4}$$

Use inverse Laplace Transform to explore the response in the time domain.

$$F(t) = F_{ss}(V)\,L^{-1}\left(\frac{1}{0.7s+1}\cdot\frac{1}{s}\right) = F_{ss}(V)\left(1-e^{-\frac{1}{0.7}t}\right) \tag{5}$$

With the desired time constant 0.7, a comparison between the identified model and step response is demonstrated as follows.

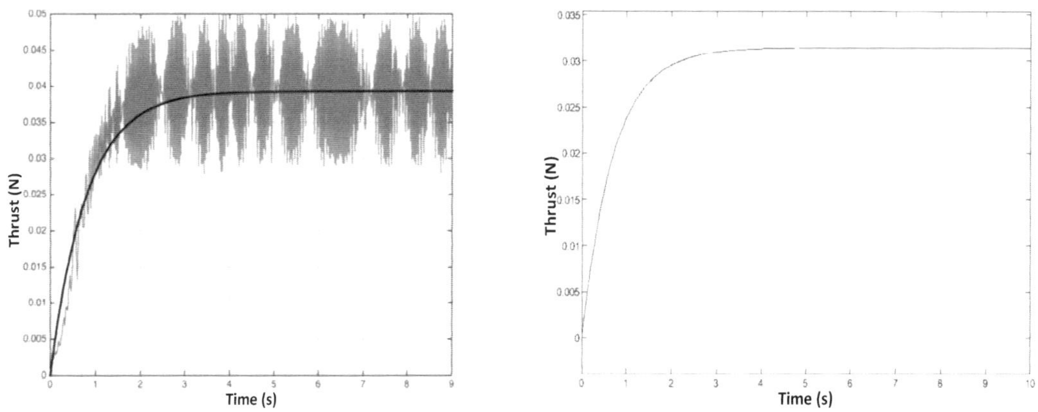


**Figure 11.** The identified model with 12v and the step response with the time constant

## *2.2.3 The rigid body dynamics*

This section is to model the mechanical structure of the helicopter, with three axes, two fans and a weight. The coordinates are defined as E, Ψ, Θ, which represent the angle of elevation, pitch and travel separately. When the angles of elevation and pitch equal to zero, that is E=Ψ=0, the state of the helicopter is called arms horizontal. The overview of the helicopter structure is shown as figure 12.

IOP Conf. Series: Materials Science and Engineering 1242 (2022) 012016 doi:10.1088/1757-899X/1242/1/012016

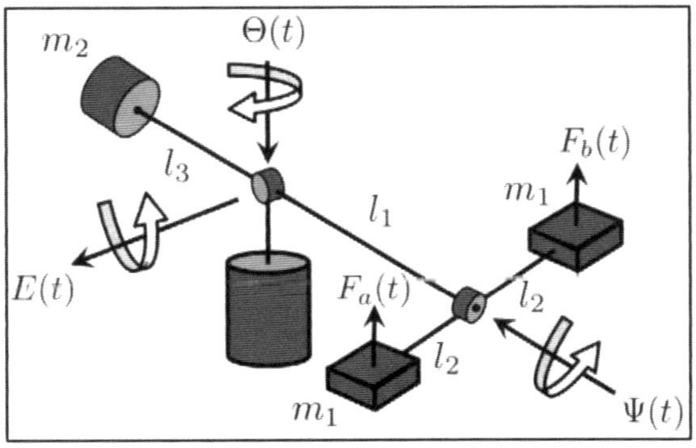


**Figure 12.** The structure of helicopter [5]

where the three arrows mean the increasing direction of each angle, the variables and their values are explained as follows:

Masses:

- Fan mass $m_1 = 0.0505$kg
- Weight mass $m_2 = 0.1$kg

Lengths:

- Helicopter arm to elevation axis $l_1 = 0.11$m
- Fan centre to pitch axis $l_2 = 0.07$m

Weight to elevation axis $l_3 = 0.108$m [5]

As mentioned above, the helicopter system is a complicate system, and its mechanical structure is also complex. In this case, the rigid body dynamics can be obtained using the Euler-Lagrange Equation.

$$\frac{\partial^2 L(t)}{\partial t \partial \dot{q}_i(t)} - \frac{\partial L(t)}{\partial q_i(t)} = Q_i(t) \tag{6}$$

where $q_i$ are the generalised coordinates of the system, $Q_i$ is the generalised forces which can express the non-conservative forces of the system, $L(t)$ is the Lagrangian of the system at time $t$, it can be calculated with the kinetic energy $T(t)$ and the potential energy $V(t)$ of the system as:

$$L(t) = T(t) - V(t) \tag{7}$$

In the helicopter system, the generalised coordinates, as well as the corresponding velocities can be defined as below:

$$q(t) = \begin{bmatrix} E(t) & \Psi(t) & \Theta(t) \end{bmatrix}^T \tag{8}$$

$$\dot{q}(t) = \begin{bmatrix} \dot{E}(t) & \dot{\Psi}(t) & \dot{\Theta}(t) \end{bmatrix}^T \tag{9}$$

The generalised forces can be obtained through force analysis, the force analysis diagram of three directions are shown severally.

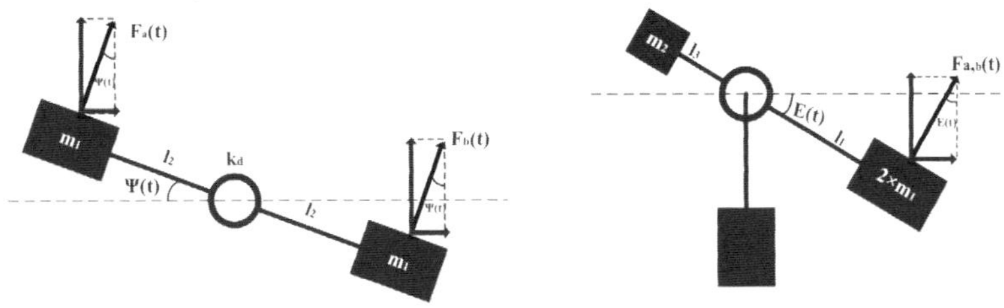


(1) (2)

**Figure 13.** The generalised forces analysis

where $F_{a,b}$ is the thrusts of two fans, $k_s$ = 0.0015 kg m$^2$s$^{-2}$, is the rotational spring, $k_d$ = 0.0005 kg m$^2$s$^{-1}$, is the damping coefficients pertaining to rotation about the pitch axis.

From figure 13 (1), despite the angle Ψ, the forces from fans are always vertical to the axis $l_2$, the generalised force on the pitch can therefore be described as:

$$-k_d\dot{\Psi}(t)+l_2(F_a(t)-F_b(t)) \tag{10}$$

As to the force on elevation, it can be concluded in figure 13 (2) that the force is to make the axis elevate, so it should be vertical to the horizontal line, that is $F_{a,b}\cdot \cos\Psi$, so the force is:

$$l_1(F_a(t)+F_b(t)\cos\Psi(t)) \tag{11}$$

The force on travel contributes to the horizontal rotation of the helicopter, so the force should be in the same direction with the horizontal line, i.e. $F_{a,b}\cdot \sin\Psi\cdot \cos E$. Therefore, the generalised force is:

$$-l_1(F_a(t)+F_b(t))\sin\Psi(t)\cos E(t) \tag{12}$$

Besides, the kinetic energy of the helicopter can also be defined using the forces analysis diagram, with some minor approximations. The kinetic energy is given by:

$$\begin{cases} T_E=\frac{1}{2}J_E\dot{E}^2, & J_E=2m_1l_1^2+m_2l_3^2 \\ T_\Psi=\frac{1}{2}J_\Psi\dot{\Psi}^2, & J_\Psi=2m_1l_2^2 \\ T_\Theta=\frac{1}{2}J_\Theta\dot{\Theta}^2\cos^2E, & J_\Theta=J_E \end{cases} \tag{13}$$

where $T_E$, $T_\Psi$, $T_\Theta$ represent the kinetic energy of the elevation, pitch, travel axis, $J_E$, $J_\Psi$, $J_\Theta$ represent the moments of inertia of these three axes. The entire kinetic energy is the sum of these, which is given by:

$$T=T_E+T_\Psi+T_\Theta \tag{14}$$

In terms of the potential energy, in the helicopter, it is stored as the sum of the gravitational potential energy and spring energy in the pitch axis, it can be defined as:

$$V=(2m_1l_1-m_2l_3)g\sin E+\frac{1}{2}k_s\Psi^2 \tag{15}$$

where g is gravitation acceleration.

Inserting (13), (14), (15) into (7), the Lagrangian can be calculated:

$$L(t)=\frac{1}{2}J_E(\dot{E}^2+\dot{\Theta}^2\cos^2E)+\frac{1}{2}J_\Psi\dot{\Psi}^2-(2m_1l_1-m_2l_3)g\sin E+\frac{1}{2}k_s\Psi^2 \tag{16}$$

Next, inserting (10), (11), (12), (16) into (6), and simplify the Euler-Lagrange equation, therefore the expressions for $\ddot{E}$, $\ddot{\Psi}$ and $\ddot{\Theta}$ can be gained as follows.

Elevation axis:

$$\ddot{E}(t)=\dot{\Theta}^2(t)\cos E(t)\sin E(t)+\frac{(F_a(t)+F_b(t))l_1\cos\Psi(t)-(m_2l_3-2m_1l_1)g\cos E(t)}{J_E} \tag{17}$$

Pitch axis:

$$\ddot{\Psi}(t)=-\frac{1}{J_\Psi}(l_2(F_a(t)-F_b(t))+k_s\Psi(t)+k_d\dot{\Psi}(t)) \tag{18}$$

Travel axis:

$$\ddot{\Theta}(t)=-\frac{l_1(F_a(t)+F_b(t))\sin\Psi(t)}{J_\Theta\cos E(t)}+2\dot{\Theta}(t)\dot{E}(t)\tan E(t) \tag{19}$$

## 3. The linearisation of the helicopter model

So far, the helicopter system has been built as a non-linear model completely. The model can reflect the behaviour of a twin-rotor helicopter with different input voltages, and the internal signal can also be measured. In order to realise offset free reference tracking, controllers need to be applied [8]. Hence, the model needs to be applicable for controller design, which means the governing, nonlinear ordinary differential equations, as well as the models of power electronics, fan dynamics obtained in the previous sections, need to be linearised as a linear, time-invariant state-space model, in the form of:

$$\begin{cases} \dot{x} = Ax + Bu \\ y = Cx + Du \end{cases} \quad (20)$$

Before deriving the state-space model, an operating point should be determined at the beginning. In the case of the helicopter, a proper operating point should be able to keep the helicopter to have a steady hover in the horizontal position, that is, the angles and the velocities should equal to 0:

$$\mathrm{E} = \Psi = \Theta = \dot{\mathrm{E}} = \dot{\Psi} = \dot{\Theta} = 0 \quad (21)$$

At the same time, the value of the voltage that contributes to achieving keeping hover is called the equilibrium voltage $\mathbf{U_e}$. Due to the dynamics of this model, it is impossible to calculate an accurate value to make the helicopter to keep steady, but the value can be determined in experiments. The typical open-loop response of elevation with double input voltages $U_e$ should be similar as:

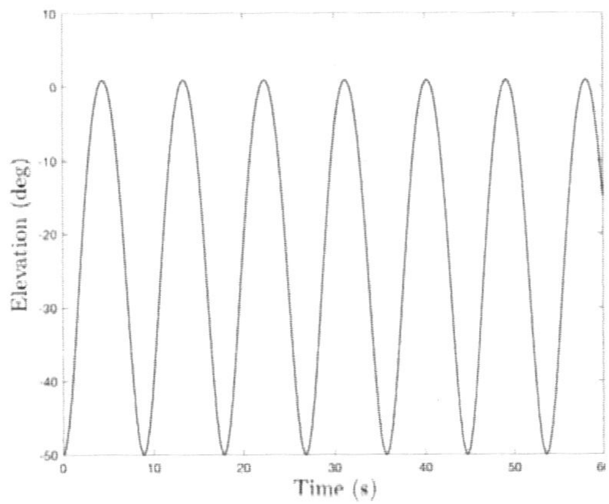


**Figure 14.** The typical response of elevation with both input $U_e$ [5]

At first, set $U_e$ to 2.5, the response is like:

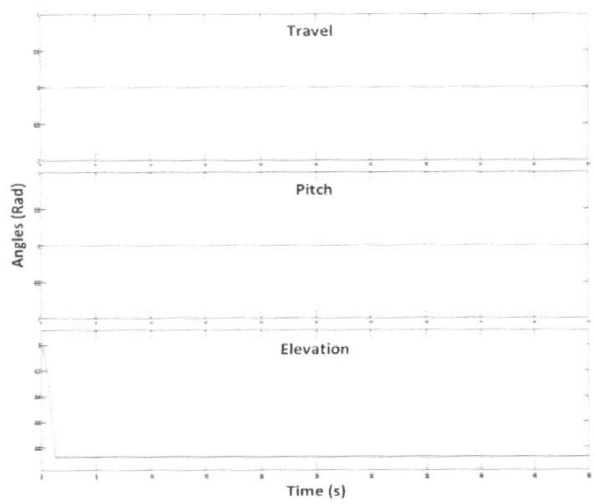


**Figure 15.** The response of axes with $U_e$=2.5

It can be seen that the helicopter fails to gain any elevation, so the value of $U_e$ should be increased until the voltage can drive the helicopter to saturate at the upper elevation limit. After adequate attempts, a satisfying equilibrium voltage $U_e$ is chosen as 5.7965, under this voltage, the response of three axes are shown as below:

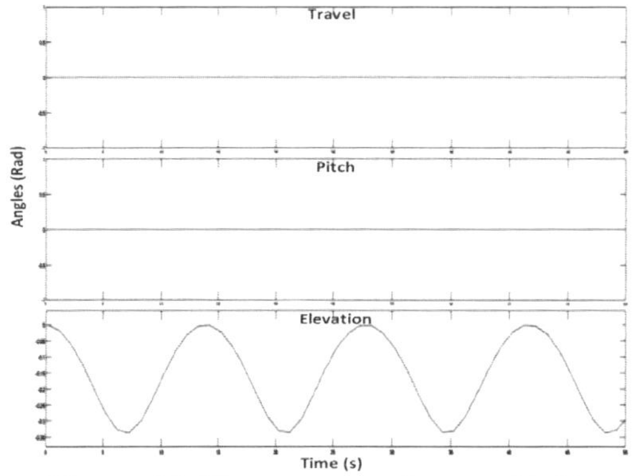


**Figure 16.** The response of axes with $U_e$

Therefore, the states, controls, and measurements of the linearised system can be defined corresponding to disturbances away from the operating point as:

$$x=\begin{bmatrix}\varepsilon\\ \psi\\ \theta\\ \dot{\varepsilon}\\ \dot{\psi}\\ \dot{\theta}\end{bmatrix}=\begin{bmatrix}E-E_e\\ \Psi-\Psi_e\\ \Theta-\Theta_e\\ \dot{E}-\dot{E}_e\\ \dot{\Psi}-\dot{\Psi}_e\\ \dot{\Theta}-\dot{\Theta}_e\end{bmatrix},\quad u=\begin{bmatrix}u_a\\ u_b\end{bmatrix}=\begin{bmatrix}U_a-U_{a,e}\\ U_b-U_{b,e}\end{bmatrix},\quad y=\begin{bmatrix}\varepsilon\\ \psi\\ \theta\end{bmatrix} \tag{22}$$

where **x** is the states, **u** is control, and **y** is the measurements of the linearised model. The matrices of the state-space model A, B, C, D can be obtained through the Jacobian linearisation method, that is:

$$A_{i,j}=\left.\frac{\partial f_i(X,U)}{\partial U_j}\right|_{X_e,U_e},\quad B_{i,j}=\left.\frac{\partial f_i(X,U)}{\partial U_j}\right|_{X_e,U_e},$$

$$C_{i,j}=\left.\frac{\partial f_i(X,U)}{\partial U_j}\right|_{X_e,U_e},\quad D_{i,j}=\left.\frac{\partial f_i(X,U)}{\partial U_j}\right|_{X_e,U_e}, \tag{23}$$

where the X, U, Y represent the states, controls and measured outputs of the model separately, which are:

$$x=\begin{bmatrix}E\\ \Psi\\ \Theta\\ \dot{E}\\ \dot{\Psi}\\ \dot{\Theta}\end{bmatrix},\quad u=\begin{bmatrix}U_a\\ U_b\end{bmatrix},\quad y=\begin{bmatrix}E\\ \Psi\\ \Theta\end{bmatrix} \tag{24}$$

The way that the states of the nonlinear system X are regarded as the functions exerting on inputs U, can be described as a vector of ordinary differential equations as:

$$f(X,U)=\begin{bmatrix} f_1(X) \\ f_2(X) \\ f_3(X) \\ f_4(X,U) \\ f_5(X,U) \\ f_6(X,U) \end{bmatrix}=\begin{bmatrix} \dot{E} \\ \dot{\Psi} \\ \dot{\Theta} \\ \ddot{E} \\ \ddot{\Psi} \\ \ddot{\Theta} \end{bmatrix} \tag{25}$$

In addition, the fan dynamics can also be linearised as an asymptotic linear function, the image of function is given by:

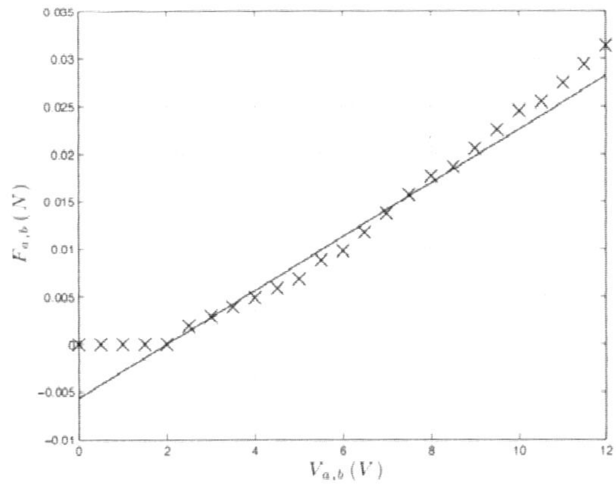


**Figure 17.** The asymptotic linear function of fan dynamics [5]

From figure 17, the intercept can be observed as -0.0057, at the same time, the slope can be calculated as 0.0028. Hence, the linearised fan dynamics can be described as:

$$F_{a,b}(t) \approx \alpha V_{a,b}(t) + \beta \tag{26}$$

where $\alpha$=0.0028 kg m $s^{-2}V^{-1}$, $\beta$ = 0.0057 kg m $s^{-2}$.

The parameters occurred so far can be summed on a table as: Based on (23), (25), (26), and table 2, the model of the helicopter can be linearised using the Jacobian method, leads to:

**Table 2.** The summary of parameters

| $k_s$/kg $m^2s^{-2}$ | $k_d$/kg $m^2$ $s^{-1}$ | $l_1$/m | $l_2$/m | $\alpha$/kg m $s^{-2}V^{-1}$ | $\beta$/kg m $s^{-2}$ | $k_a$ | $U_e$/V | $J_E$/kgm$^2$ | $J_\Psi$/kg m$^2$ | $J_\Theta$/kg m$^2$ |
|---|---|---|---|---|---|---|---|---|---|---|
| 0.0015 | 0.0005 | 0.11 | 0.07 | 0.0028 | -0.0057 | 1.2 | 5.7965 | 0.0024 | $4.95\times10^{-4}$ | 0.0024 |

$$\begin{cases} \dot{x} = \hat{A}x + \hat{B}u \\ y = \hat{C}x + \hat{D}u \end{cases} \tag{27}$$

where $\hat{A}, \hat{B}, \hat{C}, \hat{D}$ are:

$$\hat{A}=\begin{bmatrix} 0 & 0 & 0 & 1 & 0 & 0 \\ 0 & 0 & 0 & 0 & 1 & 0 \\ 0 & 0 & 0 & 0 & 0 & 1 \\ 0 & 0 & 0 & 0 & 0 & 0 \\ 0 & -\frac{k_s}{J_\Psi} & 0 & 0 & -\frac{k_d}{J_\Psi} & 0 \\ 0 & -\frac{2l_1(\alpha k_\alpha U_e+\beta)}{J_\Theta} & 0 & 0 & 0 & 0 \end{bmatrix}$$

$$\hat{B}=\begin{bmatrix} 0 & 0 \\ 0 & 0 \\ 0 & 0 \\ \frac{l_1\alpha k_\alpha}{J_E} & \frac{l_1\alpha k_\alpha}{J_E} \\ \frac{l_2\alpha k_\alpha}{J_\Psi} & -\frac{l_2\alpha k_\alpha}{J_\Psi} \\ 0 & 0 \end{bmatrix} \tag{28}$$

$$\hat{C}=\begin{bmatrix} 1 & 0 & 0 & 0 & 0 & 0 \\ 0 & 1 & 0 & 0 & 0 & 0 \\ 0 & 0 & 1 & 0 & 0 & 0 \end{bmatrix}$$

$$\hat{D}=\begin{bmatrix} 0 & 0 \\ 0 & 0 \\ 0 & 0 \end{bmatrix}$$

## 4. Data Acquisition

Until now, the modelling of the helicopter system and the deriving of a linearised state-space mode have been completed. Before designing and applying controllers, data acquisition is necessary, because all control systems implemented nowadays are based on computer control, data acquisition is convenient for a computer to process [9]. It is used to not only converting the signal and data from continuous to discrete, then back to continuous when output data, but also being a converter between analogue and digital. To be specific, the ADC implements sampling and quantizing operation, with a zero-order-hold to convert the signal to constants, whereas the DAC implements saturation and quantizing operation The overview of the structure is:

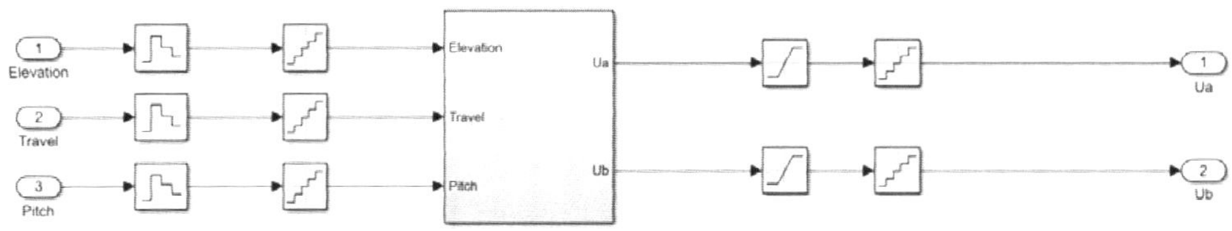


**Figure 18.** The structure of data acquisition.

where the input is the angels of three axes from the system, the output is the voltages from the controller which is to adjust the system states.

## 5. The design of controller

*5.1 The basic framework*

The basic control implementation in Simulink is shown as:

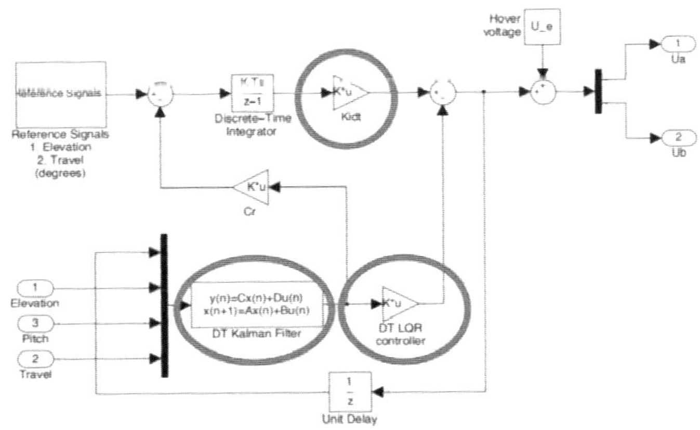


**Figure 19.** The basic control structure [5].

The figure 19 shows the fundamental control framework in 'MyDAQ Discrete Time Dynamic'. It is a conventional integral feedback control, with a Kalman filter added behind the input angles.

The performance of the system can be evaluated through inputting a group of reference values and observing the variations of three degree-of-freedoms. For example, the response with reference input 0.07, 0.08 is shown in figure 20.

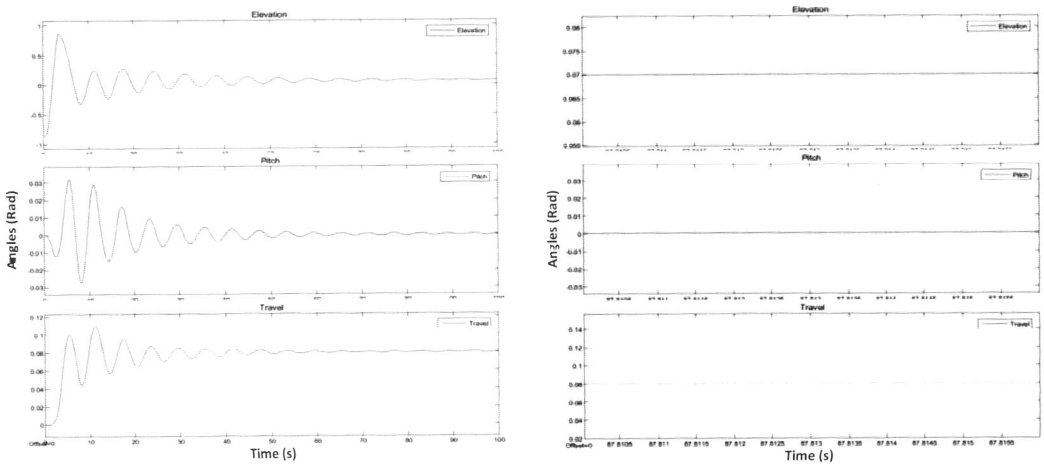


**Figure 20.** The response with reference input 0.07, 0.08.

From figure 20, it can be seen that the steady-state values are at the vicinity of 0.07, 0, 0.08, which means the system is stabilised and can realise the offset free reference tracking with the designed controllers. But as the performance in figure 20, the most significant characteristics are large settling time, as well as overshoots. Besides, the system cannot keep static after arriving reference values, it still has tiny vibrations around the values. Further, the performance will get worse with the increasing of reference inputs, like figure 21.

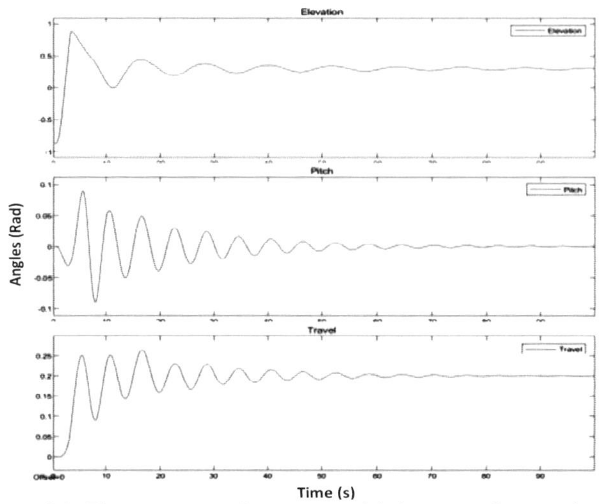


**Figure 21.** The worse performance with larger reference inputs.

From figure 20 and figure 21, it can be observed that the velocity variations of elevation, pitch, and travel are fierce. It is because of the existence of a signal delay in fan dynamics, the system cannot react to the adjusting voltage in time, so the incomplete response would be adjusted by the next control step [7]. Therefore, if the penalties of these states can be enlarged, the settling time would decrease visibly.

In addition, a cost function is defined to evaluate and compare the effectiveness of control in the form of:

$$V = \left(y - y_{ref}\right)^T Q \left(y - y_{ref}\right) \tag{29}$$

where Q is set as unit matrix here. The cost function of feedback integral control is displayed as figure 22.

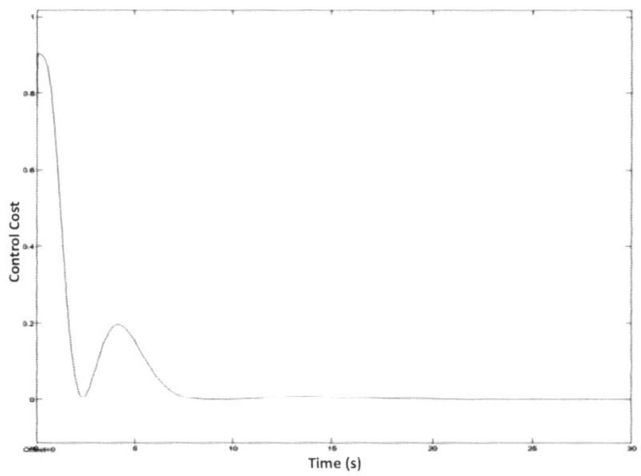


**Figure 22.** The cost function of integral feedback control with reference inputs 0.07 and 0.08.

### *5.2 The design of the advanced controller*

In the advanced control framework, the PID controller and model predictive control (MPC) controller are proposed. Compared with integral and feedback control, the MPC controller can obtain the optimal input within the constraints in every step, through measuring the output timely and solving the optimisation problems online. This can solve the issues caused by the fan dynamics delay efficiently. Besides, a PID controller is set in the control structure to enhance the control robustness, and improve the performance further.

The whole system contains three parts, PID controller, MPC controller and the plant. The PID controller is used in the outer-loop to enhance the response time, while MPC is designed in the inner-loop to improve the control effectiveness. The structure of the advanced control system is illustrated in figure 23.

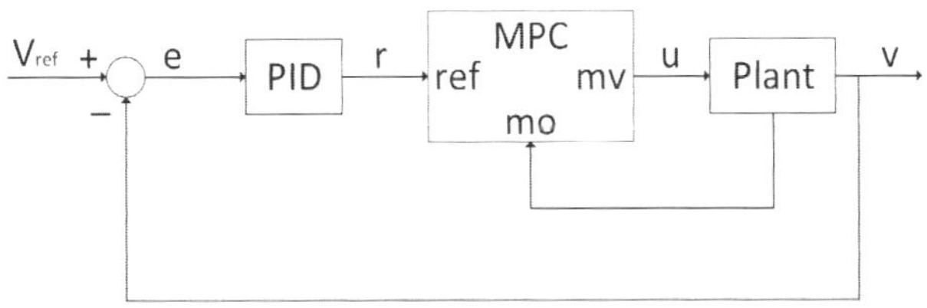


**Figure 23.** The control structure.

The design of MPC parameters is based on the linearised state-space model as well, with two inputs and three outputs. For this reason, the reference input $V_{ref}$ is a vector that contains target values of elevation, pitch and travel. The parameters in figure are:

*e*: the difference between actual and reference angles.

*r*: is the input of MPC controller, u is the adjusting voltage output from MPC controller.

*v*: is the response of three angles from the system.

The Simulink model based on figure 23 is given by:

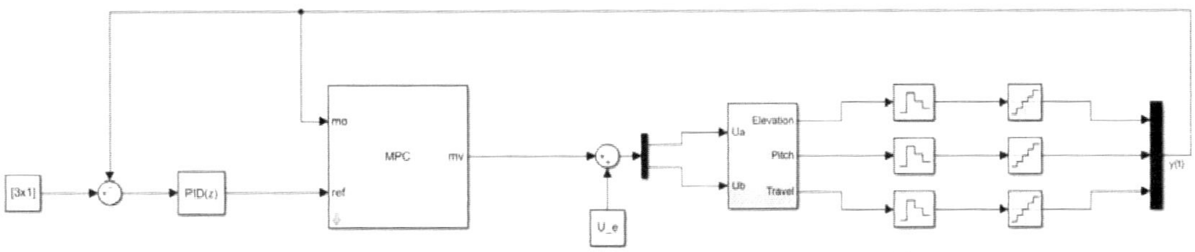


**Figure 24.** The control structure Simulink model.

where $U_e$ is the equilibrium voltage.

In terms of the design of MPC controller, there are three parameters need to be determined: prediction step N, weight Q, R, and penalty P.

In general, increasing prediction step N can improve the closed-loop performance since $K_N$ approaches to $K_\infty$ as N is close to $\infty$, but increase the computational burden. As to the performance, increasing N cannot decrease settling time, but can reduce the control cost. On the contrary, increasing R would enlarge settling time as well.

Based on the theory, design and test MPC parameters in Matlab, MPC designer. According to the response shown in chapter 4, it can be decided that the weight of elevation and travel should be relatively large, since the two angles are the main control objectives, and they could be uncontrollable when facing huge target values. Besides, N should be set with a large value, because one of the improvement aims is to reduce the cost. After some trials, the ideal values of weight for elevation, pitch and travel are set as 18, 10, 25, for input are set as 2.5 to improve its stability. Set N as 100, the principle of MPC control is shown below:

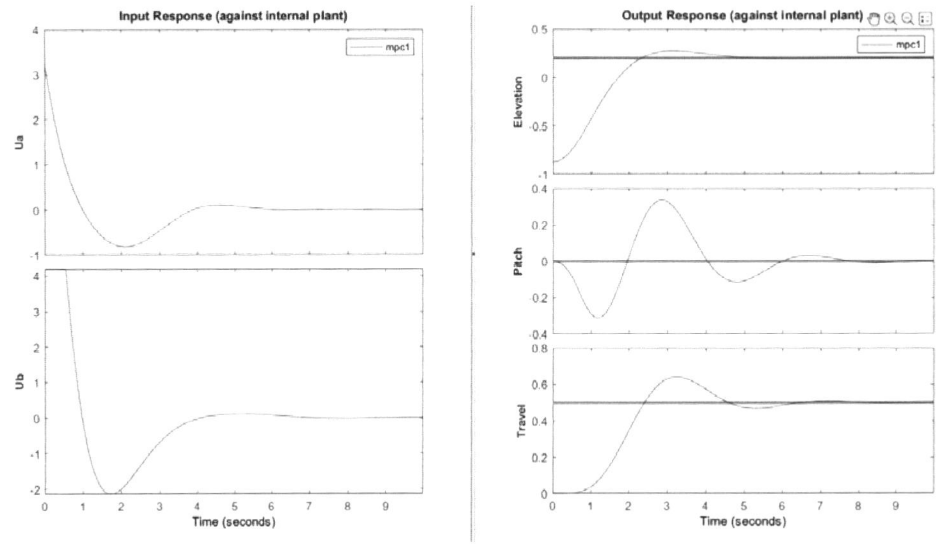


**Figure 25.** The principle of MPC controller

*5.3 The design of PID controller*

In practice, different values of proportion, integral and derivative would exert different impacts. For the proportion, it can decide the velocities of decreasing error, which means it can reduce the response time. But a large value of proportion might cause the increasing of overshoot, so it should be set as the lowest value of the set satisfying the aims. The integral part can eliminate the steady error, increase the response speed, but it would increase the vibration so that the settling time might be longer. Therefore, the value of integral should also be set as the lowest. In addition, derivative can decrease overshoot, but a larger value might cause more vibrations rather. So, it should be set as the highest value of the set satisfying the aims.

Based on the analysis, the final parameters of proportion, integral and derivative are determined as:

**Table 3.** The values of the PID controller for large angles.

| Proportion | Integral | Derivative |
|---|---|---|
| 1.25 | 0.3 | 0.85 |

**Table 4.** The values of the PID controller for small angles.

| Proportion | Integral | Derivative |
|---|---|---|
| 0.4 | 0.3 | 1.5 |

*5.4 The performance of the system*

With the designed PID and MPC controllers, the control structure has been completed. At first, input the same small angles 0.07, 0, 0.08 with figure 20, the response is displayed as

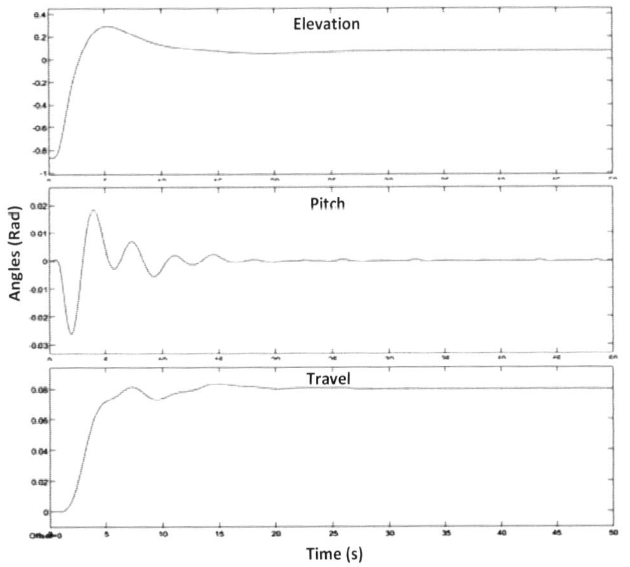


**Figure 26.** The performance with small input 0.07, 0.08.

It can be observed that the settling time is less than 20 seconds now, compared with the basic control method, the speed of system response has been increased significantly. What is more, the overshoot of elevation can be calculated as $(0.2916-0.07)/(0.07+0.8727)\times 100\%\approx 23.51\%$. In the comparison of 43.37% in the basic controllers, it is reduced distinctly.

Next, input a group of large reference values of elevation, pitch and travel as 0.25, 0, 0.5 to explore the performance.

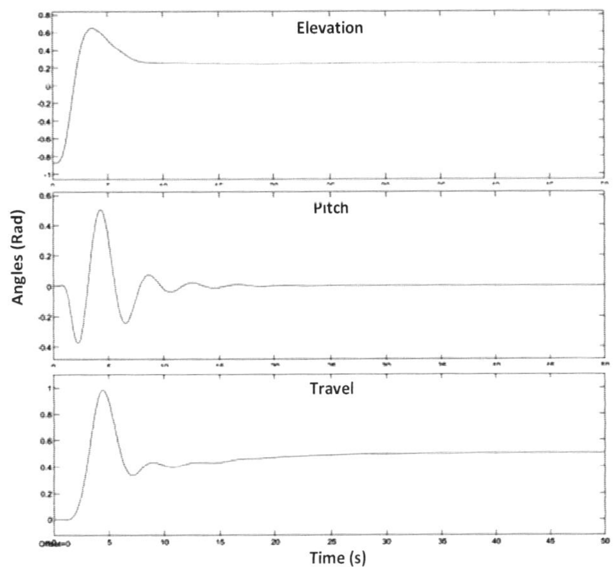


**Figure 27.** The performance with large input 0.25, 0, 0.5.

Compared with figure 21, it can be seen that the three angles track the target accurately. Especially for elevation, it can be controlled to arrive the reference input as well. At the same time, the settling time is also decreased within 30 seconds, and the overshoot of elevation can be calculated as $(0.648-0.25)/(0.25+0.8727)\times 100\%\approx 35.45\%$.

As to the control cost, it can be illustrated through plot the cost function. Input the same reference values 0.07, 0.08 for elevation and travel, the cost function of the MPC control structure and the cost function of the integral feedback control structure can be compared as follows:

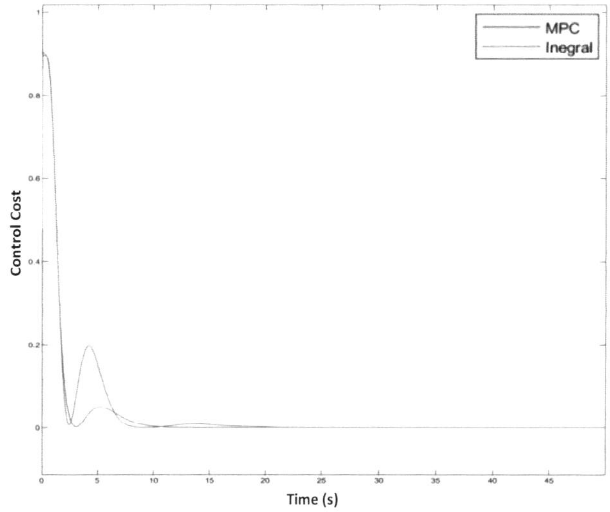


**Figure 28.** The comparison between MPC and integral control

From the figure, it can be seen that the convex curve of MPC control is lower than the one of integral control on the whole. So, it can be concluded that the PID-MPC control structure has better control effectiveness.

*5.5 The alternative scheme*

An additional scheme was proposed when designing a PID controller. It was found that if a PID controller just controls one angle, like elevation, the helicopter can be controlled relatively satisfying with one group of PID parameters, no matter with large angles or with a small angle. The control structure is illustrated below:

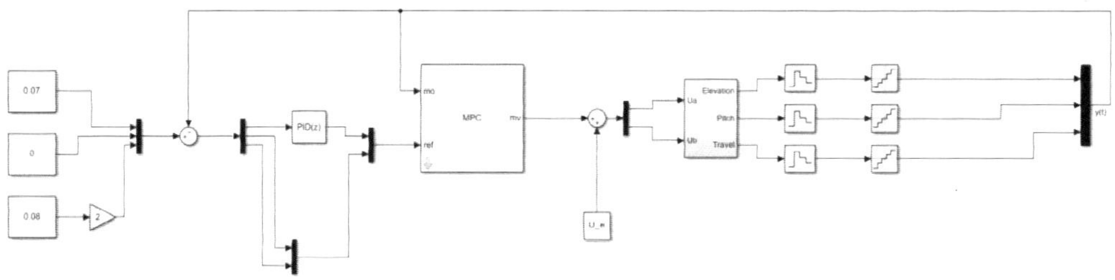

**Figure 29.** The structure of single turning with PID controller

In MPC controller, the weights of output are set as 12, 120, 400, and of input are set as 2, the prediction step is 100. The parameters of PID are set as:

**Table 5.** The values of the PID controller

| Proportion | Integral | Derivative |
|---|---|---|
| 7 | 0.75 | 12 |

Test the controller with two groups of reference input as 0.07, 0, 0.08; 0.25, 0, 0.5, the responses are displayed as follows.

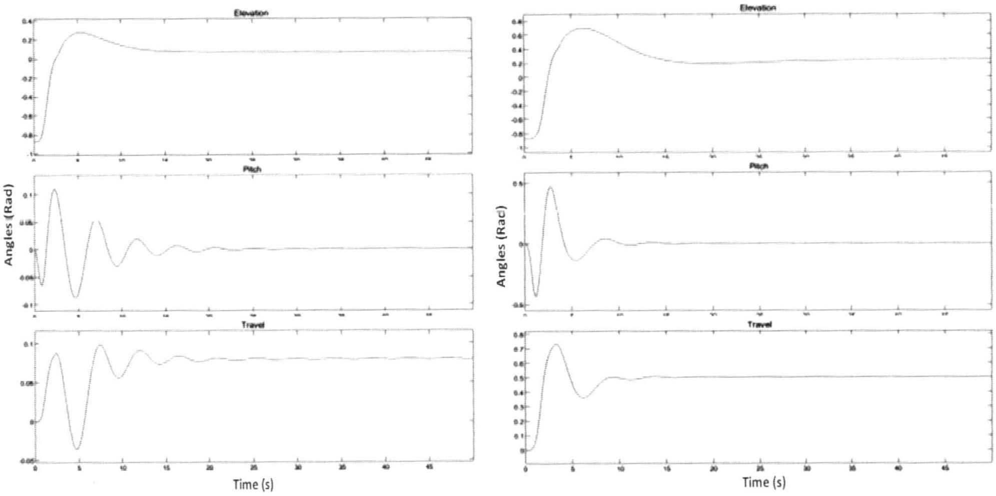


**Figure 30.** The response with input 0.07, 0, 0.08, and 0.25, 0, 0.5.

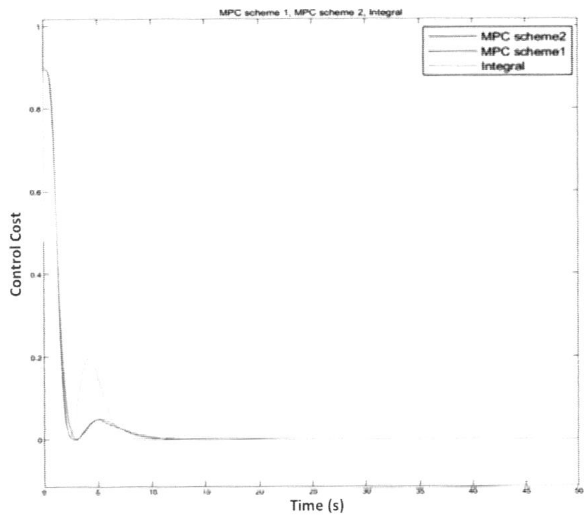


**Figure 31.** The cost functions of scheme 1, scheme 2, and the basic control.

The control cost of this scheme is slightly lower than the scheme discussed before, and both of them are lower than the basic control structure significantly. Therefore, this scheme is superior to the basic structure as well.

## 6. Conclusion

In this paper, we first discussed the characteristics of the architecture helicopter, and illustrated the difficulties of modelling and controlling the system since the nonlinearity and uncertainty. According to control theories, the helicopter can be modeled using simplification, system identification, and Euler-Lagrange method, to obtain the system dynamics. It is followed with linearisation, the Jacobian method was applied to derive a state-space model for control design, so we showed a linear, time-invariant model in R6. Based on the dynamics and state-space model, we developed and applied an advance control structure with PID and MPC controller. Comparisons between the advanced control and basic control effects showed that the performance in response speed, elevation overshoot, controllability, and control cost were improved significantly. The precision of the identified model, as well as the control performance like settling time and overshoots need to be optimised in further study in future. The application of the advanced controller will improve the performance of UAVs used in the field of architecture.

**References**

[1] Khan A.Q and Iqbal N, 2004. Modeling and Design of an Optimal Regulator for Three Degree of Freedom Helicopter/Twin Rotor Control System. , pp.45–50

[2] Chemachema, Mohamed Zeghlache, Samir, 2015. Output Feedback Linearization Based Controller for a Helicopter-like Twin Rotor MIMO System. *Journal of Intelligent & Robotic Systems*, **80**(1), pp.181–190

[3] Le Tri, Quang Lai, Ying-Chih, 2017. System Identification and Control Design of an Unmanned Helicopter Using a PI-MPC Controller **187**(1), p.12020

[4] Alam, M S & Tokhi, M O, 2007. Modelling of a twin rotor system: A particle swarm optimization approach. Proceedings of the Institution of Mechanical Engineers. Part G, *Journal of aerospace engineering*, **221**(3), pp.353–375

[5] Bryn J, ACS336 course materials

[6] Robinson F.V.P., 1997. Power electronics converters, applications and design. *Microelectronics Journal*, **28**(1), pp.105–106

[7] Wada N et al., 2008. Tracking control of a twin-rotor helicopter model under thrust constrains using state-dependent gain-scheduling and reference management pp.143–149

[8] Nguyen Tung The-Lam & Kim, Sam-Dong, 2017. Nonlinear modeling method for DC trans-conductance determination of GaAs MHEMTs. *Microwave and Optical Technology Letters*, **59**(1), pp.131–134

[9] Johnson C R 1985. Computer-controlled systems: Theory and design: Karl J. Åström and Björn Wittenmark. , **21**(6), pp.746–748

# Study on flexural behavior of graphene oxide recycled concrete beams

**K Guo[1] *, Y Z Yang[1], J H Zhou[1] and X Y Wu[1]**

[1] School of Mechanical Engineering, Shenyang Jianzhu University, Shenyang, China

* 43404844@qq.com

**Abstract**. In order to study the flexural behavior of graphene oxide (GO) recycled concrete beams, 6 GO recycled concrete beams were studied, and observe its failure mode and process. The effects of GO content, oxygen content of GO and replacement rate of recycled aggregate on the flexural behavior of the test beams, including cracking load and ultimate load, concrete strain and steel strain, mid-span deflection, were analyzed. The test results show that the cracking load and ultimate load of recycled concrete beams with GO increase to a certain extent with the increase of GO content. And the higher the oxygen content of GO, the higher the ultimate bearing capacity of GO recycled concrete beams. With the increase of replacement rate of recycled aggregate, the cracking load and ultimate load of the test beams decrease.

## 1. Introduction

With the continuous development of modern society, more and more buildings have been demolished, and how to use these waste concrete has become the focus of attention and research at home and abroad [1-3]. It has been proved to be an effective way to solve the problem of waste concrete by crushing, washing and screening waste concrete to form recycled aggregate, and then replacing part or all of natural aggregate with these recycled aggregate to form recycled concrete [4]. Many scholars at home and abroad have found that with the increase of replacement rate of recycled aggregate, the compressive strength of recycled concrete decreases correspondingly, which may be due to the large internal gap of recycled aggregate, which is easy to produce stress concentration, and there are many weak interface transition zones in recycled concrete, resulting in the low compressive strength of recycled concrete [5-7]. Zhou [8] and others found that compressive strength and splitting tensile strength of recycled concrete decreased with the increase of replacement rate of recycled aggregate. Nanometer materials are materials with unique properties, which have large specific surface area, small size effect, quantum effect, tunnel effect and interface effect [9]. Adding nanometer materials to concrete to improve its performance is a hot spot in concrete research at present.

In recent years, nanometer materials have been widely used in cement-based composites [10,11]. GO is a typical two-dimensional nanoscale material, and it is also the lead compound for preparing GO by chemical reduction-oxidation [12,13]. GO contains a large number of oxygen-containing functional groups on its surface, which make the distance between graphite sheets enlarged and have certain hydrophilicity. The literature on GO modified cement-based composites has proved that GO can significantly enhance toughening effect [14-16]. Lv, Wang, Cao and others [17-19] show that adding GO can effectively improve the tensile strength and flexural strength of cement-based composites, and improve the microstructure. Li [20] and others found that GO can improve the crack resistance of cement-based composites and the splitting tensile strength of cement-based composites like fibers. Li, Luo and others [21] found that the addition of GO into cement-based composites can accelerate the hydration process of cement and promote the formation of high-density C-S-H, and the tight bonding between more cement hydration products can inhibit the formation and development of micro-cracks, thus improving the fracture performance and toughness of cement-based composites.

At present, scholars at home and abroad focus on the properties of recycled concrete as material, while the research on recycled concrete structures or components is relatively few. In this experiment, 6 GO recycled concrete beams were designed and manufactured for flexural behavior test. The effects of GO content, oxygen content and replacement rate of recycled aggregate on the failure mode, cracking load, ultimate bearing capacity and mid-span deflection of the beams were analyzed.

## 2. Experiment

### *2.1. The test beams preparation*

In order to study the influence of GO content, oxygen content of GO and replacement rate of natural aggregate by recycled aggregate, 5 GO recycled concrete beams and 1 concrete beam with natural aggregate were designed according to the principle of orthogonal test. The content of GO is 0, 0.03% and 0.06% of cement weight, the oxygen content of GO is 20% and 40%, and the replacement rate of recycled aggregate is 0, 50% and 100%. The test beams are numbered as NC for natural aggregate concrete, NR for recycled aggregate concrete with 50% substitution rate and RC for recycled aggregate concrete with 100% substitution rate. The percentage of GO content in the concrete at each aggregate replacement rate was 0, 0.03%and 0.06% by mass of cement, and the oxygen content of RN-6-20% of the test beam was 20%, and the oxygen content of the rest was 40%. The clear span of the test beams is 2100mm, and the cross-section size is 150mm×300mm. The longitudinal tensile reinforcement and erection reinforcement are hot rolled ribbed HRB400 steel bars, and the stirrups are HPB300 steel bars. The test beams design parameters are shown in table 1.

**Table 1.** Test beams design parameters.

| test beam | Cement ($kg/m^3$) | Sand ($kg/m^3$) | Rubble ($kg/m^3$) | recycled aggregate ($kg/m^3$) | Water ($kg/m^3$) | GO ($kg/m^3$) |
|---|---|---|---|---|---|---|
| NC-0 | 39.22 | 59.32 | 114.35 | 0 | 0 | 0 |
| NC-6 | 39.22 | 59.32 | 114.35 | 0 | 0 | 0.023532 |
| RN-6-40% | 39.22 | 59.32 | 57.17 | 57.17 | 1.04 | 0.023532 |
| RN-6-20% | 39.22 | 59.32 | 57.17 | 57.17 | 1.04 | 0.023532 |
| RN-3 | 39.22 | 59.32 | 57.17 | 57.17 | 1.04 | 0.011766 |
| RC-6 | 39.22 | 59.32 | 0 | 114.35 | 2.08 | 0.023532 |

### *2.2. Materials*

Ordinary Portland Cement 42.5 grade is used, the mixing water is city tap water, the fine aggregate is natural river sand, and the coarse aggregate is continuous graded natural crushed stone with particle size of 5-20 mm and continuous graded recycled coarse aggregate with particle size of 5-20 mm. The physical properties of recycled aggregate are shown in table 2. GO is scientific research grade GO slurry (mass fraction is 1%), and the basic parameters are shown in Table 3. Longitudinal tensile bars and auxiliary steel bars adopt hot rolled ribbed HRB400 grade bars, and stirrups adopt HPB300 grade bars.

**Table 2.** Physical properties of recycled aggregate.

| Water absorption rate (%) | Apparent density (g/cm3) | Crushing index (%) |
|---|---|---|
| 2.3 | 2.5378 | 16.3 |

**Table 3.** Basic parameters of GO.

| Thickness (nm) | Single layer diameter (μm) | Strippable rate (%) | Oxygen content (%) | Carbon content (%) | Sulfur content (%) |
|---|---|---|---|---|---|
| 11 | 0.2-10.0 | >95 | 40.0 | 42.7 | <2.1 |
| 1 | 0.2-10.0 | >95 | 20.0 | 42.7 | <2.1 |

### *2.3. Dispersed graphene oxide*

Although many oxygen-containing groups in the internal structure of GO can make GO more soluble in water, flocculation will occur when GO is directly added into cement, which will adversely affect its dispersion in cement-based composites and directly affect the strengthening effect of GO on cement-based composites. Sometimes the strength is even lost due to matrix defects due to agglomeration. Therefore, in order to uniformly disperse GO in cement and prepare GO recycled concrete with good performance, some measures need to be taken. Magnetic stirring and ultrasonic dispersion are common methods for dispersing GO.

(1) Magnetic stirring

The magnetic heating stirrer is used to heat and stir the polycarboxylate superplasticizer, GO slurry and water uniformly, with the rotating speed of 1500r/min and the heating and stirring time of 10min.

(2) Ultrasonic dispersion.

Ultrasonic instruments can transform ordinary oscillation into mechanical oscillation with higher frequency. So that ultrasonic waves can be scattered in the solution and generate positive radiation. Due to the flow, a large number of small bubbles with diameters between 50 μm and 500 μm are oscillating, and these bubbles are generated in the negative pressure area and propagate along the ultrasound. In the positive pressure region, if the sound pressure increases to the limit, these bubbles will increase rapidly and then decrease rapidly. In this process, bubbles will have shock waves, so as to remove dirt, disperse and separate solid particles.

In this experiment, the Jietuo ultrasonic cleaner was used to ultrasonically disperse the solution obtained after magnetic stirring. The heating temperature was controlled at 20°C, and the dispersion time was 20 minutes each time. Finally, the dispersed GO dispersion was obtained.

### *2.4. Test scheme*

In order to eliminate the influence of shear stress on the normal section of the test beam, this experiment adopts the loading at three equal points loading, and the load is symmetrically applied to the test beam through the distribution beam, thus forming a pure bending section of 600 mm in the middle of the span, and the two ends of the specimen are simply supported. The hydraulic jack is used to apply the vertical load through the distribution beam. The test is preloaded first, which can be divided into three stages, and the loading value of each stage is 5% of the estimated ultimate load. Uninstall the device after normal debugging, and then perform hierarchical formal loading. The loading experiment adopts the method of loading step by step, and the loading is stopped after the test beam is damaged.

After the maintenance of the test beam is completed, the concrete surface is polished and leveled, and then five resistance strain gauges are pasted along the height direction of the section to measure the strain of the concrete in the pure bending section. Displacement gauges are arranged at both ends of the beam and at the loading point to measure the deformation of the test beam. Displacement meters are arranged at the supports and loading points at both ends of the beam to measure the deformation of the specimen. The arrangement of strain gauges and displacement meters is shown in figure 1. Automatic data acquisition system is used to measure the mid-span deflection and deformation of concrete and reinforcement.

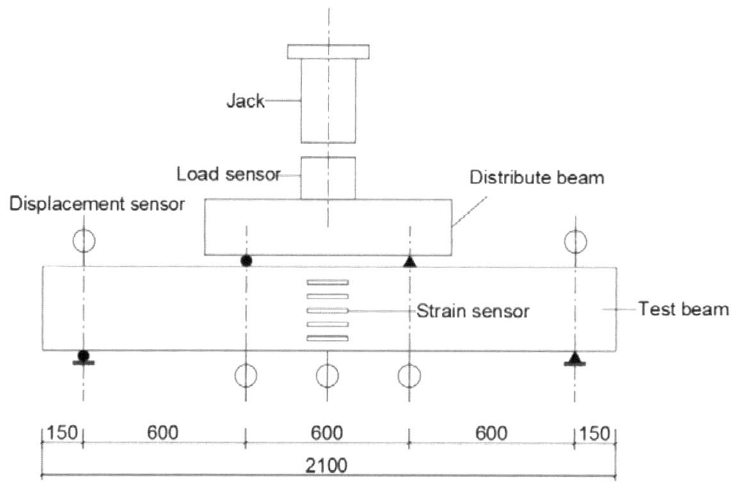


**Figure 1.** Bending test instrument of GO recycled concrete beam.

## 3. Experimental results

### *3.1. Testing phenomena*

6 GO recycled concrete beams were designed and manufactured in this experiment. The cracking load, ultimate load and load when inclined cracks appeared are shown in table 4.

**Table 4.** The cracking load, inclined cracks load and ultimate load test results of beam.

| test beam | cracking load (kN) | inclined cracks load (kN) | ultimate load (kN) |
|---|---|---|---|
| NC-0 | 33 | 120 | 217 |
| NC-6 | 36 | 118 | 220 |
| RN-6-40% | 32 | 120 | 216 |
| RN-6-20% | 29 | 120 | 212 |
| RN-3 | 27 | 105 | 208 |
| RC-6 | 24 | 100 | 198 |

All 6 test beams are crushed and destroyed by concrete in compression zone, and the test phenomena of all test beams are similar. Take the test beam RN-6-40% as an example. During the test, it was found that the test beam cracked when the load was 32 kN, and the crack width was about 0.1 mm. With the continuous loading, many vertical cracks are added in the pure bending section, which continuously develops along the height.

At the same time, the inclined section cracking. With the further increase of load, the mid-span crack stops extending or slowly extends, but the crack width increases sharply. Then the concrete near the loading point appears transverse cracks, and finally the concrete in the middle of the span is crushed and swelled, and the test beam is damaged.

The damage pattern is shown in figure 2.

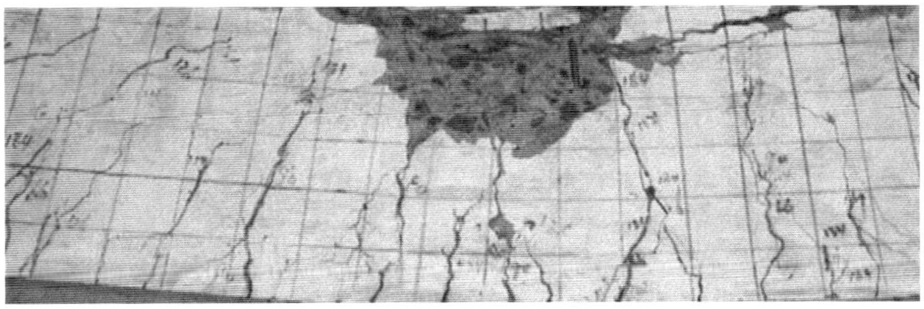

**Figure 2.** Failure modes of the test beam RN-6-40%.

### *3.2. Load-deflection curve*

The load-deflection curve of the specimen at the mid-span is shown in figure 3. Before reaching the yield point, the mid-span deflection of the 6 test beams changes with the load in the process of bending loading, and the mid-span deflection increases linearly in direct proportion to the load. Before the concrete cracks, the deflection increases slowly with the load, at this time, the stiffness of the test beam is large, and the bending moment of the beam is jointly borne by the concrete in the tensile zone and the tensile reinforcement, and the 6 test beams show the same characteristics. After the concrete cracks, the concrete in the tensile zone quits work, the stiffness of the test beam decreases, and the growth rate of mid-span deflection accelerates. With the further increase of load, the deflection of the test beam increases continuously until the beam is destroyed.

Compared with NC-0 and NC-6, the yield load of NC-6 is obviously higher than NC-0, and the ultimate load of NC-6 is increased by 1.38% compared with NC-0. After the yield load, the mid-span deflection of the test beam NC-6 doped with GO increases slowly. Therefore, the test beam doped with GO has better bearing capacity and ductility. Compared with the test beams RN-6-20% and RN-6-40%, the load deflection curves are roughly the same. The yield load of test beam RN-6-40% is slightly higher than that of test beam RN-6-20%, and the ultimate load of test beam RN-6-40% is increased by 1.89% compared with RN-6-20%. Therefore, the test beam with oxygen content of 40% has better bearing capacity. Compared with NC-6, RN-6-40% and RC-6, the yield load of NC-6 is higher than that of RN-6-40% and RC-6, and the ultimate loads of RN-6-40% and RC-6 are reduced by 1.82% and 10% respectively compared with NC-6. After the yield load, the mid-span deflection of RC-6 grows fastest, followed by RN-6-40%, and the mid-span deflection of test beam NC-6 grows slowest. It can be seen that the addition of recycled aggregate will reduce the bearing capacity and ductility of the test beams, and with the increase of aggregate replacement rate, the bearing capacity of the test beam will decrease. Compared with RN-3, the yield load of RN-6-40% is slightly higher than that of RN-3, and the ultimate load of RN-6-40% is 3.85% higher than that of RN-3. After the yield load, the mid-span deflection of the test beam RN-6-40% increases more slowly than that of RN-3. Therefore, with the increase of Go content, the bearing capacity and ductility of the test beam will increase.

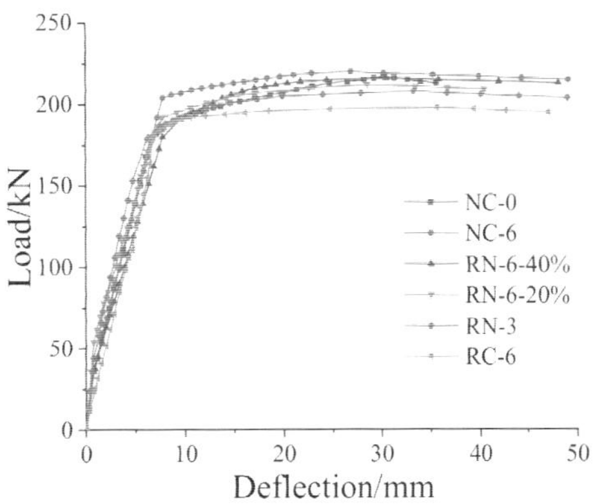


**Figure 3.** Load-mid-span deflection.

### *3.3. Cracking load and ultimate load*

It can be seen from figure 4 and figure 5 that adding 0.06% GO can increase the cracking load of the test beam by 9.09% and the ultimate load by 1.38%. Compared with the test beam with 20% oxygen content, the cracking load of the test beam with 40% GO increased by 10.34%, and the ultimate load increased by 1.89%. The cracking load and ultimate load of the test beams with 50% and 100% recycled aggregates decreased by 11.11% and 33.33% respectively, and decreased by 1.82% and 10% respectively. Compared with the test beam with 0.03% GO content, the cracking load of the test beam with 0.06% GO content is increased by 11.11%, and the ultimate load is increased by 3.85%. Therefore, increasing the content of GO and the oxygen content of GO can effectively improve the bearing capacity of recycled concrete test beams. However, with the increase of recycled aggregate content, the bearing capacity of recycled concrete test beams decreases.

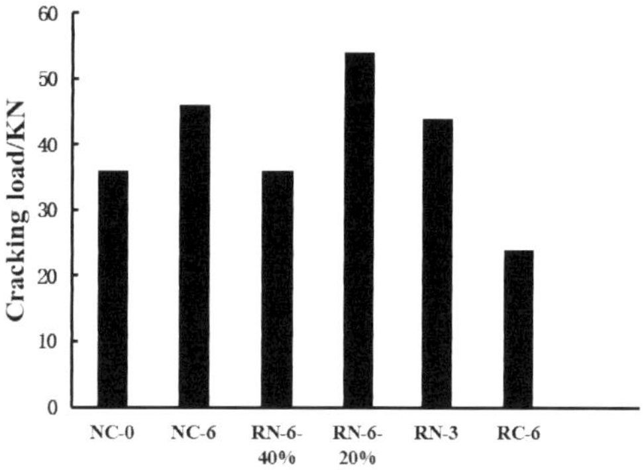


**Figure 4.** Cracking load.

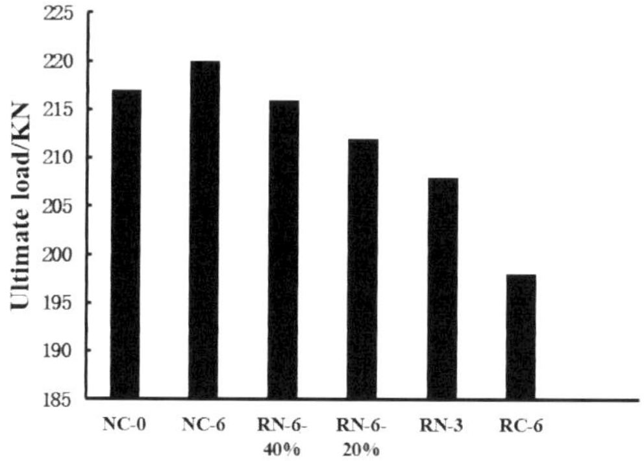


**Figure 5.** Ultimate load.

### *3.4. Reinforcement strain*

The Strain curves of longitudinal reinforcement of test beams under flexural loading is shown in figure 6. In the early stage of test beams loading, the longitudinal reinforcement strain of all test beams increases linearly and proportionally with the increase of load. After cracking, the concrete in the tensile zone gradually quits work with the increase of load, and the steel bars gradually assumed the main load.

The strain variation laws of the longitudinal reinforcement of the 6 test beams were basically the same during this period. As the load continues to increase, when the load increases to the yield stress of the longitudinal reinforcement, the slope of the strain curve of the longitudinal reinforcement obviously increases, and the strain of the reinforcement increases rapidly.

Through the analysis of the test results, it can be seen that the strain variation laws of the longitudinal reinforcement in the 6 test beams are similar in general, which can better reflect the stress and deformation characteristics of the test beams in each stage of the loading process. After the loading reaches the yield stage, the increasing trend of longitudinal reinforcement strain of the test beam NC-6 is lower than that of the test beam NC-0. The RN-6-40% of the test beam was roughly the same as the RN-6-20%, and the increasing trend of the longitudinal reinforcement strain of the test beam NC-6 was more obvious than that of the test beam RN-6-40%. The longitudinal strain of rN-6-40% of the test beam is higher than that of RC-6.

Therefore, the increase of recycled aggregate replacement rate will reduce the adhesion between concrete and steel bars, and the adhesion between concrete and steel bars will also increase with the increase of Go content. These results indicate that the adhesion between concrete and reinforcement decreases with the increase of replacement rate of recycled aggregate, and the adhesion property between concrete and reinforcement increases with the increase of GO content.

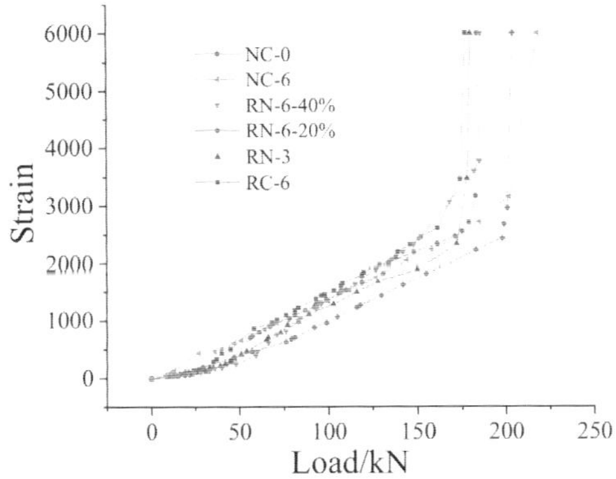


**Figure 6.** Strain-load variation of mid-span reinforcement.

### *3.5. Concrete strain*

Figure 7 shows the distribution of concrete strain along the section height of the test beam under different load levels. Under any specific load, the average strain curve of concrete in the mid-span section of 6 test beams is approximately linear distribution, and the strain at any point on the section is approximately proportional to the distance from this point to the neutral axis. Moreover, the factors such as the replacement rate of recycled aggregate, oxygen content of GO, GO content have no obvious influence on the strain distribution law of the mid-span section of concrete beams, and the strain distribution of the mid-span section in the process of flexural failure basically conforms to the stress plain section assumption.

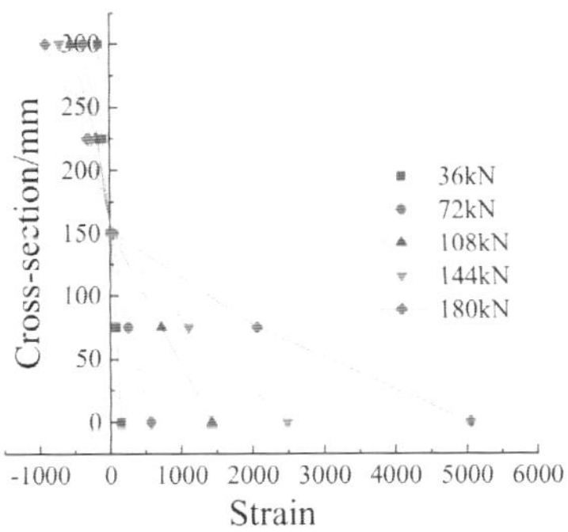


(a) NC-0

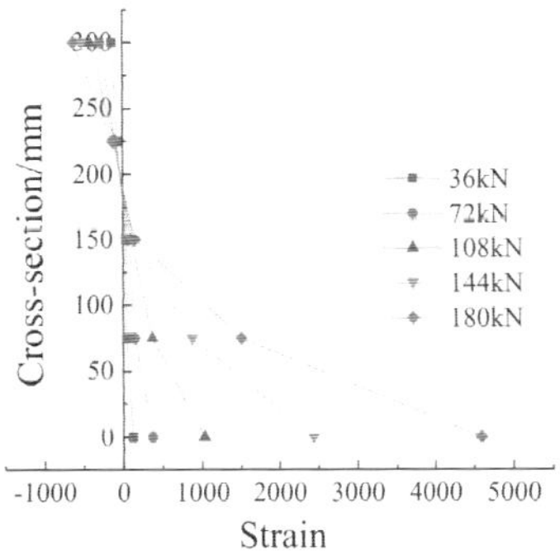


(b) NC-6

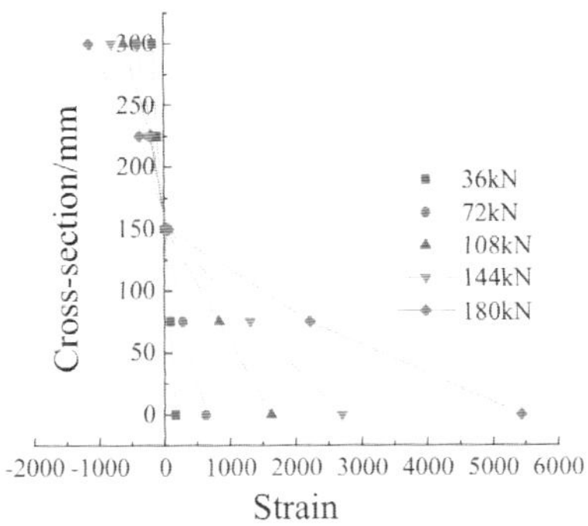


(c) RN-6-40%

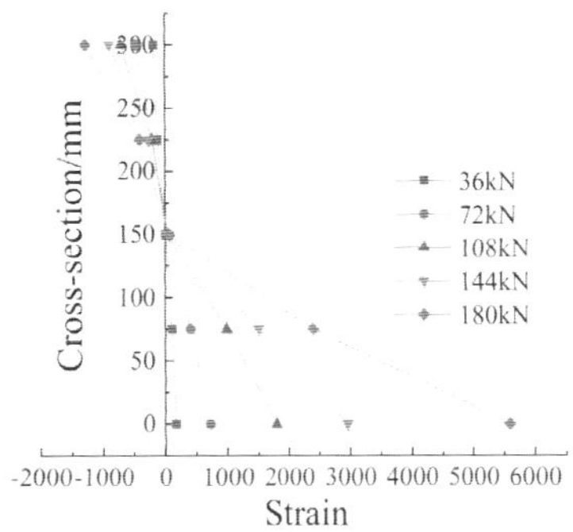


(d) RN-6-20%

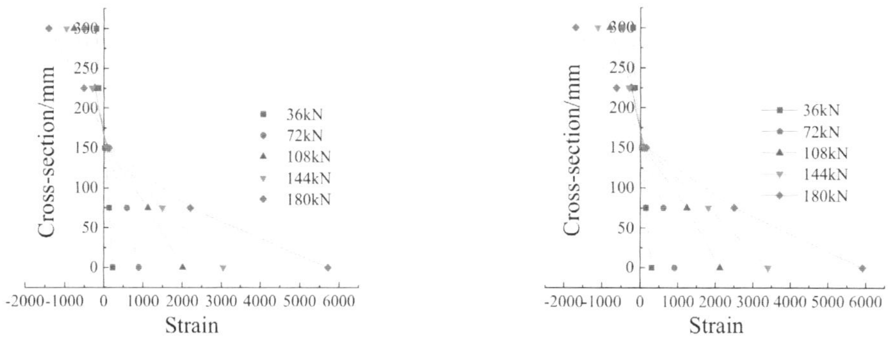


(e) RN-3 (f) RC-6

**Figure 7.** Average strain distribution in mid-span section.

## 4. Conclusion

6 GO recycled concrete beams were designed and manufactured in this experiment. The influence of GO content, oxygen content of GO and recycled concrete content on flexural capacity of recycled concrete beams with GO was studied. The characteristic load, strain, deflection and crack propagation of each test beam are analyzed, and the conclusions are as follows:

(1) The addition of GO can obviously improve the performance of recycled concrete and improve the mechanical properties of recycled concrete. When the content of GO is 0.06%, the cracking load and ultimate load are increased by 9.09% and 1.38% compared with the test beam without GO, and by 11.11% and 3.85% compared with the test beam with 0.03% GO, while the cracking load and ultimate load of the test beam with 40% GO oxygen content are increased by 10.30% compared with the test beam with 20% oxygen content. With the increase of the content of GO, the cracking load and ultimate load of the test beams are increased to some extent. The addition of GO can improve the bearing capacity of recycled concrete test beams, and within limits, with the increase of GO, the bearing capacity of the test beams will be improved. Moreover, with the increase of the content of GO, the adhesion between concrete and reinforcement also increases.

(2) With the increase of recycled aggregate content, the mechanical properties of recycled concrete decrease. Compared with the test beam without recycled aggregate, the cracking load of the test beams with 50% and 100% recycled aggregate replacement rate decreased by 11.11% and 33.33% respectively, and the ultimate load decreased by 1.82% and 10% respectively. The cracking load and ultimate load of the test beam are reduced, so it can be concluded that the increase of recycled aggregate will reduce the bearing capacity of recycled concrete test beam. Moreover, with the increase of recycled aggregate content, the adhesion between concrete and reinforcement decreases.

(3) According to the distribution diagram of concrete strain along the section height of the test beams under different load levels, it can be seen that the replacement rate of recycled aggregate, the oxygen content of GO, the content of Go and other factors have no obvious influence on the strain distribution law of the GO recycled concrete beams, and the strain distribution of the concrete beam basically conforms to the stress plain section assumption in the process of flexural failure.

**Acknowledgment**
This project is sponsored by Scientific Research Project of Liaoning Provincial Department of Education (LJKZ0565).

**References**

[1] Zhou W, Chen J and Lu H 2009 Present situation and countermeasures of construction waste resource recovery Building Technology **40(8)** 741-744
[2] Du T, Li H and Qin Y 2002 Discussion on the future development of recycled concrete Concrete **(4)** 44+49-50

[3] Xiao J, Li W and Fan Y 2012 An overview of study on recycled aggregate concrete in China (1996-2011) Construction & Building Materials **31(6)** 364-383
[4] Xiao J Z, Li J B and Lan Y 2003 Latest development and review of recycled concrete technology Concrete **(10)** 17-20,+57
[5] Chen F, Wu K and Ren L J 2019 Internal curing effect and compressive strength calculation of recycled clay brick aggregate concrete Construction and Building Materials **12(11)** 1815
[6] Guo Y X, Li Q Y and Yue G B 2018 Calculation of compressive strength of recycled concrete considering coarse aggregate quality and substitution rate Journal of Building Structures **39 (4)** 153-15
[7] Huo H Y, Fan C C and Chen A Y 2017 Influence of recycled aggregate with different strength grades on basic mechanical properties of recycled concrete Concrete **(2)** 60-62, p65
[8] Zhou J H, He H J, Meng X H and Yang Y Z 2010 Test of basic mechanical properties of recycled concrete Journal of Shenyang Jianzhu University **26(3)** 467-467
[9] Bai C L 2002 Nanotechnology and its developing prospect Advanced Materials Industry **51(1)** 37-37
[10] Fan J, Xiong G J and LI G Y 2014 Research progress and development trend of carbon nanotube cement-based composites Material Guide A: Review **28(6)** 142-148
[11] Dong J M, Liu C and Long S Z 2012 Effects of nano-Si02 on microstructure and properties of cement-based materials under different dispersion conditions Journal of Building Materials **15(4)** 490-493
[12] Ahmad H, Fan M and Hui D 2018 Graphene oxide incorporated functional materials: a review Composites Part B: Engineering **145** 270-280
[13] De Silva K K H, Huang H H and Joshi R K 2017 Chemical reduction of graphene oxide using green reductants Carbon **119** 190-199
[14] Reddy Indukuri C S, Nerella R and Chand Madduru S R 2019 Effect of graphene oxide on microstructure and strengthened properties of fly ash and silica fume based cement composites Construction and Building Materials 229
[15] Xu G, Du S and He J L 2019 The role of admixed graphene oxide in a cement hydration system Carbon **148** 141-150
[16] Birenboim M, Nadiv R and Alatawna A 2019 Reinforcement and workability aspects of graphene-oxide-reinforced cement nanocomposites Composites Part B: Engineering **161** 68-76
[17] Lv S H, Sun T and Ma Y J 2013 Control and toughening effect of nanometer graphene oxide on hydration crystal structure of cement composites Concrete **(11)** 1-6
[18] Wang Q, Wang J and Lv C X 2015 Effect of graphene oxide on microstructure and mechanical properties of cement-based composites New Carbon Materials **30(4)** 349-356
[19] Cao M L, Zhang H X and Zhang C 2015 The effect of graphene on mechanical properties and microstructure of cement paste Journal of Harbin Institute of Technology **47(12)** 26-30
[20] Li X Y, Lu Z Y and Chuah S 2017 Effects of graphene oxide aggregates on hydration degree, sorptivity, and tensile splitting strength of cement paste Composites Part A: Applied Science and Manufacturing **100** 1-8
[21] Li X and Luo S R 2021 Fracture properties of graphene oxide reinforced cement composites Journal of Composite Materials **38(2)** 618-620

# Numerical case studies about two-dimensional SHS joints with symmetrical and asymmetrical top-connection

**P Heinemann[1*], D N Isopescu[1] and S G Maxineasa[1]**

[1] Gheorghe Asachi Technical University of Iași, Faculty of Civil Engineering and Building Services, Department of Civil and Industrial Engineering, Iași, Romania

* patrick.heinemann@tuiasi.ro

**Abstract.** Hollow section steel profiles are widely used for industrial purpose. Next to circular hollow section, which are set to for example offshore constructions, squared hollow section profiles have many advantages. The profiles are lightweight, have got a high rotational stiffness and in contrast to circular profiles, squared profiles can be welded more easily without having multiplanar welding radii. The weakest point of columns with two or more members is the node. Due to the geometrical heterogeneity and the imperfection due to the process of welding it is a complex part to calculate.
This paper is about a numerical case study of a full-overlapped squared hollow section joint. A focus is the influence of the inclination angle on the maximum stresses and deflection in the ultimate limit state. The analysis is split into symmetrical and asymmetrical structures. The aim is to figure out the influence of the inclination angle on the stability of the joint.

## 1. Introduction

Hollow section steel profiles are widely used for industrial purpose. One of the most common steel profiles is the squared hollow section profile. The advantages are lightweight, a high rotational stiffness and in contrast to circular profiles, squared profiles can be welded more easily because there are no multiplanar welding radii. The weakest part for vertical columns with hollow section profiles is the node. Due to the geometrical heterogeneity and the process of welding the calculation of the node is the most complex part. Much research was done to that topic. The current standards, like the Eurocode 3 [1] or CIDECT [2] have got limitations regarding the geometry. Inclination angles smaller than 30° are offset the standard [3].

The following paper is about two numerical case studies of a uniplanar full-overlapped SHS steel joint on one hand with symmetrical and on the other hand with asymmetrical brace connection. In studies about multiplanar nodes (Heinemann et al. [4]) complex three-dimensional stress distribution in the welding line was analysed. There is significant difference between the ultimate load values in uniplanar KT-joints and corresponding values in two-planar connections [5]. In uniplanar systems, like the joint in this paper or in studies with different profile types (Heinemann et al. [6], Heinemann et al. [7]) the stress distribution is in a symmetrical way and has got the three main types of stresses: Nominal, geometrical and local stress (Saini [8]). The column is part of a roof construction. The load scenario of a roof construction are mainly constant loads, like dead load or snow. The load case of the numerical system is generated as compression load for both braces. A focus is given on the von-Mises stresses and deformation of these joints in the ultimate limit state. Aim of this study is to figure out the influence of the inclination angle and symmetry on the resistance of the joint.

## 2. Numerical Model

The following paragraph is about the description of the numerical study of uniplanar steel Y-joints with top-connection. The aim is to figure out the maximum load capacity of the welding line and of the joint itself. For the numerical studies the software ANSYS [9] is used.

The studies of 6 models can be split into two parts. First there are the results of the squared-hollow-section profiles with a symmetrical geometry. The inclination angle is set to 30°, 45° and 60°. The second part is about SHS profiles with an asymmetrical shape. The inclination angles are set to 30° 45°, 30°-60° and 45°-60°. The joints are completely overlapped with a negative gap. Because of the different inclination angle, it is not realistic to fix the length of the gap. There are also mean variations when the edge distance is fixed. However, the method the get the most precise comparable results is to fix the inner distance of the height. Figure 1 visualises the distance z, the negative gap g and the edge distance. The main geometric parameters are important and the most influencing part of the resistance of a joint [10]. The parameters are defined for the symmetrical cases in table 1. All geometrical parameters were listed by Heinemann et al. [11].

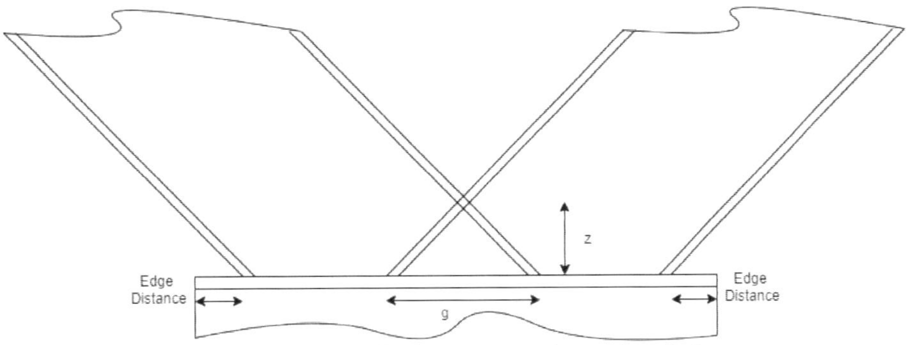


**Figure 1.** Geometrical Parameters in sectional plane

A special focus is given on the deflections. The load case is constant for all models. The compression force affects the bottom of the chord, while the top of the braces is set as pinned. The force is iterated to the maximum resistant stress of 235 N/mm². So, the analysis is in the linear behaviour of the material. By increasing the load, cracks can arise.

**Table 1.** Geometrical parameters.

| Angle | Gap (g) [mm] | Edge Distance [mm] | Height (z) [mm] |
|---|---|---|---|
| 30° | 89.9 | 6.4 | 29.3 |
| 45° | 58.0 | 20.4 | 29.0 |
| 60° | 20.78 | 15.0 | 28.7 |

Due to the complexity of crack propagation, which was analysed in different papers (Heinemann et al. [12], Atteya [13], Almukhtar [14]), this paper is about the linear elastic behaviour. Fatigue strength, like it was analysed by Remes [15] is not focused in this paper. This load case fits better to the most common laboratory testing machines. In case of roof structures, the load case of top compression is more realistic.

The chord is a SHS 120 x 120 x 3 mm and the braces have a cross section of 50 x 50 x 2.5 mm. Both profiles are defined in the standard EN 10219-2 [16]. Rammohan [17] concluded, that the diameter of the central cross section around the weld increases due to buckling regardless of the thickness of the hollow cylinder in case of CHS double V-butt joints. Because of the large contact area

in the model of 30°-60°, the parameter z cannot be constant because of geometrical reasons. The horizontal shifting in the dimension of a few millimetres is neglected. The length of the chord is set to 350 mm, the length of the braces is 150 mm. For a better connection between the braces and the chord, a 5 mm top plate is located at the top of the chord. It is necessary to get a base for the welding line. The welding line is a 3 mm fillet weld, which connects the braces on one hand among each other and on the other hand to the chord. Saini [8] explained, that the weld geometry does not need to be modelled when there is no high local bending. In this study the welding line is modelled, because of the small stress peaks, which arises at the corners of the weld. The material is given with the S 235 steel, which is one of the most common steels in Europe. Heinemann et.al [18] and Younise et al. [19] did numerical studies about the influence of materials regarding joints. It was figured out, that the difference between steel types is small compared to non-steel types.
Due to the same diameters of the braces, it is not possible to weld the side flanges. Because of that a chamfer of 2 mm is set to the ends of the braces. To connect the three members, a 3 mm fillet weld including a triangular cross section is modelled (figure 2).

## 3. Case studies: The welded joint of a three-pipe system with squared cross section

### *3.1. Symmetrical model*

#### *3.1.1. Load scenario*

The first analysis is about a full-overlapped two dimensional SHS joint with a symmetrical shape. The inclination angles are set to 30°, 45° and 60°. Figure 2 visualises the general model, in this case 45° and the load case. Much research was done to the different load cases. Islamovic [20] did studies about joints with bending moment.

The load scenario and the boundary conditions are visualised in figure 2. An axial compression force affects the bottom of the chord. The value of the forces is iterated to the yield stress of the material. Table 2 shows the maximum compression forces, which affects the chord, depending on the inclination angle. The braces ends are set as pinned in axial direction. Horizontal deflections are blocked.

**Table 2.** Applied forces – symmetrical model.

| Inclination Angle [°] | Iterated Force [kN] |
|---|---|
| 30 | 10.45 |
| 45 | 10.86 |
| 60 | 9.27 |

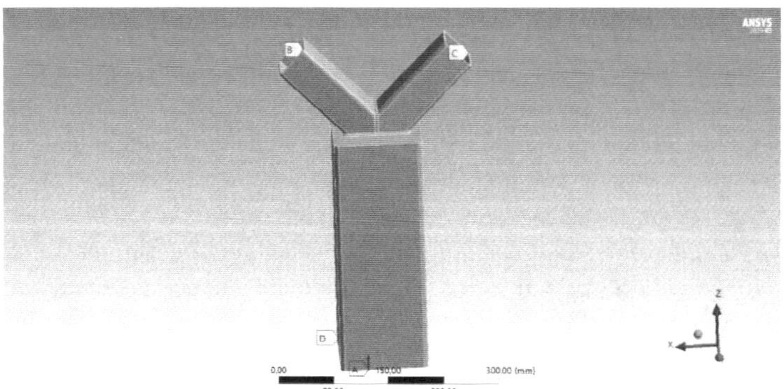


**Figure 2.** System and Load case SHS.

*3.1.2. Mesh*

The mesh with the three-dimensional solid elements can be seen in figure 3.

**Figure 3.** Mesh SHS symmetrical.

The mesh is given with 10 mm tetrahedral solid elements with quadratic element shape function. The refinement in the welding line area is 1.3 mm. The connection between the braces and the chord is set as frictionless. So, the force can only be transferred through the welding line.

The shape of the welding line can be seen in this picture, too.

*3.1.3. Inclination angle influence*

The distribution of the von-Mises stresses on the surface of the braces and welds is printed in figure 4. The maximum stresses arise in the edges of the welding line, due to the geometrical heterogeneity. Figure 5 visualises a sectional plane. There the maximum stresses are printed through the cross section of the profile.

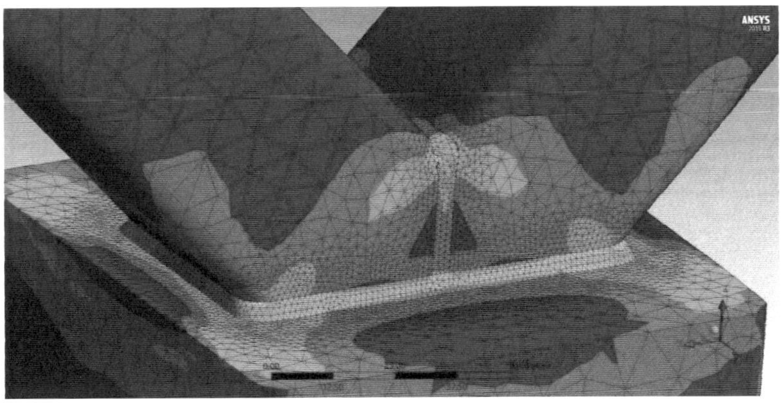


**Figure 4.** Von-Mises Stress.

The load is iterated in the way, the maximum stress is equal to the yield stress. The deflection of the 30°, 45° and 60° model is visualised in figure 6-8. In case of the braces and welding the maximum deflections arise in the 30° model.

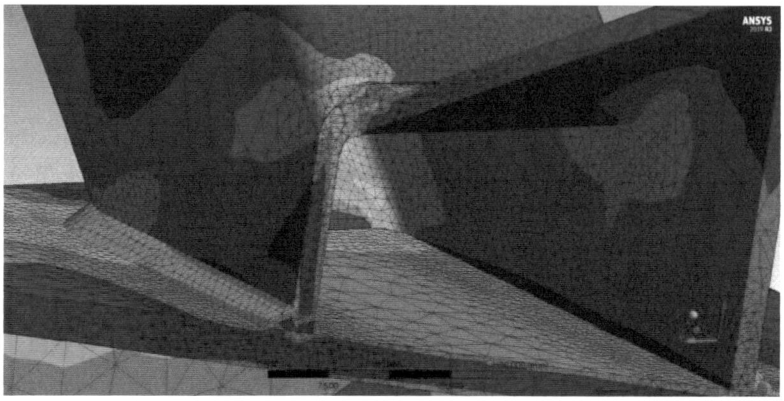


**Figure 5.** Von-Mises Stress detailed.

**Figure 6.** Deflection 30°.

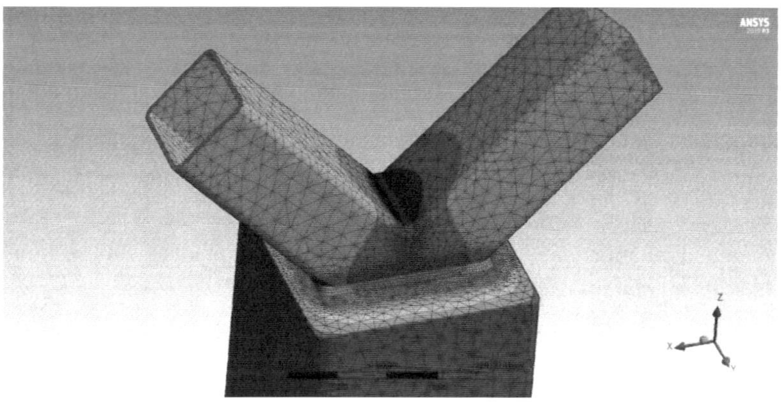


**Figure 7.** Deflection 45°.

**Figure 8.** Deflection 60°.

The maximum compression force in the ultimate limit state for the symmetrical SHS joint is printed in figure 9, the results of the deflection is shown in figure 10. While the 30° and 45° models have an increasing behaviour for both parameters, the 60° model has got a low maximum compression force and deflection respectively.

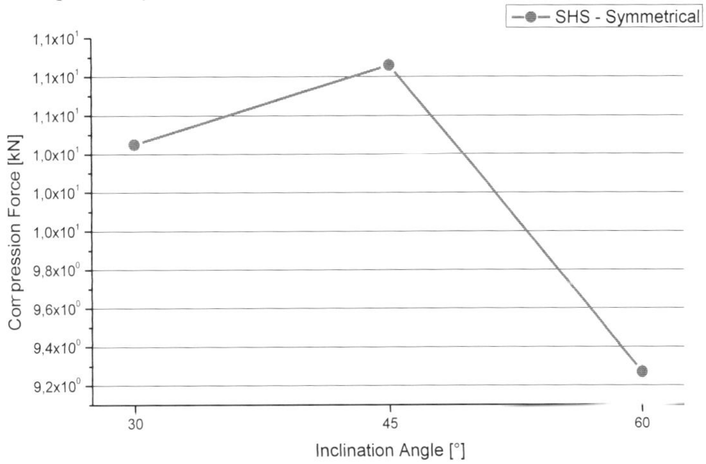


**Figure 9.** Compression Force SHS symmetrical.

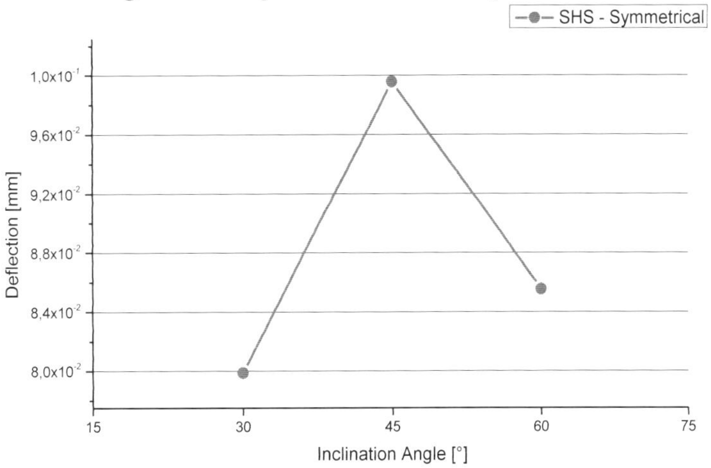


**Figure 10.** Deflection SHS symmetrical.

### *3.2. Asymmetrical Model*

#### *3.2.1. Load scenario and mesh*

The second study is about steel SHS joints with an asymmetrical shape. The inclination angles are 30°-45°, 30°-60° and 45°-60°. Due to the different contact area for these braces, the edge distance is not symmetrical (Fig. 11). Also, this figure shows the general system and the load scenario.

The pinned boundary conditions together with the axial displacement is dependent on the inclination angle. The load case, mesh and analysation process are equal to paragraph 3. Table 3 shows the iterated maximum axial compression forces, which are applied to the chord.

**Table 3.** Applied forces – asymmetrical model.

| Inclination Angle [°] | Iterated Force [kN] |
|---|---|
| 30 | 9.44 |
| 45 | 9.52 |
| 60 | 9.99 |

#### *3.2.2. Inclination angle influence*

The distribution of the von-Mises stresses can be seen in figure 12. The results are comparable to the results of the symmetrical SHS model.

The maximum stresses arise in the edges of the welding line of the brace-to-chord and the brace-to-brace connection. The maximum deflections are printed in figure 13-15. The highest deflections for the braces are at the end of the 30° brace. Overall, the chord has got the highest deflection.

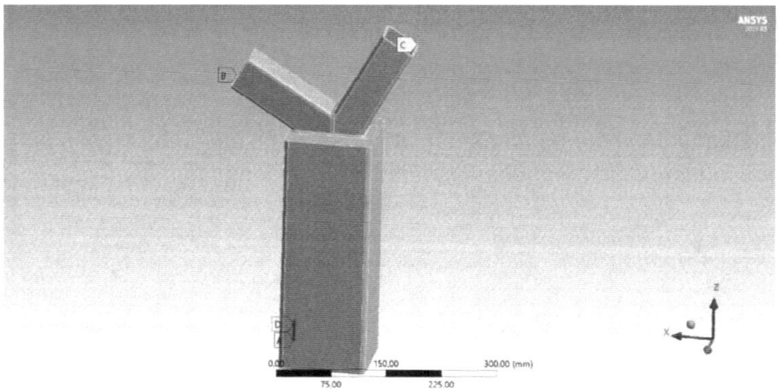


**Figure 11.** Model and Load Case.

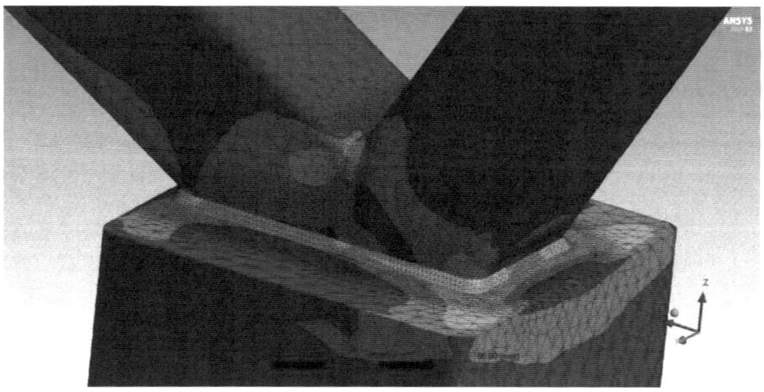


**Figure 12.** Von-Mises stress.

**Figure 13.** Deformation 30°-45°.

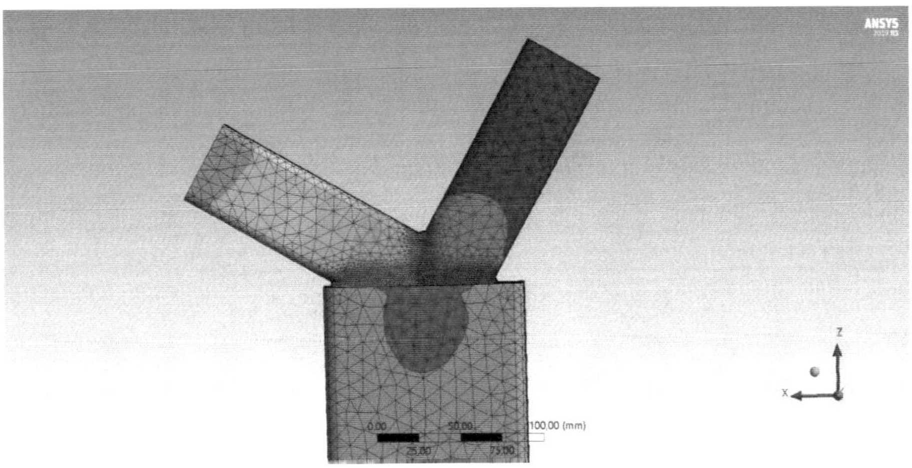


**Figure 14.** Deformation 30°-60°.

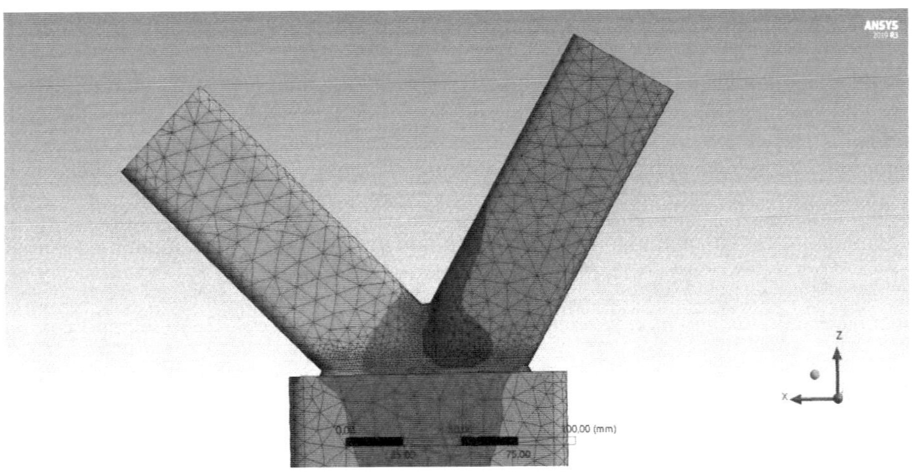


**Figure 15.** Deformation 45°-60°.

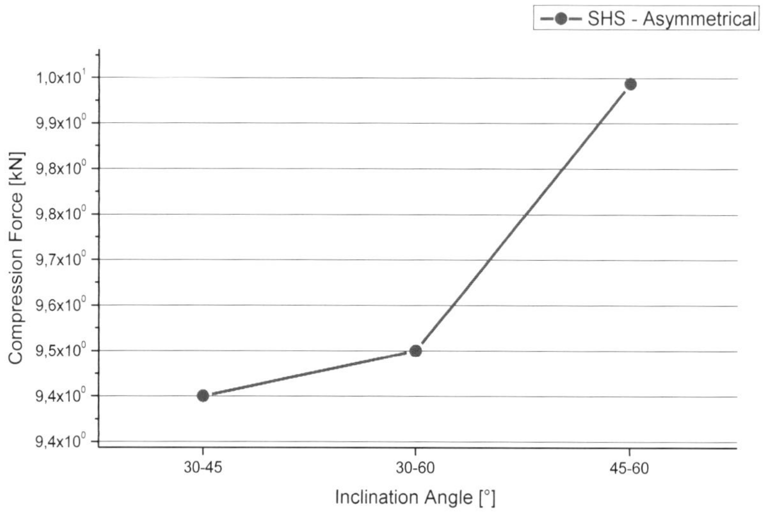


**Figure 16.** Max. Compression Force SHS Asymmetrical.

Figure 16 shows the results for the maximum compression force in the ultimate limit state for the SHS joint with asymmetrical shape. The shape of the graph has got an increasing trend for plain to steeper inclination angle. So, when there is a joint with a steeper inclination angle, a higher force can affect the bottom of the chord.

The deflection is visualised in figure 17. The trend is increasing, too. The shape is similar to the compression force. The 30°-45° model has got the smallest deflection and compression force, while the 45°-60° model has got the highest compression force and deflection. But the range of the results are close together with a delta factor of 1.06 and 1.18 respectively.

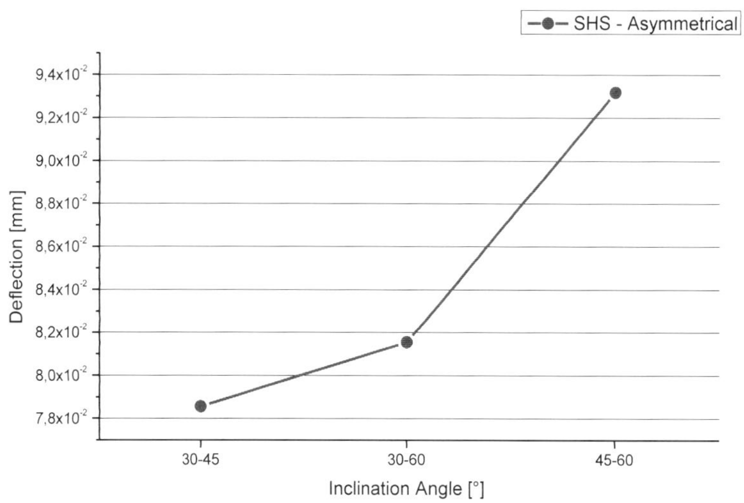


**Figure 17.** Deflection SHS Asymmetrical

## 4. Conclusion

In this paper two case studies of symmetrical and asymmetrical full overlapped squared hollow section joints were presented. The distributions of the von-Mises stresses and deflections are similar to both systems. The results of the maximum compression force and deflection have got nearly the same range, but the shape of the graphs is partly different. Especially the results for the 60° symmetrical model are not fitting to the trend of the asymmetrical model. In this case the pair of the steepest inclination angle has got the highest joint resistance. In future studies more inclination angles and different diameters of the steel profiles can be analysed to verify the shape of the results.

## References

[1] DIN EN 1993-1-8, Eurocode 3 2010-12 Bemessung und Konstruktion von Stahlbauten – Bemessung von Anschlüssen. *Deutsches Institut für Normung*

[2] Wardenier J, Kurobane Y, Packer J A, Dutta D and Yeomans N 1991 Design guide for circular hollow section (CHS) connections under predominantly static loading, *CIDECT series 'Construction with hollow sections'* Serial No. **1** TÜV-Verlag, Köln, Germany

[3] Jurcikova A and Rosmanit M 2012 FEM model of joint consisting RHS and HEA profiles *Procedia Engineering* No. **40** pp. 183-188

[4] Heinemann P, Isopescu D N and Maxineasa S G 2021 Numerical case study about three-dimensional CHS joints with overlapped top-Connection [manuscript submitted for publication] *Universitatea Technica de constructii Bucuresti*, Faculty Construction and Installation, TU Iasi

[5] Azari-Dodaran N and Ahmadi H 2019 Static behaviour of offshore two-planar tubular KT-joints under axial loading at fire-induced elevated temperatures, *Journal of Ocean Engineering and Science*, S2468-0133(No. **19**)30079-8

[6] Heinemann P, Isopescu D N and Maxineasa S G 2021 Numerical case studies about two-dimensional CHS joints with symmetrical full-overlapped top connection [Unpublished manuscript], Faculty Construction and Installation, TU Iasi

[7] Heinemann P, Isopescu D N and Maxineasa S G 2021 Numerical case studies about two-dimensional CHS joints with asymmetrical full-overlapped top connection [manuscript submitted for publication] *Materials today: Proceedings*, Faculty Construction and Installation, TU Iasi

[8] Saini D S, Karmakar D and Ray-Chaudhuri S 2016 A review of stress concentration factors in tubular and non-tubular joints for design of offshore installations, *Journal of Ocean Engineering and Science* No. **1** pp. 186-202

[9] ANSYS Inc. Products 2019 R3 © 2006-2019

[10] Espinosa J M, Schafhirt S and Muskulus M 2017 Mesh Parametric Study for Fatigue Assessment of tubular K-joints using Numerical Methods, *Energy Procedia* No. **137** pp. 514-522

[11] Heinemann P, Isopescu D N and Maxineasa S G [in press] 2021 Case Studies on Finite Element Modelling of Welded Joints *Buletinul Institutului Politehnic Din Iasi*

[12] Heinemann P, Isopescu D N and Maxineasa S G 2020 Study on the Modelling of crack propagation in the joints of tubular steel elements *Buletinul Institutului Politehnic Din Iasi* Volume **66** (70) No. 2 Constructii Architectura

[13] Atteya M, Mikkelsen O, Lemu H G 2019 *State-of-the-art of the crack propagation modelling in tubular joints*. IOP Conf. Series: Materials Science and Engineering **700** (2019) 012035

[14] Almukhtar A M 2011 *Fracture simulation of welded joints*. Nova Science Publishers (**2011**) ISBN 978-1-61470-799-8

[15] Remes H, Gallo P, Jelovica J, Romanoff J, Lehto P 2020 *Fatigue strength modelling of high-performing welded joints*. International Journal of fatigue **135** (2020) 105555

[16] EN 10219-2, 2019 Cold formed welded steel structural hollow sections – Part 2: Tolerances, dimensions and sectional properties. *Deutsches Institut für Normung*

[17] Sampath S S, Rammohan N and Shetty R 2016 Experimental analysis of stresses in weld metal subjected to tensile load, *World Journal of Engineering* **13/4** pp. 281-287

[18] Heinemann P, Isopescu D N and Maxineasa S G 2021 The influence of materials on the behaviour of joints with multiple bar connections, *IOP Conference Series: Materials Science and Engineering* **1138** (2021) 012023

[19] Younise B, Rakin M, Gubeljak N, Medo B Sedmak A 2016 Numerical prediction of ductile

fracture resistance of welded joint zones. *Procedia Structural Integrity* **2** 753-760
[20] Islamovic F, Muratovic P, Gaco D, Kulenovic F 2009 *Bend testing of the welded joints.* 7$^{th}$ International Scientific Conference on Production Engineering

# The stone cantilevers rehabilitation of heritage buildings

**A V Lapuste**[1*] **and B Marton**[1]

[1] Technical University of Cluj Napoca, Faculty of Civil Engineering, Romania

* adina.lapuste@mecon.utcluj.ro

**Abstract**. The aim of this article is to analyse multiple solutions for the stone cantilevers rehabilitation commonly found in heritage buildings. The stone cantilevers are arranged in the inner courtyards of historical buildings and ensure horizontal circulation. Throughout the article will analyse two categories of situations encountered in the rehabilitation practice: for the first situation the cantilever rehabilitation can be done after the stone plate demolition, and for the second situation the cantilever rehabilitation must be done without having the option of demolishing the flat plates. The difficulty of the intervention is given by the fact that the fixed ending restrain of the cantilever was ensured during the execution technology, the cantilever fixed in the masonry being ensured during the building of the wall. The rehabilitation of the stone cantilevers is conditioned by the fact that the fixed ending restrain point cannot be rebuild on the initial principles. This fact requires the adoption of rehabilitation principles based on a different approach to ensure the fixed end point. The proposed solutions solve both issues ensuring the required load capacity of the cantilever and ensuring the required level of fixed end restrain

## 1. Introduction

The aim of this article is to analyses multiple solutions for the stone cantilevers rehabilitation commonly found in heritage buildings [1]. The stone cantilevers are arranged in the inner courtyards of historical buildings and ensure horizontal circulation (figure 1). Throughout the article will analyses two categories of situations encountered in the rehabilitation practice: for the first situation the cantilever rehabilitation can be done after the stone plate demolition, and for the second situation the cantilever rehabilitation must be done without having the option of demolishing the flat plates. The difficulty of the intervention is given by the fact that the fixed ending restrain of the cantilever was ensured during the execution technology, the cantilever fixed in the masonry being provided during the building of the wall. The rehabilitation of the stone cantilevers is conditioned by the fact that the fixed ending restrain point cannot be rebuild on the initial principles [2-5]. This fact requires the adoption of rehabilitation principles based on a different approach to ensure the fixed end point. The proposed solutions solve both issues ensuring the required load capacity of the cantilever and providing the required level of fixed end restrain [6,7].

## 2. Rehabilitation solutions

For the first situation there is no possibility of removing the stone plate and the rehabilitation must be done with the plates in position, it is considered two possible rehabilitation options. The first option (figure 3) is considered two L profiles disposed on one side and the other side of the cantilever stone being fixed one by the other through blots that crosse the stone cantilever holes. The profiles are introduced in the masonry wall passing on the other side of the wall. For this situation it is possible to drill holes in wall to introduce the steel profile and to ensure fixed restrain, similar with the cantilever fixed restrain. There is no collapse risk when drilling the holes in the walls as the holes are made on one side and the other side of the cantilever. The fixed end restrain is ensure outside the wall. When fixing the profiles, to obtain a good contact between the steel profiles and the stone plates will introduce grout or mastic. A perfect contact between them can be ensure only by injecting the grout or mastic [9,10].

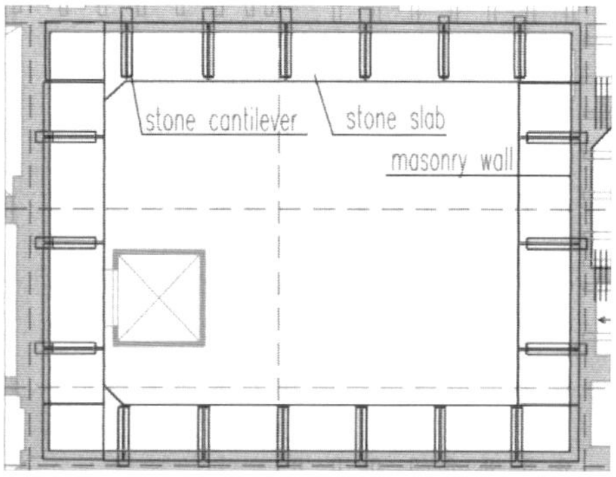


**Figure 1.** Building's plan [8].

**Figure 2.** Building's overview [8].

This solution has been analyzed, and the investigations made, showed that the console is developed differently - it appears wider inside than on the outside. It could have been applied if the two corners had been fixed to the console stone to the other end. In this situation will consider the next 2 solutions analyzed and proposed.

This solution (figure 3) is one that corrects especially the cantilever detachment tendency caused by water infiltrations or by material degradation over the years and improves the fixed end restrain. The cantilever fixed restrain is ensure on the cantilevers width and by introducing the steel profiles it is provide additional contact area.

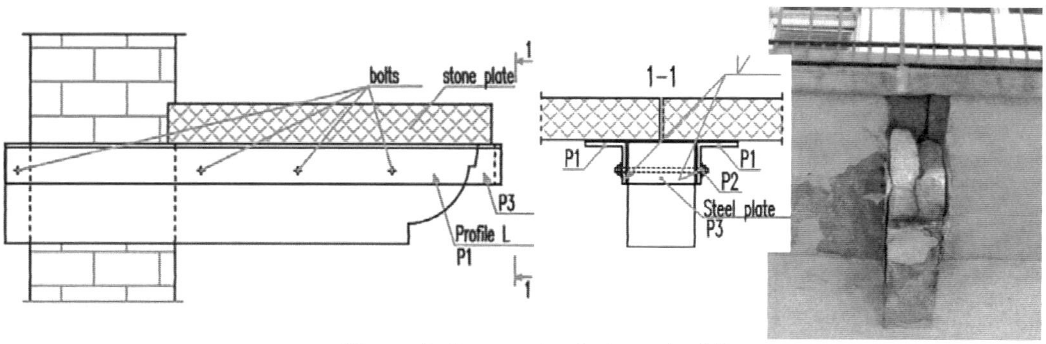


**Figure 3.** Proposed solution - 1.a [8].

Regarding the second option (figure 4), the principles are similar to the first option, the cantilever rehabilitation is provided by two L profiles located on one side and the other side of the cantilever, but in certain situations it is very difficult to drill the wall next to the cantilever as usually the stone cantilever is not carved in the interior of the wall as well as in the exterior of the wall, the seen part of the cantilever. Due to this fact the steel profile cannot be fixed right next to the cantilever to create a single straight element, therefore the first option in no longer valid. In this situation will introduce the profile on a shorter length of the wall, 10-15cm, and the profile must be anchored in the wall. The tension stresses of the cantilever are taken by rod and tube placed on the upper side of the element. The upper profile connects the two degraded areas and resists tension stresses, similar with a rebar disposed on the upper side of the cantilever. This way of reinforcement becomes active by the fact that the profile is linked in the cantilever points, providing the tension stresses transfer; the steel profile is being fixed at one end. For a better behavior that can be controlled, a threaded rod was introduced and anchored on the other side. The tube can adapt easily as can be molded very well on the support conditions, providing a good contact between the steel tube and the masonry wall.

The slides between this reinforcement and the console are taken over by normal flat rods (threaded rods inserted from one profile to the other by drilling the stone).

The malfunctions of the cantilevers were caused by the access of water in the space between two stones of the corridor, which stagnated there permanently and gradually made its place in that space. The quality of the stone in the consoles was weaker, the stone was more porous, with lower tensile strength. By tightening with metal profiles and by injection, the continuity was restored, and the degradation process was further prevented from spreading. Once the stones have been removed, regardless of the chosen option, will ensure proper insulation at the top.

This solution (figure 4) was adopted because the profiles are introduced in the wall only 10-15cm, thus the restrain cannot be considered as a fixed one, in these conditions the profiles provide the tension resistance of the element.

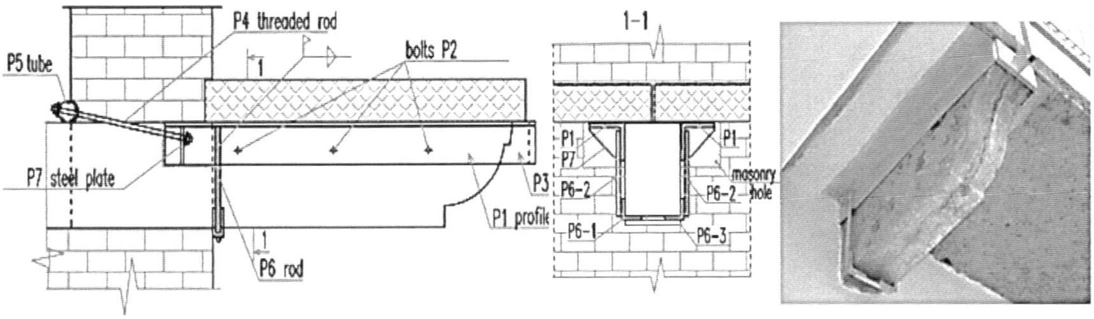


**Figure 4.** Proposed solution - 1.b [8].

For the second situation (figure 5) there is the possibility of removing the stone slab and instead of 2L profiles disposed below the slab, will use one T steel profile disposed on the stone cantilever and subsequent will place the slab. In this situation the challenge is to design the steel profile as being the only resistance element capable of resisting the gravity load provided by the slab. The cantilever could have a higher degree of degradation thus the aim is not to consolidate the cantilever to provide the initial capacity of it but to add a resistance element capable of providing the initial capacity alone. Thus, it is mandatory to ensure a fixed end restrain of the element. In the T profile is a hole in which it is introduced a steel tube anchored in the masonry wall to a certain depth. The threaded rods are fixed in position by making a slit in masonry. The perfect contact between the steel profiles and masonry is provided only by injecting the grout or mastic between elements.

The consolidation solutions were analyzed to integral resist permanent and temporary loads. The aim was to ensure a controllable contact between the steel profiles and the elements with which is in contact. The compression is taken over by the stone material of the cantilever, and the tension is taken by the rod anchored to the wall. The local crushing stone was checked, which, in wall, no longer had the level of degradation that was present in the stone coming out of the console, because the stone in the wall was not subject to moisture and other external factors.

The threaded rod placed at the end was designed to resist tensile force and the pressure on the masonry was verified because it enters and reaches a circular steel profile. Thus, the pressure distribution on the masonry generated by the action of the threaded rod was verified.

The cantilever is fixed in the wall, the bending moment being taken by a couple formed by the compressed area - the T-shaped metal profile at the end of the stone console (local compression verification) and the threaded rod that prevents the lifting of the cantilever, and which is anchored in the masonry, with the solution presented and verified by calculation [8-10].

Technologically - a hole is made in the wall just above the beam. The hole can be made because there are no actions on the cantilever. The hole is made square, up to the other side, with a height equal to the height of the T profile plus 3-4 cm, and the width is equal to the width of the console.

The matting is made at the top, because at the bottom the console is well supported by masonry. At the top, a continuous support is provided by using a cement mortar or a suitable mastic placed before or by injection-after. The same operation is performed on the part that is embedded in the wall on the inside and which is responsible for taking over the snatching force. The piece is U-shaped, and when it is in position, a picture mortar is placed on top of it. It is placed in position next to the rods, and when they are tightened, the excess mortar pushes back, and the system starts working.

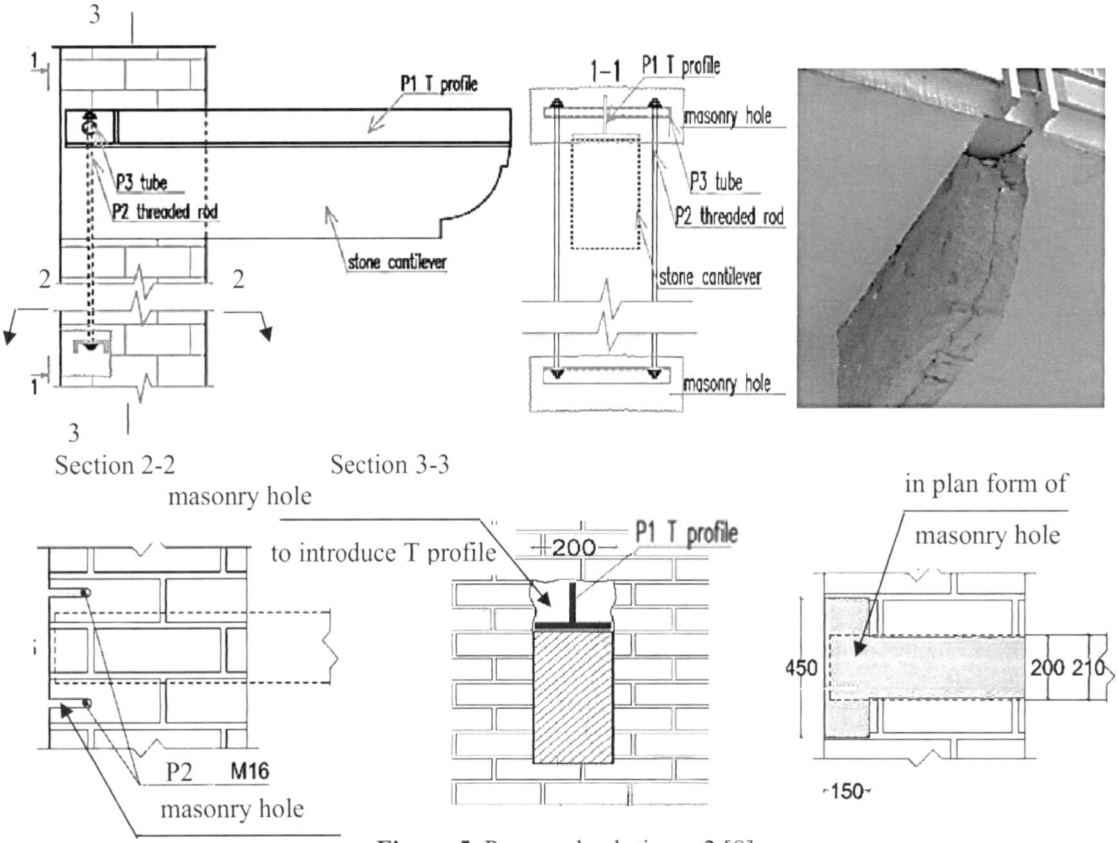


**Figure 5.** Proposed solution – 2 [8].

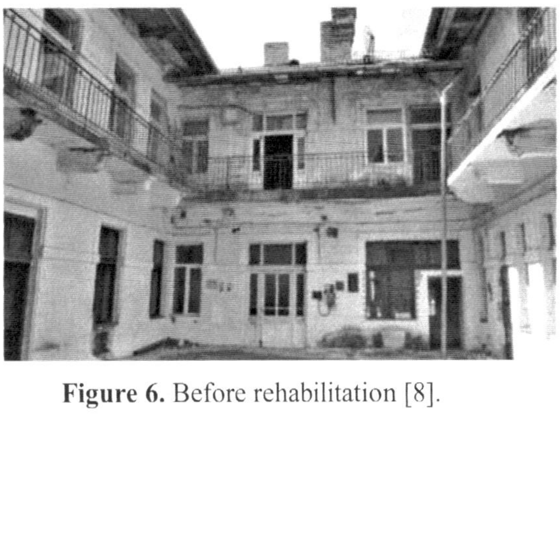

**Figure 6.** Before rehabilitation [8].

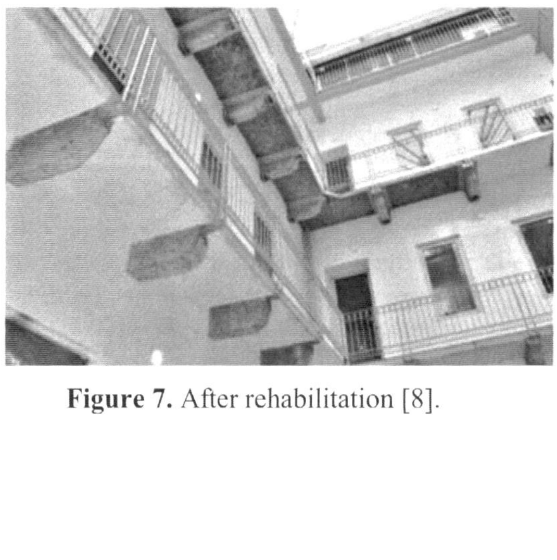

**Figure 7.** After rehabilitation [8].

## 3. Conclusions

The goal of this research was to give answers to a current issue that frequently appears in rehabilitation practice. A current issue that generates several complications when involves heritage buildings [11,12].

The analyze was to solve the stresses transfer from the stone cantilever to the new resistance elements: first it is mandatory to transfer the stresses from the cantilever to the new elements and second to correct the cantilever degradations over the years.

The steel profiles will correct and will take over the functioning deficiencies. Thus, the new elements must have a presence as discreet as possible and must be as much as possible in accordance with the aesthetics of the building. The priority is to have a solution that is efficient with minor modifications of the buildings image. The research gives answers to all three situations depending on the challenge characteristics. From all three solutions, the third solution (figure 5) after the rehabilitation and relocation of the circulation area, the intervention cannot be seen, is a more technical solution with a more desirable effect. For heritage buildings it is desirable that the rehabilitation interventions will not significantly change the buildings image. The first step when choosing the best solution is to establish whether the slabs can or cannot be removed during intervention. If this option is possible, that the third solution will be chosen. If not, will choose one of the first two solutions depending on the occurred situation. Most likely, the second option will be applied, which also solves the issue of stone uniformity for the part fixed in the masonry wall, where the stone is not sufficiently carved. The first option is flawless if the stone cantilever is perfectly carved along its entire length [8].

## References

[1] Lourenço P B 2013 *Conservation of Cultural Heritage Buildings: Methodology and Application to Case Studies* (Rev ALCONPAT) p 101

[2] Šekularac N Risti´c N.D Mijovi´c D Cvetkovi´c V Bariši´c S Ivanovi´c-Šekularac J 2019 *The Use of Natural Stone as an Authentic Building Material for the Restoration of Historic Buildings in Order to Test Sustainable Refurbishment: Case Study Sustainability* (Natural Stone and Architectural Heritage) p 11

[3] B Fitzner K Heinrichs & D La Bouchardiere 2000 *Damage index for stone monuments* (Proceedings of the 5th International Symposium on the Conservation of Monuments in the Mediterranean Basin) p 70

[4] B Fitzner & K Heinrichs 2002 *Damage diagnosis on stone monuments – weathering forms damage categories and damage indices Understanding and managing stone decay* (Proc Int Conf on stone weathering and atmospheric pollution network) p 26

[5] Bernd Fitzner 2004 *Documentation and Evaluation of Stone Damage on Monuments* (10th International Congress on Deterioration and Conservation Of Stone Stockholm) p 677

[6] Malcata M Ponte M Tiberti S Bento R Milani G Failure 2020 *Analysis of a Portuguese Cultural Heritage Masterpiece: Bonet Building in Sintra* (Engineering Failure AnalysisVolume 115)

[7] Giordano E Ferrante A Clementi F Milani G Formisano A 2019 *Cultural Heritage and Earthquake: The Case Study of San Francesco's Church in Amandola* (Proceedings of the IMEKO TC4 International Conference on Metrology for Archaeology and Cultural Heritage MetroArchaeo Florence Italy) p 37

[8] Uctra Expert SRL 2016 Consolidation design of the stone cantilever Heritage Building *Piata Unirii nr13 Timisoara*

[9] Eurocod 3 Steel Structures Design

[10] Eurocod 6 Masonry Structures Design

[11] Günçe K 2016 *Adaptive reuse strategies for heritage buildings: A holistic approach* (Sustainable Cities and Society Volume 26) p 91

[12] Mija Milić Mislav Stepinac 2021 *Assessment and Rehabilitation of Culturally Protected Prince Rudolf Infantry Barracks in Zagreb after Major Earthquake* (Assessment Reconstruction and Decision Procedures for the Preservation of Existing Structures after Earthquakes) p 508

# Study on quality detection of stone surface

**W Liu[1]*, M Zhao[1,2] and H Shang[1]**

[1] College of Arts and Information Engineering, Dalian Polytechnic University, Dalian City, China
[2] Shenyang Jianzhu University, Shenyang City, China

* lwb18040109652@163.com

**Abstract.** With the continuous improvement for the safety of stone installation, the quality detection of its machined surface is also improving. But its surface crack was always a difficult problem in detection. In recent years, with the rapid development of image processing technology a new method to solve this problem can be realized. In this paper, CCD camera was used to detect the surface crack and roughness of stone slabs. The research about image correction and image enhancement was carried, so as to eliminate noise and protect the crack edge in the greatest extent. The improved threshold segmentation algorithm was used to segment image of linear fracture filtering. Mathematical morphology was applied to refinement of single pixel of crack in order to gain crack size. The CCD camera and micro lens are assembled as image sensors to enlarge the surface of stone and take a sample by many taking pictures. The surface profile of the tested stone slab is obtained by reconstruction method, and the surface roughness of stone slab is finally obtained by analysis and calculation. Through this research, a fast detection technology was realized, which solves the problems of low processing efficiency and high cost in production. The result shows certain theoretical value and practical application significance.

## 1. Introduction

In recent years, the development trend of stone processing equipment is to develop in the direction of high efficiency, high precision, high automation and low consumption. However, at present, the stone industry is still in the manual detection stage in terms of product detection, resulting in very low stone production rate, processing accuracy and installation safety. If automatic detection is adopted, the efficiency can be greatly improved and the processing cost can be reduced at the same time [1]. Based on digital image processing technology, this paper uses CCD camera to detect the surface crack and roughness of stone slabs. The digital image of stone slabs is obtained by CCD camera. According to the characteristics of crack image, the image correction and image enhancement are studied [2]. Then the linear crack filter image is segmented by the improved threshold segmentation method. Finally, the crack is refined by single pixel using mathematical morphology to obtain the crack size [3]. The CCD camera and the magnifying lens are assembled as an image sensor to take a sequence of microscopic images, extract the clear areas in the image and fuse them into a full focus image, then the depth information of different areas of the image is obtained by the focus measure operator, the surface profile of the tested stone slab is obtained by reconstruction method, and the surface roughness of stone slab is finally obtained by analysis and calculation [4-6]. A fast detection technology is realized so as to solve the problems of low processing efficiency, time consuming and laborious in production engineering.

## 2. Experimental Platform

The detection system shown in figure 1 is mainly used to detect the surface cracks of large slabs [7-9]. The micro detection system is mainly used to detect the surface roughness of large stone slabs. The measurement principle is shown in figure 2 [10,11]. In this experiment the measurement of stone surface roughness is to obtain through the surface roughness measuring instrument, as shown in figure 3 [12,13].

**Figure 1.** Crack detection experimental device.

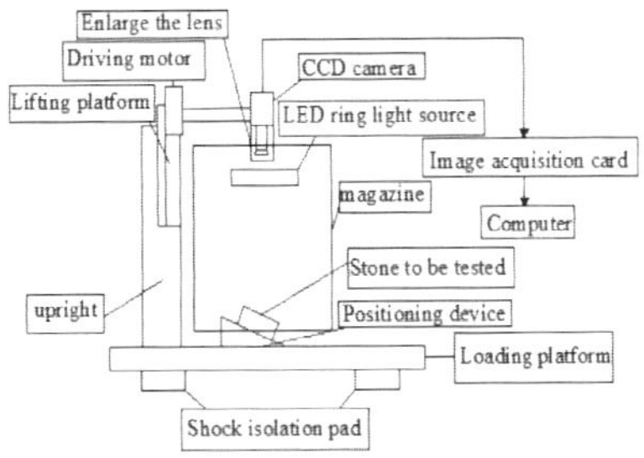


**Figure 2.** Roughness detection and measurement principle.

**Figure 3.** Roughness profiler.

## 3. Defect detection

According to the characteristics of crack image on the surface of large stone slabs, the flow chart of crack recognition algorithm adopted in this paper is shown in figure 4.

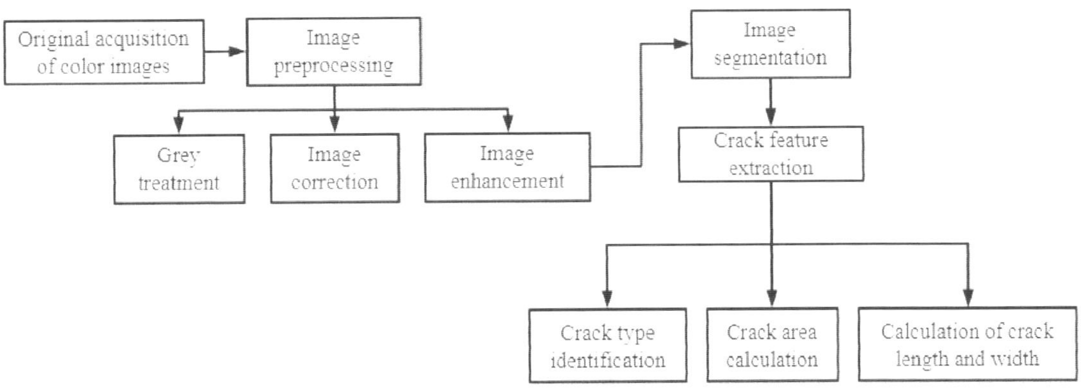


**Figure 4.** Flow chart of crack identification algorithm.

### *3.1. Image preprocessing*

The preprocessing of crack image is divided into three parts: crack image graying, crack image correction and crack image enhancement. The purpose of image processing is to obtain high-quality images, and wavelet transform can improve the shortcomings of classical interpolation algorithm. Therefore, this paper uses the improved interpolation algorithm to correct the fracture image by combining wavelet transform with traditional interpolation algorithm [14]. The bilinear interpolation algorithm which can make the interpolated image smoother and the algorithm complexity is also lower is selected to be combined with wavelet transform, so as to obtain the required background image by interpolation and to correct the original fracture image. The flow chart of the improved interpolation algorithm is shown in figure 5.

The comparison of MATLAB simulation results between the improved correction algorithm and the correction algorithm based on bilinear interpolation is shown in figure 6.

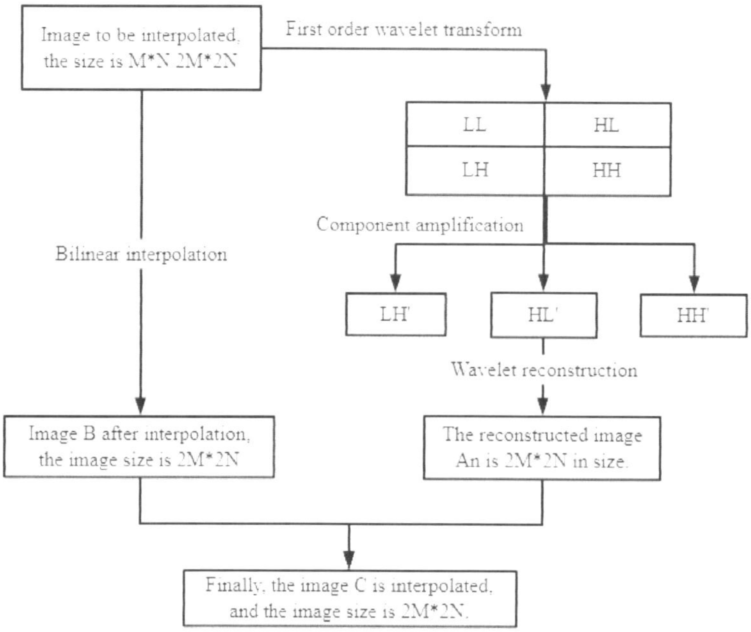


**Figure 5.** Flow chart of improved interpolation algorithm.

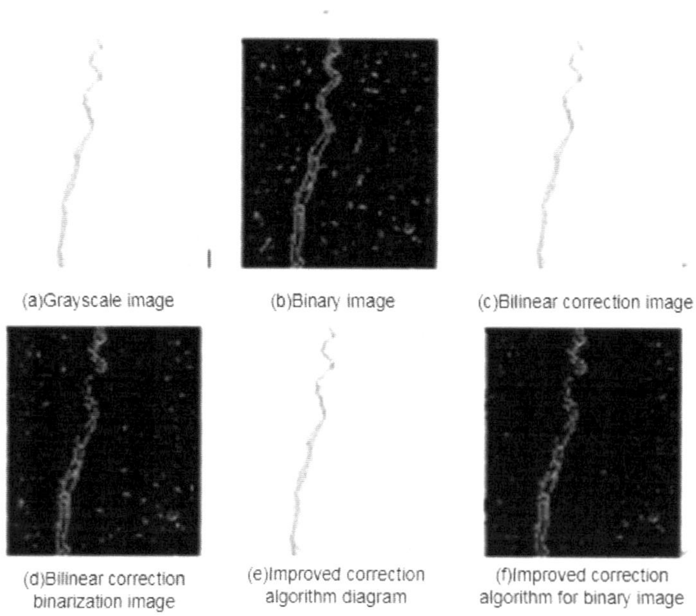


**Figure 6.** Comparison image of MATLAB simulation experiment results.

### *3.2. Crack type analysis*

Through a large number of experiments, the threshold used in this paper is t = 10. If the crack type analyzed in the first step is a linear crack, it is also necessary to analyze whether the crack is a transverse crack or a longitudinal crack. The analysis method is to compare the projection sizes of the crack on the x-axis and y-axis respectively. If the projection length of the crack on the x-axis is greater than its projection length on the y-axis, it can be determined that the crack is a transverse crack. On the contrary, if the projection length of the crack on the X axis is less than its projection length on the Y axis, the crack is a longitudinal crack. The analysis process of crack type is shown in figure 7.

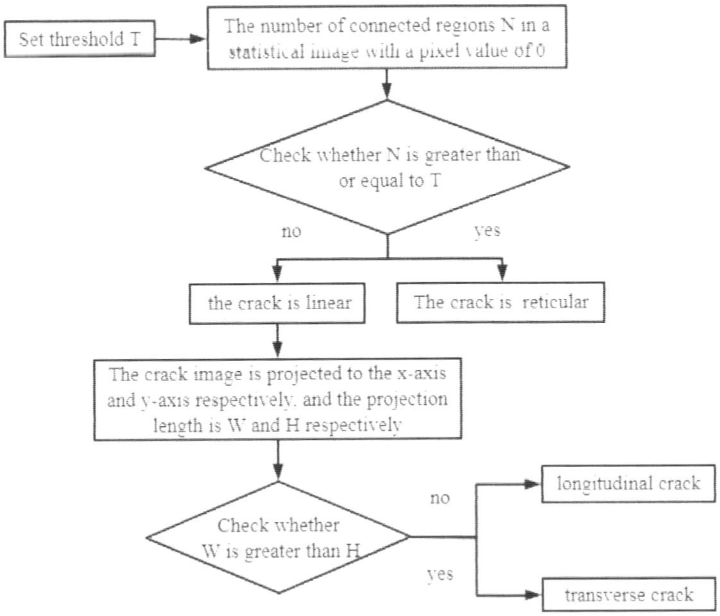


**Figure 7.** Analysis process of crack type.

### *3.3. Calculation of crack length and width*

The crack skeleton is extracted based on the maximum disk method, and the image thinning is mainly for the binary image. In this paper, the look-up table method is used for thinning. The judgment basis for the retention or removal of a point is the 8 adjacent points, as shown in table 1.

**Table 1.** The status and judgment method of 8 adjacent points.

| 8 adjacent points of a point | Retention or removal |
|---|---|
| 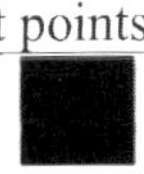 | This point cannot be removed, it is an internal point |
|  | This point cannot be removed, it is an internal point |
|  | This point cannot be removed, and the removal may disconnect the connected part |
| 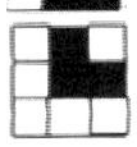 | This point can be removed, it's not a skeleton |
| 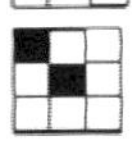 | This point cannot be removed. It is the endpoint of the line |

For the refined crack image, it is necessary to count the number of pixels of the crack target (that is, the total number of white pixels in the statistical image). Assuming that the total number is m, the actual length of the crack is $L=m\mu$, where L represents the actual length corresponding to each pixel ($\mu \approx 0.92$mm). The area calculation method of linear crack is very simple. It is to count the number of pixels with pixel value of 0 in the binary inverse color image($N_1$), then calculate the proportion of these pixels in the whole image(N) and then multiply this proportion by the actual area($S_1$) size of the large in the image to obtain the linear crack area, The calculation formula is $S=S_1(N_1/N)$. After calculating the area (S) and length (L) of the crack, the calculation of the width (W) of the crack (because the width of the crack in each place is inconsistent, the average width is calculated here) is as follows: W=S/L. Selected in this paper 150× 150mm stone slabs with cracks was divided into 8 crack pictures with good quality for experiment. Relevant experimental results are shown in table 2.

**Table 2.** Experimental data of crack feature extraction.

| Name | Crack type | Judgment type | Crack length/mm | Residual error (v) | Square of residual error ($v^2$) |
|---|---|---|---|---|---|
| Img1-1 | longitudinal | longitudinal | 149.951 | 0.311 | 0.096721 |
| Img1-2 | longitudinal | longitudinal | 150.104 | 0.464 | 0.215296 |
| Img1-3 | longitudinal | longitudinal | 150.095 | 0.455 | 0.207025 |
| Img1-4 | longitudinal | longitudinal | 149.862 | 0.222 | 0.049284 |
| Img1-5 | longitudinal | longitudinal | 149.857 | 0.217 | 0.047089 |
| Img1-6 | transverse | transverse | 150.008 | 0.368 | 0.135424 |
| Img1-7 | transverse | transverse | 149.924 | 0.284 | 0.080656 |
| Img1-8 | transverse | transverse | 149.907 | 0.267 | 0.071289 |

### *3.4. Results and analysis*

According to the comparison image of the MATLAB simulation experiment results in figure 6, the middle part of Figure a is bright, while the two sides are dark, which has a certain impact on the crack image. There are also cracks in the binarized image, this is because the middle light makes the gray value of some crack target pixels larger, and these pixels are wrongly judged as the background during binarization, this has a bad impact on the subsequent feature extraction. Figure c is the effect image after the crack image is corrected by using the image correction algorithm based on bilinear interpolation. It can be seen that the fracture of the crack is reduced, in other words, the number of pixels misjudged is less than the image corrected. The binarization result shown in Figure f is better than that in Figure d, and the number of pixels misjudged is further reduced. Experiments show that it is necessary to correct the crack image, and the improved image correction algorithm is effective for the crack image.

According to the crack data extracted from the experiment in table 2, the error is analyzed. From table 2, it can be concluded that $\sigma = 28.701$, then $3\sigma = 86.103$, which meets the principle of $|\sigma| < 3\sigma$, so there is no gross error, and the experimental results are ideal.

## 4. Roughness Detection

### *4.1. Stylus contact surface roughness measurement experiment*

According to GB/T1031, the sampling length was selected as 0.25mm, and the evaluation length was 1.25mm. The measurement results are shown in table 3.

**Table 3.** Roughness measurement values of stone slab.

| Serial number | Ra1 | Ra3 | Ra3 | Ra4 | Ra5 | mean value |
|---|---|---|---|---|---|---|
| Slab A | 0.0267 | 0.0250 | 0.0274 | 0.0292 | 0.0252 | 0.0267 |
| Slab B | 0.0351 | 0.0877 | 0.0153 | 0.0391 | 0.1358 | 0.0626 |
| Slab C | 0.0447 | 0.1031 | 0.0236 | 0.0643 | 0.0242 | 0.0519 |
| Slab D | 0.0169 | 0.0164 | 0.5391 | 0.0367 | 0.0281 | 0.1275 |

### *4.2. Non-contact surface roughness measurement experiment*

#### *4.2.1. 3D surface reconstruction experiment*

The large stone slabs to be tested was placed on the loading platform, and a magnifying lens of 400 times was selected to assemble with CCD camera as an image sensor [15]. The computer control is the stepper motor to make the image sensor move upward along the Z-axis and record the step size. The image sensor is used to take continuous photos on the surface of the measured large stone slabs. The sequence of microscopic images captured are transferred to the computer by the image acquisition card for image processing, as shown in figure 2. Setting μ as the step size and the image sequence as j (j=1, 2, 3, …, n), If a certain region on the j image of the sequence image is clear, then the height of the area z-axis on the focusing plane is j×μ.

According to the above process, sequence photo sampling was carried out for the large stone slabs. Figure 8 and 9 are the collected part of sequence diagram on large stone slabs.

**Figure 8.** Sequence Diagram 10.

**Figure 9.** Sequence Diagram 20.

**Figure 10.** Fusion image.

Matlab software is used to fuse the sequence phase images. In order to better extract the characteristic information of stone surface, the image was preprocessed. The fused image is shown in figure 10, and figure 11 is the height index map of the measured stone slab image reconstructed. The 3D reconstruction results are shown in figure 12.

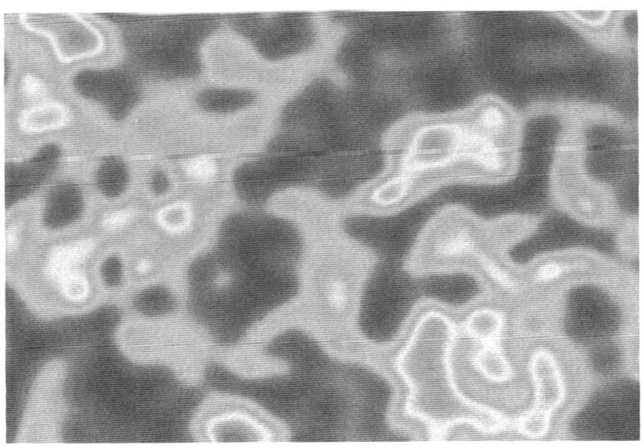

**Figure 11.** The height map after the reconstructed.

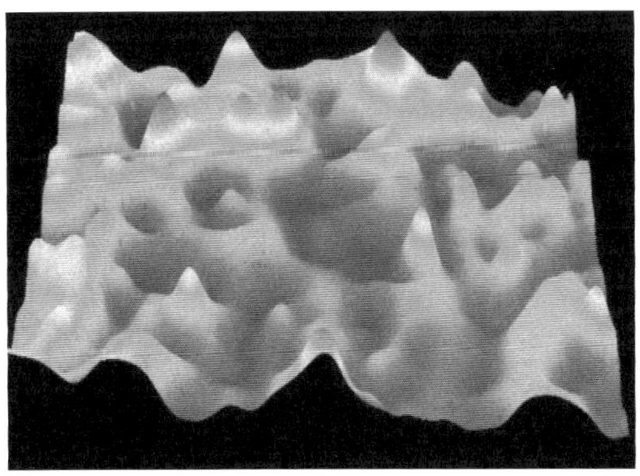

**Figure 12.** 3D reconstruction image of stone slabs.

#### *4.2.2. Results and analysis*

In order to test the accuracy and reliability of the experimental method in this paper, the surface roughness of the same stone slab was detected by a contact desktop roughness profiler. Table 4 lists the comparison and relative errors of a set of roughness profilometer measurements with the experimental data.

**Table 4.** Comparison table of parameter values.

| Stone sample | Slab A | Slab B | Slab C | Slab D |
|---|---|---|---|---|
| Profilometer measurement (μm) | 0.0277 | 0.0656 | 0.0519 | 0.0525 |
| The measured value in this experiment (μm) | 0.0313 | 0.0608 | 0.0564 | 0.0581 |
| The relative error (%) | 11.50 | 7.89 | 7.98 | 9.64 |

It can be seen from the experimental data that there is little difference between the measured value of this test and profilometer. The relative error of slab A is the largest and that of slab B is the smallest. The average relative error of the four slabs is 9.25%.

It can be seen from the comparison results of table 4, The change trend of measured and calculated roughness values of the four slabs is the same, the relative error is around 10%. The measured results are satisfied. It can be seen that the surface roughness measurement method based on sequence microscopic image fusion 3D reconstruction proposed in this paper is accurate and reliable, it provides a basis for the measurement of 3D roughness.

It can be seen from the data that the measured value in this experiment is relatively large, mainly because the contact pin of the roughness profiler has a certain size, so it is difficult to collect the contour at the lower part of the surface of the large stone slab under experiment. However, these contours can be detected in this experiment.

## 5. Conclusion

In the crack detection, the image is preprocessed firstly, and the combination of wavelet transform and traditional interpolation algorithm is proposed to realize the correction of crack image. In this paper, the common filter Windows are combined into a centrosymmetric template to enhance the crack image, so as to eliminate the noise and protect the crack edge to the greatest extent. Then, the improved threshold segmentation algorithm is used to segment the linear crack filter image. Finally, the length and area of cracks are calculated by counting the number of target pixels of cracks. In the detection of surface roughness, the CCD camera and micro lens are assembled as image sensors to enlarge the surface of stone and take a sample by many taking pictures. The surface profile of the tested stone slab is obtained by reconstruction method, and the surface roughness of stone slab is finally obtained by analysis and calculation. Finally, the comparative experimental results show that the average relative error between the measured value of the test and the measured value of the profiler is 9.25%, which shows that the application of this method for surface quality detection is reliable. Through this research, a fast detection technology was realized, which solves the problems of low processing efficiency and high cost in production. The result shows certain theoretical value and practical application significance.

## References

[1] Han Y F and Shi P F. 2005 Surface Defect Detection Based on Image Processing Techniques *J Measurement and control technology*,**24 (9)**:15-18

[2] Paul B., Ravishankar R and Martin C, et al. 2003 Automatic Defect Classification for Semiconductor Manufacturing *J. Machine Vision and Application.*, **9**:201-214

[3] Zhao H Y and Feng X H. 2000 Analysis of ceramic tile surface quality based on machine vision *J. Journal of Xian University of Architecture & Technology*, **1**:43-46

[4] Liu H. 2012 Research on 3D reconstruction of workpiece surface crack based on digital image processing *D. Changchun University of Science and Technology*

[5] Yu H B, Zhao R C and Wang B. 2004 Overview of 3D Surface Reconstruction Technology based on Shadow. Computer *J Engineering and Applications*, **40 (10)**: 30-33

[6] Qian Y, Yin W Q and Lin X Z, et al. 2004 Overview of 3D surface reconstruction technology based on shadow *J. Computer Engineering and Applications*, **40 (10)**: 30-33

[7] Zhao M, Peng X C and Shao M, et al. 2005 Research on virtual technology of optimized stone processing *J. Journal of Shenyang Jianzhu University: Natural Science*, **(2)**:160-163
[8] Tsai D M and Huang T Y. 2003 Automated surface inspection for statistical textures *J. Image and Vision Computing*, **21(4)**:307-323
[9] Kumar, Ajay and Pang, et al. 2002 Defect Detection in Textured Materials Using Optimized Filters *J. IEEE Transactions on Systems, Man & Cybernetics: Part B*, **32(5)**:553
[10] Chen G B. 2007 Research on 3D Reconstruction method based on defocus microscopic Image *D. Zhejiang University of Technology*
[11] Zhou X L, Zhang Y and Jiang N D, et al. 2014 Research on 3D Image Reconstruction of Asphalt Pavement Micro-texture Based on Depth of Focus *J. Journal of Highway and Transportation Technology*, **31(9)**: 6-12
[12] Jiang Z G, Shi W H and Han D B, et al. 2004 Microscopic 3D imaging system based on focusing synthesis *J. CT Theory & Application Research*, **13(4)**: 9-15
[13] Le J, Yang J and Wang J, et al. 2010 Three-dimensional Surface Roughness Detection Based on Multi-Layer Sequence Image *J. Chinese Journal of Scientific Instrument*, **9**: 2036-41
[14] Yang Y, Lv Z Y and Yan X P, et al. 2006 Microscope-based measurement of abrasive surface topography *J. Lubrication Engineering*, **7**: 110-112
[15] Dong W P, Sullivan P J and Stout K J. 1994 Comprehensive Study of Parameters for Characterising Three-dimensional Surface Topography IV: Parameters for Characterising Spatial and Hybrid Properties *J. Wear*, **178(1-2)**:45-60

IOP Conf. Series: Materials Science and Engineering 1242 (2022) 012021 doi:10.1088/1757-899X/1242/1/012021

# Hempcrete - modern solutions for green buildings

**M L Lupu[1*], D N Isopescu[1], I-R Baciu[1], S G Maxineasa[1], L Pruna[1] and R Gheorghiu[2]**

[1] "Gheorghe Asachi" Technical University of Iasi, Faculty of Civil Engineering and Building Services, Iasi, Romania
[2] SC Cronos Consulting SRL, Iasi, Romania

* mariuslucianlupu@gmail.com

**Abstract.** It is widely established that the building industry has a negative impact on the environment and a significant influence on the phenomena that contribute to climate change. Traditional construction materials, such as cement, contribute considerably to environmental pollution. Given the enormous quantity of energy and materials used by the construction sector, this industry must adopt more sustainable practices. Nowadays, an increasing number of natural building materials are used in the structural component or the insulation of buildings. As a result, natural construction materials may be a superior alternative to accomplish this goal. This article discusses the features and applications of hempcrete in the building industry. Hempcrete is a sustainable material composed of industrial hemp, lime as a binder, and water. Due to hemp's porous structure, it has deformation capacity, sound-absorbing qualities, better hygrothermal properties than conventional concrete, and, depending on the proportions of hemp, lime, and water, fire resistant capabilities due to the presence of lime.

## 1. Introduction

In comparison to other key sectors of human activity, the building industry has a substantial impact on the environment via the use of natural resources and energy. These consumptions result in pollutants being emitted and non-renewable natural resources being depleted. To address these issues, it is vital to limit energy usage and to utilize materials that have a low environmental effect [1].

Combating climate change has been a main priority for a multitude of UN organizations. The New Urban Agenda, which was endorsed in October 2016 at the United Nations Habitat III conference in Quito, Ecuador, was mostly focused on the sustainability and energy consumption of built environments in order to enhance lives and battle climate change [2]. As a result, interest in green buildings will spike in the next years. The World Green Building Council defines a 'green' building as "a building that, in its design, construction or operation, reduces or eliminates negative impacts, and can create positive impacts, on our climate and natural environment" [3]. Latest developments in green building design integrate passive as well as active technology designed to increase a building's energy efficiency. While adopting such energy-efficient construction is certainly advantageous, using conventional building materials has a negative impact on the environment. To minimize building energy usage, phase change materials, passive house approaches, and other comparable strategies are being investigated [4, 5]. On a related note, scientists have been researching a relatively new concept known as geoengineering, which entails the intentional, large-scale modification of the environment in order to mitigate certain symptoms of anthropogenic climate change [6]. It was proposed that this environmental manipulation may be accomplished via the use of carbon dioxide reduction (CDR) measures and albedo modifications (AM). Reforestation and afforestation, bioenergy with carbon capture and storage (BECCS), and accelerated weathering are just a few of the $CO_2$ reduction techniques that have proved the potential to capture or remove several gigatons of $CO_2$ from the atmosphere, thereby minimizing the effects of anthropogenic climate change [7, 8]. Thus, commercial cultivation of fast-growing crops/plants such as hemp, sisal, jute, flax, kenaf, and hazelnut must be considered.

Meanwhile, scientists began exploring the possibilities of biomass products as construction materials. Hempcrete, hemp lime, lime-hemp concrete (LHC), or hemp concrete is a kind of cellulose aggregate concrete made from hemp hurds or shivs with a lime-based binder. Hemp concrete is a subset of a broader category of building materials that incorporate bio-based aggregates and a mineral binder. These materials are referred to as cellulose aggregate concrete (CAC) [8], bio-aggregate-based building materials [9], agro-concretes, or vegetal concretes [10]. This article disseminates information on the several features of hempcrete and its uses in building construction that has been investigated and collated via a study of pertinent research published by different writers.

## 2. Mechanical, thermophysical and acoustic properties

Hempcrete is a composite material made up of three components: hemp, binder, and water. Hemp aggregates are the non-fibrous portion of the hemp stem, referred to as "shiv" or "hurd", which exhibits a unique porosity.

The binder is typically lime-based, however some formulations (for example, combining pozzolanic material and hydraulic binder) are utilized to enhance hemp lime's effectiveness. Hempcrete, commonly known as hemp lime, has pores of varying sizes [11]:

- Macropores (up to 1 cm in diameter) caused by the irregular arrangement of hemp shiv in the mix;
- Mesoporores (0.1 to 1 mm) between shiv and binder (trapped air);
- Micropores between hydrates in the binder matrix (less than 0.01m);

This wide range of porosity, which is unique to hemp lime, has a significant effect on the material's thermal, acoustic, and hygrometric properties. Obviously, the microstructure of hemp lime varies with the ratio of hemp to binder; porosity varies with the origin and manufacture of shiv; and density varies with the ratio of hemp to binder and then with the kind of application (walls, roofing, roughcast) and installation method (manually, by projection machine, projection distance). As shown in table 1, a variable density value has an effect on thermal and mechanical performance [12].

**Table 1.** Mechanical and thermal properties of hempcrete as a function of the ratio of hemp to limestone [12].

| Application | Shiv : binder proportion (by mass) | Target density [kg/m$^3$] | Typical ultimate compressive strength [N/mm$^2$] | Typical thermal conductivity λ [W/mK] |
|---|---|---|---|---|
| Roof | 1:1 | 220 | 0.05 | 0.06 |
| Wall | 1:1.5 | 275 | 0.11 | 0.06 - 0.09 |
| Wall | 1:2 | 330 | 0.22 | 0.09 - 0.115 |
| Wall | 1:2 compressed | 440 | 0.35 | 0.115 |
| Floor | 1:3 | 500 | 0.8 | 0.13 |
| Floor | 1:4 | 600 | 1.15 | 0.14 |
| Pre-cast Structural | 1:4 compressed | 600-1000 | 2 - 6 | 0.14 - 0.27 |

Several compression tests demonstrate this material's exceptional ductility, which enables large deformations and load absorption without delicate fractures or macroscopic cracks. However, compression resistance approaches very low levels, necessitating the use of other structural forms (timber, steel or concrete structures).

Numerous variables have an effect on mechanical characteristics [13]: chemical structure of the binder; composition of the mixture [14, 15] (an increase in lime concentration increases resistance but decreases non-fragile behavior); aging (mechanical performances increase during the first two years); treatment conditions (in shifting humidity levels); shiv nature and dimension (resistance increases with larger but also decreases with longer particles); degree of compactness (greater resistance but lower porosity); thermal properties with increased compacting strength); particle orientation in the mix for anisotropy [15, 16]. In general, it is feasible to consider compression resistance as follows:

$$\sigma_{max} < 1\,MPa \tag{1}$$

Regarding dimension variations, retreat deformations are seen at 20°C, 50% HR, and a rate of about 3 mm/m after 75 days of drying [17]. Due of hempcrete's high porosity, it is very responsive to liquid water. A water absorption test [18] demonstrates hemp's rapid absorption capacity: its degree of saturation is more than 95% after ten minutes of immersion. The exchange of water between hemp and lime is critical throughout the component mixing process, since hemp transfers water to lime during the hydration phase. It is critical to monitor water dosage carefully, since an excessive amount may result in some excellent yields during the drying process.

The hygrothermal and acoustic properties of a material are highly dependent on its composition (formulation and material matrix) and porosity [19]. With regards to thermal characteristics, it's worthwhile to highlight a few key terms: thermal conductivity (λ), specific heat (c), and thermal diffusivity (α). The first one describes a material's capacity to conduct heat. It is dependent on the type and content of the mix, the installation, and the presence of water in the material in hempcrete. In general, it is feasible to examine the following:

$$\lambda = 0.06 \div 0.2\,W/(mK) \tag{2}$$

Specific heat enables the quantification of an element's ability to absorb or release energy due to thermal changes, — in other words its capacity to store heat. It was determined or quantified in a variety of research using various characteristics such as density, the presence of water, and the moisture formulation [20]:

$$c = 940 \div 3000\,J/(kgK) \tag{3}$$

Thermal diffusivity is a property of materials that indicates their capacity to convey temperature fluctuations; it is determined by their density, conductivity, and capacity.

$$\alpha = 0.98 \cdot 10^{-7} \div 3.02 \cdot 10^{-7} m^2/s \tag{4}$$

The excellent acoustic behavior of hempcrete was shown in acoustic experiments. Due to the material's unique porosity, it exhibits a high absorption coefficient ($\alpha_a$). It was determined in relation to various frequency and mix formulations [21].

$$\alpha_a > 0.9\ (100Hz \div 200\ Hz) \tag{5}$$

$$\alpha_a > 0.6\ (400Hz \div 500\ Hz) \tag{6}$$

$$\alpha_a > 0.6\ (1000Hz \div 2000\ Hz) \tag{7}$$

## 3. Comfort aspects

Currently, comfort issues are quite widespread. Even though structural aspects are critical for human safety, how people feel at home or at work has an effect on their productivity. A discomfort sensation might result in health complications, discontent, or energy waste.

Thus, it is essential to consider not only the mechanical properties of materials but also their thermal, hygrometric, and acoustic properties.

*3.1. Acoustic characteristics*

Noise protection (from the external environment or adjacent rooms) and noise absorption are critical in both residential and commercial structures. Absorption is mostly determined by porosity. In the case of hempcrete, some measures are made based on material qualities and structural systems [22].

Cerezo's research reveal that hempcrete, due to its high porosity, has absorption coefficient values between 0.3 and 0.9 at frequencies of 100 - 200 Hz [12] . These values are especially interesting when contrasted with those of other more conventional materials, such as cellular concrete (porosity between 0.2 and 0.3), which has a mostly closed porosity.

The formulation and thickness of samples may be exploited to enhance their acoustic qualities.

With a large lime dose, the absorption peaks are shifted to lower frequencies, limiting the global amount of absorption. Additionally, increasing the thickness results in a reduction of the absorption peaks without a significant change in absorption.

Bütschi et al. also conducted studies on phonic insulation in hempcrete bricks with a thickness of 31 cm and a density of roughly 700 kg/m$^3$ [23]. Acoustic intensity levels range from 43 dB when using raw hempcrete to 47 dB when using plaster on two wall facing.

Another research revealed that hempcrete absorbs sound due to its porosity. Gle et al. have also demonstrated that altering the ingredients and manufacturing technique of hempcrete may affect its acoustic absorption [24].

*3.2. Hygrothermic characteristics*

Thermal comfort is one of the most key design parameters nowadays due to its impact on energy waste: high-performance buildings and user well-being enable energy load reduction and improved health. Several studies on the hygrothermic behavior of hemp lime [25, 26] demonstrate the difficulties of the issue because to the complexity of the physical transfer of water vapour and liquid water, making it critical to accurately define the material and do more research on the subject. Phase altering nonlinear occurrences are related with energy release and absorption. These factors enhance insulation performance by maintaining a steady ambient temperature.

The Bath University study demonstrated significant stability of conditions in an experimental building constructed of hempcrete and timber (called HemPod) in concepts of Relative Humidity and Temperature when comparison to a mineral wool-insulated timber-framed building, indicating that hempcrete can help reduce daily interior Relative Humidity variations [27].

Hempcrete is known to have phase change qualities as a consequence of its latent heating effects, these are a consequence of the material's high heat capacity paired with its permeability and high moisture absorption capacity. The connection between relative humidity and thermal capacity of a hempcrete wall is seen in figure 1. Thus, whereas thermal bridges created by mortar joints boost heat fluxes, they are reduced by moisture absorption during periods of high humidity. Moreover, the quantity to which moisture migration suppresses is unknown [28, 29].

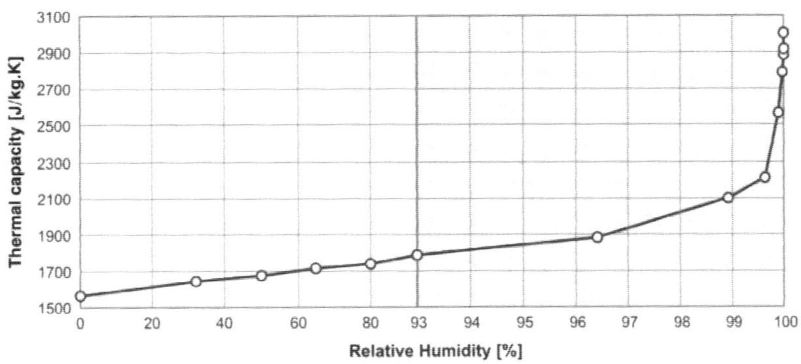


**Figure 1.** The relation between Relative humidity of hempcrete and its thermal capacity [30].

*3.3. Reaction to fire*

Fire response is essential to evaluate since it related to building safety, which has an effect on construction permits and insurance. This is, though, a relatively unexplored feature of hempcrete, with a small number of investigations. However, the research have unanimously acknowledged hempcrete's superior fire resistance [31]. Fernea et al. [32] discovered that hemp concrete passed the fire test standards in their examination of the material's response to fire. The investigated samples were constructed as an acoustic absorption medium using lime and water. Three samples were investigated with a constant volume of lime and a variable hemp/lime ratio. The ratios were 1:1, 2:1, and 3:1, respectively. The sample with a 2:1 ratio was used as a control/reference. The water-lime ratio was maintained at 1:2 throughout. No information was provided regarding the curing period. The samples were subjected to a fire reaction test in accordance with SR EN520/A1:2010 [31].

As part of their multi-criteria research, Sassoni et al. investigated the fire resistance of a hemp composite. The samples were classified as low density (LD; 330 kg/m$^3$) or medium density (MD; 640 kg/m$^3$). The fire resistance test was done in accordance with EN 13823, and the LD samples passed with a class 'C' rating, while the MD samples passed with a class 'B' rating [15].

## 4. Environmental studies

Environmental issues arise all across the agriculture and construction processes. Concerning the agricultural process, hemp cultivation may have a negative influence on the environment owing to the use of nitrogenous fertilizer and transportation (see table 2). Performance of hemp hurd may be enhanced by minimizing nitrogen fertilizer use and cultivating hemp types that make the greatest use of available nitrogen. Reduced travel distance for the hemp straw would also enhance its environmental performance; in the case of France, transportation issues are overcome by widespread hemp production across the country. Concerning the construction process, the research emphasizes the beneficial influence on the greenhouse effect of a hempcrete wall's potential to operate as an overall carbon sink for at least 100 years [33]. The manufacturing of lime-based binder is the single largest contributor to greenhouse gas emissions, non-renewable energy usage, photochemical ozone generation, and resource depletion. Transportation is the primary contributor to ozone layer degradation and the second largest contributor in terms of effects on non-renewable energy use and the greenhouse effect. Reduced greenhouse gas emissions from the production of lime are contingent on its manufacturing sector. Reduced transportation time between the facility manufacturing the lime binder and distributors would also boost the construction stage's total potential effect. Additionally, the environmental performance of end-of-life products may be enhanced with the inclusion of recycling choices or recovery methods [34]. Hempcrete has a remarkably low ecological impact compared to other building materials since it can retain roughly 35 kg of $CO_2$ per square meter of wall (25 cm thickness) during a 100-year period.

**Table 2.** Environmental impacts of hemp culture - Life Cycle Assessment [34].

| Impacts | Production of raw materials | | Construction | Use | End of life | Transport | Total |
|---|---|---|---|---|---|---|---|
| | Hemp hurd | Other materials | | | | | |
| Resource depletion (kg Sb eq) | $2.8 \cdot 10^{-2}$ | $7.7 \cdot 10^{-2}$ | $1.2 \cdot 10^{-3}$ | 0 | 0 | $2.6 \cdot 10^{-2}$ | $1.3 \cdot 10^{-1}$ |
| Atmospheric acidification (kg $SO_2$ eq) | $5.1 \cdot 10^{-2}$ | $4.8 \cdot 10^{-2}$ | $1.3 \cdot 10^{-3}$ | 0 | 0 | $5.1 \cdot 10^{-3}$ | $1.1 \cdot 10^{-1}$ |
| Greenhouse effect 100 years (kg $CO_2$ eq) | -45.9 | 23.1 | 0.2 | -13.6 | 0 | $6.7 \cdot 10^{-1}$ | -35.5 |

| | | | | | | | |
|---|---|---|---|---|---|---|---|
| Destruction of $O_3$ layer (kg CF-11 eq) | $7.1 \cdot 10^{-7}$ | $3.3 \cdot 10^{-6}$ | $3.4 \cdot 10^{-7}$ | 0 | 0 | $5.7 \cdot 10^{-6}$ | $9.9 \cdot 10^{-6}$ |
| Formation of photochemical $O_3$ (kg $C_2H_4$ eq) | $7.1 \cdot 10^{-4}$ | $4.2 \cdot 10^{-3}$ | $5.0 \cdot 10^{-5}$ | 0 | 0 | $3.8 \cdot 10^{-4}$ | $5.4 \cdot 10^{-3}$ |
| Nonrenewable energy (MJ) | 52.3 | 265.8 | 19.9 | 0 | 0 | 56.3 | 394.2 |
| Air pollution ($m^3$) | 674 | 207.2 | 19.9 | 0 | 0 | 128.2 | 1024 |
| Water pollution ($m^3$) | 4.3 | 2.2 | $6.1 \cdot 10^{-2}$ | 0 | 0 | $1.1 \cdot 10^{-1}$ | 6.7 |
| Generation of waste (kg) | 6 | n.a. | 0.9 | 0 | 98 | n.a. | 104.9 |

## 5. Manufacturing

Hemp hurd, which are generally free of fibrous layers, are treated in a hammer mill to reduce them to fragments no bigger than 40 ± 5 mm in their widest diameter [35, 36].

**Figure 2.** Molded hempcrete [35].

Sinka et al. utilized hemp hurds with an average length of roughly 15 mm and the longest hurd particle measuring 55 mm in their investigations [36]. Alternatively, some commercial methods use the decorticator, which simultaneously separates the fiber from the whole hemp straw and reduces it to minute particles. The binder is mostly composed of hydrated lime with a little amount of pozzolanic material or a commercial hydraulic lime-based binder. Hempcrete walls may be constructed on-site in two ways: by pouring the mixture into a temporary formwork and manually tamping it (see figure 2), or by spraying the material onto the target location using the projection method (see figure 3). Both processes result in low compaction, and there is no control over the material's maturation conditions.

**Figure 3.** Sprayed hempcrete [37].

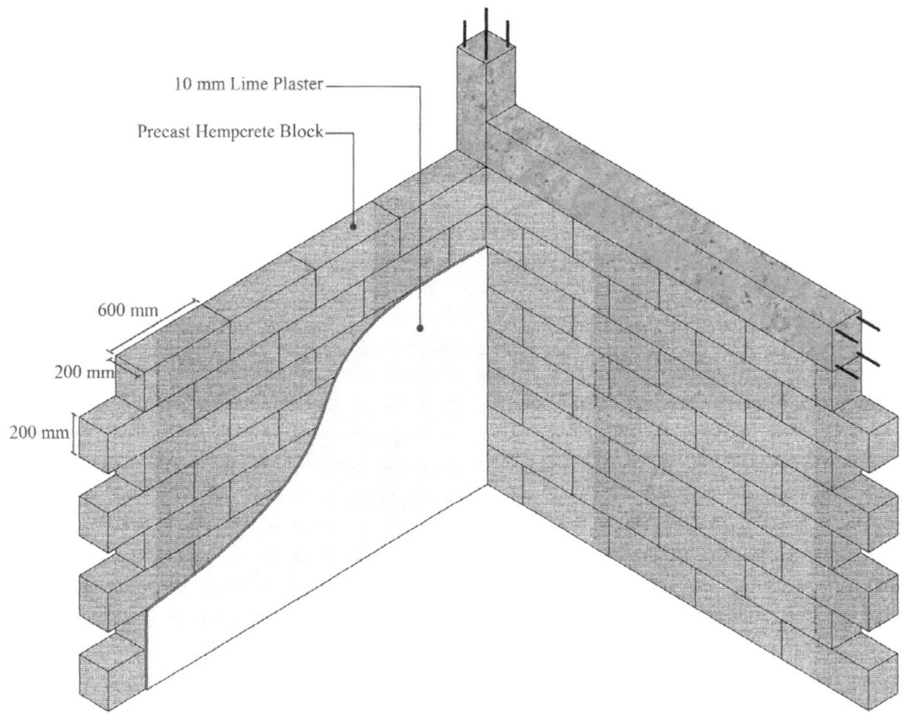


**Figure 4.** Wall assembly made of large hempcrete blocks [30].

Recent developments in the industry show a higher interest in pre-cast blocks rather than cast hemp concrete walls [38, 39]. Some innovations also help with stacking the blocks without mortar joints through physical interlocking [40]. Figure 4 illustrates how pre-cast blocks are used to build infill walls using conventional masonry in a reinforced cement concrete frame [30]. Some of the advantages of pre-cast blocks are:

- controlling the compression applied to the mixtures ensures uniformity in mechanical strength and performance;
- they may be cured individually and then transported to the site, removing seasonal limits from the building process.

Numerous studies have shown that hempcrete outperform equivalent conventional materials not just in terms of embodied energy, but also in terms of net $CO_2$ emissions during the average building's whole life cycle. This is mostly due to hemp's active carbon sequestration throughout the growth period, as well as its rusticity, which requires little agronomic input. Additionally, the minimal number of operations necessary to manufacture the wall (electricity is required to combine hemp and lime) affected the outcomes favorably. Gradual carbon sequestration by lime carbonation after construction, as seen in previous studies, might further enhance the carbon balance, and this could be the subject of future study [41, 42].

## 6. Conclusion

This research enables some thoughts on the real-world uses of hempcrete in relation to envelope technology. Although the material cost is two to three times that of standard insulating materials (such as mineral wool), it is suitable when considering sustainability and hygrothermic performance. Aspects of hemp concrete that are critical include practice and technique.

The hemp variety is determined by climate and producer . With comprehensive control across the whole manufacturing process, it would be successful in developing a hemp variety with very well dimensions and quality features. Given the friable nature of hemp, this may not be easy.

Concerning implementation, projection equipments are not standardized, and it is critical to instruct employees prior to applying the material on the construction site, since material qualities might vary depending on the projection method, distance, and compactness. The mixing formulation demands extra attention throughout the installation phase and for achieving the project's desired thickness. However, the ease of installation enables rapid construction and opens up several application domains. Without a defined standard, the quality of building components is determined by the installation's best practices. In terms of mechanical properties, hemp lime cannot be considered a structural component due to its insufficient compressive strength (1 MPa).

To achieve good thermal performance and a low U value hempcrete walls must be thicker (30 cm). However, it is essential to keep in mind that this is the ultimate thickness, in contrast to other insulating materials that are linked to substructures, and that the benefits of interior thermal comfort are achieved not only via thermal resistance, but also through hygrometric behavior. Since a result of these benefits, hempcrete may be an intriguing alternative to typical insulating materials, as it enables the structure to breathe.

Another advantage is the broad range of applications in both new and existing buildings, but there are a few limitations in terms of building typologies. The height of the building is limited if it is entirely composed of hempcrete and timber structure (a steel or concrete structure has more environmental consequences); the significant thickness of the walls consumes useful surface; in terms of visual characteristics, hempcrete surface, even with smooth work, retains a "rustic" appearance. A viable option is to use an indoor hempcrete plaster in residential structures to enhance their interior acoustic performance.

In terms of energy consumption, a repair of an envelope component with a hempcrete coating results in a decrease of consumptions in particular types of residential buildings due to the hygrothermal properties of hempcrete.

Environmental examination of a 1 $m^2$ hempcrete wall with a timber framework (26 cm in thickness) reveals that the effect of the hemp material is much less than that of other construction methods. In this instance, hempcrete is a wise option.

The following benefits summarize the usage of hempcrete in buildings: it is a light weight material with low conductivity; the porous structure allows for the stabilization of indoor air temperature and humidity regardless of external variations; it has great acoustic properties; it has a lower carbon footprint when compared to traditional construction materials; hemp hurd is more resistant to biological deterioration than other building materials.

Even though hempcrete has superior technical characteristics and superior performance in terms of interior comfort, several practical difficulties remain unresolved. Currently, construction aplications are constrained, and they must be enhanced by more research and technological advancements.

**Acknowledgements**
This paper was elaborated with the support of the "Eco-innovative Products and Technologies for Energy Efficiency in Constructions – EFECON" research grant, project ID P_40_295/105524, Program co-financed by the European Regional Development Fund through Operational Program Competitiveness 2014-2020.

**References**

[1] Maxineasa S G, Isopescu D N, Baciu I-R, Lupu M L and Dragan T C 2020 IOP Conf. Ser.: Mater. Sc*i. Eng.* **789** 012039
[2] Habitat III 2017 *Third United Nations Conference on Housing and Sustainable Urban Development* 1-10
[3] WGBC 2021 https://www.worldgbc.org/what-green-building. (Accessed 1 December 2021)
[4] Laaouatni A, Martaj N, Bennacer R, El Omari M and El Ganaoui M 2017 *Enrgy Proced* **139** 744 749
[5] Pitts A 2017 *Sustainability-Basel* **9** 272
[6] Sandler T 2017 *Rev Int Organ* **13** 105-125.
[7] Sovacool B K 2021 *Energy Strateg Rev* **35** 100656
[8] Jami T, Rawtani D andAgrawal Y K 2016 *Emerg Mater Res* **5** 240-247
[9] Amziane S and Collet F 2017 Bio-aggregates Based Building Materials: RILEM State- Of-The-Art Reports *Springer Netherlands pp 149-186*
[10] Amziane S and Sonebi M 2016 *Tech Lett* **1** 31
[11] Collet F, Bart M, Serres L and Miriel J 2008 *Constr Build Mater* **22** 1271-1280
[12] Cerezo V 2005 *PhD Thesis* INSA de Lyon
[13] Arnaud L and Gourlay E 2012 *Constr Build Mater* **28** 50-56
[14] Tronet P, Picandet V, Lecompte T and Baley C 2011 *JNC17* 161.
[15] Sassoni E, Manzi S, Motori A, Montecchi M and Canti M 2014 *Energ Buildings* **77** 219–226
[16] Elfordy S, Lucas F, Tancret F, Scudeller Y and Goudet L 2008 *Constr Build Mater* **22** 2116-2123
[17] Hustache Y 2008 Synthèse des connaissances sur les bétons et mortiers de chanvre 104
[18] Tran Le A D 2010 *PhD Thesis* Université de Reims
[19] Ecomat 2013 *Construction and bioresources* 60
[20] Collet F and Pretot S 2014 *Constr Build Mater* **65** 612-619
[21] Contrada F, Kindinis A and Dony A 2015 *CIGOS-PARIS*
[22] Glé P, Gourdon E and Arnaud L 2012 *Constr Build Mater* **37** 801–811
[23] Bütschi P-Y, Deschenaux C, Miao B and Srivastava N-K 2004 *Can. J. Civ. Eng.* **31** 526–529
[24] Glé P, Gourdon E and Arnaud L 2011 *Appl. Acoust.* **72** 249-259
[25] Tran Le A D, Maalouf C, Mai T H, Wurtz E and Collet F 2010 *Energ Buildings* **42** 1797–1806
[26] Oumeziane YA 2012 *AUGC-IBPSA*
[27] Lawence M, Fodde E, Paine K and Walker P 2012 *Constr Build Mater* **36** 270-275
[28] Ahlberg J, Georges E and Norlen M 2014 *Examensarbete 15 hp*
[29] Simpson S 2004 *MSc thesis* University of East London
[30] Jami T, Karade S and Singh L P 2019 *J Clean Prod* **239** 117852
[31] Murphy F, Pavia S and Walker R 2010 *BCRI Bridge* 431-438.
[32] Fernea R, Tamas-Gavrea D R, Manea D L, Rosca I C, Aciu C, Munteanu C 2017 *Procedia Eng* **181** 1005-1012.
[33] Daly P, Ronchetti P, Woolley T 2009 *Epa Strive* 126
[34] Pretot S, Collet F, Garnier C 2014 *Build Environ* **72** 223-231
[35] Hempcrete Cymru 2021 https://www.hempcretecymru.com/what-is-hempcrete (Accessed 10

November 2021)
[36] Sinka M, Sahmenko G, Korjakins A 2014 *J Sustain Archit Civ Eng* **8**
[37] Mukherjee A 2013 *MSc thesis* Queen's University Canada
[38] IsoHemp https://www.isohemp.com/en/hempcrete-blocks-naturally-efficient-masonry. (Accessed 14 November 2021).
[39] American Lime Technology http://www.americanlimetechnology.com/tradical-hembuild/. (Accessed 14 November 2021).
[40] Vicat 2018 https://www.vicat.com/Vicat-Group/News-from-the-Group/Byosis-Interlocking-hempcrete-blocks. (Accessed 15 Noiembrie 2021).
[41] Florentin Y, Pearlmutter D, Givoni B and Gal E. 2017 Energ Buildings **156** 293–305
[42] Arrigoni A, Pelosato R, Melia P, Ruggieri G, Sabbadini S and Dotelli G 2017 J Clean Prod **149** 1051–1061

# The use of perlite in civil engineering applications

**S G Maxineasa[1]*, D N Isopescu[1], M L Lupu[1], I-R Baciu[1], L Pruna[2] and C Somacescu[3]**

[1] "Gheorghe Asachi" Technical University of Iasi, Faculty of Civil Engineering and Building Services, Department of Civil and Industrial Engineering, Romania
[2] "Gheorghe Asachi" Technical University of Iasi, Faculty of Civil Engineering and Building Services, Department of Graphic Engineering, Romania
[3] SC PROCEMA PERLIT SRL, Ilfov, Romania

* sebastian.maxineasa@tuiasi.ro

**Abstract**. In the last years, research applications in the construction sector have taken a different turn due to rapid urbanization and the necessity to protect the environment and create a balance between the natural and built environments. Taking into account the fact that the thermal efficiency of buildings has an important influence over the environmental performances of the construction sector, the civil engineering specialists have tried to develop and use different new solutions in order to improve this essential building characteristic. Thus, different new materials have been used with the goal of reducing the overall thermal transmittance of the envelope elements and of the total energy consumption needed for heating and cooling. Perlite is one of the materials that have been considered for developing new energy efficient construction applications. The main goal of the present paper is to present all the relevant solutions that have been developed and used in the construction sector that consider perlite as a component part. The present study is the first part of a more extensive research program that takes under evaluation the possibility of using perlite as a thermal insulating material for different constructions.

## 1. Introduction

Thermal insulating plasters have emerged as an attractive option for insulating existing wall structures, particularly older masonry structures, where refurbishment interventions may require the replacement of damaged plasters. Intensive research is being conducted to optimize the thermal conductivity and environmental impact of these materials (combination of lightweight aggregates, binders and additives) [1].

Building energy consumption has overtaken industrial and transportation energy consumption to become one of society's three largest energy consumers [2]. In China, buildings use 40% of total energy [3,4]. Thermal insulation materials used in or on a building's interior and exterior walls are an efficient approach to minimize energy consumption [5–7]. This material is meant to prevent interior building heat from escaping to the outside, or to achieve the reverse, by modifying the structure of the building's exterior enclosure in order to preserve the inside temperature of the building. Thermal insulation materials are often used in three categories: organic, inorganic, and thermal insulation-mortar. Organic thermal insulation materials include polyurethane, polyethylene, and polystyrene, which are all extensively utilized for their superior thermal insulation qualities and inexpensive cost [8,9]. They do, however, burn easily and emit hazardous gases throughout the combustion process. As a result, extensive usage of these materials increases the danger of building fires and has a negative impact on the environment [10–12].

The most of the world's building stock was constructed prior to 1970, when there were no regulations demanding energy-efficient construction. Additionally, the primary energy consumers were associated with civil use, with space heating/cooling accounting for the lion's share of overall energy consumption [13]. The most essential component was the installation of opaque components, which contributed to the building's increased energy loss [14]. As a result, retrofitting existing buildings with energy efficient components is a viable method of reducing the building's overall energy use. Given the predominance of wood in the built environment, thermal insulating plasters may be considered suitable for energy retrofits. These materials are similar to ordinary plasters except for the absence of sand and the addition of a variety of lightweight aggregates (LWA) that contribute to the reduction of heat conductivity.

Perlite is a siliceous volcanic rock with a trace of mixed water (between 2 percent and 6 percent). Perlite ore is roughly 75% SiO2, with oxides of A1 (14.8%), K (4.8%), Na (2.9%), Ca (0.9%), Mg (0.1%), R2O3 (1.5%), and bound water (3.0%). When presized crude perlite particles are burned to temperatures more than 870°C (1600°F), they expand infinitely to create a network of sealed glassy cells resembling popcorn. This explains expanded perlite's exceptional insulating and lightweight qualities. While crude perlite is translucent to light grey to glossy black in hue, expanded perlite is snowy white to greyish white in color. Expanded perlite is a form of synthetic pumice that offers many benefits over natural pumice. Its characteristics may be controlled by adjusting the duration and temperature of calcination to achieve specified grain size and porosity [15].

Perlite is a completely eco-friendly substance. Perlite is now employed in a range of applications, including horticulture, agriculture, pisciculture, refractory and cryogenic uses, filtration and disinfection, and, lastly, building.

The major goal of this study is to outline all significant perlite-based solutions that have been developed and used in the construction sector. The present study is the initial stage in a bigger research program that will examine the thermal insulating properties of perlite in a range of designs.

## 2. The areas of applications for perlite

Expanded perlite is a white, ultra-lightweight aggregate with particles ranging from extremely fine powder to up to 6 mm in diameter. It is inorganic, inert, pH neutral, biocompatible, and free of asbestos. Additionally, it is a completely eco-friendly substance. Today, perlite is utilized in a wide number of applications, including horticulture, agriculture, pisciculture, refractory and cryogenic uses, filtration and disinfection, and building.

It has exceptional thermal insulation qualities throughout a broad temperature range, from cryogenic applications operating at minus 273°C (absolute zero) to refractory applications operating at temperatures above 1000°C. Additionally, because of its highly adsorbent surface and low bulk density, it serves as an excellent carrier or low-cost filler for a broad range of chemical compositions.

Perlite is utilized as a heat insulator in refractory and cryogenic applications because of its outstanding thermal characteristics. It is crucial in filtration and decontamination operations since it is a superior filter and absorbent material that is both recyclable and ecologically friendly [15].

However, due to its excessive water retention capacity, soil loosening, and airing qualities, it is most usually utilized in agricultural and horticultural operations. Additionally, it is employed as a substrate in aquaponic (totally ecological) and hydroponic vegetable gardens, where it is used in place of mineral wool cubes [16-17].

In recent years, substrate and hydroponic systems have been developed to boost production uniformity and to make crop upkeep and harvesting simpler, leading in greater crop yields and quality. Additionally, these technologies help in isolating and minimizing disease issues, which are becoming increasingly difficult to manage in soil-grown strawberries due to the prohibition or difficulty of soil disinfection [18].

Hydroponics, or "soilless" plant culture, presupposes that the roots of plants are immersed in a nutrient solution composed of water and nutrients or planted in an inert and porous environment (mineral wool, clay, perlite, vermiculite, coconut fibers, etc.) that retains the nutrient solution, allows for root oxygenation, and supports the plants' growth (figure 1). These systems may cultivate a wide range of plants, including green vegetables, tomatoes, cucumbers, peppers, and strawberries.

**Figure 1**. The use of perlite in agriculture - hydroponic crops [19].

Hydroponic systems are an integral part of every plant production. Commercial hydroponics for greenhouse crop production are already available in the form of nutrient film methods and deepwater culture systems specifically suited for small crops such as leafy greens [19]. Seedlings and transplants, as well as fruiting crops, have been produced using substrate culture. Aeroponics is another exciting way of plant cultivation. Today's hydroponic systems are adapted and modified from greenhouse crop production techniques. Thus, the essential ideas and understanding of nutrient control in hydroponics are same [20].

Whatever hydroponic approach is utilized, it is vital to keep in mind that plant roots need three things: water, nutrients, and oxygen. Hydroponic system differences arise from the method in which these three components are supplied to plant roots.

Hydroponic fertilizers must be water soluble and include all of the components necessary for plant growth in a range of combinations that are suitable for each crop and developmental stage. A good hydroponic fertilizer must include nitrogen, potassium, phosphorus, calcium, magnesium, sulfur, iron, manganese, copper, zinc, molybdenum, boron, and chlorine. It is necessary to use hydroponic-specific fertilizers, which are readily available at specialty retailers. Temperature and pH of the fertilizer solution must be controlled to fulfill the unique needs of each crop. The water used to prepare the nutrient solution must be of excellent quality, filtered, and tested for mineral salts, bacteria, and other pollutants.

Aquaponics (figure 2) is a hybrid of hydroponics (growing plants without soil and using chemical fertilizers) and aquaculture (fish farming) (growing fish). Aquaponics is a self-contained system made up of two subsystems (hydroponics and aquaculture). They are held together by water [21]. Water is crucial for delivering nutrients from fish excrement to plants. Thus, the plants expand with the support of these nutrients, the nutrients are absorbed into the water, and the water is cleaned in this way, allowing the fish to grow in a healthy environment.

Aquamony is a very effective way of generating a great quantity of food in a little amount of area via the use of a waste-to-resources ecosystem. Beneficial bacteria convert fish waste into chemical compounds (nutrients) that are readily absorbed by plants (vegetables, fruits, etc.) in a recirculating system through the filtering system [22].

Due to the granular structure with a high porosity, it retains up to 35% to 40% of the water it contains, releasing it to the soil when the plant needs it. It is recommended for planting shrubs, trees, flowers, roses, and vegetables when mixed with the substrate to a maximum of 30%. When the seedling is planted, a handful of perlite is added to the bottom of the pit to support in root growth and to help keep moisture in the soil throughout the summer's dry periods.

**Figure 2**. Aquaponic system [21].

## 3. Perlite - in the building industry

Perlite is a crucial component in construction and building. It is used in conjunction with cement, bitumen, lime, gypsum, or other binders to impart desirable qualities to concrete or mortar. Aggregates provide volume stability, resistance to wear and erosion, and other required physical attributes to final products. There are two types of lightweight aggregates: natural and manufactured. Diatomite, pumice, volcanic scoria, sawdust, volcanic cinders, and tufts are natural lightweight aggregates; expanded shale, slate, perlite, and vermiculite are manufactured lightweight aggregates. All of these substances are often utilized in mortar and concrete. These lightweight man-made aggregates arise during the heating process when the aggregate loses its crystallization water and expands. Apart from these aggregates, solidified blast furnace slag is utilized in the manufacture of lightweight or reinforced concrete [23,24].

Expanded perlite has a broad range of applications in the construction, industrial, chemical, horticultural, and petrochemical sectors. Perlite is a common horticultural material in the United States owing to its capacity to promote soil aeration and drainage. Perlite keeps moisture and nutrients on the surface, which roots and cuttings may take advantage of [25].

Standards have been developed for a range of industrial uses, including thermal insulation for calcium silicate blocks and pipelines, structural insulating boards, and asbestos-cement-perlite composite boards.

Due to the low unit weight of expanded perlite (EP), it provides substantial advantages in terms of heat and sound insulation in the building sector. EP, on the other hand, degrades the physical and mechanical qualities of concrete owing to its high potential for water absorption. As a result, it is anticipated that coating the EP with a polymer would significantly reduce water absorption, thereby improving the EP's mechanical and physical properties [26].

Numerous studies have been conducted to highlight the benefits of adopting this kind of material in the building sector. According to a study conducted by the International Association of Perlite [27], several brick walls with two large voids of varying thicknesses were filled with perlite; the walls filled with perlite increased thermal transmittance between 30% and 65% (depending on the wall thickness); it was also discovered that the density of the perlite filling had no effect on the thermal transmittance $U$ except in a small proportion of 5%; additionally, the difference found of the level of the pearlite layer was not affected by the density of the perlite filling [28].

Thermal insulating plasters are formed by combining a range of various kinds of Light Weight Aggregates (LWA) into a matrix of cement, gypsum, natural hydraulic lime, or hydraulic lime. Mineral aggregates, including perlite, vermiculite, diatomite, zeolite, and pumice, are a common kind of LWA.

Numerous investigations on plasters comprised of these diverse aggregates have been undertaken. L.M. Silva et al. [29] performed research in 2009 on a number of perlite and vermiculite-based plaster compositions. The research established that increasing the mineral concentration decreased the mechanical characteristics of the material. Another research [30] examined how cement mortars interact with expanded perlite, polystyrene, expanded glass, and hollow microspheres. Another conclusion of this research is that polystyrene and perlite were the LWAs with the most significant influence on mechanical qualities. S. Abidi et al. [31] created a biphasic gypsum, perlite, and vermiculite plaster system. His computational and experimental studies indicated that the inclusion of a mineral aggregate enhanced porosity and lowered heat conductivity. Thermal conductivity was lowered by 25% when perlite and vermiculite were used: 0.23 W/mK for vermiculite and 0.16 W/mK for perlite (from a starting value of 0.50 W/mK).

Perlite is often used in building sector, because of its outstanding thermal insulation characteristics and low weight. Perlite improves both fire and acoustic protection in addition to its thermal insulating properties. When pearly rock is used to manufacture building materials, the resulting product is lightweight, fire resistant, and possesses exceptional thermal insulation properties, making it ideal for roof tiles, partitions, and exterior walls that require a high level of fire resistance, as well as areas with high humidity. Additionally, the perlite enables the walls to breathe, creates a natural microclimate, is allergen-free, and retains its distinctive characteristics in perpetuity.

Perlite is processed without the use of chemicals; it is chemically inert, and no by-products are produced. Perlite is non-toxic and harmless to humans, animals, and marine life. At the end of its useful life, perlite is returned to the ecocycle in its entirety. Perlite is also non-polluting in terms of water and soil, since it contains no heavy metal compounds or toxic organic pollutants. Perlite is a form of natural glass and as such is chemically inert with a pH of roughly 7 [32].

Perlite is a limited resource. The world's perlite reserves are estimated to be 700 million tonnes. Armenia has confirmed perlite resources totaling 150 million cubic meters, while projected resources total up to 3 billion cubic meters. With a specific density of 1.1 ton/m$^3$, Armenia's confirmed reserves total 136 million tons. Other reported reserves include Greece's 120 million tonnes, Turkey's, the United States of America's, and Hungary's approximately 49-57 million tonnes. The world's perlite production, led by China, Turkey, Greece, the United States of America, Armenia, and Hungary, totaled 4.6 million tonnes in 2018 [33,34].

The following are some of the benefits of using perlite:

- low weight;
- ease of handling, mixing, and shipping;
- ease of commissioning;
- exceptionally low humidity (1%);
- excellent heat insulator, acoustical, and fireproof.

There are numerous well-conducted studies of workers exposed to perlite ore and expanded perlite in the workplace, some of which included workers exposed to perlite dust at levels above the current OEL. Taken together, these findings provide compelling evidence that occupational exposure to perlite dust has relatively few adverse health effects; there is little (if any) pneumoconiosis, lung function decline, or respiratory symptoms [35, 36]. The morbidity study's reported findings contradict the negative findings of several studies of workers occupationally exposed to perlite. While environmental exposure occurs 24 hours a day, seven days a week, whereas occupational exposure occurs only during work hours, mining dust concentrations in the workplace are almost certainly significantly higher than those in the environment [36].

The objective of this project is to illustrate how expanded perlite may be utilized to produce lightweight, thermally insulating, sound-absorbing, and fire-resistant floor supports (figure 3).

*Pearlic screeds* are an unstructured floor support system made up of pearly screeds. The pearly hoes are used to distribute the weight evenly over the floorboards, strengthen their thermal resistance, and offer suitable fire protection for all kinds of floorboards. These materials may be reinforced dispersively or with concrete steel, and they can be applied manually or mechanically. Additionally, they offer the benefit of human or automated final polishing [37]. These pearlescent screeds are recommended for the following applications: Interior spaces supported by reinforced concrete, wood, or profiled sheet flooring. External terraces constructed of slope-light concrete that doubles as a waterproof membrane.

*Perolytic iso-pat system with semi-wet screed* is a multilayer floor support system comprised of a granular, dry bed of expanded perlite covered by a thin layer of weakly reinforced semi-wet screed [37]. This floor support system can be used for a variety of purposes, including thermal and sound insulation of floors, reducing the load on concrete, sheet, or wood flooring by two to four times, implementing an underfloor heating system, and economically and quickly replacing the double-layered system composed of dense screed and extruded polystyrene.

**Figure 3**. The use of perlite in the field of construction - different types of pearlic screed [37].

*Iso-perolytic bed system with dry screed* is a two-layer flooring support consisting of a granular, dry bed of expanded perlite covered by two rows of rigid plates [37,38].

Taking the preceding parts into account, the following features of this product can be asserted:

- Provides insulation and soundproofing for floors installed over concrete, wood, or profiled sheet flooring while maintaining an A1 fire rating.
- Thermal insulation of concrete slabs over the ground and at the foundations of houses;
- Complete compatibility with underfloor heating elements;
- Virtually unlimited dimensional stability over time and at temperatures up to $500^0$C;
- Easy integration of electrical, water, and heating networks when applied;
- The ability to glue the waterproofing membrane with both adhesive and flammable glue.
- Due to the granular perlite's capacity to entirely fill the area intended for it, it inhibits the construction of thermal or acoustic bridges (such as those made by heat-insulating materials in the shape of plates).

The floor supported by *a wood framework filled with insulating granular perlite* is useful for thermal insulation and soundproofing floors, as well as lowering the weight of concrete, sheet, or wood flooring by up to tenfold. Among the unique characteristics of this kind of floor support are the following:

- It insulates and isolates the sound of floors installed over concrete, wood, or profiled sheet flooring. It requires no drying time and can be laid immediately after the tiles are installed. It thermally insulates concrete slabs above ground, at dwelling foundations. It thermally insulates the floors between the attic and the top floor of a home.
- Due to the very low density of the perlite bed (35 kg/m$^3$ – 230 kg/m$^3$) and the wooden structure, the system is unsuitable for rat and insect nesting.

## 4. Future research

Due to the wood-like qualities of composite materials comprised of lightweight, insulating, and acoustic perlite, they work as a wood replacement. Perlite products may put little stress on existing walls and building structures due to their low weight and fire resistance. Perlite aggregate plaster resists fire and improves the structure's resilience to cracking.

Due to the fact that the core material contributes for 40% of the overall cost of a fumed silica-based vacuum insulation panel (VIP) [37,38], recent years have seen a rise in research into possibly lower-cost core materials [39–45]. Expanded perlites, which now cost roughly a sixth of the price of fumed silica, have been discovered as prospective options for core materials. Apart from its usage as a VIP core, evacuated perlites may be utilized directly as insulation in a range of applications, including thermal storage tanks and cold storage boxes.

Although perlite and fumed silica have a similar chemical makeup (over 90% $SiO_2$) and are both categorized as powder insulants, their heat conductivities are significantly different. This is because their particle organization order (geometric structure) and structural parameters such as particle size, pore size, and porosity are unlike [46].

Surprisingly, the majority of these factors are controllable during the perlite manufacturing process (particle/pore size, porosity) or the insulation manufacturing process (particle/pore size, porosity) (pressure, density). It is critical to understand the quantitative effects of these factors on the thermal conductivity of evacuated perlite in order to build high-performance, cost-effective VIPs.

Following a review of the specialized literature on perlite and its application in the field of construction, several research directions regarding the use of this material as a heat-insulating material for timber frame type constructive structures have been identified; therefore, it is desired to expand on the study presented in this article and also to find thermal insulation solutions that are appropriate for our country's requirements [47-50].

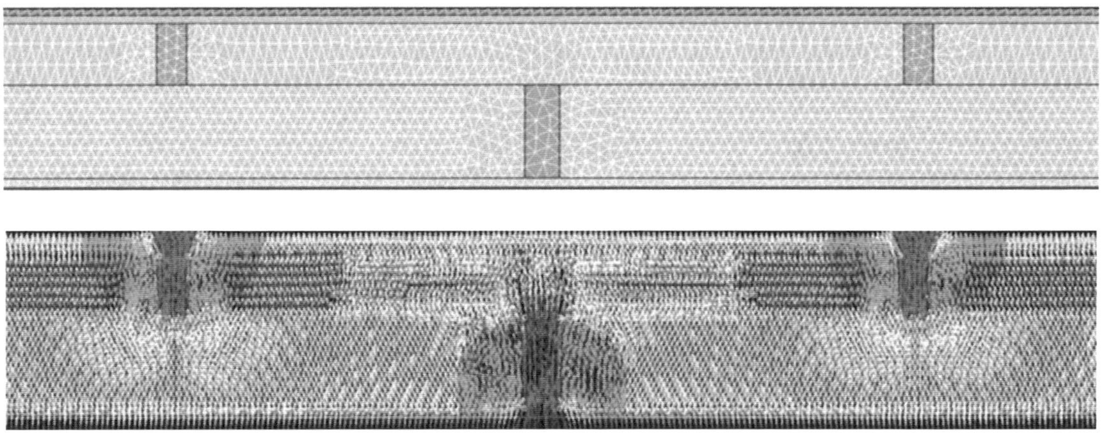

**Figure 4.** Thermal bridge analysis for a timber frame structure with perlite thermal insulation layer.

To continue this inquiry based on the early findings, more research will be undertaken using thermal simulations conducted using the RDM 6.17 software. The major purpose of this study will be to investigate and compare the thermal characteristics connected with the different forms of thermo-insolation that are available for this structure. This study also includes an examination of thermal bridges in relation to a wood frame construction with a perlite layer of thermal insulation (figure 4).

## 5. Conclusions

Perlite is a versatile and ecologically friendly mineral that is mined and processed in a manner that has a negligible effect on the environment. Additionally, the green community views perlite-enhanced products—such as lightweight insulating concrete roof decks—as a high-performance solution drawn from a practically infinite natural resource.

Globally, the building sector utilizes more than 50% of expanded perlite, according to data. Perlite is widely used in the construction industry as an aggregate in lightweight insulating concrete and plaster; as a loose fill insulation material for concrete masonry blocks, cavity walls, and residential homes; and in a variety of specialty applications such as fireproofing sprays, chimney fillers, interstitial floors, and acoustical sprays.

Civil engineering professionals have attempted to design and implement novel methods to enhance this critical building feature, recognizing that thermal efficiency has a substantial impact on the construction industry's environmental performance.

Taking into account the information offered in this article, it can be concluded that perlite may be utilized in a variety of ways in the area of building, most notably as a heat-insulating material.

**Acknowledgments**
This paper was elaborated with the support of the "Ecoinnovative Products and Technologies for Energy Efficiency in Constructions – EFECON" research grant, project ID P_40_295/105524, Program co-financed by the European Regional Development Fund through Operational Program Competitiveness 2014-2020.

**References**

[1] Fenglio E, Fantucci S, Serra V, Carbonaro C and Pollo R 2018 *Energy & Buildings* **179** 26-38.
[2] E.N.E.A. 2017 *(RAEE) 2017: analisi e risultati delle policy di efficienza energetica del nostro Paese*.
[3] U.S. Department of Energy 2014 (energy.gov).
[4] Geng Y, Ji W, Lin B, Hong J and Zhu Y 2018 *Energy Build.* **172** 181–191, https://doi.org/10.1016/j.enbuild.2018.04.047.
[5] Berardi H 2017 *Resour. Conserv. Recy.* **123** 230–241, https://doi.org/10.1016/j.resconrec.2016.03.014.
[6] Zhao H-X and Magoulès F 2012 *Renewable Sustainable Energy Rev.* **16** 3586–3592, https://doi.org/10.1016/j.rser.2012.02.049.
[7] Sadineni S B, Madala S and Boehm RF 2011 *Renewable Sustainable Energy Rev.* **15** 3617–3631, https://doi.org/10.1016/j.rser.2011.07.014.
[8] Huang H, Zhou Y, Huang R, Wu H, Sun Y, Huang G and Xu T 2020 *Sustainable Cities Soc.* **52** 101840, https://doi.org/10.1016/j.scs.2019.101840.
[9] Liu L, Zou S I, Li H, Deng L U, Bai C, Zhang X, Wang S and Li N 2019 *Energy Build.* **201** 19–36, https://doi.org/10.1016/j.enbuild.2019.07.037.
[10] Aditya L, Mahlia T M I, Rismanchi B, Ng H M, Hasan M H, Metselaar H S C, Muraza O and Aditiya H B 201 *Renewable Sustainable Energy Rev.* **73** 1352–1365, https://doi.org/10.1016/j.rser.2017.02.034.
[11] Zhu Q W, Wu F D and Zhao J P 2012 *New Build. Mater.* **39** 12–16, https://doi.org/CNKI:SUN:XXJZ.0.2012-06-004.
[12] Yang H, Jiang Y, Liu H, Xie D, Wan C, Pan H and Jiang S 2018 *J. Colloid Interface Sci.* **530** 163–170, https://doi.org/10.1016/j.jcis.2018.06.075.
[13] Wang F, Wei H, Liu C, Sun H, Zhu Z, Liang W and Li A N *React. Funct. Polym.* **143** 104345, https://doi.org/10.1016/j.reactfunctpolym.2019.104345
[14] Abu-Jdayil B, Mourad A-H, Hittini W, Hassan M and Hameedi S 2019 *Constr. Build. Mater.* **214** 709–735, https://doi.org/10.1016/j.conbuildmat.2019.04.102
[15] Singh M and Garg Mridul 1991 *Construction & Building Materials* **5** 75-81
[16] Papadopoulos A, Bar-Tal A, Silber A, Saha U K and Raviv M 2008 *Soilless Culture – Theory and Practice* 505-543
[17] Grillas S, Lucas M, Bardopoulou E and Sarafopoulos S 2001 *ISHS Acta Hort.* **548** 105-113 10.17660/ActaHortic.2001.548.10
[18] Gül M, Yoruk M A, Hayirli A, Turgut L and Karaoglu M 2005 *J. Appl. Poult. Res.* **14** 217-225
[19] Masabni J and Niu G 2022 *Plant Factory Basics, Applications and Advances* 153-166
[20] S. Pedersen and T. Wik 2020 *Aquacult. Eng.* **89** 102059
[21] Wongkiew S, Hu Z, Chandran K, Lee J W and Kumar Khanal S 2017 *J. Aquac. Eng. Fish.*

*Res.* **76** 9-19 10.1016/j.aquaeng.2017.01.004
[22] Kataki R, Pandey A, Khanal S K and Pant D 2020 *Current Developments in Biotechnology and Bioengineering* 475-494, 10.1016/B978-0-444-64309-4.00020-9
[23] Rashad A M 2016 *Construction and Building Materials* **121** 338-353.
[24] Bergej B 2019 Eds *Elsevier* 1-421
[25] Swapna Mukherjee 2011 *Springer* 425–489, http://dx.doi.org/10.1007/978-94-007-1162-4_17
[26] Akyuncu V and Sanliturk F 2021 *Journal of Building Engineering* **38** 102182
[27] Perlite Institute 2018 (https://www.perlite.org/library/)
[28] Türkmen I and Kantarci A 2006 *Indian J. Eng. Mater. Sci.* **13** 247-252
[29] Silva L M, Ribeiro R A, Labrincha J A and FerreiraV M 2010 *Cement Concrete Composit.* **32** 19–24. https://doi.org/10.1016/j.cemconcomp.2009.07.0 03
[30] Lanzón Torres M and García-Ruiz P A 2009 *Cement Concrete Composit.* **31** 114–119. https://doi.org/ 10.1016/j.cemconcomp.20 08.11.0 03
[31] Abidi S, Nait-Ali B, Joliff Y and Favotto C 2015 *Composit.* **68** 392–400. https://doi.org/ 10.1016/j.compositesb.2014.07.030
[32] Reformex 2020 (https://www.reformex.ro/), Accessed 4 December 2021).
[33] M. Bennett, Shawnna 2019 *U.S. Geological Survey, Mineral Commodity Summaries*.
[34] USGS 2019 (www.usgs.gov)
[35] Chatzi L, Prokopakis E, Tzanakis N, Alegakis A, Bizakis I, Siafakas N and Lionis C 2005 Chest 217 (1) 372-378 10.1378/chest.127.1.372
[36] Perlite Institute 2011 (www.perlite.org.)
[37] Procema 2020 (https://www.procema-perlit.ro/)
[38] Builtspace 2021 (https://www1.builtspace.com/)
[39] Alam M, Singh H, Suresh S and Redpath D A G 2017 *Appl. Energy* **188** 1–8, https://doi.org/10.1016/j.apenergy.2016.11.115
[40] Almeida FA, Corker J, Ferreira N, Neto M A, Fan M, Beyrichen H and Caps R 2017 *Ciencia Tecnol. Dos Mater.* **29** 151–156, https://doi.org/10.1016/j. ctmat.2016.10.002
[41] Alam M, Singh H, Brunner S and Naziris C 2014 *Energy Build.* **69** 442–450, https://doi.org/10.1016/j.enbuild.2013.11.027
[42] Chang B, Zhong L and Akinc M 2016 *Vacuum* **131** 120–126, https://doi.org/10.1016/j. vacuum.2016.05.027
[43] Li C D, Li B, Pan N, Chen Z, Saeed M U, Xu T Z and Yang Y 2016 *Energy Build.* **125** 298–309, https:// doi.org/10.1016/j.enbuild.2016.05.013
[44] Li C D, Saeed M U, Pan N, Chen Z F and Xu T Z 2016 *Mater. Des.* **107** 440–449, https://doi.org/10.1016/j. matdes.2016.06.071
[45] Zhuang J, Ghaffar S H, Fan M and Corker J 2017 *Compos. B Eng.* **127** 215–221, https://doi.org/10.1016/j.compositesb.2017.06.019
[46] Nemanic V and Zumer M 2015 *Energy Build.* **90** 137–141, https://doi.org/10.1016/j. enbuild.2015.01.012
[47] Chen Z, Chen Z, Yang Z, Hu J, Yang Y, Chang L, Lee L J and Xu T 2015 *Energy* **93** 945–954, https://doi.org/10.1016/j. energy.2015.08.105
[48] Tang G H, Bi C, Zhao Y and Tao W Q 2015 *Energy* **90** 701–721, https://doi. org/10.1016/j.energy.2015.07.109
[49] Maxineasa S G, Isopescu D N, Baciu I-R, Lupu M L and Dragan T C 2020 *IOP Conference Series Materials Science and Engineering* **789** 012039
[50] Baciu I-R, Isopescu D N, Taranu N, Lupu M L and Maxineasa S G 2020 *IOP Conference Series Materials Science and Engineering* **916** 012004

# Studies on sliding stabilization solutions of a built site in a residential area

**M Măntulescu[1*] and I Tuns[1]**

[1] Transilvania University of Brasov, Faculty of Civil Engineering, Romania

* mmantulescu@unitbv.ro

**Abstract**. The paper presents a case study performed on a site for residential buildings, arranged by terracing on a sloping ground. The built site, subject to investigations, shows an obvious state of degradation downstream, manifested by the development of concentric cracks, penetrated deep, as a result of displacements/rotations produced in the massif of soil susceptible to sliding. The analyzed comparative solutions aim at achieving mechanical barriers developed in dep, taking the tensile efforts through reinforced soil works and improving the shear resistance of the soil.

## 1. Introduction

For the execution of the parking platform related to the apartment buildings in a residential area of Brasov city, figure 1, consolidated filling was executed, supported on the downstream side by a structure made of reinforced soil.

**Figure 1**. General view of the parking.

### *1.1. Technical characteristics*

The parking lot platform has a length of about 196.00 meters and a width of about 19.00 meters (Situation plan, figure 2). The initial terrain had a uniform slope with an inclination of about 27 - 30°, an inclination that is still present on the ground around the platform under study. Upstream of the platform we find two apartment buildings with basement + ground floor + 12 floors and a small construction with basement + ground floor, located between the two blocks. The parking platform connects access to the basement of the apartment buildings. The platform is limited by a slope with an angle of about 79° and a level difference of about 4.00 - 6.00 meters, between the upper face of the parking lot and the surface of the natural land downstream.

IOP Conf. Series: Materials Science and Engineering 1242 (2022) 012023 doi:10.1088/1757-899X/1242/1/012023

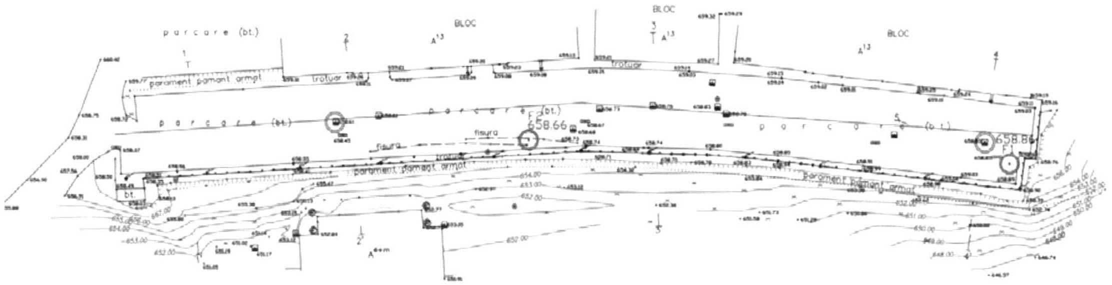


(a)

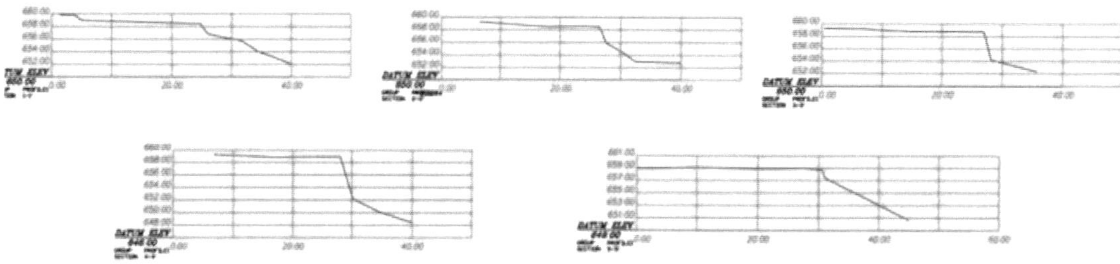

(b)

**Figure 2**. (a) Situation plan; (b) profiles of the parking area.

*1.2. General presentation of the existing situation data*

The platform was made of mechanically stabilized soil. The filling material used was obtained, according to the initial design and field observations, from the excavations carried out for the apartment buildings from the neighborhood. The reinforced soil massif, according to the project, has a variable height between 4 and 6 meters, consists of compacted elementary layers with a thickness of 50 cm. The filler material indicated in the design consists of a clay with a friction angle of 11.7° and a cohesion of 47.4 kPa.

**Figure 3**. Destroyed parament.

The reinforcement was made with Tencate Rock Pec 150 geogrid and Geocomposit 402 C geocomposites. The "protection" of the façade was made with two rows of welded nets with 10 cm meshes between which granular material was placed.

The entire car park shows obvious signs of failure of the substrate, with possible consequence on the stability of the site.

The arguments behind this statement are as follows:

- the facing consists of two rows of welded nets at a distance of approx. 30 cm, with 10 cm mesh. It is destroyed in several places and the material between the nets (fine, scattered soil), is mostly washed (see figure 3).

- the curb was detached from the road on a considerable portion, 50 - 70 m, in a semicircular shape, with a maximum arrow of 10 - 12 cm, which indicates a displacement of the support structure in the direction of overturning (see figure 4).

**Figure 4**. Displaced curb.

Concentric semicircular cracks appeared in the rigid system of the road, indicating again a displacement (at least in the sense of settlement), of the terrain (see figure 5).

**Figure 5.** Semicircular cracks in road.

The rainwater collection system is compromised; all sewer manholes are clogged and the piping is certainly displaced. This is deeply damaging to the supporting structure, which is thus additionally loaded with a saturated material and the mechanical properties are consequently reduced.

### *1.3. Geological-geomorphological context*

The studied location belongs from a geomorphological point of view to the piedmont on the southern slope of Mount Tâmpa. The piedmont was formed by the twinning of manure cones, resulting from the deposition of disintegrated material and transported by torrents. As a result, these proluvial deposits have a cross-stratification in which they alternate layers with very varied granularity (rubble, clays, sandy clays, dusty sands and gravel) [1]. A diluvial clay layer lies over them, which tapers downstream.

The bedrock, the Jurassic Tithonian-Barremian reef limestone, is found at depths of tens of meters, perhaps over a hundred meters.

The geological map of the studied area is presented in figure 6.

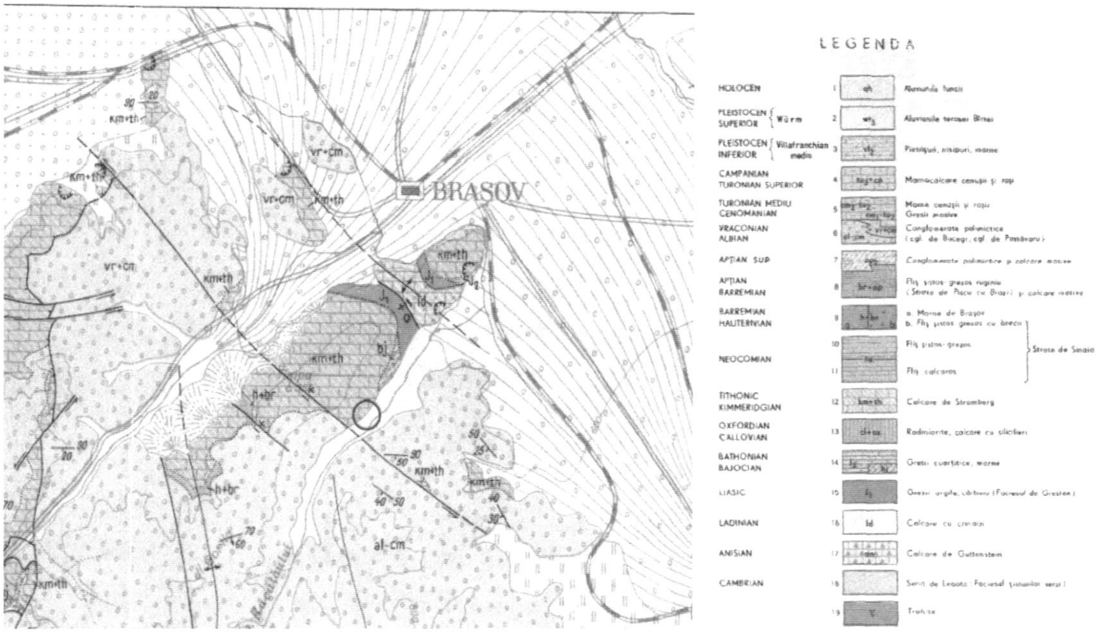


**Figure 6.** Geological map of the area.

## 2. Geotechnical investigation of the site and the analyses of the proposed consolidation solutions

Soil investigations was made by drilling with continuous coring and collecting undisturbed samples and also by dynamic penetrations. The results of the investigations revealed a complex lithology, in which cohesive soil alternates with non-cohesive material. According to the groundwater level also, the slip surfaces are situated at 6 and 9 meters depth from the ground level at that time. Here are the results of the most significant boreholes:

Borehole - F1

0.00 - 0.20 m concrete

0.20 - 1.10 m broken stone filling;

1.10 - 8.00 m semi-cohesive filling made of silty clay with fragments of limestone, thick plastic, reddish;

8.00 - 11.00 m silty clay, hard plastic, reddish;

F2 Borehole - F1

0.00 - 0.20 m concrete;

0.20 - 1.60 m broken stone filling;

1.60 - 6.50 m filling of reddish silty clay with intercalations of limestone with a frequency of about 0.50 meters;

6.50 - 10.00 m silty clay, hard plastic, reddish, with intercepted limestone intercalations at 7.10 and 8.00 meters respectively;

Aspects concerning the shape of the samples taken are presented in the figures 7a, 7b, and a view of the drilling rig used, is shoved in the figure 7c.

a) b) c)

**Figure 7.** a, b - Samples of soil investigations; c - The drilling rig on the site.

Slope stability was determined with Geo 5 software by analytical methods. Presuming only that the road will be restored, the safety factor was inacceptable (less than one). The next step was calculated by adding a wall of piles, for which the dimensions were adjusted until a safety factor over 1.5 has been obtained. The length of the piles was calculated on the ground pressure plus traffic surcharge, without counting the passive resistance of the soil below the road. This lack of resistance was counted because the sliding phenomenon is much larger and it is possible to go on in the future. The system with a sheet of piles was conceived according to [2], and over it a blanket was made of reinforced soil.

Aspects concerning the profile networks used in Geo 5 software and the sliding mass highlighting are showed in figure 8.

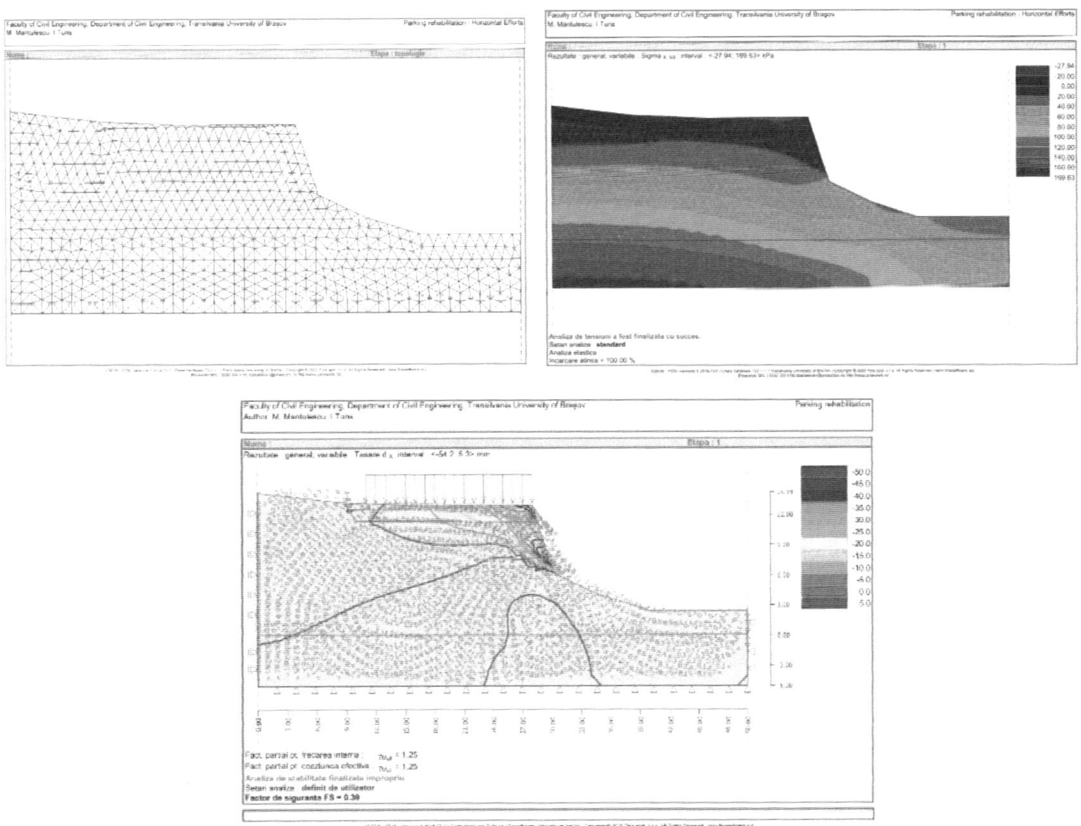

**Figure 8**. The networks of profile with sliding mass highlighting.

IOP Conf. Series: Materials Science and Engineering 1242 (2022) 012023 doi:10.1088/1757-899X/1242/1/012023

Given the fact that the structure of the road is rigid, we consider that the existing cracks in the form of an arch are due to the horizontal efforts that are induced by the displacement of the foundation ground consisting of reinforced soil.

The theoretical displacement of this massive support would be, as seen in figure 9, of 33.7 cm, but it is limited by the friction with the rigid concrete structure and by the layers that form the road structure.

Part of this movement, around half of it, was unevenly consumed along the structure. This can be seen at the contact between the curb and the road.

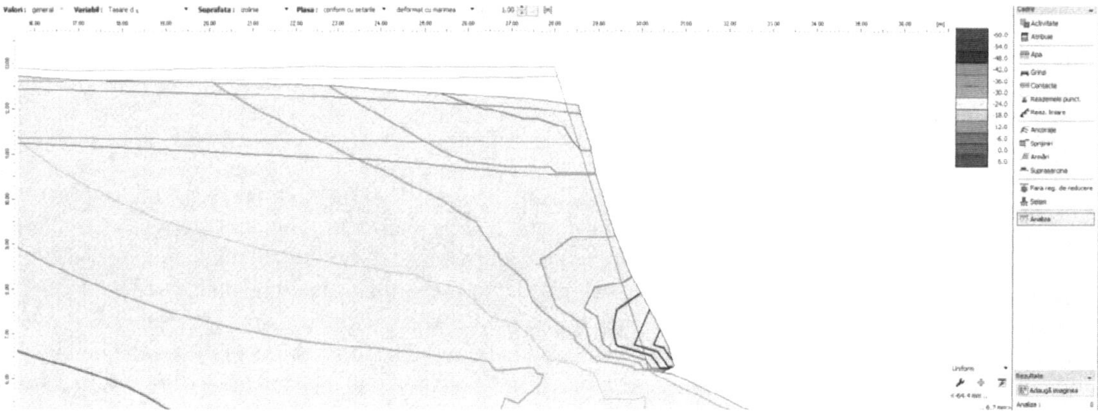

**Figure 9**. The deformed massif with the horizontal displacement contours (in millimeters) calculated by FEM analyses (GEO5).

Based on the aspects presented, the following variants are proposed as solutions to strengthen the retaining wall:

**Table 1**. Proposed variants with advantages and disadvantages.

| 1. Proposed variant | 2. Advantages | 3. Disadvantages |
|---|---|---|
| A. Sheet of drilled piles | Safe solution without total disruption of parking functioning | Expensive and less environmentally friendly solution |
| B. Retaining wall with buttresses | Safe solution without total disruption of parking functioning | Very expensive solution, less ecological and with a poor aesthetic |
| C. Grouting jet in the existing structure | Discreet solution that allows a free aesthetic arrangement | Less reliable solution because the injected solution will propagate on predetermined planes, without absolute control |
| D. total restoration of the support structure with reinforced soil | Safe solution provided correct execution | Solution that involves decommissioning the entire parking lot over a period of several months. At the same time, it will be necessary to create a platform for the storage of the filling material which implies higher costs and legal difficulties. |

IOP Conf. Series: Materials Science and Engineering 1242 (2022) 012023 doi:10.1088/1757-899X/1242/1/012023

VARIANT A

Making a sheet of drilled piles, located at the top of the existing retaining wall, at approx. 1.00 m from the facing of the existing retaining wall. The sheet will be made of tangent piles. The piles will be secured at the top with a crown beam, located below the current elevation of the platform, and the sidewalk can be made as a console of the crown beam.

The sheet of piles will be calculated when pushing the ground using the characteristics of the geotechnical study, without taking into account the existing retaining wall of reinforced soil [3].

It is recommended to use the recoverable metal pipe technology of execution.

The solution also involves a local stability analysis of the existing retaining wall, given the load by the machine used.

VARIANT B:

Execution of a reinforced concrete retaining wall with buttresses, executed downstream of the existing retaining wall and subsequent execution of the filling between the facing of the existing retaining wall and the elevation of the new retaining wall.

The solution involves the use of digging machines to build the retaining wall, which cannot work on some areas of the alignment.

VARIANT C:

Execution of a reinforcement with resistance net and perforated anchors in the filling of the platform. An anti-erosion and facing protection system will be provided. The solution requires a deeper understanding of the existing geotechnical parameters of the filling material. Given the location conditions, the technical expert recommends option A or C to consolidate the existing retaining wall, choosing which will be more economically advantageous and taking into account resistance / stability.

In both variants, repair works will be carried out for the concrete platform of the parking lot and the rainwater collection systems will be cleared / replaced.

VARIANT D:

Total restoration of the support structure with reinforced soil.

VARIANT E:

Construction of a retaining wall downstream of the existing retaining wall. Foundations on reinforced concrete piles, executed at the level of the parking platform and reinforced concrete elevation + protective parapet. After the construction of the new retaining wall, a filling will be made between its reinforced concrete elevation and the facing of the existing retaining wall.

The solution involves an analysis of the stability of the existing retaining wall, at the load given by the machine used, given that the conformation of the terrain does not allow, along its entire length, the execution of piles in the downstream area of the existing wall.

A situation plan marking the proposed work area is presented in figure 10.

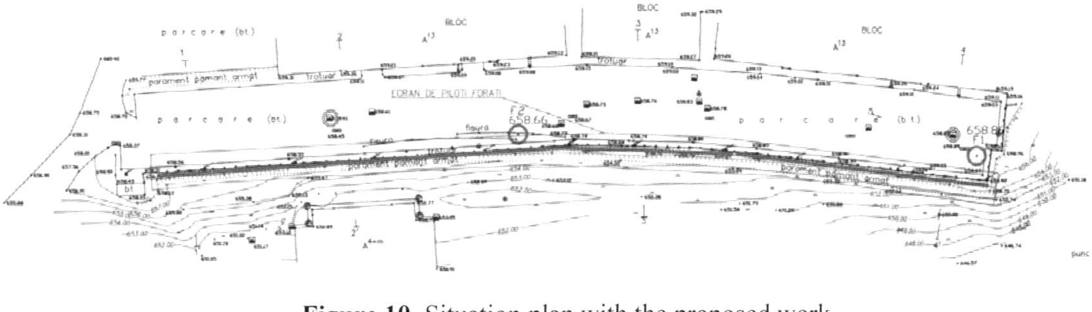


**Figure 10**. Situation plan with the proposed work.

To accommodate additional loads that were induced to execution by downstream compaction, a row of piles was provided. These were embedded in the base rock layer with an average length of 12.00 meters (from West to East - 13.00; 11.00; 9.00; 6.00 m) with 0.80 m diameter, at a distance of at least 80 cm apart. The so described position of the piles is also in agreement with the [4] report's conclusions. The continuity of the piles will be checked by with down-hole method, according [5].

Because of the unpredictability of the events, a monitoring program is needed to be set up to raise alerts of potential displacements that go over the reasonable limits.

The support structure was calculated in accordance with the provisions of Eurocode 7 (geotechnical design) and with the Romanian norm NP 122: 2010 (Determination of the characteristic and calculation values of the geotechnical parameters) [6] as follows:

**Table 2** – Used partial factors

| Partial fact. for Material (M) - Status Project. Permanently | | |
|---|---|---|
| partially fact for internal friction: | $\gamma_\phi$ = | 1.25 [–] |
| partially fact for effective cohesion: | $\gamma_c$ = | 1.25 [–] |
| partially fact for undrained shearing resistance: | $\gamma_{cu}$ = | 1.40 [–] |

| Fact. partial. for. actions (A) Status Project. permanently | | | | | |
|---|---|---|---|---|---|
| | | Status STR | | Status GEO | |
| | | Unfavorable | Favorable | Unfavorable | Favorable |
| Standing actions: | $\gamma_G$ = | 1.35 [–] | 1.00 [–] | 1.00 [–] | 1.00 [–] |
| Variable actions: | $\gamma_Q$ = | 1.50 [–] | 0.00 [–] | 1.30 [–] | 0.00 [–] |
| Act. from water: | $\gamma_w$ = | - | - | 1.00 [–] | |

**Filling made of red silty clay with limestone intercalations**

| | | | |
|---|---|---|---|
| Volumetric weight: | $\gamma$ | = | 20.00 kN/m$^3$ |
| Internal friction angle: | $\varphi_{ef}$ | = | 11.10 ° |
| soil cohesion: | $c_{ef}$ | = | 10.00 kPa |
| Volumetric weight saturated | $\gamma_{sat}$ | = | 22.00 kN/m$^3$ |

**Silty clay reddish, firmly**

| | | | |
|---|---|---|---|
| Volumetric weight: | $\gamma$ | = | 19.70 kN/m$^3$ |
| Internal friction angle: | $\varphi_{ef}$ | = | 3.00 ° |
| soil cohesion: | $c_{ef}$ | = | 80.00 kPa |
| Volumetric weight saturated | $\gamma_{sat}$ | = | 20.50 kN/m$^3$ |

**Tithonian limestone**

| | | | |
|---|---|---|---|
| Volumetric weight: | $\gamma$ | = | 25.00 kN/m$^3$ |
| Internal friction angle: | $\varphi_{ef}$ | = | 29.00 ° |
| soil cohesion: | $c_{ef}$ | = | 150.00 kPa |
| Volumetric weight saturated | $\gamma_{sat}$ | = | 25.50 kN/m$^3$ |

In according with GEO 5 software, the calculation scheme for sizing the retaining structure is represented in figure 11 and the diagrams of the pressure on the structure, displacing, bending moments and shearing, in figure 12.

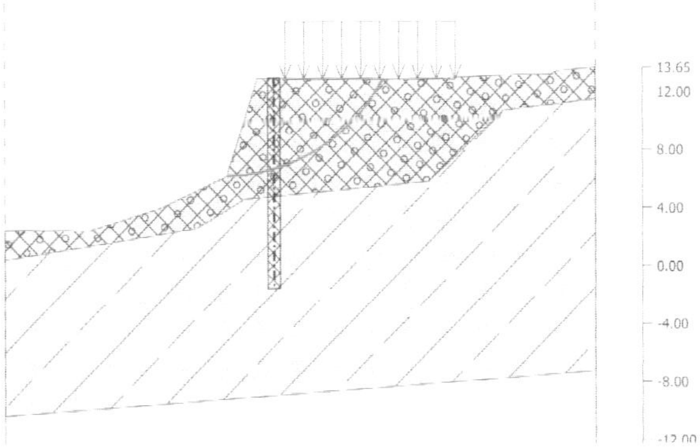


**Figure 11.** Dimensions in meters of the retaining structure.

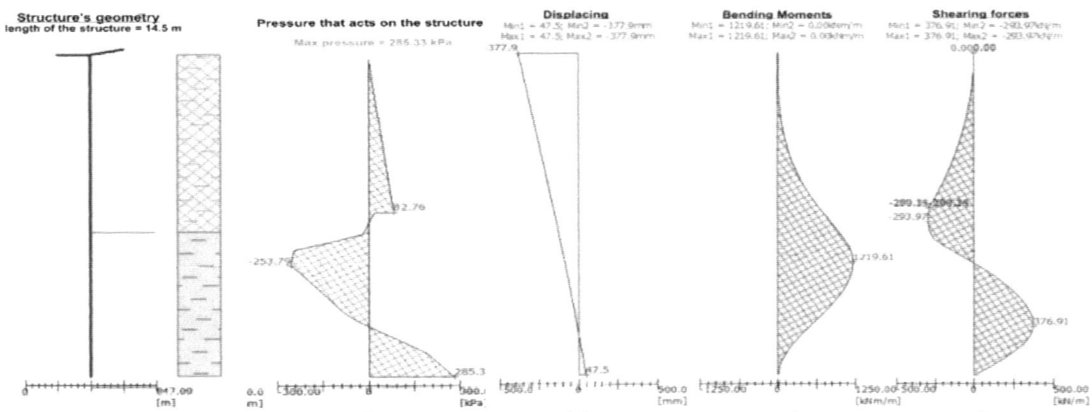


a) b) c) d)

**Figure 12.** Horizontal pressure (a) on the structure, displacing in millimeters (b), bending moments (c) in kN and shearing forces (d) in kPa on the designed structure.

The calculation scheme for verifying of the retaining structure and dimensions of the front face (reinforced soil with aesthetic role), in according with GEO 5 software, is presented in figure 13.

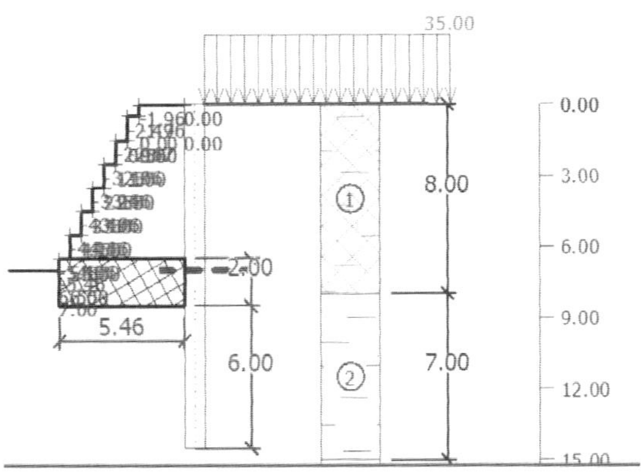


**Figure 13.** The retaining structure sketch for verifying it according to [7] and [8] with dimensions in meters of the front face, the surcharge at top in kPa and the soil layers dimensions that act on it.

In figure 14 it can be seen that the dimensions of the estimated horizontal displacing for a 8 meters high is according to the calculus made with GEO5. It is obvious that the first situation (a) for the soft clay fits to the analyzed situation.

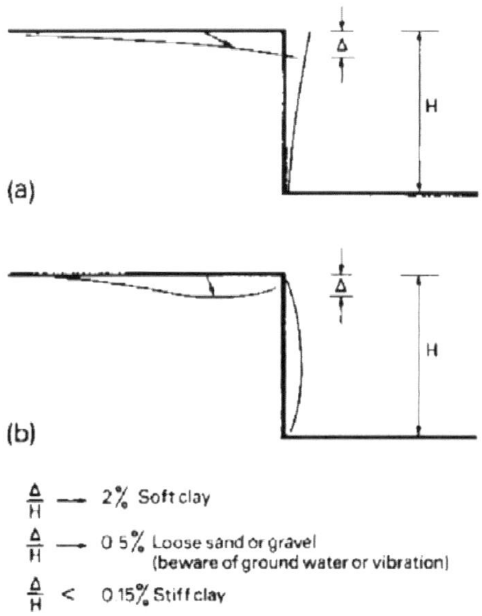


**Figure 14.** The diagram of the displacement beneath the loaded area [9].

## 3. Conclusions

Considering, on the one hand, the fact that the car park cannot be taken out of service for a long period of time, and on the other hand, the cost and safety, the sheet of piles are the best solution.

The consolidation work has an average total length of 199 m, being made up of 5 reinforced concrete rigid rafts and 224 drilled piles, Ø 0.88 m, respectively: the T1 and T2 rigid rafts, 94 piles with length of 13.00 m, the T3 rigid raft, 53 piles with length of 11.00 m, the T4 rigid raft, 38 piles with a length of 9.00 m, and the T5 rigid raft, 39 piles with a length of 6.00 m The reinforced concrete rafts have a height (thickness) of 1.20 m, and the width between 2.40 m – 3.60 m.

The drilled piles are arranged in a single row, the distance interaxle being 0.88m, the same as the diameter; they represent a screen of tangent piles.

In order to achieve an aesthetic landscape, it was proposed that the support structure would be masked by a reinforced soil structure that will be executed along the entire work, with different sectional dimensions.

The work will be executed in steps of 0.50 meters high with a retreat of 0.50 meters to 1.00 meters high (see figure 13).

At the base of the reinforced soil structure, a walled gutter with raw stone will be made.

## References

[1] Tuns I, Mantulescu M and Tamas F 2009 Analysis on the degradation state of the site of certain residential buildings *Bull. Transilv. Univ. Brasov. Eng. Sci.* **Series I** 379

[2] NP 123 2010 Normativ privind proiectarea geotehnică a fundațiilor pe piloți (Normative on geotechnical design of foundations on piles)

[3] Tuns I and Mantulescu M 2017 Assessment of Structural Foundation's Contribution to the Stability of a Site Susceptible to Sliding *Procedia Eng.* **181** 76–83

[4] Poulos H G 1995 Design of reinforcing piles to increase slope stability *Can. Geotech. J.* **32** 808–18

[5] NP 137 2014 Normativ pentru evaluarea in situ a rezistentei betonului din construcțiile existente (Normative for the assessment of in situ resistance of concrete in existing constructions)

[6] NP 122 2010 Normativ privind determinarea valorilor caracteristiceşi de calcul ale parametrilor geotehnic (Determination of the characteristic and calculation values of the geotechnical parameters)
[7] GP 093 2006 Ghid proiectarea structurilor de pământ armat cu materiale geosintetice și metalice (Guide to the design of reinforced soil structures with geosynthetic and metallic materials)
[8] Lin H, Ni L, Suleiman M T and Raich A 2015 Interaction between Laterally Loaded Pile and Surrounding Soil *J. Geotech. Geoenvironmental Eng.* **141** 04014119
[9] Burland J B, Simpson B and St. John H D 1979 Movements Around Excavations in London Clay. *Proc. Eur. Conf. Soil Mech. Found. Eng.* **1** 13–29

IOP Conf. Series: Materials Science and Engineering 1242 (2022) 012024 doi:10.1088/1757-899X/1242/1/012024

# Numerical studies regarding the behaviour a large span roof under seismic load when using anti-seismic devices

**C Mojolic**[1*]

[1] Structural Mechanics Department, Faculty of Civil Engineering, Technical University of Cluj-Napoca, Romania

* cristian.mojolic@mecon.utcluj.ro

**Abstract**. The study presents the data obtained on a large span structure equipped with passive seismic isolators, placed at roof level, when subjected to multiple seismic loads. Three distinct passive isolation systems are taken into consideration to reduce the seismic response of the structure. The first case uses Friction Pendulum Systems (FP), the second uses a combination of High Damping Rubber Bearing (HDRB) and low friction pot-bearing systems, and the third implies the use of Lead Rubber Bearings (LRB) and low friction pot-bearing systems. Using seven sets of accelerograms, recorded and generated, the values for the main seismic response parameters are extracted in to underline the efficiency of the proposed systems. The results obtained in terms of natural vibration period, accelerations, bending moment, base shear are compared.

## 1. Introduction

The research focuses on the effect of base isolation on a large span roof structure situated in Romania, in the influence zone of Vrancea seismic area., which is responsible for 90% of the seismic energy released in the country. The studied building is a 9.4 m height, one story structure, with concrete columns and steel roof. The 12 round columns, with a diameter of 110 cm, that are supporting the roof, are and situated 12 m apart and are considered fixed at ground level. The roof has a span of 40 m and a total surface of 2400 m$^2$, with 6 main steel trusses with a maximum height of 3.5 m. The roof is stiffened in the longitudinal direction by using lattice beams, located in the nodes of the main beams The design of the structures considers the main trusses pinned to the top of the concrete columns. All the elements of the structure are calculated in accordance with the Romanian design codes.

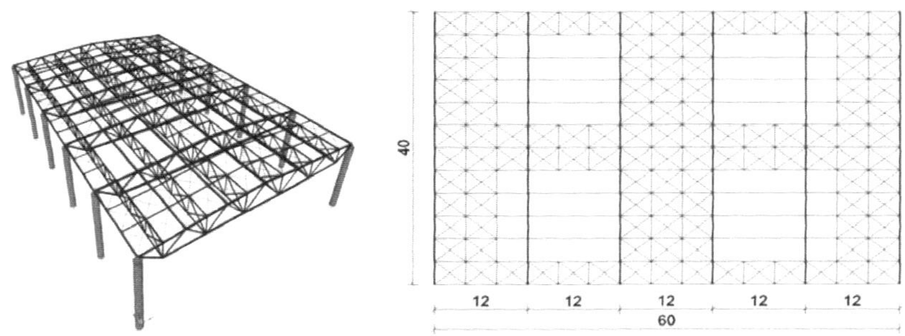


**Figure 1.** Structure layout.

To reduce the seismic impact on the reference structure, three types of passive base isolation solutions are proposed in this study. The considered seismic load is applied by accelerogram and response spectrum methods, according to the design code P100-1/2013 [1]. The response of the structure in the four cases is studied and the results in terms of natural vibration period, accelerations, bending moment and base shear are compared.

## 2. Seismic isolation

The base isolation operation of a structure implies decupling the whole structure or a part of that structure, by adding a series of elements on the entire surface of the isolation level. The wanted effect is that all the deformations that can appear due to the seismic action on the structure, will be at the isolators level, so that the rest of the structure will behave as a rigid body.

Usually, the position of passive seismic isolators is above the infrastructure level, but when most of the structure's weight is represented by the steel roof, the proposed solution involves changing the position of the isolation to the top of the columns. This means decupling the roof from the concrete poles. Three different types of seismic isolation were considered for this study.

### *2.1. Friction Pendulum System (F.P.)*

The first seismic isolation proposed, implies the use of Friction Pendulum Systems (FP). The radius of the spherical sliding surface is 4 m, with a designed displacement of 0.458 m.

Using equation 1 from below [2][3], a value of 0.194 kN/mm was obtained for the effective stiffness.

$$k_{eff} = k_1 + \frac{Q}{u_{max}} = W\left(\frac{1}{R} + \frac{\mu_f}{u_{max}}\right) \tag{1}$$

The effective damping for this system is $\xi_{eff}$ = 19%. To ensure a constant pressure between the PTFE plate and the slider, the dependence of the coefficient of friction on the velocity of sliding must be taken into consideration. The relationship between the coefficient of friction and the velocity of sliding at the interface can be described by the exponential function (2) [2].

$$\mu_f = \mu_{max} - (\mu_{max} - \mu_{min}) \cdot e^{-\alpha|\dot{u}|} \tag{2}$$

In the previous equation, $\mu_{min}$ and $\mu_{max}$ are the values of the coefficient of friction at very small and very high velocities of sliding, $\alpha$ is a parameter that controls the intensity change for the friction coefficient, and $\dot{u}$ is the sliding velocity. To simplify the modelling of the velocity dependence of friction for response history analysis, a fixed value of the ratio of $\mu_{min}$ to $\mu_{max}$ was chosen ($\alpha$ = 0.2), thus obtaining $\mu_{min}$ = 0.015 and $\mu_{max}$ = 0.055. In general, the FP systems cannot be used together with other passive isolators such as HDRB and LRB because the displacement is not just horizontal but also vertical, as the slider reaches the top end of the curved plate. The F.P. are placed at the top of each column as shown in figure 2-a.

### *2.2. High Damping Rubber Bearing (HDRB) and low friction pot-bearing systems*

The second seismic isolation solution uses High Damping Rubber Bearings-HDRB and pure friction elements (PF), multidirectional pot-bearing elements that are characterized by a very low friction coefficient. The energy dissipation of the pot-bearing devices was not taken into consideration for further computation. To be able to find the characteristics of the devices, a certain period of the isolation system was targeted. Using this period for the isolation system the effective stiffness was determined using equation (3) [4].

$$T_{eff} = 2\pi\sqrt{\frac{W}{k_{eff} \cdot g}} \tag{3}$$

The resulted value for the stiffness is 2.88 kN/mm. Four isolators were considered enough to satisfy the imposed isolation requirements. The main characteristics of the chosen HDRB isolators are: diameter (D) = 950 mm, shape factor (S) = 21.4, $K_{eff}$ = 0.72 kN/mm, $F_y$ = 120 kN. A special attention was given to the positioning of the isolators under the roof, so that the vibration modes would not be affected, and the effects of torsion would be reduced to minimum (figure 2-b).

### *2.3. Lead Rubber Bearings (LRB) and low friction pot-bearing systems*

The third proposed solution involves using Lead Rubber Bearing (LRB) isolators instead of HDRB, maintaining the same positions and the same number for the isolators as the previous solution. The main characteristics of the resulting LRB isolators are: diameter (D) = 750 mm, diameter of the lead core (d) = 110 mm, shape factor (S) = 22.6, $K_{eff}$ = 0.70 kN/mm, $F_y$ = 155 kN.

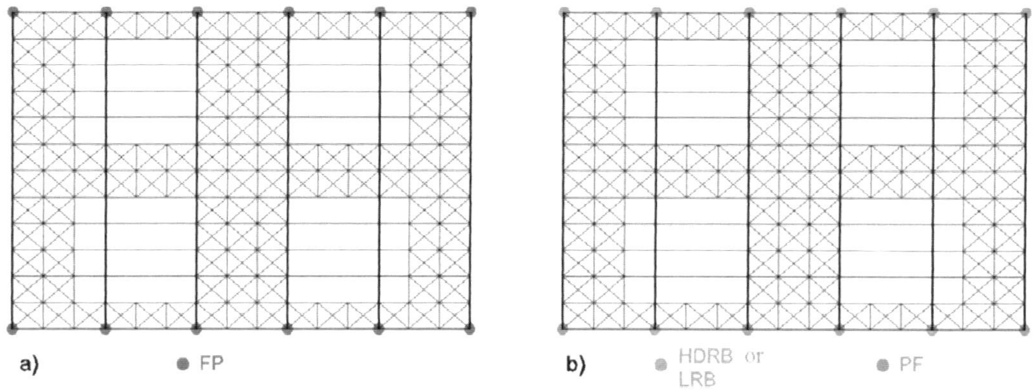


**Figure 2.** Position of the isolators.

## 3. Seismic load

The structure is situated in a seismic area with a peak ground acceleration (PGA) of 0.3g and $T_c$ = 1.6s for MRI = 225 years (20% probability of exceedance in 50 years). The seismic load considered for this study involves using both accelerograms and response spectrum, in accordance with the Romanian seismic design code P100-1/2013.

7 sets of ground motions were used, three recorded accelerograms of the main Romanian earthquakes (Vrancea 77, Vrancea 86 and Vrancea 90) and four sets of artificially generated motions (A1, A2, A3, A4).

All the three recorded ground motions (Vr77, Vr86, Vr90) were scaled according to the response spectra with with SeismoMatch software.

The artificial ground motions were generated using SeismoArtif software and a 40 second time frame was considered for each one. And were also scaled to match the response spectrum for the location of the structure (figure 3).

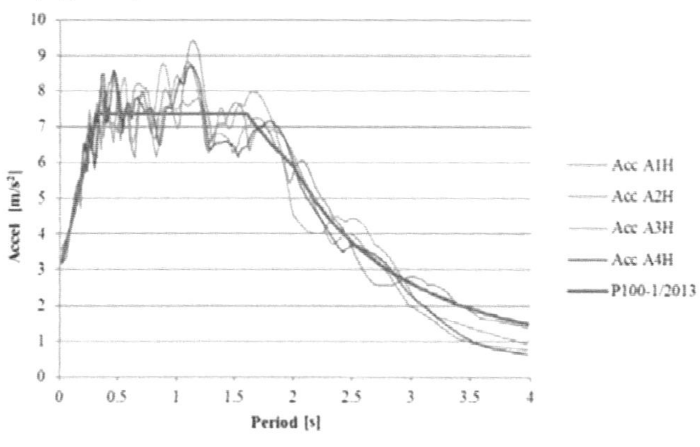


**Figure 3.** Response spectrum for artificial ground motions.

The accelerograms used in the study are compatible with the elastic response spectrum for the chosen location of the structure for a dumping ratio of 5%.

Considering that large span structures are sensitive to vertical displacements, the vertical component for each ground motion, recorded and generated, was considered.

According to the basic requirements for base insulation set out in P100-1/2013, insulation devices must be made with a higher degree of safety than that used to design the structure, the seismic action applied to each device must be amplified by a factor of 1.2.

## 4. Structure response

A nonlinear time-history analysis (fast Nonlinear Analysis - FNA) in SAP200 was performed on the structure, using all 7 sets of ground motions, both in horizontal and vertical direction.

The isolators were defined as link elements, considered as finite elements with a nonlinear behavior.

Hysteresis loops (load vs. displacement) are very important graphs when analysing the seismic behaviour of structures with base isolation systems under earthquake excitations. The displacements of the isolator in both directions are also relevant and verifications were made to ensure that they do not exceed the displacement capacity provided by the design.

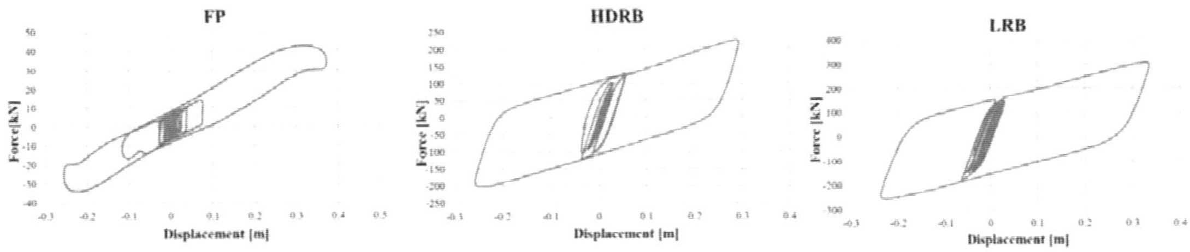


**Figure 4.** Hysteretic loops of the isolators during Vr77 earthquake.

As a result of the modal analysis, it was observed that the first two vibration modes follow translations in X and Y directions and the third is a torsional mode. By placing the isolation level immediately under the roof, besides the changes in the roof elements, important modifications occur on other structural elements situated above the isolation level [5].

## 5. Results

One of the main consequences of using base isolation devices is the increase of the natural vibration period of the structure. The increase of the normal period results in a reduction of the dynamic amplification factor. To be considered efficient, following the seismic isolation of the structure, the obtained main period must be at least three times higher than the period of the non-isolated structure. This condition is fulfilled in all three cases presented above. The period of the structure increases from 0.77s when no isolators are used to 2.89s when using FP, 2.78s for HDRB and 3.07s in the case of LRB isolation.

**Table 1.** Accelerations on isolated structures from ones on the basic structure.

| Isolator type | Vrancea 77 | Vrancea 86 | Vrancea 90 | A1 | A2 | A3 | A4 |
|---|---|---|---|---|---|---|---|
| FP | 18% | 10% | 16% | 16% | 13% | 14% | 12% |
| HDRB | 28% | 25% | 29% | 23% | 24% | 24% | 25% |
| LRB | 28% | 25% | 29% | 22% | 24% | 24% | 25% |

A significant reduce in accelerations was recorded when using seismic isolators, with values that vary from 72% to 90% from the ones on the uninsulated structure. The values for each isolation type and for every seismic load are presented in table 1.

Another important parameter of the seismic response of the structure is the reduction of the bending moment and shear force. A significant decrease of the maximum bending moment and shear force at the bottom of the concrete columns is obtained, especially when using friction pendulum systems. Also, the displacement at the top of the columns is reduced by 79% when using FP, 39% for HDRB and 50% for LRB.

The total base shear of the structure is significantly reduced for each seismic action and each isolation case, with values that vary from 79% to 88%.

**Table 2.** Maximum bending moment at the bottom of the concrete column.

| Seismic action | $M_{max}$ [kNm] | | | |
|---|---|---|---|---|
| | Basic structure | FP | HDRB | LRB |
| Vr77 | 4536 | 912 | 2690 | 1944 |
| Vr 86 | 4719 | 621 | 2233 | 1872 |
| Vr 90 | 4848 | 865 | 2763 | 2551 |
| A1 | 4765 | 697 | 2125 | 1671 |
| A2 | 4847 | 636 | 2356 | 1914 |
| A3 | 4679 | 722 | 1886 | 1690 |
| A4 | 4698 | 523 | 2177 | 1733 |

It can be concluded that the effect of using passive seismic isolation solutions for large span structures has a positive effect on the behavior of the structure under seismic loads. All the important parameters that characterize the response of the structure show a large improvement in terms of displacement, shear force, acceleration, bending moment and shear force.

## References

[1] Technical University of Bucharest and Minister for Regional Development and Public Administration 2013 *Seismic design code - Part I – Design provisions for buildings, code P100-1/2013* (Bucharest)

[2] Cheng F, Y Jiang H and Lou K 2008 *Smart Structures Innovative Systems for Seismic Response Control* ed Taylor& Francis Group

[3] Constantinou M, Whittaker A, Kalpakidis Y, Fenz D and Warn G 2007 *Performance of Seismic Isolation Hardware under Service and Seismic Loading* Technical Report MCEER-07-0012 (Buffalo, New York) p 20

[4] Naeim F and Kelly J 1999 *Design of Seismic Isolated Structures- From Theory to Practice* ed John Wiley & Sons

[5] Mojolic C and Petrina T 2017 AIP Conference Proceedings **1863**, 560037

# Validation of elastic wave arrival detection method based on use of sparse matrix computation

**K Nakamura[1*], Y Kobayashi[1], K Oda[1], S Shigemura[1] and K Ikebata[1]**

[1] Department of Civil Engineering, College of Science and Technology, Nihon University, Tokyo, Japan

* nakamura.katsuya@nihon-u.ac.jp

**Abstract**. In the field of non-destructive investigations, the arrival time of the elastic waves is the important parameters to visualize the locations of the failures and/or elastic wave velocity distributions that are generally used for the soundness evaluation. The accuracy of the visualization depends on the accuracy of the arrival times, and it is demanded that the arrival times are detected automatically with required accuracy from huge amount of the elastic wave measurement data for the practical applications. In this paper, a method of the arrival time detection is proposed on the basis of the sparse matrix computation. The root mean square voltages in the noise of the wave form, is used as an index to separate the elastic waves between the noise and the signal, and the boundary of the index between the noise and the signal in the wave forms is determined in the sparse matrix. Therefore, the arrival times located in the boundary of the noise and it is detected by sparse matrix computations. The proposed method is verified on the model tests and compared with the conventional method, and it was confirmed that the proposed method detects more arrival time accurately in comparison with the conventional method.

## 1. Introduction

The elastic wave velocity distributions are attractive to evaluate the soundness of the existing structures or investigate the mechanism of the progress failure inside the materials. The non-destructive inspections based on use of the elastic wave velocity distributions in which the examples are elastic wave tomography [1] and Acoustic Emission tomography [2], have been applied to identify the elastic wave velocity distributions of the targets by inverse analysis techniques [3]. Moreover, if the elastic wave velocity distributions are identified by the tomography, the accurate arrival times which are parameters of inverse analysis, is required. The arrival times are the boundary of the index between the noises and the signals in the wave forms which are measured from the surface of the targets.

The visual confirmation in which is one of the detection methods, works well to tell different between the noises and the signals, and detect the practically arrival times. However, the huge amount of the elastic waves is generally required to clearly identify the elastic wave velocity distributions by the tomography. Therefore, it is demanded that the arrival times are detected automatically with required accuracy from huge amount of measurement data for the practical applications. Due to this demand, AR-AIC [4] have applied to detect the arrival times. For example, the locations of failures in specimens, are visualized by detected the arrival times [5, 6]. However, if AR-AIC is applied to the wave form which are formed large different amplitudes and low S/N signals, AR-AIC is expected to detect the arrival times included huge errors, because of the algorism expects that signals are propagated with regularity. The object of developing AR-AIC is to detect the arrival times of seismic waves. The propagations of waves from sources of seismic to sensors, are longer than the propagations used in the non-destructive inspections based on use of the elastic wave velocity distributions. It is easy to set the analysis areas consisted regularity waves. Therefore, the algorism of AR-AIC is not required to consider the analysis areas mixed the other waves such as shear waves, surface waves and the different waves contributed other events. In addition, setting the analysis areas consisted regularity waves, are required the empirical judgments, and it is the difficulty of automatically setting that in the tomography.

In this paper, a method of the arrival time detection is proposed based on the root mean square voltage in the noises and the sparse matrix computations. The root mean square voltages in the noises of the wave form, is used as an index to separate the elastic waves between the noises and the signals if the noises are measured before conducting the model test or the monitoring of the existing structures. The boundary of the index between the noises and the signals in the wave forms is determined in the sparse matrix. The proposed method is verified on the model tests in which the elastic waves are generated on the aluminum plates to detect the arrival times, and the accuracy of the arrival times investigated on the proposed method and the conventional method which is AR-AIC.

## 2. The proposed arrival detection method

### 2.1. The discriminant functions

The wave forms are consisted two of the parts which are the noise parts and the signal parts. The arrival time of the elastic wave is the point where is the boundary of the index between the noise parts and the signal parts. If the noise parts and the signal parts, are discriminated, it is the potential that the arrival times are detected from the wave forms in which the signals are formed by the large different amplitudes and low S/N signals. The root mean square voltage in the noises is approximately constant under the same measurement conditions, it is expected to be the criterion of detecting the arrival time. In order to separate to noises and signals, the discriminant function in which is Gaussian function based on use of the root mean square voltage in the noises, apply to the wave forms. The root means square voltage in the noises $V_n$, are defined as

$$V_n = \sqrt{\frac{1}{T_m}\int_0^{T_m} y^2(t)\,dt} \tag{1}$$

where $y(t)$ is the wave form, $T_m$ is the duration on the noises. The plot of $V_n$ used the example of the measured noises with $T_m$, are shown in figure 1. According to the figure 1, $V_n$ show that the constant values of root mean square voltage, if $T_m$ is above 1.0 ms. It is noted that $V_n$ work to be the criterion of detecting the arrival time if $T_m$ are the sufficient duration. Furthermore, if $V_n$ is computed by the noises measured with conducting model tests or health monitoring, it is expected that $V_n$ are the irregular values. Thus, $V_n$ require to be computed by the noises before measuring the elastic waves. The discriminant function $\phi_i$ at each amplitude $i$, is defined as

$$\phi_i = exp\left\{-\frac{(|y_i| - V_n)^2}{2s^2}\right\} \tag{2}$$

where $|y_i|$ is the amplitude of the wave form, $s^2$ is sensitivity parameter of the discriminant function based on use of the variance of the noises. The difference between $V_n$ and $|y_i|$, is clearly evaluated if $s^2$ is sufficient small. However, it is expected the miss detection of the accurate arrival time because of detecting the weak reflected waves or surface waves located in the wave form if $s^2$ is smaller than the statistical variability of the noises. In order to demand the accuracy of the arrival time, the deviation of the noises $s$ is defined as

$$s = 1.65\sqrt{\frac{1}{m}\sum_{j=1}^{m}(|y_j| - V_n)^2} \quad (3)$$

where $m$ is the number of the data same as time step of $T_m$, $j$ is the amplitude number of the noises. In Gaussian distribution, the size of $s$ is including approximate 90 % of the variability in the noises, if $s$ is computed by the product of the standard deviation of the noises and 1.65. It is expected to prevents the miss detection of the arrival time by $s$ computed by equation (3), because of considered the demanded statistical variability. The example of discriminating the wave form between the signals and the noises by $\phi_i$, is shown in figure 2. The output of $\phi_i$ is from1.0 to 0. If $|y_i|$ approximate $V_n$, the output is 1.0. Theoretically, the outputs of $\phi_i$ can transform the wave forms to step function that the noise is 1.0 and the signal is 0. Thus, the output of the discriminant functions can be defined as a step function included errors. Therefore, the arrival times are located in the boundary of step function and detected if the optimized step function is computed by the results of sparse matrix computations.

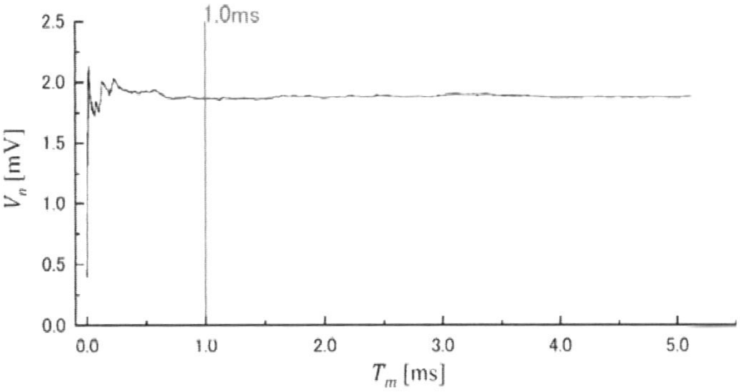


**Figure 1.** The relation of $V_n$ and $T_m$.

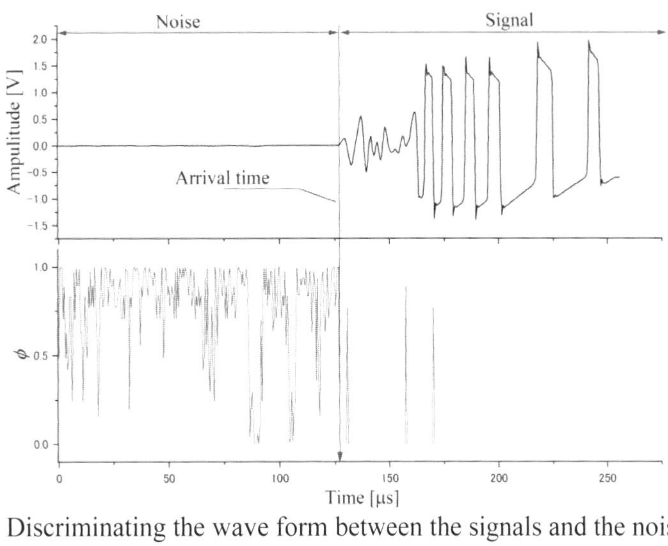


**Figure 2.** Discriminating the wave form between the signals and the noises by $\phi_i$.

### 2.2. The sparse matrix computation

The sparse matrix computation is defined as

$$\min_{\Phi'}\left\{\frac{1}{2}\|\Phi - A\Phi'\|^2 + \lambda\|\Phi'\|\right\} \tag{4}$$

where $\Phi$ is the vector of discriminant function, $A$ is the stair matrix, $\Phi'$ is the vector of correction value, $\lambda$ is the tuning parameter to control a sparse characteristic of $\Phi'$. The elements of $\Phi$ is the output of $\phi_i$. Thus, the number of the elements is same number as the discretized amplitudes. The equation (4) is defined as the least absolute shrinking and selection operator (LASSO) regression [7] and it is computed by alternating direction method of multipliers (ADMM) [8], because of generally using to sparse matrix computations. The columns of $A$ defined as the equation (5), is the discretized step function.

$$A = \begin{bmatrix} 1 & 1 & \cdots & 1 & 1 \\ 0 & 1 & \ddots & 1 & 1 \\ 0 & 0 & \ddots & 1 & 1 \\ \vdots & \vdots & \ddots & \vdots & \vdots \\ 0 & 0 & \cdots & 0 & 1 \end{bmatrix} \tag{5}$$

The ideal situation is that one optimal step function is determined by the product of $A$ and $\Phi'$ which is the results of the equation (4) and the boundary of the optimal step function is the index of the arrival times. However, $\Phi$ include errors and $\Phi'$ have the multiple elements which is not 0. Thus, the optimal step function which the boundary of 1.0 and 0, is not determined. In order to determine the optimal step function, the candidates of optimized step function are computed by the product of $\Phi'$ and the columns of $A$, and the optimized function is selected from the candidates. The candidates of optimized step function $\Psi_i$ is defined as

$$\Psi_i = D_i\phi_i' \tag{6}$$

where $D_i$ is the columns $i$ of $A$, $\phi_i'$ is the element of $\Phi'$. $D_i$ is defined as

$$D_i^T = [\alpha_0 \quad \alpha_1 \quad \cdots \quad \alpha_i \quad \cdots \quad \alpha_{n-1}] \tag{7}$$

$$a_j = \begin{cases} 1, j \leq i \\ 0, j > i \end{cases} \tag{8}$$

$$0 \leq j < n \tag{9}$$

where $j$ is the number of the element of $D_i$, $n$ is the total data of the discretized amplitudes. Moreover, the optimized $\Psi_i$ is selected by the average errors $\delta_i$ definde as

$$\gamma_i = \|\Phi - \Psi_i\|_1 \tag{10}$$

$$\delta_i = \frac{\gamma_i}{n} \tag{11}$$

where $\gamma_i$ is the 1-norm of the errors. The optimized $\Psi_i$ is selected if the minimu $\delta_i$ is comuputed by the equation (11). It is noted that $i$ is the location where is the end of noises in the wave form. Thus, $i+1$ is the arrival time. In additon, the eaxmple of the arrival time detection by $\Psi_i$, is shown in the figure 3. Although, $\phi_i$ is approximate 1.0 in the noise parts, and 0 in the signal parts, and the boundary of the noises parts and signal parts are not clear because of what the noises include the weak reflected waves or surface waves in figure 2, the boundary of the noises part and signal part which is the arrival time, is cleared by optimized $\Psi_i$ in figure 3.

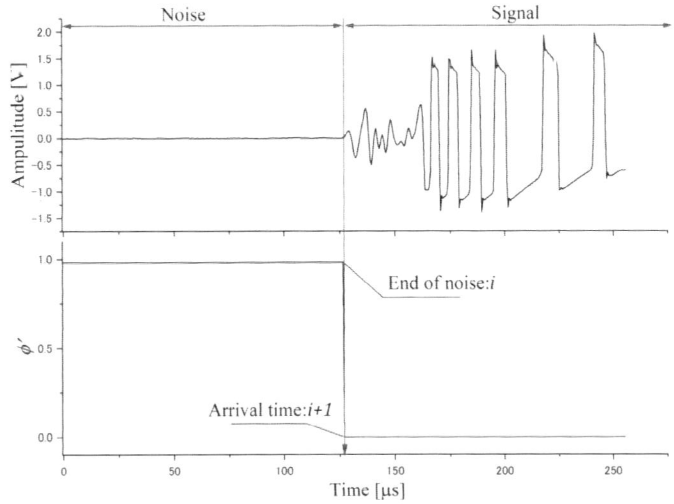


**Figure 3.** The outline of the arrival time detection by optimized $\Psi_i$.

## 3. Validation of the proposed method by the model test

### *3.1. The experimental details and the initial conditions of analyses*

In order to validate the performance of the proposed method, the elastic wave measurements are conducted based on using sensors located on the aluminium plate which is the thickness of 5 mm shown in figure 4. According to the Huygens principle, the elastic waves generated on the uniform elastic wave velocity distribution, such as a metallic material, propagates with linear paths. Therefore, the elastic wave velocity can be computed by the quotient of the linear distance between the sensors located in the propagation direction and the travel time computed by the detected arrival times. In this study, the performance of the proposed method is validated based on the computed elastic wave velocities. According to figure 4, the propagations which are ch1-ch2 and ch1-ch3, is expected to show the uniform elastic wave velocity distribution, if the accurate arrival time is detected in each the sensor. Moreover, the elastic waves measured in ch3, are hugely attenuated because of the diffraction and it is considered the difficulty to detect the accurate arrival time by AR-AIC. Generally, there is the difference in the amplitudes between shear waves and elastic waves, and it contribute the large difference if the wave forms are attenuated. It is expected that AR-AIC detect the arrival time of shear wave if AR-AIC apply to the wave form measured in ch2. Thus, the proposed method is validated if the accurate arrival times of the elastic waves measured in ch2, is detected, and it shows the similar elastic wave velocity as the propagation in straight.

The elastic waves are artificially generated by Pencil lead breaks in the sources point show in figure 4. Pencil lead breaks generally is applied to the studied of Acoustic Emission (AE). Furthermore, AE is the elastic wave generated by the occurrence of cracks. Pencil lead breaks is a familiar method in this filed, because of generating the one elastic wave in each of the test [9, 10]. The measurement conditions are shown in table 1, and refereed the measurement case applied pencil lead breaks to the metallic materials [11]. Operating frequency of the sensors is 35-100 kHz. Measured elastic wave is amplified 60 dB by the preamp and sampling frequency is 2 MHz. Therefore, arrival times include errors of minimum 0.5 μs. In addition, the measurement was started if the amplitude exceeded the threshold in which the 35 dB, and the waveforms measured for 255.5 μs, were stored in the measurement PC.

Moreover, the pre-trigger time is set 127.5 μs to record the waveform until the amplitude of elastic wave exceeded the threshold.

In order to conduct the comparison with the performance of the proposed method and AR-AIC, the proposed method and AR-AIC apply to the measured waveform based on use of the parameters shown in table2. In the initial condition of AR-AIC, the autoregressive coefficients of noise and signal are respective 10. Moreover, the duration of $V_n$ is set 1023.5 μs and the tuning parameter $\lambda$ is $10^{-3}$ as initial condition of the proposed method.

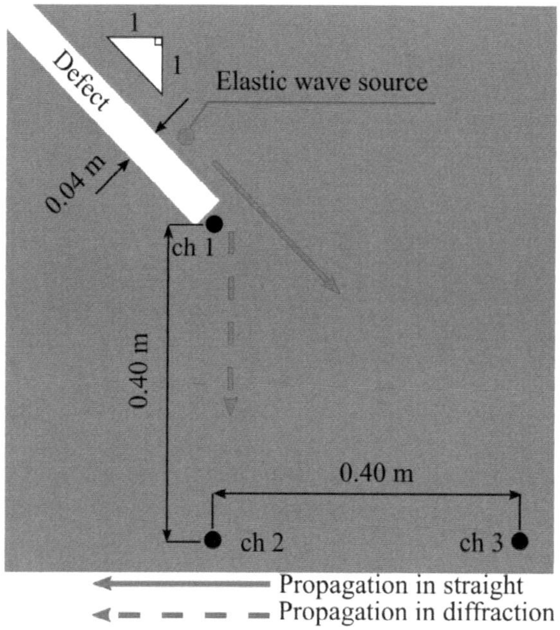


**Figure 4.** The outline of the elastic wave measurement and Pencil lead breaks.

**Table 1.** The measurement conditions of the model test.

| | |
|---|---|
| Operating frequency [kHz] | 35-100 |
| Preamplifier gain [dB] | 60 |
| Sampling frequency [MHz] | 2 |
| Threshold of measurement trigger [dB] | 35 |
| Pre-trigger time [μs] | 127.5 |
| Wave length [μs] | 255.5 |

**Table 2.** The initial conditions of analyses.

| | | |
|---|---|---|
| AR-AIC | Autoregressive coefficients of noises | Autoregressive coefficients of signals |
| | 10 | 10 |
| Proposed method | Duration on noises [μs] | Tuning parameter |
| | 1023.5 | $10^{-3}$ |

*3.2. Results*

The average elastic wave velocities computed by 3 measured elastic waves, are shown in table 3. The average elastic wave velocities in which is the distance of ch1-ch3, are not clear differences between results of AR-AIC and the proposed method. The results are noted that the proposed method detects the practical arrival times of the elastic waves propagated in straight measured in ch1 and ch3, as the conventional method. On the other hand, the results of AR-AIC and the proposed method in the distance

**Table 3.** The average of elastic wave velocities.

| | | AR-AIC | Proposed method |
|---|---|---|---|
| ch1-ch3 | Average [m/s] | 5537.0 | 5510.2 |
| | Standard deviation [m/s] | 29.4 | 36.4 |
| ch1-ch2 | Average [m/s] | 2461.9 | 5357.4 |
| | Standard deviation [m/s] | 31.8 | 34.5 |

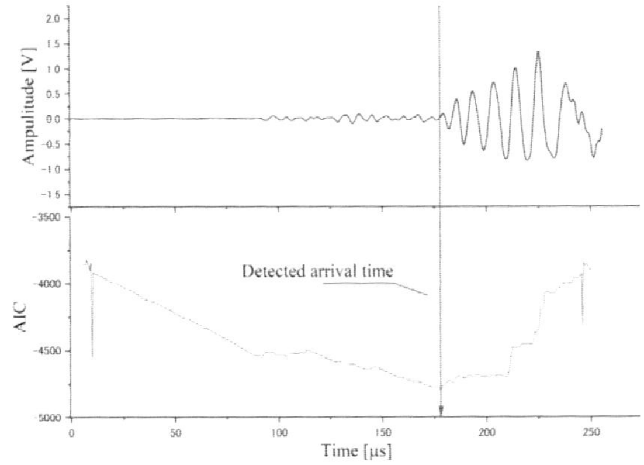


(a) wave1

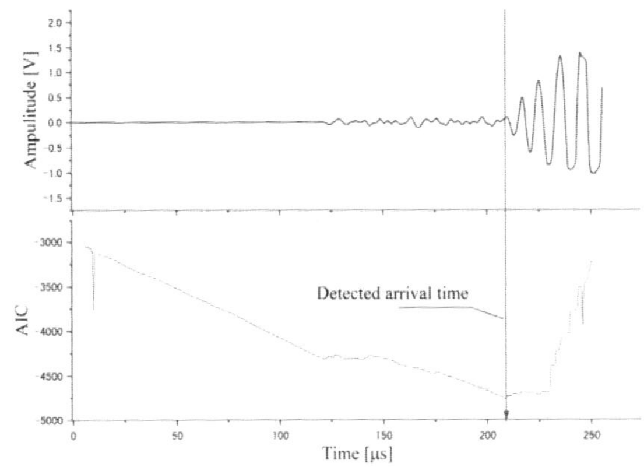


(b) wave2

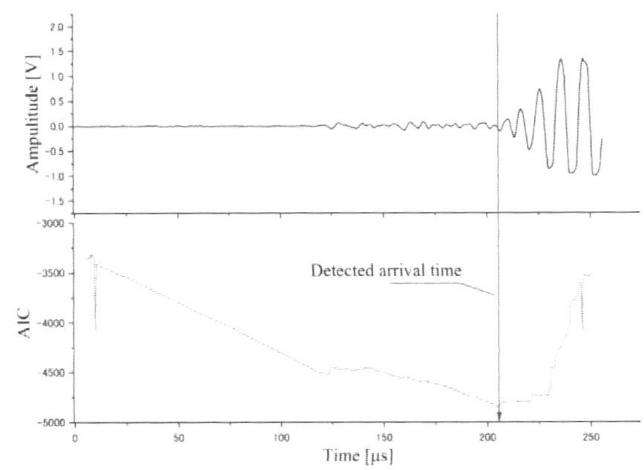


(c) wave3

**Figure 5.** The results of AR-AIC.

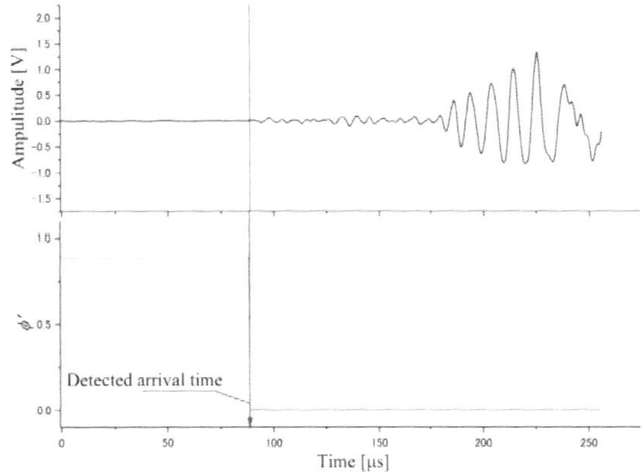


(a) wave1

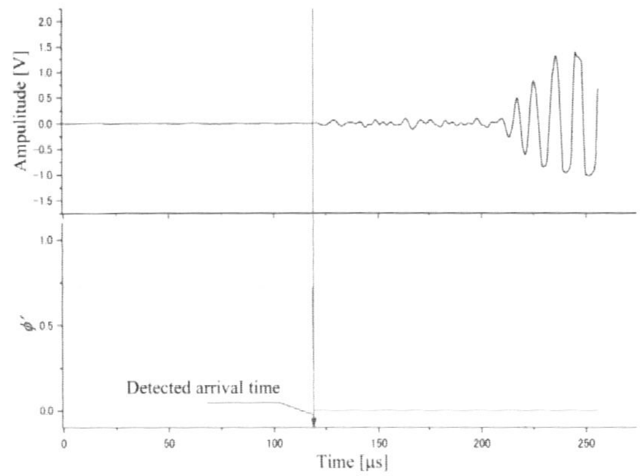


(b) wave2

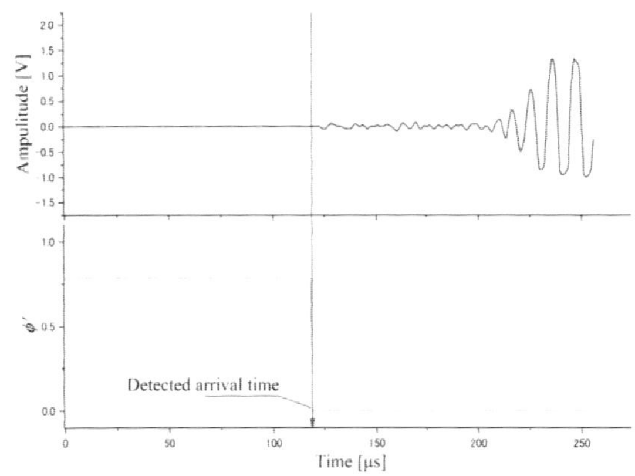


(c) wave3

**Figure 6.** The results of the proposed method.

of ch1-ch2, show the difference of the average elastic waves. It is expected that the detected arrival times are different between the results of AR-AIC and the proposed method in ch2. In addition, the defects are not located the distance of ch1-ch2 and it is expected that the elastic wave velocity are identical as ch1-ch2, and AR-AIC detects the times in which is not the accurate arrival time in ch2. The elastic waves and the results of arrival detections are shows in figure 5, 6. In figure 5, 6, there are in comparisons with results of AR-AIC and the proposed method in the identical event of Pencil lead breaks. According to figure 6, AR-AIC does not detect the arrival time of the waves firstly arrived at the ch2, and is detected time in which the signal part has different S/N. In the algorism of AR-AIC, the signals are considered continuous functions and it selects the boundary of the signals if the signal functions are optimized. Thus, the boundary which optimizes the signal functions, is the arrival times. However, if the signal includes multi waves which is large different S/N, AR-AIC detects the boundary of large different S/N to shrink the errors of the signal function, and it is difficult to detect elastic waves. According to figure 6, it is northly that the proposed method detects the arrival time of the elastic wave expected by the visual confirmations. In order to discriminate the signals, the proposed method uses the root mean square voltage in the noises. Therefore, the proposed method does not contribute to decreasing the accuracy of the arrival time detection if the signals includes multi waves. In identifications of non-uniform elastic wave velocity distributions by the tomography, it is considered that the elastic waves propagated with refractions and diffractions, are measured. Therefore, it is required to detect the arrival time of the elastic waves consisted low S/N signals and large different amplitudes waves as figure 5, 6. It is implied that elastic wave velocity distributions identified by the tomography, are improved by the practical arrival time detected by the proposed method.

## 4. Conclusions

In this paper, the proposed method is validated by performing the model tests on the defected aluminium plate to develop a method for detecting practical arrival time of the elastic wave. The conclusions that are obtained in this study are written below.

- According to table 3, the average elastic waves in the conventional method which is AR-AIC, are shown difference between propagation in straight and diffraction. However, the results of the proposed method show the uniform velocity distributions in ch1-ch3 and ch1-ch2, and it is expected that the proposed method detects the accurate arrival times.
- According to figure 6, the proposed method detects the arrival time of the elastic wave from the wave form in which AR-AIC is difficult to detect the accurate arrival times. It is implied that using the practical arrival time detected by the proposed method, contributes to improve the identification of elastic wave velocity distributions by the tomography.

In this study, the applicability of the proposed method to the tomography is shown. However, the contribution of the proposed method to improve results of the tomography is still unknown. Although, the initial conditions used in the model tests, contribute to detect the accurate arrival time, it is required to continue investigations of suitable initial conditions based on using other conditions. Therefore, it is necessary to continue to study the applicability of the proposed method in the tomography.

## References

[1] Sassa K 1988 Suggested for seismic testing within and between boreholes *International journal of rock me-chanics and mining sciences & geomechanics abstracts* **Vol.25** No.6 pp.447-472
[2] Kobayashi Y and Shiotani T 2016 Computerized AE tomography: Innovative AE and NDT techniques for onsite measurement of concrete and masonry structures *State-of-the-Art Report of the RILEM Technical Committee 239-MCM* Springer pp.47-68
[3] Nakamura K, Kobayashi Y, Oda K, Shigemura S and Ikebata K 2021 Validation of AE tomography based on model tests of soil specimen with heterogeneous elastic wave velocity distribution *IOP Conference Series: Materials Science and Engineering* **Vol.1138** No.012033
[4] Takanami T and Kitagawa G 1988 A new efficient procedure for the estimation of onset times of seismic waves *Journal of Physics of the Earth* **Vol.36** No.6 pp.267-290
[5] Mao W, Goto S and Towhata I 2020 A study on particle breakage behavior during pile penetration process using acoustic emission source location *Geoscience Frontiers* **Vol.11** pp.413-427
[6] Ohno K, Uji K, Ueno A and Ohtsu M 2014 Fracture process zone in notched concrete beam under three-point bending by acoustic emission *Construction and Building Materials* **Vol.67** pp.139–145
[7] Tibshirani R 1996 Regression shrinkage and selection via the lasso *Journal of the Royal Statistical*

*Society. Series B(Methodological)* **Vol.58** No.1 pp.267-288
[8] Boyd S, Parikh N, Chu E, Peleato B and Eckstein J 2010 Distributed Optimization and Statistical Learning via the Alternating Direction Method of Multipliers *Foundations and Trends in Machine Learning* **Vol. 3** No. 1 pp.1–122
[9] Qiu X, Wang Y, Xu J, Xiao S and Li C 2020 Acoustic emission propagation characteristics and damage source localization of asphalt mixtures, *Construction and Building Materials* **Vol.252** No.119086
[10] Tai J, Liu X, Wang X, Shan Y and He T 2021 An adaptive localization method of simultaneous two acoustic emission source based on energy filtering algorithm for coupled array signal, *Mechanical Systems and Signal Processing* **Vol.154** No.107557
[11] Ai L, Soltangharaei V, Bayat M, Greer B and Ziehl P 2021 Source localization on large-scale canisters for used nuclear fuel storage using optimal number of acoustic emission sensors *Nuclear Engineering and Design* **Vol.375** No.111097

# Flexible steel structures used for ski tunnels in Parang Mountains, Romania

**D Negrea[1*] and Ch Cazacu[2]**

[1] Viacon Romania SRL, Brasov, Romania
[2] Transilvania University of Brasov, Faculty of Civil Engineering, Romania

* doina.negrea@viacon.ro

**Abstract.** The main idea of this project was to create an access road for the tourists and also to improve the winter sports conditions, especially the ski, in the beautiful Parag Mountains. It was a very challenging project, as the road takes until 1600 m at alpine elevation. One of the structures was delivered and assembled in November 2019, in 3 weeks of hard weather conditions, with a 5 working men team.

## 1. Introduction

A tunnel can be defined as a closed or roofed structure carrying a road through, or under an obstacle. This obstacle may be anything in the path of a preferred road alignment such as a mountain, a body of water, a building, or a complete development [1].

As a short history, tunnels have been known since ancient times. The first tunnels were natural, formed in different caves. They had variable sections and were found in limestone rocks.

It is found that the first tunnel was made by the Egyptians and the Babylonians, 4000 years ago. This tunnel served the purpose of connecting two buildings in Babylon. The connection was from the royal palace to the temple. The length of this tunnel was found to be 910 m, which was brick-lined. In Europe, the first tunnel was built by Roman Emperor Claudius in A.D. 54 for carrying spring water through the Apennine Mountains. Its length was 5.8 km. It was completed in 12 years by 30,000 laborers.

The development of first railway tunnel in the world was constructed to connect Switzerland and Italy in 1895 - 1922. The shape of the cross-section was a horseshoe, its length was 19.82 km. The first highway tunnel was constructed in Hungary in 1851-1853 and its length was about 350 m. The first underground tunnel was constructed in 1843, by Sir Marc Brunel. The first navigation tunnel was constructed in France in 1679. Its length was about 160 m. America's first vehicular tunnel was built in 1866 which is known as Washington S.T. Tunnel constructed under the Chicago River.

During the second half of the twentieth century, the development and progress of explosives and sophisticated equipment were flourished, which made the tunneling process more feasible.

The tunneling helps to make work better for different approaches under different circumstances. It is hence dependent on the experience and the knowledge of the engineer. Mainly the tunneling works are carried out by civil engineers [2].

Nowadays, tunnels can be grouped into categories according to their purpose and the place where they are built: railway tunnels - railway tunnels (single or double track) and subways, Road tunnels - tunnels for public roads and car traffic, Hydrotechnical tunnels (galleries) with free-flow or pressurized pipes, tunnels for navigation and floating, tunnels – aqueduct, protection tunnels and tunnels for the mining industry. Also, according to the geographical area where the tunnels are built, they are divided into Mountain tunnels, tunnels in cities, and tunnels under the bottom of the water.

We can also Classify the tunnels based on surrounding material tunnels in mud-stone rocks, tunnels in sandy soil, tunnels in hard stone rocks, tunnels undersea or river bed, and open-cut tunnels [2].

Tunnels have improved the connection of regions and have been used lately as a catalyst for the economic development of previously isolated regions. Conventional tunnel construction has always been a significant engineering challenge. Success depends not only on careful planning and execution but also on ensuring that there is an optimal interaction between processing technology and the specialized chemistry of support systems. Depending on the rock and the soil encountered, problems can occur during excavation, even with the best possible planning, high-performance technologies, and optimal construction execution [3].

In the selection of the best and most efficient solution for any tunnel, there are key points that have to be considered: The cross-section with length, Waterproofing type, nature of rocks, Equipment, Time, Energy, and Cost Estimation.

The design concept and construction of the tunnels must follow the steps from the logic scheme (figure 1).

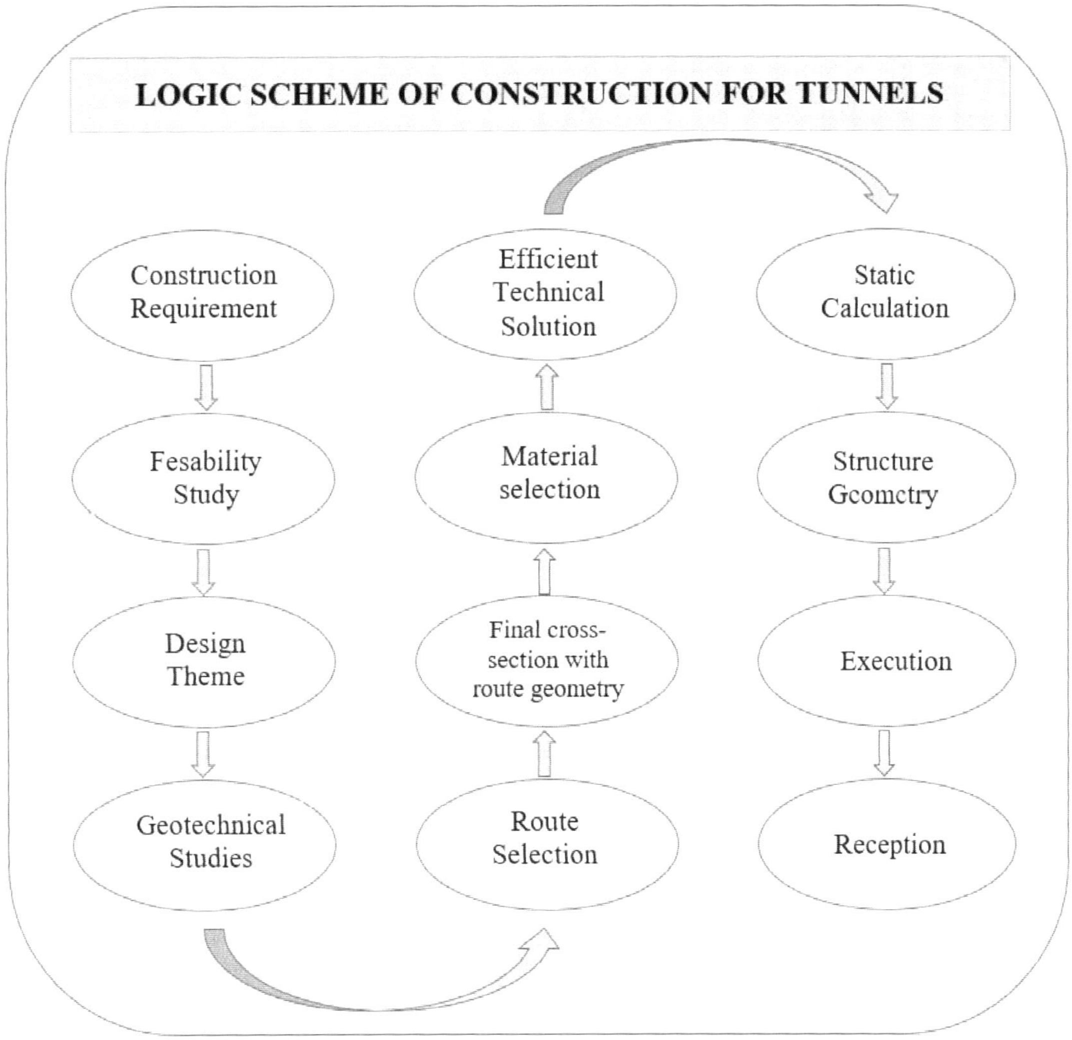


**Figure 1.** Logic scheme of construction for tunnels.

The Parang Mountains are located in the Southern Carpathians, in the Petrosani region, between Jiu, Strei and Olt rivers, in the south-west of the Parang Sureanu Lotrului mountain chain. Their length is over 50 kilometers and wide approximative 25 kilometers. Parang Mare peak has a maximum altitude of 2519 meters. This mountain is crossed by the highest tourist route in Romania, the Transalpina, with spectacular serpentines of 150 kilometers and a beautiful view. The most known tourist resort is Ranca, where visitors can have a dream vacation in the middle of nature [4].

**Figure2.** Parang Mountains [4].

For the project from Parang Mountain, there was necessary to find an unconventional technical solution, to provide both functions: road access for vehicles and ski slope. The ski tunnels can be classified into 2 types: indoor and outdoor. The tunnel which has been built on this project is considered an outdoor tunnel and facilitated with more than one purpose: Access for vehicles inside the tunnel and a ski slope on the top of the tunnel, which makes this tunnel more productive and unconventional.

As we mentioned the types of tunnels in earlier descriptions, those all belong to the conventional category (railway tunnels, road tunnels, hydro-technical tunnels, etc.). Considering all the characteristics of this project, this tunnel falls into the unconventional category. By dint of this unconventional tunnel, there is no possibility to build it conventionally. The reason that makes this project special with an unconventional tunnel it was necessary to build a tunnel to the surface of the road, not to make a tunnel throw the mountain (for example to use the NATM method).

Since it was unconventional tunnels, was highly important to find the best solution in all the stages of these tunnel buildings. The selection of Steel structure played a very important role to keep the quality of structure as well as the costing of the project on an efficient level.

The route has been made in such a way considering the comfort and optimum timesaving. They make alpine areas more attractive as a habitat for people and wildlife animals. Also, underground traffic diversion conserves natural resources and prevents Noise pollution. In developing urban areas, these underground tunnels guarantee eco-friendly nature and efficient mobility. As a pioneer in the construction of bridges, tunnels, and the protection of the environment, the ViaCon Group recognized early on the potential of tunnel construction. By participating in important infrastructure projects, the company had the opportunity to leave its mark, becoming a top specialist in this field.

## 2. Case study

A similar project was realized in Algeria, in 2015, railway tunnel, with the function of the tunnel above the railway line. The structures had the following parameters: a Supercor structure with a span of 12.02 m, rise of 12.02 m, bottom length of 180.08 m, and top length of 138.62 m.

**Figure 3.** Railway tunnel Algeria (Viacon Poland case study).

**Figure 4.** Railway tunnel Algeria (Viacon Poland case study).

The expansion process of infrastructure in Romania is growing significantly under the government initiative known as the PNDL Program, which supports infrastructure building with many aspects as well as helping to solve the lack of roads and ski slopes.

PNDL is a multiannual financing program, coordinated by the Ministry of Regional Development and Public Administration, which has as its general objective the equipping of the administrative-territorial units with all the technical-urban endowments, of educational, health and environmental infrastructure, sports, social-cultural and touristic, administrative and access to the communication ways to ensure an attractive investment climate for the localities of Romania.

The beneficiaries of the tunnel projects have applied under PNDL Program in terms of the transformation of the existing road into a multi-functional road that will provide the access to vehicles as well as facilitated with ski-slope.

### *2.1. Design*

The present-day design consideration includes the long-term environmental impact and also the life-cycle cost. Hence more importance is given in current design codes to the durability of structures [3].

The design stage of the project is one of the most important stages and so it requires a lot of attention.

*2.1.1. Parameters of the Multiplate VM35 structures.* The parameters for the steel structures were established after was collected all the information from the site. The major dimension of the structure consists of 8.55 meter span, a rise of 5.98 meter, bottom, and top lengths are 50.87 meters and 35.35 meters respectively, S355 steel with a total structural weight of 70 tones per each tunnel.

Due to the durability of corrugated steel sheet structures and lower weight with good strength, can be installed quickly and without difficulties using light equipment. The structures/flexibles are designed in such a way as to distribute the external loads in the embankment around them and towards the foundations. That is why the installation must be carried out with special care to ensure their correct functioning.

A well-positioned corrugated steel structure, installed on a properly prepared foundation, properly assembled, and surrounded by a backfill made of carefully compacted stable material will work efficiently and correctly throughout the entire service life for which it was designed.

The flexibility of the structure allows a high tolerance in terms of dimensional or locational changes that in the case of rigid structures often cause cracks

The Euro Codes and National Annex standards have been considered for all the design and parametrical aspects; MATHCAD has been used for static calculation of the structure [5].

### *2.2. Execution*

The project execution started in autumn 2018. This project was designed to help to develop the respective area, in the following ways:

- widening of the roadway by 4 meters;
- pour asphalt level;
- building culverts;
- support walls.

The principal idea of this project is to create an easy access road for the tourists and also to enhance the conditions for winter sports, especially ski, in the beautiful Parang. This road is until the 1600 alpine elevation making it a very challenging project, one of the structures was delivered and assembled in November 2019 within 3 weeks in hard weather, with a team of 5 people.

**Figure 5**. First steel tunnel (author's photo).

The second structure was delivered and assembled in May 2020, also in 3 weeks was already assembled. Was necessary to use a crane for 40 tones and an arm of 25 m to realize the assembly.

**Figure 6.** Second tunnel assembly in May 2020 (author's photo).

After the structures were assembled, the construction team must make the overflow around and cover the structures with compacted ballast, following the instruction from ViaCon Romania SRL.

*2.2.1. Assembly of the steel structures.* Plates are delivered to the construction site in bundles. Each bundle contains several plates with maximum numbers. The weight of a single bundle does not exceed 4 metric tons. The delivery includes a Euro pallet - containing fasteners, anchor bolts, base channels, the assembly drawing (figure 7), and the assembly kit (figure 8) [6].

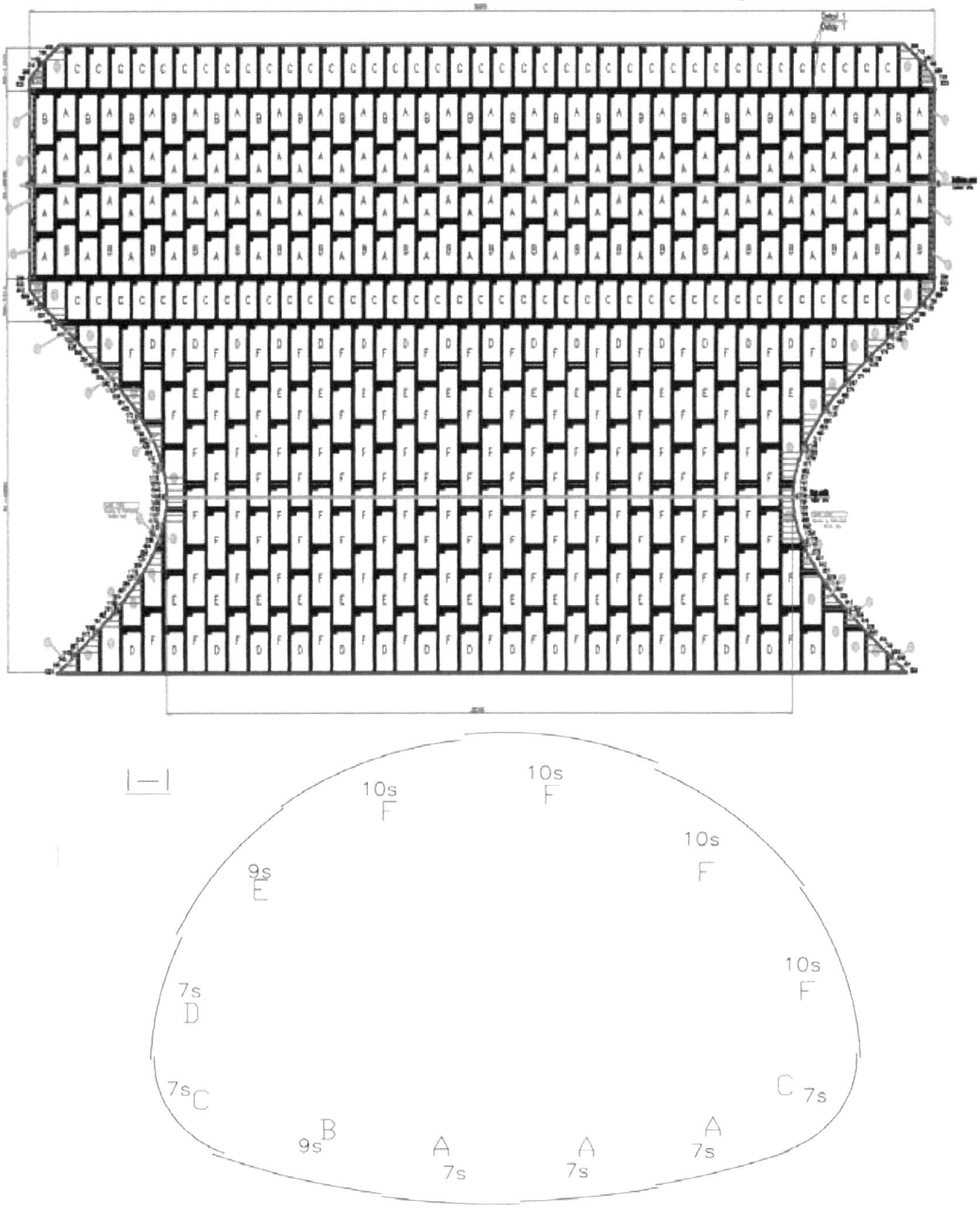


**Figure 7.** Assembly drawing of Multiplate VM35.

In figure 7, all plate types are marked with a letter (A, B, C, etc.) or a number (1, 2, 3, etc.). All plates marked with the same letter are identical and can be put in any place of the structure where such plate type is planned according to the drawing. All numbered plates have unique numbers and must be assembled exactly there where are they marked on the drawing [6].

Also, to each drawing, there is attached a table with types of plates and their heat numbers.

The total number of plates for one structure was 477 pieces.

The plates type A was marked with black paint and is 126 pieces. The plates type B was marked with red paint and is 42 pieces. The plates type C was marked with yellow paint and are 78 pieces. The plates type D was marked with blue paint and are 34 pieces. The plates type E was marked with green paint and are 30 pieces and the plates type F was marked with orange paint and are 119 pieces.

The plates relate to high resistance galvanized bolts and nuts M20 class 8.8.

The green lines from the assembly drawing are the axis of the structure (bottom axis and top axis).

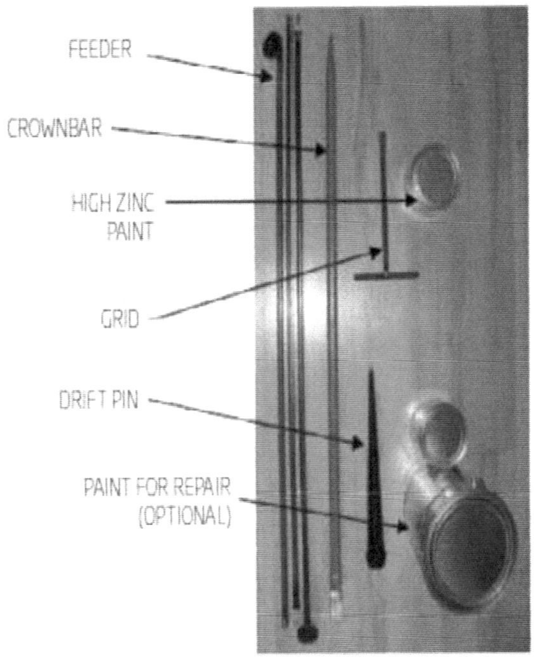


**Figure 8.** Assembly kit.[6]

The assembly kit is supplied with the structure (figure 8). Is necessary to have the required accessories with the team, such as a feeder to match the holes of the plates, and high zinc paint to repair the layer of zinc if will be scratched.

The assembly crew should be equipped with the following additional equipment:

- crane with an adequate load capacity;
- telescopic handler;
- slings and belts for plates transport;
- impact and hand wrenches;
- ladders, scaffolding, basket lifts;
- belts and tension chains;
- hammers min. 3 kg (used to adjust holes, using the drift pins).

Is essential to know the size of the structure, the real situation from the construction site, and the purpose of the structure, the type of necessary equipment and tools should be agreed upon individually with ViaCon Technical Department.

*2.2.2. Rules of assembly.* Assembly of the structure should be started from the bottom plates (figure 9) [6].

**Figure 9.** Assembly of bottom plates (author's photo).

During assembly of the bottom part, it is necessary to provide the mounting space under the plates to put in the bolts. The bottom axis should be controlled all the time at the stage of assembly of the bottom. After assembling the bottom plates, we assemble the side plates (figure 10) [6].

**Figure 10.** Assembly of side plates (author's photo).

After assembling the side plates, we proceed to assemble the top plates (figure 11). Top plates can also be pre-assembled in small sections to give ease in the assembly with side plates. During assembly it is advisable to not tighten the bolts to the maximum defined torque (they should be tightened firmly) - this will be helpful to avoid problems with the upcoming activities of mounting the next plates (rings). Tightening the bolts to the required torque shall be done after assembling the entire structure and before the backfilling. Nuts can be located inside or outside the structure. Their location does not affect the "work" and bearing capacity of the structure [6].

For practical reasons, it is good to place all the nuts in the bottom part of the structure on the inside, while in the rest of the structure on the outside (from the backfill side). Access to the nut placed from outside at the area of the bottom plate is limited and does not allow for properly tightening the bolts [6].

**Figure 11.** Assembly of top plates (author's photo).

During assembly, the dimensions of the structure have been checked on an ongoing basis. Structures with larger cross-sections deformed under their weight, which makes assembly more difficult. For structures with a span of more than 4 m, like in our case, it is recommended that the assembly of all cross-sectional sheets is carried out by just a few meters. This applies to the side plates with the bottom and side panels with the upper ones [6].

Additionally, the mounting belts should be used to maintain the nominal dimensions of the cross-section. After assembling the structure, it is necessary to repair the damage to the anti-corrosion coating that occurred during the transport and assembly of the structure [6].

*2.2.3. Control of the bolt tightening torque and the structure shape.* After the assembly, the bolt tightening torque should be checked. Required tightening torque: 360 Nm (minimum) for the structure with a span bigger than 7m, in our article the span is 8.55 m [7].

Tightening of the bolts has been checked using a torque wrench. 5% of the total number of bolts should be checked. A minimum of 95% of the bolts tested have to meet the requirements regarding the size of the torque. The inspection is carried out on randomly selected bolting connectors, located evenly around the structure.

During the backfilling of the structure, the tightening of the bolting connectors has been checked at a random pattern.

The site engineer is responsible to check the dimensions of the structure during its construction. During the assembly of large span profiles, there is a natural thing that the structure deflects under its weight, as a result of which the structure may reach a lower height and a greater span. During backfilling, the structure crown point will be picked up [7].

When the backfill reaches the level of the crown point, the process of uplifting shall be finished. This process should be controlled by performing necessary measurements so that the tolerance of the cross-section is maintained [7].

The contractor was obliqued to present a report of the behaviors of the structure after assembly and during backfilling and the report of bolting connectors tightening.

*2.2.4. Soil preparation.* It is necessary to follow the instructions, from the technical data sheet, when the foundation under the construction is realized, is indicated to use gravel-sand mixtures with fraction 0 - 45 mm, uniformity coefficient $Cu \geq 4.0$, curvature coefficient $1 < Cc < 3$, and water permeability k > 6 m/day [7].

The compaction index Is min.= 0.98, according to the standard Proctor test. For the structures with a span bigger than 4,0 m, like in our case, the foundation was contoured in the way corresponding to the shape of the bottom plate and reached the level of side plates (a few centimeters below the connecting place of side plates with bottom plates). In any case, particular attention shall be paid to the densification of the foundation aggregate in the area of the side and bottom plates connection place. The thickness of the aggregate foundation was 25 cm. Regardless of whether the aggregate foundation is flat or contoured, its upper layer (5 - 10 cm) must consist of uncompacted gravel and sand bedding so that it can freely adapt to the corrugation of the plates [8].

Aggregate located directly next to the structure shouldn't contain grains larger than 32 mm. Before assembly of the structure, the delineation of the structure axes, ordinates of foundation, shape, and density index need to be checked. The correctness of the substrate preparation should be verified and accepted by the responsible person before starting assembly works of the structure [9].

*2.2.5. Final work.* The project was timely finished and is fully functional since September 2021. It was a big challenge to finish this project within timeline and with quality because of adverse weather conditions and the access was challenging for all the team.

In the Parang Mountains, the weather was bad with stormy wind resulting in lots of difficulties to handle the plates in these adverse conditions.

**Figure 12.** The final stage of the first tunnel (internet photo newspaper "Stirile Transilvaniei").

**Figure 13.** Drone view of two tunnels and slope ski (internet photo newspaper "Stirile Transilvaniei")

## 3. Conclusions

The design of the project for the construction of the two steel structures analyzed in this paper raised a significant series of difficulties in terms of consolidating and improving the existing road, joining the new extension functions: access for the cars, and also for the realizing of the sloped sky, working with the team in an area with high slope, difficult access of construction equipment due to the narrow area and on the slope.

Currently, the construction of the tunnels works was finished. The beneficiary was asked to periodically, at intervals of one year, make a current inspection of the entire tunnels, to verify the behavior over time of the constructions. If it is proved that the executed works will not show damages, the adopted solutions can be used for other similar constructions.

The steel structures have a series of advantages in comparison with the classic structures (concrete, precast):

- simple design (few details regarding the execution),
- easy and rapid assembly,
- reduced execution time,
- can be put into position without stopping the traffic, with few equipment and workers.

Each structure was assembled in 3 weeks, with a team of five workers. Was necessary to use a crane for 40 tones and an arm of 25 m to realize the assembly.

Tunnels can be extended in length if future needs dictate. The juncture between old and new sections of the tunnel can be accomplished using the same type of structure (type Multiplate, profile VM35). Care must be taken during any future modifications to keep loads and soil envelope support for the tunnels balanced and uniform. Any tap-ins or similar modifications must be done in a manner to keep the entire tunnel periphery symmetrically loaded and supported and to maintain the stability of the soil envelope [10].

**Acknowledgment**
The work presented in this paper was realized by ViaCon Romania, with big support from Strabag Romania. We are so grateful and sending our best regards to the investors: Hunedoara County Council and Petrosani City Hall.

**References**
[1] Amundsen F H and Solvik O L 1995 *Classification of Tunnels, Existing Guidelines, and Experiences, Recommendations*
[2] Sharma A K 2021 *Tunnel Construction- A Practical Approach Demystified* (BFC Publications)
[3] Dobrovolski M and Cazacu C 2010 Importance of execution technologies in tunnels constructions **CIBv 2010**
[4] www.romania-travel-guide.com, accessed on October 2021
[5] The European Union 2006 EN 1993-1-1 (2005) (English): Eurocode 3: Design of steel structures - Part 1-1: General rules and rules for buildings *Concr.* **40**
[6] Internal documents (ViaCon Group)
[7] www.geosegypt.com (internet source), accessed on October 2021
[8] Foley A H and Olsen F 1960 Project EHV-Preliminary Corona Investigations: The Effect of Harmonics on Corona Losses *Trans. Am. Inst. Electr. Eng. Part III Power Appar. Syst.* **79** 310–5
[9] Terzaghi K 2007 Earth Pressure on Temporary Supports in Cuts, Tunnels, and Shafts *Theoretical Soil Mechanics* pp 182–215
[10] www.conteches.com (internet source), accessed on September 2021

# Study on the corrosion and frost – thaw behavior of the concretes with polystyrene granules

**A Panțiru[1], C Cădere[1], L I Diaconu[1*], M Bărbuță [1], M Rujanu [1] and A C Diaconu [2]**

[1] Technical University "Gheorghe Asachi" from Iassy, Faculty of Civil Engineering and Building Services, Romania

[2] Romanian Association of Earthquake Engineering

* livia-ingrid.diaconu@academic.tuiasi.ro

**Abstract.** The study was conducted in the Laboratories of the Faculty of Civil engineering and Building Services, department of Concrete, Materials, Technology and Management, Technical University "Gheorghe Asachi" from Iassy. The main objective of the experimental research from this paper is the analysis of some types of concretes prepared with polystyrene granules and fly ash, aiming to obtain a material which is corresponding to the structural requirements, to be energetically efficient and to be ecological. Using these components, the durability characteristics (frost – thaw and corrosion), the influence of the polystyrene granules on the environment and the mode in which the materials properties are maintained in time under environmental factors, were experimentally determined. The concretes were prepared with polystyrene granules and fly ash from CET Holboca Iassy, Romania. The fly ash was utilized in proportion of 10% and 20% from the cement amount and the polystyrene granules replaced some sorts of the natural aggregate. The types of concretes are CPOL1, CPOL2, CPOL3 and CPOL4.

## 1. Introduction

The building materials and the construction industry is trying to accomplish the durable development criterions byproducts using, in order to reduce to minim the impact on the environment [1, 2]. The concrete is a widely utilized in construction, presenting a great durability potential, taking into account that it can be combined with different types of byproducts, such as fly ash, silica or furnace slag. The byproducts can be used as admixtures in concretes to improve the physico – mechanical durability, thermo – physical, acoustical characteristics. The thermo power plant ash, silica or furnace slag are only some of the byproducts added in concrete to obtain concretes with superior characteristics in comparison with the usual concretes, the byproducts can replace different dosages of cement or aggregates [3, 4]. The aggregates can be replaced in proportions of 100% with byproducts, polystyrene granules etc.

The effect of the polymer granules in concretes composition are depending on the type of the concrete (cement concrete or polymeric concrete), the dosage and the type of the polystyrene and on the other admixtures which are introduced [5].

The first studies on the utilization of the polystyrene granules were conducted from 1973 [6]. In the following decades, by many experimental and theoretical researches, the concrete with polystyrene granules (CPOL) was studied. The replacing of the natural granules with recovered or artificial aggregates, in concrete's manufacturing will lead to an increase of the building materials industry sustainability.

The polystyrene concrete is consisting in cement, sand, natural aggregates and polystyrene granules, which can be utilized as admixtures or aggregates replacement [7]. The polymer concrete is a light weight concrete, in the case when a single aggregate sort is used, even when many sorts are used. Function of the aggregate dosage, also structural concretes can be obtained, presenting densities greater than 2000 kg/m$^3$, but the mechanical strengths are lower than the traditional concretes [8].

Many experimental studies were conducted, trying to improve the characteristics of the concrete with polymer granules [9-11]. The studies regarding the compositions and the mechanical and durability characteristics conducted on concrete with polystyrene granules, highlighted a series of properties specific to these types of concretes. Factors such as: reduced density, high strength / weight ratio, reduced thermal coefficient, good fire behavior, increase durability etc. are considered advantages for these types of concretes instead to the traditional concrete. Using these concretes, the elements and structures weights are reduced and their cross section as well. Elements and structures with great apertures can be realized, which are presenting a good behavior to seismically type loads [9].

The concrete's corrosion represents a physico – chemical degradation process under aggressive chemical actions, whose effects are depending on its composition and the compactness, the intensity, nature and duration of the chemical actions.

In this paper are presented the results of the experimental studies conducted on concretes with polystyrene granules subjected to frost – thaw and aggressive chemical agents.

## 2. Experimental program

### *2.1. Mixes composition*

For the preparation of the concrete with polystyrene granules, a witness mix (CCI) was used, containing the following materials:

- cement type CEM II 42,5 - 360 kg/m$^3$;
- river aggregates: sand sort 0 – 4 mm - 803 kg/m$^3$;
- sand sort 4 - 8 mm - 384 kg/m$^3$;
- sand sort 8 - 16 mm - 558 kg/m$^3$;
- water 172 l;
- Glenium Sky super plasticizer / water reducing additive in a dosage of 1% from cement's amount

In the experimental program, a series of cement concrete mixes were prepared, where the cement was replaced with 10% (BC10) and 20% (BC20) with thermo power plant ash (FA) from CET Holboca Iassy. The receipts are presented in table 1.

**Table 1.** Witnesses receipts BC10 and BC20 with thermo power plant ash.

| Receipt | Cement [kg/m$^3$] | Water [l/m$^3$] | Thermo power plant ash [kg/m$^3$] | Sand [kg/m$^3$] | Aggregate [kg/m$^3$] | Aggregate [kg/m$^3$] | Additive [kg/m$^3$] |
|---|---|---|---|---|---|---|---|
| BC | 360 | 172 | 0 | 803 | 384 | 558 | 3.6 |
| BC10 | 324 | 172 | 36 | 803 | 384 | 558 | 3.24 |
| BC20 | 288 | 172 | 72 | 803 | 384 | 558 | 2.88 |

### *2.2. Experimental set up*

To determine the durability characteristics, the strength to aggressive chemical agents (concretes corrosion) and the frost – thaw strength, samples with fly ash and polystyrene granules were realized and analyzed.

Throughout the experimental analysis are determined the durability characteristics, the influence of the polystyrene granules on the material and the mode of maintaining in time of the materials under the action of the environmental factors.

*2.2.1. Study of the influence of the chemical agents on the durability of concretes with polystyrene granules.* The testing process of the samples for accelerated corrosion was performed by introducing the samples in a (KOH) and (HCl) solutions. The solution of KOH was used in concentrations KOH 30 g/l and KOH 100 g/l and the HCl solution in 3% and 18% (prepared from concentrated HCl having the density of 1.19 g/ml. The analysis for the study of the corrosion strengths of the materials were conducted on two samples from each concrete receipt for each concentration. After 12 days, when the samples were preserved in KOH (figure 1) and KCl (figure 2) solutions, were analysed to observe the visible modifications from the surfaces which came into contact with the chemical agent and the masses of the samples before and after the chemical reactions occurred, were determined by weighting [12].

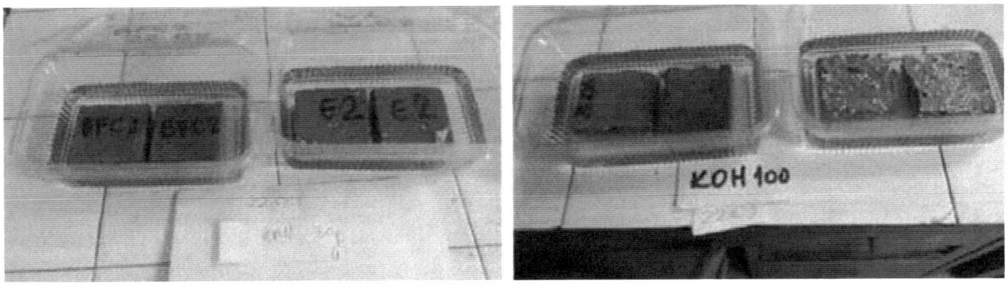


**Figure 1.** Testing of the samples for the concrete's corrosion determination – (KOH 30g/l).

**Figure 2.** Testing of the samples for the concrete's corrosion determination – (HCl 30g/l).

*2.2.2. Frost thaw strength (gelivity) determination.* According to SR 3518-2009 [12] the frost – thaw of the concrete can be determined by a destructive method. The variation of the compression strength is determined.

Cubic samples having the side of 100 mm were realized (figure 3)., prepared according to SREN 12390-2/2002 [13].

**Figure 3.** Samples subjected to frost – thaw.

For the samples which are subjected to frost – thaw cycles and for the witness samples, prepared from the same concrete and preserved in the same conditions, the compression strength is determined in order to assess the strength loss [14].

The samples are introduced in water bath at a temperature of (20±5) $^0$C 4 days before testing, to saturate. The water is added in the bath up to ¼ from the sample's heights; after 24 hours the water is added up to ½ and, after other 24 hours up to ¾ and after 3 days the water has to be with at least 20 mm above the samples. They are preserved another 24 hours for saturation.

For testing, for frost – thaw, the samples are introduced in a refrigerating chamber at a temperature of (-17±2) $^0$C, where are maintained for 4 hours where the witnesses samples are preserved in water.

After 4 hours the samples are taken off from refrigerating chamber and are introduced in the water bath for another 4 hours. After the frost thaw - cycles it is performed the strength loss on the samples subjected to compression testing on three subjected samples and three of the witness samples.

To verify the gelivity degree G50 the samples are subjected to 50 frost - thaw cycles and then the compression strength is determined and is compared to the witness samples figure 4.

The strength loss, according to the standard (η) is calculated with the following formula:

$$\eta = \frac{f_n - f_f}{f_n} \times 100 \quad (1)$$

where:
$f_n$ – compression strength determined on witness, samples (N/mm$^2$)
$f_f$ – compression strength determined on samples which are subjected to frost – defrost cycles, (N/mm$^2$)

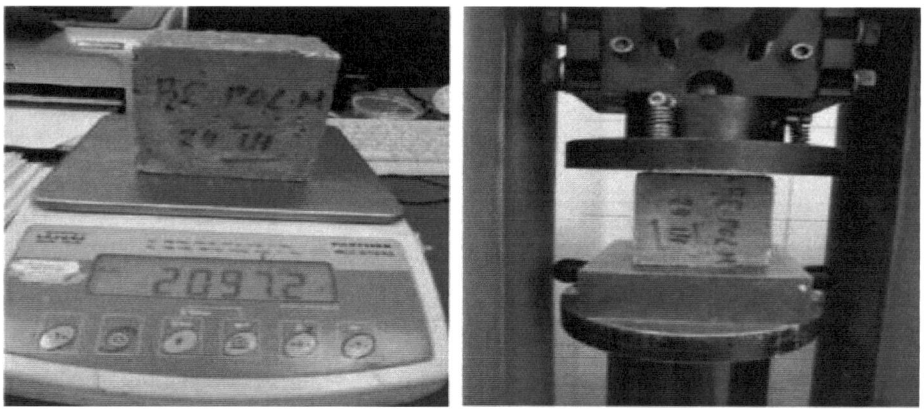

**Figure 4.** Compression testing of the samples subjected to 50 frost thaw cycles and samples density.

It is considered that the concrete is resistant to a maxim number of frost – thaw cycles without a reducing of the compression strength greater than 25%.

The gelivity testing were performed on samples realized from the following types of concretes:

- BCPOL1, 10 % fly ash as cement replacement and the sand was replaced in a proportion of 40% with polystyrene granules, amount equivalent in volume;
- BCPOL2, 10% fly ash as cement replacement and the sort 4 - 8 mm was replaced in a proportion of 60% with polystyrene granules, amount equivalent in volume;
- BCPOL3, prepared with 10% fly ash as cement replacement, the sand was replaced with 10% polystyrene granules with sizes 0-4 mm and the aggregate sort 4 – 8 mm was replaced in a proportion of 10% with polystyrene granules with the dimensions of 4 – 8 mm;
- BCPOL4, prepared with 20% fly ash as cement replacement, the sand was replaced with 20% polystyrene granules of dimension 0-4 mm and the sort 4 – 8 mm was replaced in a proportion of 10% with polystyrene granules having the dimensions of 4 – 8 mm;

The determinations of the frost – thaw strength of the concrete samples where the aggregates were replaced with polystyrene granules were conducted according to the current standards [15-19].

The samples are verified before the testing beginning in order to not present degradations. The frost – the cycles are performed according to the standard, observing if in time degradations are appearing.

The visible degradations from the following categories are noted:

- surface swellings;
- cavities with diameter of 5 mm;
- cracks.

The assessment of the degradations due to the frost – thaw cycles are performed only for uncut sides of the samples. If the samples are not presenting deterioration from the mentioned above categories, the testing is considered finished. In the case when deteriorations can be observed, the compression strength of the samples which were subjected to frost – thaw cycles and of the witnesses samples, as well.

## 3. Results and discussion

### *3.1. Influence of the chemical agents on the durability of concretes with polystyrene granules (concrete's corrosion)*

As can be observed from the figure 5 can mentioned the following issues:

- the samples which were introduced in KOH, in both of the concentrations are presenting an appropriate behaviour from chemical strength's point of view;
- the samples subjected to HCl action suffered evident degradations instead the samples subjected to KOH action;
- the mass losses are greater in the case of the samples subjected to HCl attack instead of the samples subjected to KOH.
- For the sample introduced in HCl solution with concentration of 18% the material's decay by the total decalcification of the cement stone and modification in polystyrene granules structure can be observed.

**Figure 5.** Samples degradation.

### *3.2. Frost thaw strength (gelivity)*

The results for the gelivity testing are presented in the table 2.

**Table 2.** Witnesses receipts BC10 and BC230 with thermo power plant ash.

| Sample | Compression strength loss η % |
|---|---|
| CPOL1 | 14.38 |
| CPOL2 | 17.74 |
| CPOL3 | 13.05 |
| CPOL4 | 3.18 |

The exposure of the concrete with thermo – power plans ash and polystyrene granules to the accelerated action of the KOH and HCI are showing the following:

- The samples subjected to the action of the KOH solutions of 30 g/l and 100 g/l did not present visible degradations, which could affect the cement stone, only low mass losses, due to the degradation of a thin layer from the sample's surface;
- The samples subjected to the action of the HCI solutions of 3% did not present important degradations of the cement stone, which was destroyed by soaking and the concrete became brittle.
- Taking into account that the testing mode is tough as duration and concentration (12 days and very high concentrations of the chemical agents, can be appreciated that the concrete with thermo – power plant ash is presenting a good behaviour to the action of the chemical agents;
- According to the standard SR EN ISO 10545-13:2017 [12] for the HCl 3% and KOH 30 g/l (low concentrations - L), the concrete with thermo power plant ash is situating in the superior class LA(V), providing that is not presenting visible effects. For the KOH 100 g/l of (high concentrations – H), the concrete with thermo power plant ash is situating in the superior class HA(V) providing that visible effects don't exist and for the HCl of 18% solutions (high concentrations – H) the concrete with thermo power plant ash is situating in the superior class HA(V) providing partial or entire loss of the original aspect.
- After subjecting to frost – thaw cycles the concretes with thermo power plant ash and polystyrene granules suffered strength losses between 3,18% and 17,74% and before the compression testing, the samples were free eye analyzed and deformations, swellings, cracks, spallings or delamination cannot be observed. All the samples are presenting a good behavior to frost – thaw, the compression strength decrease is below 25% for all types of concretes. The concretes gelivity is G50.

## 4. Conclusions

In the study, polystyrene granules wastes were used for preparing concrete with substitution of cement and aggregates in different percentages and the durability characteristics were analysed. The dosage of polystyrene granules is influencing the behaviour of experimental concretes under accelerate aggressive agents and under frost-thaw cycles. The solutions of KOH and HCl used as corrosive agents had attacked the concrete, visible degradations were observed. After 50 cycles of frost-thaw the loss of compressive strengths was below 25% that indicate that experimental concrete with polystyrene granules has the gelivity G50.

The concrete obtained by using polystyrene granules as substitution of aggregates had a good behavior under corrosion and frost – thaw cycles and it can be used for obtaining concrete blocks which are presenting a better thermal conductivity than the traditional concrete.

## References

[1] Aciu C, Manea, D. et.al. 2015 Recycling of polystyrene waste in the composition of ecological mortars. In: 8th International Conference Interdisciplinarity in Engineering, INTER-ENG 2014, 9-10 October 2014, Tirgu Mures, Romania

[2] Ioniuc I, Diaconu L I et.al. 2016 The flying ash taken by wet process from cet holboca Iassy– major pollution source, The 9th International Conference INTER-ENG 2015, pp. 413-418, Procedia Technology vol. 22, Tîrgu Mureș, România

[3] Diaconu LI, Plian D et.al. 2019 The long term behaviour of some fly ash concrete mixes subjected to dissolving-levigation corrosion, Romanian Journal of Materials, pp. 267-273, vol.49, nr.2

[4] Bărbuta M, Bucur R et.al. 2016 Combined effect of fly ash and fibers on properties of cement concrete, 9th International Conference Interdisciplinarity in engineering, INTER-ENG 2015, pp. 280 – 284, Procedia Technology, vol. 181, Târgu Mureș, Romania

[5] Muntean R, Cazacu C et.al. 2016 Practical applications of dispersely reinforced concrete with polypropylene fibers beams, 9th International Conference Interdisciplinarity in engineering,

INTER-ENG 2015, pp. 285-292 Procedia Engineering, vol. 181, Târgu Mureș, Romania
[6] Cook D J 1973 Expanded polystyrene beads as lightweight aggregate for concrete, Precast Concrete, vol. IV, pp. 691-693
[7] Cook D J 1983 Expanded polystyrene concrete, concrete technology and design, New Concrete Materials," Surrey Univ. Press, pp. 41-69
[8] Sayadi A, Tapia J A et.al. 2016 Effects of expanded polystyrene (EPS) particles on fire resistance, thermal conductivity and compressive strength of foamed concrete" Constr. And Building Materials, vol. 112, pp. 716-724
[9] Chen B, Liu J 2004 Properties of lightweight expanded polystyrene concrete reinforced with steel fiber," Cement and Concrete Research, vol. 34, pp. 1259-1263,
[10] Chen B, Liu J 2010 Experimental study of lightweight expanded polystyrene aggregate concrete containing silica fume and polypropylene fibers," Journal of Shanghai Jiaotong University, vol. 15, nr. 2, pp. 129-137
[11] Shi W, Miao L, Luo J, Wang J and Chen Y 2015 Durability of Modified Expanded Polystyrene Concrete after," Hindawi Publishing Corporation
[12] SR EN ISO 10545-13: Ceramic tiles - Determination of chemical resistance
[13] SR EN 3518: Tests on cement concretes. Determination of the freeze-thawing restance by measuring the variations of the resistance strength and/or of the dynamic relative elastic modulus
[14] SR EN 12390-1: Testing hardened concrete Shape, dimensions and other requirements for specimens and moulds
[15] SR EN 12390-2: Testing hardened concrete - Part 2: Making and curing specimens for strength tests
[16] SR EN 12390-3: Testing hardened concrete Compressive strength of test specimens
[17] SR EN 3518: Tests on concretes. Determination of freeze-thawing resistance by measuring the variation of the resistance strength and/or of the dynamic relative elastic modulus
[18] SR EN 12390-4: Testing hardened concrete- Part: Compressive strenght- Specification for testing machines
[19] SR EN 3518: Tests on concretes. Determination of the freeze-thawing restance by measuning the variations of the resistance strength and/or of the dynamic relative elastic modulus

# The impact of building occupant habits on heating energy consumption: a case study

**S G Petre[1*], D N Isopescu[1] and M Pruteanu[1]**

[1] "Gheorghe Asachi" Technical University of Iaşi, Faculty of Civil Engineering and Building Services, Romania

* sergiu.petre@tuiasi.ro

**Abstract**. The effort to reduce the consumption generated by the use of buildings is one of the important objectives of today's global society, involving authorities that regulate the energy performance of buildings, manufacturers of materials, specialists involved in the design and last but not least the occupants of buildings. The habits of the occupants of buildings regarding the indoor comfort temperature have a direct influence on the amount of energy consumed. The paper aims to highlight whether the results expected by applying the values recommended in the regulations are in line with the occupant's habits. Based on a case study, the energy requirement for heating was assessed using different indoor temperatures, which were established in relation to the temperatures monitored in situ. The inside temperature values were monitored in three buildings, located in the third climatic zone of Romania. The article presents the results of the case study, emphasizing that indoor temperature control and monitoring are of high importance in the desired expectations for energy performance of buildings, to meet the sustainability objectives. Knowing the mutual influence between energy consumption and thermal comfort demands, building designers can find solutions to achieve energy conservation goals.

## 1. Introduction

Increasing the energy efficiency of buildings is a key point in reducing global warming, and to achieve this, efforts must be made by all professionals involved in the building sector, but at the same time, building occupants have an important role to play. Several authors [1-5] note that the behaviour of occupants directly influences the energy consumption in buildings. Considering this influence and that in European Union, in 2019, 64% of final energy consumption in the residential sector was generated by heating of the spaces [6] it can be said that, some significant reductions in consumption can be made with a better understanding from the occupants, on how they can participate in reducing the total energy consumption. To reduce the difference between actual and theoretical consumption and to be able to forecast accurate consumption, it is also very important that the specialists involved in calculating the energy performance of buildings use scenarios as close as possible to the habits of the occupants.

The purpose of the article is to present to the specialists involved in the design of buildings, the influence of the variation of interior temperatures on the energy consumption, necessary for heating. A two part case study was conducted to present this. The first part of the study consists in the monitorization of the indoor temperature in three spaces, in different apartments, to observe occupant habits in terms of thermal comfort. The second part of the case study presents the variation of the heating energy consumption determined in the case of a typical apartment of the existing built environment in the north-eastern part of Romanian. The specific conditions of the third climatic zone of Romania have been selected, where the outside temperatures have a standard average value for the winter season. The results highlight a substantial increase in energy consumption for each degree of interior temperature increase when the structure of the envelope elements is kept constant.

## 2. Case study

To establish the influence of building occupant habits in energy consumption, the energy requirement for heating was determined for an apartment part of a condominium located in Romania's third climate zone, when different indoor calculation temperatures were taken into account.

The analysis was done for a typical apartment of the Romanian existing built environment, with thermal protection solutions that do not meet the actual standard requirements related to energy consumption. In this case, the exterior and staircase walls have a layer of expanded polystyrene with a thickness of 5 cm, applied 10 years ago. This is the only thermal protection solution, considering that no thermal protection solutions have been implemented at the lower floor over the unheated basement. The heating is made with steel radiators connected to a gas central heating boiler and the room temperature is controlled with a single thermostat. The floor plan of the apartment is presented in figure 1.

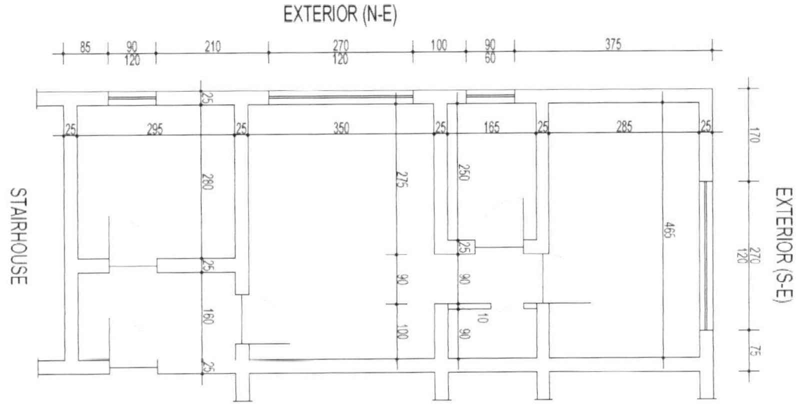


**Figure 1.** Floor plan of the apartment.

In situ measurements have been carried out to determine the thermal performance of existing thermal insulating materials. Measurements have been made with a portable system for the measurement of heat transfer properties of materials Isomet 2114 Applied Precision [7], presented in figure 2.

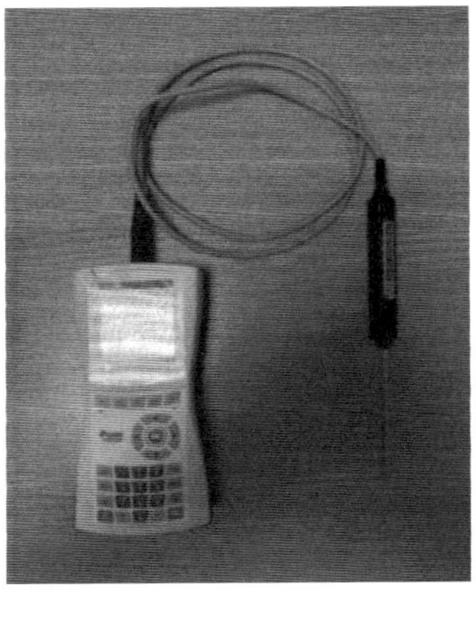

**Figure 2.** Portable Heat transfer analyser – Isomet 2114 Applied Precision.

### *2.1. Habits of the building occupants on thermal comfort*

To observe the habits of the occupants, temperature monitoring was made in three different apartments. In the first two apartments, measurements were made in bedroom spaces, mentioned below as Bedroom 1 and Bedroom 2. In the third apartment, the monitorization was made in an office room. The monitorization was realized with three temperature data loggers which, for two months of winter, have made temperature recordings at a regular 15-minute interval. The type of sensor used is LogTag Recorders HAXO-8 [8] and is shown in figure 3.

**Figure 3.** Data logger – LogTag Recorders HAXO-8.

A summary of the minimum, maximum, average values and the standard indoor temperature according to the methodology for calculating the energy performance of buildings in Romania, for the three locations, is presented in table 1.

**Table 1.** Summary of temperature values [°C].

| | Minimum | Maximum | Average | Standard indoor temperature |
|---|---|---|---|---|
| **Bedroom 1** | 17.90 | 21.60 | 19.80 | 20.00 |
| **Bedroom 2** | 20.30 | 26.10 | 22.80 | 20.00 |
| **Office** | 16.00 | 29.30 | 25.20 | 20.00 |

Differences may be observed between the standard indoor temperatures and the mean of the indoor measured temperatures, the greatest difference between them being in the case of the office, where the mean of the measured values is 26.00% higher than the standard value. A higher temperature can also be observed in bedroom 2 where the average temperature recorded was 14% higher than the standard one. In the case of bedroom 1, the difference is the lowest, only 1% and this time the average temperature recorded is lower than the standard indoor temperature.

Whether it is a temperature higher or lower than the standard temperature, the large variability between users is also noted by researchers from Greece [3], England [5] or China [9]. Therefore, for professionals involved in identifying solutions to increase the energy performance of buildings, the variation of indoor temperatures within such extensive limits becomes an uncertainty in determining the actual energy consumption for heating. And the question that results is how correct is the determination of the parameters that establish the energy performance of buildings in the real conditions, identified in their exploitation.

### *2.2. Heating energy consumption for different interior temperatures*

The determination of the energy consumption can be evaluated by calculation based on the structure of envelope elements and the thermal conductivity of the layers that make up these structural elements. Given the age of the thermal insulation layer, it was chosen to identify by in situ measurements its thermal conductivity.

The thermal conductivity of expanded polystyrene from the outer walls was measured. figure 4 shows the location where the measurements were taken.

**Figure 4.** The location where the measurements were taken.

In table 2 are presented the results of the measurements. In this particular case, the measured value of thermal conductivity is similar to the value prescribed by the methodology for calculating the energy performance of buildings in Romania [10-13]. For expanded polystyrene more than 10 years old the standard gives a value of 0.042 W/mK for the thermal conductivity. In the determination of thermal resistance of walls, the average measured value was considered.

**Table 2.** Results of thermal conductivity measurement.

| | Measurement 1 | Measurement 2 | Measurement 3 | Average |
|---|---|---|---|---|
| Thermal conductivity [W/mK] | 0.0414 | 0.0424 | 0.0447 | 0.0428 |

For the outer walls and the staircase walls, the structure of the element is presented in figure 5 and the determination of the thermal resistance of the element is presented in table 3.

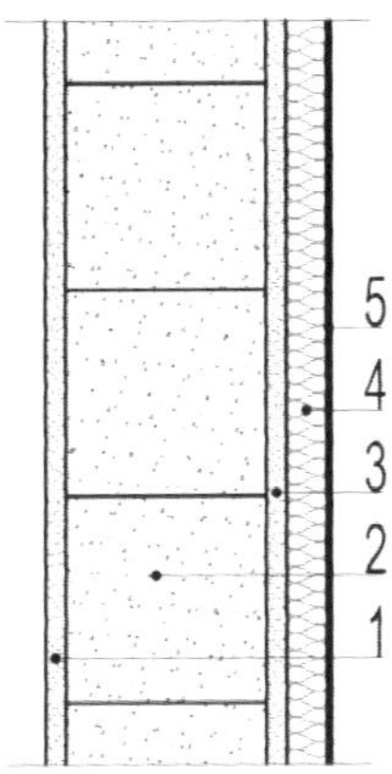


**Figure 5.** Cross-section of the wall.

**Table 3.** Thermal resistance calculation for the wall structure.

| Nr. | Layer | Thickness [m] | Thermal Conductivity [W/mK] | Layer thermal resistance [m²K/W] | Thermal resistance [m²K/W] |
|---|---|---|---|---|---|
| 1 | Internal plaster | 0.025 | 0.93 | 0.026 | 2.125 |
| 2 | Autoclaved aerated concrete | 0.25 | 0.30 | 0.794 | |
| 3 | External plaster | 0.025 | 0.93 | 0.026 | |
| 4 | Expanded polystyrene | 0.05 | 0.043 | 1.107 | |
| 5 | External mineral plaster | 0.005 | 0.93 | 0.005 | |

For the lower floor, the structure of the element is presented in figure 6 and the determination of the thermal resistance of the element is presented in table 4.

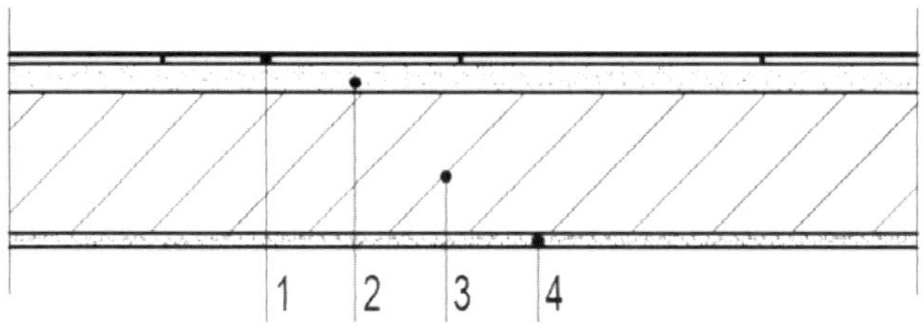


**Figure 6.** Cross-section of the lower floor

**Table 4.** Thermal resistance calculation for the lower floor structure

| Nr. | Layer | Thickness [m] | Thermal Conductivity [W/mK] | Layer thermal resistance [m²K/W] | Thermal resistance [m²K/W] |
|---|---|---|---|---|---|
| 1 | Ceramic tile | 0.01 | 1.60 | 0.006 | 0.391 |
| 2 | Cement floor mortar | 0.03 | 0.93 | 0.031 | |
| 3 | Reinforced concrete | 0.15 | 1.74 | 0.086 | |
| 4 | Plaster | 0.015 | 0.93 | 0.016 | |

The determination of the consumption level for each required temperature has been calculated according to the methodology for calculating the energy performance of buildings valid in Romania [9-12], using the equation (1):

$$Q_{fh} = Q_L - \eta * Q_g + Q_{th} = H * (\theta_{em} - \theta_i) * t - \eta * Q_g + Q_{th} \; [kWh/year] \quad (1)$$

where $Q_{fh}$ energy consumption for heating (kWh/year); $Q_L$ represents the heat loss of the building (kWh/year); $\eta$ represents usage factor (-); $Q_g$ represents heat intakes of the building (kWh/year); $Q_{th}$ represents heat loss caused by heating installations (kWh/year); H represents heat loss coefficient of the building (kW/K); $\theta_{em}$ represents the average outside temperature of the cold season (°C); $\theta_i$ represents average internal temperature (°C); t represents the duration of the heating season (h).

Consumption determinations have been made in the range 18°C to 24°C, considering that the standard indoor temperature is 20°C. The summary of the calculations is presented in table 5.

The summary of the results of the case study is presented in table 6.

The case study shows that for small temperature changes, the variation in energy consumption is considerable, for example, increasing the temperature by 10% from 20°C to 22°C, generates an increase in the consumption by about 18% or increasing the temperature by 20% from 20°C to 24°C, generates an increase in the consumption by about 40%.

**Table 5.** Summary of the calculations.

| H | $\theta_i$ [°C] | $\theta_{em}$ [°C] | t [h] | $Q_L$ [kWh/year] | η [-] | $Q_g$ [kWh/year] | $Q_h$ [kWh/year] | $Q_{th}$ [kWh/year] | $Q_{fh}$ [kWh/year] |
|---|---|---|---|---|---|---|---|---|---|
| | 18.00 | 5.09 | 5856 | 10799.60 | | 1631.07 | 9168.53 | 965.42 | 10133.95 |
| | 19.00 | 5.54 | 6144 | 11813.44 | | 1693.75 | 10119.69 | 1064.99 | 11184.68 |
| | 20.00 | 6.04 | 6432 | 12826.60 | | 1756.43 | 11070.17 | 1164.82 | 12234.99 |
| 0.14 | 21.00 | 6.99 | 6936 | 13881.21 | 1.00 | 1866.12 | 12015.09 | 1264.41 | 13279.50 |
| | 22.00 | 7.39 | 7176 | 14976.59 | | 1918.35 | 13058.24 | 1370.20 | 14428.44 |
| | 23.00 | 8.32 | 7728 | 16205.91 | | 2038.49 | 14167.42 | 1490.85 | 15658.27 |
| | 24.00 | 9.87 | 8760 | 17681.80 | | 2263.09 | 15418.71 | 1622.86 | 17041.57 |

**Table 6.** Summary of the results of the case study.

| Indoor adopted temperature [°C] | Heating energy consumption [kWh/m²year] | Standard energy consumption[a] [kWh/m²year] | Difference between the adopted and standard temperature consumption [%] |
|---|---|---|---|
| 18.00 | 186.25 | | -17.17 |
| 19.00 | 205.56 | | -8.58 |
| 20.00 | 224.87 | | 0.00 |
| 21.00 | 244.06 | 224.87 | 8.54 |
| 22.00 | 265.18 | | 17.93 |
| 23.00 | 287.78 | | 27.98 |
| 24.00 | 313.21 | | 39.29 |

[a] Standard energy consumption is the consumption obtained for the standard indoor temperature (20°C).

## 3. Conclusions

Reducing energy consumption from buildings is a continuing concern of today's society. Important efforts are being made to find effective ways of reducing consumption, whether we are talking about developing high-performance materials or building design at very high energy efficiency. Energy performance authorities are also actively working to limit consumption and develop forecasts. In this context, the users of buildings are a particularly important part of the development process and must be actively involved in its realization.

In some cases, it can be seen that the difference between standard indoor temperature by the regulations in Romania and actual operating temperatures can be as high as 26%. Differences between actual and standardized temperatures may vary greatly depending on the user and this may be caused by the lack of intelligent temperature control in different spaces, poor heating system design, by user preferences in terms of indoor comfort temperature or even depending on the economic situation of this. For the difference between the calculated and actual consumption to be as low as possible, it is necessary to implement intelligent temperature control solutions in as many buildings as possible. Good information to users on the impact of non-compliance with the recommended temperature in terms of both consumption and health of the interior environment can also have a positive impact.

The case study shows the extent to which the change in temperature influences the energy consumption for heating. It also shows, that for small temperature changes the variation in consumption is considerable. Considering the great influence that the user has on the final energy consumption, a good understanding of their habits by designers can lead to more accurate forecasts. In this respect, the authors consider that the widespread deployment of indoor temperature monitoring and the creation of current use scenarios of buildings is essential in the overall aim of reducing energy consumption.

IOP Conf. Series: Materials Science and Engineering 1242 (2022) 012028 doi:10.1088/1757-899X/1242/1/012028

## References


[1] Diao L, Sun Y, Chen Z and Chen J Modeling energy consumption in residential buildings: A bottom-up analysis based on occupant behavior pattern clustering and stochastic simulation 2017 *Energy Build* **147** 47-66

[2] Niemann P and Schmitz G Impacts of occupancy on energy demand and thermal comfort for a large-sized administration building 2020 *Build Environ* **182** 107027

[3] Dascalaki E G and Balaras C 2021 Impacts on Indoor Thermal Comfort and Heating Energy Use in Hellenic Dwellings from Occupant Behavioral Reactions *Appl Sci* **11(14)** 6254

[4] Csoknyai T, Lagardeur J, Akle A A and Horvath M Analysis of energy consumption profiles in residential buildings and impact assessment of a serious game on occupants' behavior 2019 *Energy Build* **196** 1-20

[5] Huebner G M, McMichael M, Shipworth D, Shipworth M, Durand-Dubin M and Summerfield A The reality of English living rooms – A comparison of internal temperatures against common model assumptions 2013 *Energy Build* **66** 688-696

[6] EUROSTAT – Statistics Explained https://ec.europa.eu/eurostat/

[7] Applied Precision https://www.appliedp.com/

[8] LogTag https://logtagrecorders.com/

[9] Yang L, Yan H and Lam J C Thermal comfort and building energy consumption implications – review 2014 *Appl Energy* **115** 164-173

[10] MC 001/1-2006 Metodologie de calcul al performanţei energetice a clădirilor. Partea I-Anvelopa clădirii.

[11] MC 001/2-2006 Metodologie de calcul al performanţei energetice a clădirilor. Partea II – Performanţa energetică a instalaţiilor din clădiri.

[12] MC 001/4-2009 Breviar de calcul al performanţei energetice a clădirilor şi apartamentelor. Partea a IV-a.

[13] MC 001/6-2013 Parametrii climatici necesari determinării performanței energetice a clădirilor noi și existente, dimensionării instalațiilor de climatizare a clădirilor și dimensionării higrotermice a elementelor de anvelopă ale clădirilor. Partea a VI-a.

# Research on design method of human resource system test case in construction enterprise

**Z Ping[1], W Yueyong[2], D Fangjing[1] and G Xin[2]**

[1] School of Digital Media, Liaoning Communication University, Shenyang, China
[2] School of Intelligent Engineering, Shenyang City University, Shenyang, China

* wyyg_2000@163.com

**Abstract**. In the process of human resource system test design in construction enterprises, various test design methods are often used, such as equivalence class division, boundary value analysis, causal graph, error inference and so on. How to choose the correct test design method is related to the rationality of test case design and high efficiency of test execution. Therefore, the selection of appropriate test design method plays a vital role in use case design. In this paper, the basic use of equivalence class partition and boundary value analysis and their relationship are briefly described. It is hoped that it can be used for reference in the study of human resource system test cases in construction enterprises.

## 1. Introduction

With the improvement of informatization level in various industries, software products have been widely used, providing great convenience for social production and life. Software testing is an important link in the application development process of modern software products and an important work to ensure and improve software quality [1-3].

The software testing problem based on human resource system of construction enterprise is an important task to be solved urgently. The main purpose of implementing software testing is to ensure the quality of software and ensure that the problems existing in software can be dealt with in time, so as to realize the high quality application of software. From the perspective of whether they care about the internal structure and specific implementation of software, software testing methods are mainly white box testing and black box testing [4-5]. White box test is a test based on all the covered codes, paths and conditions.

Through the test, it detects the internal performance of the product and checks whether the path in the program can complete the work as required. However, it does not test the function and is mainly used for software verification. Black box test method only checks whether functions can meet specified requirements, which is mainly used in the confirmation test stage of software, including equivalence class division, boundary value analysis, decision table and other methods [6-8]. This paper mainly introduces the research results of black box test method of human resource system of construction enterprise, and briefly introduces the equivalence class division test method of login page of human resource system of construction enterprise based on Web front-end development. To provide quality assurance for human resource system of construction enterprises, so that the system can operate safely and reliably, and greatly improve the work efficiency of relevant human resource system.

## 2. Equivalence class partition method

Equivalence class partitioning is a typical black box testing method. Equivalence class partitioning is to divide the input field of the program into several parts, and then select a few representative data from each part as test cases. The role of representative data of each class in the test is equivalent to the other values in that class, that is, if an example in a class finds an error, the other examples in that equivalence class can also find the same error; Conversely, if no error is found in one example of a class, no error is found in the other examples of that class.

### *2.1 Partitioning equivalence classes and listing equivalence classes*

Equivalence class partitioning is a typical black box testing method. Using this method to design test cases, we must first divide equivalence classes and list equivalence class table on the basis of analyzing requirements specification. All input data can be reasonably divided into several equivalence classes.

Taking one data from each equivalence class as the input condition of the test, a small amount of representative test data can be used to obtain better test results. Equivalence class partitioning is to divide the input field of the program into several parts, and then select a few representative data from each part as test cases. The role of representative data of each class in the test is equivalent to the other values in that class, that is, if an example in a class finds an error, the other examples in that equivalence class can also find the same error; Conversely, if no error is found in one example of a class, no error is found in the other examples of that class.

There are two different cases of equivalence class partitioning:

- valid equivalence class: a collection of input data that is reasonable and meaningful for the specification of a program;
- invalid equivalence classes: the exact opposite of the definition of a valid equivalence class.

When designing test cases, consider both of these equivalence classes. Because software must not only accept reasonable data, it must also be able to withstand the unexpected.

### *2.2 Principles for determining equivalence classes*

One valid equivalence class and two invalid equivalence classes can be established when the input conditions specify the range of values or the number of values.

A valid equivalence class and an invalid equivalence class can be established in the case that the input conditions specify a set of input values or a "must be" condition.

When the input condition is a Boolean, a valid equivalence class and an invalid equivalence class can be determined.

N valid equivalence classes and one invalid equivalence class can be established in the case that a set of values (assumed to be N) for the input data is specified and each input value is treated separately by the program.

In the case of rules that must be followed by input data, one valid equivalence class (conforming to the rule) and several invalid equivalence classes (violating the rule from different angles) can be established.

The equivalence class should be further divided into smaller equivalence classes if it is sure that the elements of the equivalence class are treated differently in the program.

### *2.3 Identify test cases*

From the list of equivalence classes, follow these steps to determine the test cases:

Assign a unique number to each equivalence class.

Design a new test case that covers as many valid equivalence classes as possible that have not been covered. Repeat this step so that all valid equivalence classes are covered by test cases.

Design a new test case that covers only one invalid equivalence class. Repeat this step to overwrite all invalid equivalence classes.

### *2.4 Considerations for equivalence class partitioning*

Both valid equivalence classes and invalid equivalence classes should be considered.

Divide carefully and review the division.

## 3. Boundary value analysis method

Definition: a black box test that tests the boundary values of input or output. In general, boundary value analysis is a supplement to equivalence class partitioning. In this case, the test case comes from the boundary of the equivalence class.

### *3.1 Comparison of boundary values and equivalence classes*

Equivalence class partition is to select any element from equivalence class, while boundary value analysis needs to select one or more elements that are boundary elements;

Boundary value analysis not only considers the input conditions, but also the test conditions generated by the output space.

### *3.2 The basic idea of boundary value analysis*

If the input criteria specify a range of values, the test input data should be values that just reach the range boundary and values that just exceed the range boundary.

If the input condition specifies the number of values, the maximum, minimum, one less than the minimum, or one more than the maximum are used as test data.

Apply principles 1 and 2 above for each output condition of the specification.

If the program's specification gives an ordered set of input fields or output fields, the first and last elements of the set should be selected as test cases.

## 4. Example

Taking the weight setting of independent weight correlation type in the strategic performance-performance template of human resource system of construction enterprise as an example, this use case applies the combination design of article 5 of equivalence class division method principle and Article 1 of boundary value analysis method principle above.

The weight-related independent demand in the strategic performance-performance template of human resource system of construction enterprises is described as follows: "Weight correlation is independent. If there are performance indicators and key work areas, the weight sum of performance indicators and key work areas is 100%, the weight sum of all indicators is 100%, and the weight sum of all key work items is 100%. The total evaluation result is calculated according to the weight weighting of regions."

The equivalence class classification table is as follows:

**Table 1.** Equivalence class partition table.

| Condition | Valid equivalence class | Serial number | Invalid equivalence class | | Serial number |
|---|---|---|---|---|---|
| Two weights: performance target weight and key work weight | Enter two weights a and B | A1 | Non-bivariate weight | empty | B1 |
| | | | | a | B2 |
| The weights are positive and the sum is zero | a=80，b=20 | A2 | negative | a=-1 | B3 |

| | | | | | |
|---|---|---|---|---|---|
| 100 | | | | | |
| | | | alphabet | a=F | B4 |
| | | | special symbol | a=* | B5 |
| | | | The sum of the weights is not 100 | a=80, b=21 | B6 |
| | | | | a=79, b=20 | B7 |
| One of the weights is 0 | a=0，b=100 | A3 | Not a weight of zero 0 | a=0, b=0 | B8 |
| There are two weights and: performance target weight and key work weight and | Enter two weights and A, B | A4 | Not the sum of two weights | empty | B9 |
| | | | | A | B10 |
| The sum of both weights is 100 | A=100，B=100 | A5 | The sum of the two weights is not 100 | A=101 | B11 |
| | | | | A=99 | B12 |
| | | | | B=101 | B13 |
| | | | | B=99 | B14 |

According to equivalence class partition table and boundary value analysis design, for example:

**Table 2.** Design the use case table

| **Serial** | **Valid use case input** | **Overrides valid equivalence classes** | **serial** | **Invalid use case input** | **Overrides invalid equivalence classes** |
|---|---|---|---|---|---|
| 1 | a=80，b=20 | A1, A2 | 4 | empty | B1 |
| 2 | a=0，b=100 | A1, A3 | 5 | a | B2 |
| 3 | A=100，B=100 | A4, A5 | 6 | a=-1，b=20 | B3 |
| | | | 7 | a=F | B4 |
| | | | 8 | a=* | B5 |
| | | | 9 | a=80, b=21 | B6 |
| | | | 10 | a=79, b=20 | B7 |
| | | | 11 | a=0, b=0 | B8 |
| | | | 12 | empty | B9 |
| | | | 13 | A | B10 |
| | | | 14 | A=101, B=100 | B11 |
| | | | 15 | A=99, B=100 | B12 |
| | | | 16 | A=100, B=101 | B13 |
| | | | 17 | A=100, B=99 | B14 |

The above is the method and process of designing weighted independent demand test cases in the strategic performance-performance template of human resource system of construction enterprises based on equivalence class division and boundary value analysis in the black box test method of software testing method.

## 5. Construction enterprise human resources system login home page

The following briefly introduces the process of login homepage of human resource system of construction enterprise based on Web front-end development, and the method of equivalence class division of user login function. The login page of human resource system of construction enterprise is shown in figure 1.

**Figure 1.** Human resource system login page of construction enterprise.

The implementation code is shown in figure 2 and figure 3.

```
<!DOCTYPE HTML>
<html>
<head>
<meta http-equiv="Content-Type" content="text/html; charset=utf-8">
<title>login</title>
    <style type="text/css">
        body{
            text-align: center;
            background:url("photo1.png");
            background-repeat:no-repeat;
            background-size: cover;
        }
        fieldset{
            background: #73A5AB;
            border-radius: 13px; width: 260px;height: 190px;
            margin: 0 auto;
            float: none;
        }
    </style>
</head>
```

**Figure 2.** Implementation code.

```
<body>
<form id="form1" title="User Login" action="#" method="post" >
     <fieldset>
        <legend align="left">User Login</legend>
         uername: <input type="text" name="userName" maxlength="6"/>
        <br/>
        <br/>
         password: <input type="password" name="userPsd" maxlength="8"/>
        <br/>
        <br/>
           <input type="radio" value="boy"   name="gender"/>administrator
           <input type="radio" value="girl"   name="gender" checked="checked"/>user
        <br/>
           <input type="checkbox" value="1" name="rempas" checked="checked"/>remember password
        <br/>
        <br/>
          <input type="submit" name="submit1" value="Login"/>
     </fieldset>
   </form>
</body>
</html>
```

**Figure 3.** Implementation code.

## 6. Conclusion

To sum up, based on equivalence class division and boundary value analysis of black box software testing method, the method and process of weighted independent demand test cases in strategic performance-performance template of human resource system of construction enterprises are designed.

In addition, this paper introduces the realization process of the login homepage function of human resource system in construction enterprise based on web front-end development technology, and the design of the equivalence class division method of user login function. Lay a foundation for the follow-up human resource system test case design of construction enterprises.

## References

[1] Liu Nin-xin 2009 Design and Implementation of Human Resource Management System based on JavaWeb [D]. Guangzhou: Guangdong University of Technology
[2] Ron Patton 2006 Software Testing [M]. Second Edition. SAMS Publishing
[3] Vliet H V 2000 Software Engineering-Principles and Practice [M] Second Edition.New York:John Wiley&Sons
[4] Wu Zhou 2013 Design and Implementation of Human Resource Management Information System in Sichuan Luqiao Group [D]. Chengdu: University of Electronic Science and Technology of China
[5] Zhang Ludan 2015 Design and Implementation of human Resource Management System in MCC North China [D]. Dalian: *Dalian University of Technology*
[6] Papazoglou M P and Georgakopoulos D 2013 Service-Oriented Computing:Introduction[J].Communications of the ACM,46(10):24-28
[7] Jorgensen P C 2013 Software Testing A Craftsman' s Approach[M].London:CRC Press
[8] Zhang Ping, GUAN Pingyu and Yang Jing 2011 Discussion on software Testing of Human Resource Management System [J]. *Science and Technology Information*, pp. 615

# Calculation algorithm and method for optimizing electricity consumption for the railway traction system

**A F Popov[1*] and M Fratu[2]**

[1] Transilvania University of Braşov, Department of Automatics and Technology of Information, Romania
[2] Transilvania University of Braşov, Department of Building Services, Romania

* alex.popov68@yahoo.com

**Abstract.** This article presents a calculation algorithm for establishing the electricity demand consumed by the railway electric traction system on a railway section, as well as methods for lowering power consumption and cost. In order to conceive the calculation algorithm, the following variables are taken into account: the number of trains in circulation on the section, the slope of the section, the train speed and the speed restrictions. For factors that substantially increase electricity consumption, but can be controlled, optimal values are computed using specific equations in the MATLAB program. The optimal values are implemented using the train's timetable or a railway control system (SCADA). A decrease in power consumption of 5% can be expected with this method.

## 1. Introduction

The aim of this paper is to propose a method of increasing the energy efficiency and cost-effectiveness of railway transport. This can be accomplished through the design of a calculation algorithm to establish the electricity demand for each section of the railway, allowing for the implementation of an optimization measure as well as the realization of more accurate consumption forecasts on hourly intervals for railway electric traction [1].

For the design of the calculation algorithm and the optimization method, the following aspects must be taken into account: the number of trains in circulation on the section, the section slope, the train speed and the speed restrictions, factors that will substantially increase the electricity consumption required for railway electric traction [2].

Speed restrictions are limitations on the speed of trains on distances over which it is not possible to ensure the safety of train traffic at the speed set in the timetable, due to rail defects. On a railway section there may be several sections with restricted speed. The development of speed over time in a restricted sector is described as: braking until the restricted speed is reached, travel at a constant but reduced speed and an acceleration at the exit from the restricted area until the set speed is reached. In addition to delays, speed restrictions directly cause increases in the electricity consumption by the rolling stock, proportional to the number of affected sections and the speed limit in force [3].

MATLAB parameter optimization has been used for smoothing the power fluctuation on the electrical grid, with the purpose of improving power quality – using the software to calculate optimized parameters and then implementing them [4].

The inclusion of the hourly electricity consumption on the railway section in the forecasted one eliminates the payment of the penalties perceived by the supplier to the traffic operators.

## 2. Calculation algorithm for determining the electricity demand for a railway section

We define the railway section as the distance measured in km from the injection point of a supply feeder from an electric traction substation (ETS) to the feeder area of an adjacent ETS. The length of a usual railway section varies between 25 and 35 km, and is delimited at the ends by neutral areas. Figure 1 shows the electricity supply scheme of the railway sections.

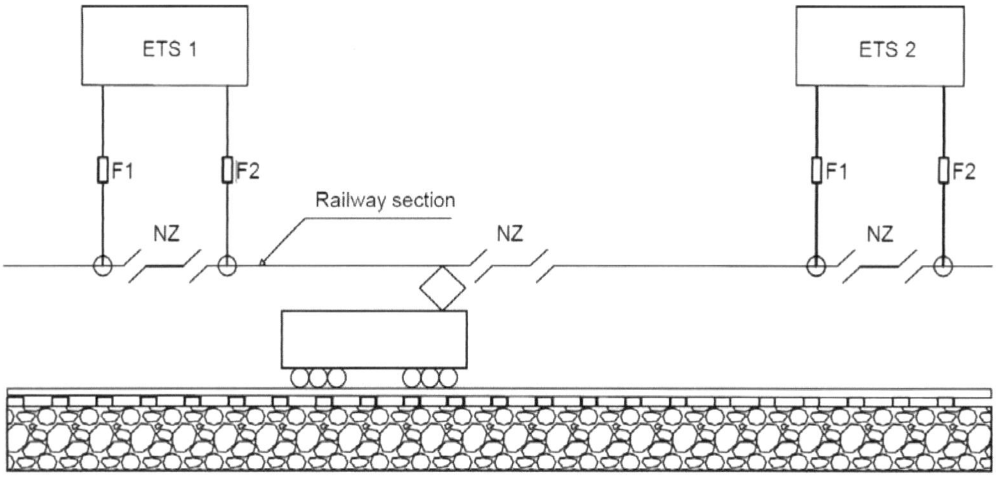


**Figure 1.** Scheme of electricity supply of the railway, delimitation of the railway section. ETS 1, 2 - electric traction substations; F1, 2- feeder; NZ- neutral zone.

The mathematical model of electricity consumption starts from the speed-time relationship, respectively from the distance-time relationship. The velocity distance curve and the velocity-time curve is a diagram showing the instantaneous velocity of the train along the ordinate and the time along the abscissa. This represents the change in speed at different times after the train starts. In the speed-time curve, the distance traveled during the given time interval is given by the area between the curve and the abscissa. Acceleration and deceleration are also given by the slope at any point on the curve to the abscissa [2,5].

Figure 2 shows the speed-time relationship for a given train, with the four distinct components: acceleration period, constant speed travel, inertial travel and braking.

The distinct components represented by the speed-time relationship in figure 2 are described by the following periods:

a. Acceleration period - the train is accelerated to maximum speed. Here, the train speed increases continuously and reaches its maximum value. The power absorbed during this period is maximum.

b. Constant travel period - the train moves at a constant speed reached at the end of the acceleration period. During this period the absorbed power is constant.

c. Inertial running period - the engine supply is interrupted at the end of the constant running period; the train is allowed to run due to its own inertia. No power is absorbed from the power supply during this period, the movement is due to the kinetic energy acquired by the wheels. The speed of the train gradually decreases due to the resistance of the train.

d. Braking period - the brakes are applied at the end of the inertial travel period to bring the train to rest [6].

The force developed by the traction unit at the level of the axles for the movement of the traction unit and its train is called traction effort, and serves the following functions:

- ensure the necessary acceleration for the mass of the train (i.e. linear and angular acceleration);

- exceeds the gravitational component of the train weight;
- oppose train resistance (eg wind, friction resistance and curve resistance).

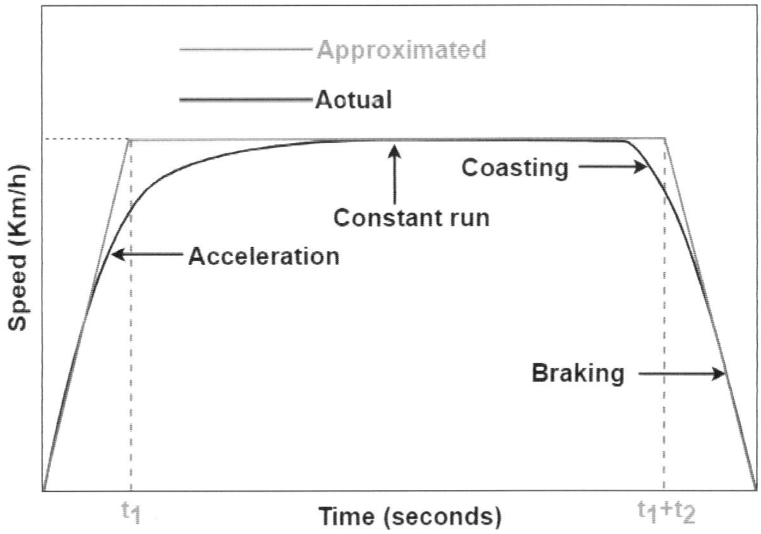


**Figure 2.** Typical speed-time curve.

Thus, the mathematical equation of traction effort for train travel can be expressed as:

$$F_e = m \cdot a + \Sigma F \tag{1}$$

where:
$F_e$ = traction force required to drive the train (Newton)
m = traction transmission mass (kg)
a = train acceleration (m/sec$^2$)
$\Sigma F$ = the sum of the forces acting against the movement of the train, is expressed by the relation:

$$\Sigma F = F_R + F_G \tag{2}$$

where:
$F_R$ = tensile force required to overcome the train's resistance
$F_G$ = tensile force required to overcome the effect of gravity defined by the relation:

$$F_G = m \cdot g \cdot \sin\Phi \tag{3}$$

g = acceleration due to gravity (m/s$^2$)
sin Φ = slope of the railway section
It is calculated based on the linearized speed-time curve the electricity consumed for towing the train during the time t1 in which the acceleration is constant and the gradient positive.

$$Et_l = \text{Force} \cdot \text{Distance} = Ft_l \cdot ½ \cdot V_m \cdot t_l \quad (\text{kWh}) \tag{4}$$

$$Ft_l = [m \cdot a \pm m \cdot g \cdot \sin \Phi + m \cdot r] \tag{5}$$

$Ft_l$ - is the effort exerted by the traction system during t1 due to acceleration, gravity and tensile strength.
$V_m$ is the maximum speed.
r is running resistance.
The positive sign indicates a positive slope that implies an upward movement, and the negative sign indicates a negative slope that implies a downward movement.

The energy consumption from time t1 to t2 is denoted by Et2 and is given by the equation,

$$Et_2 = Ft_2 \cdot V_m \cdot t_2 \tag{6}$$

where $Ft_1$ - is the effort exerted by the traction system during t2.
During constant speed, the acceleration a = 0
Therefore, the energy consumption at constant speed is:

$$Et_2 = (m \cdot g \cdot \sin\Phi + m \cdot r) \cdot V_m \cdot t_2 \tag{7}$$

Total energy consumption is

$$E_{cons.sec} = Et_1 \cdot Et_2 \tag{8}$$

$$E_{cons.sec} = m \cdot V_m \left[((a \pm g \cdot \sin\Phi + r)/2) \cdot t_1 + (\pm g \cdot \sin\Phi + r) \cdot t_2\right] \tag{9}$$

where $E_{cons.sec}$ is the energy consumed on a railway section by one train.

Equation (9) represents the total consumption of electricity for towing a train on a railway section, where the positive sign represents a positive gradient and the negative sign corresponds to a negative gradient.

The total electricity needed to tow trains on the railway section is:

$$E_{total} = \sum_{tr=1}^{n} E_{cons.sec} \tag{10}$$

The capacity to provide the necessary electricity for towing trains on the railway section depends on the short-circuit level of the power supply, the possibility to dynamically modify the supply scheme in accordance with the real-time consumption of the section, the level of control, supervision and automation of the installations that serve the electric power supply system of the railway electric traction.

## 3. Method for optimizing electricity consumption in the electric railway traction system

The total energy consumption equation defined in equation (9) is considered a function to be optimized according to certain constraints [2,7].

The optimization of electricity consumption for traction is done under the following two constraints:

a) the maximum speed ($V_m$) reached by the train does not exceed the specified value;

b) the distance traveled by the train (S) must be constant.

From a mathematical point of view, the objective function of the optimal control problem is defined as follows:

$$Min\,F(X) = E = m * V_m[\left(\frac{a \pm g \cdot sin\Phi + r}{2}\right) \cdot t_1 + (\pm g \cdot sin\Phi + r) \cdot t_2] \tag{11}$$

Constraints can be defined mathematically as follows:

$$a \cdot t_1 \leq V_m; \tag{12}$$

$$0.5 \cdot a \cdot {t_1}^2 + a \cdot t_1 \cdot t_2 = S \tag{13}$$

where S is the distance traveled by the train.

In order to calculate the total electricity consumption for traction, real values for the parameters of equation (10), the basic equation and the two constraints are entered in the MATLAB program. Using the conventional optimization method available in the MATLAB program, optimal values of the parameters corresponding to the lowest overall power consumption.

In order to validate this method, the expected power consumption based on MATLAB simulation was compared with actual power consumption measured in electric substations.

Before any optimization, the meter at ETS1 (Brașov) showed a consumption of 1416.74 kWh. The consumption calculated using equation (9) was 1394.76.

Optimized parameters were calculated based on the input values corresponding to a train on the Brașov – Predeal railway route in Romania (see table 1). This simulation showed an expected power consumption of 1320.26 kWh with the optimal values.

Actual power consumption was measured using an electronic meter (ALPHA® Plus) installed on the feeders of an ETSs that supplies the railway section.

The optimal values obtained from the MATLAB simulation were implemented by simply modifying the timetable of the train in question. As expected, by using the optimized parameters, the measured power consumption decreased to 1356.30 kWh (by approximately 5%).

The measured consumption did not significantly differ from the MATLAB estimated one, therefore the method was considered effective.

**Table 1:** Input parameters and parameters optimized by the conventional method.

| Input parameters | Value | Optimized parameters | Value |
|---|---|---|---|
| M (t) | 400 | Acceleration a (m/s$^2$) | 0.2 |
| $V_m$ (m/s) | 50 | Acceleration $t_1$ | 490 |
| sinΦ | $3*10^{-3}$ | Free run period $t_2$ | 3700 |
| r (N/kg) | 4628 | Energy consumption (kWh) | 1320.36 |

A more effective implementation of the method of optimizing the electricity consumption for the railway traction could be done in a monitoring, control and data acquisition system (SCADA). MATLAB software can be set up to provide the SCADA system with optimal control values for a specific section [8].

There is research that showed a potential of 14.3% reduction in power consumption using a different mathematical model, however, that reduction was achieved using regenerative braking, making the results not directly comparable [9].

## 4. Cost-effective electricity consumption for the electric rail traction system

The current method of electricity cost calculation for railway traction at national level (according to local legislation) is that the consumer (CFR SA) forecasts the hourly / daily / monthly consumption for each electric traction substation, transmits the forecast to the energy supplier electricity, and the supplier is entitled to charge penalties for forecast errors (see table 2) [10].

**Table 2.** Calculation of forecast error penalties for individual ETSs.

| | Amount (MWh) | Tariff (Lei/kWh) | Tariff (%) | Value (Lei) |
|---|---|---|---|---|
| Peak energy underconsumption | 92.438 | 0.5057 | 15 | 7011.88 |
| Exceeding peak energy | 20.152 | 0.5057 | 25 | 2547.72 |
| Energy remains underconsumption | 163.57 | 0.2588 | 15 | 6349.79 |
| Energy left over | 32.66 | 0.2588 | 25 | 2113.10 |
| Empty energy underconsumption | 206.154 | 0.200 | 15 | 6184.62 |
| Empty excess energy | 38.114 | 0.200 | 25 | 1905.70 |
| Total value not included in the forecast | | | | 26112.81 |
| Penalty percentage 100% | Total penalty value | | | 26112.81 |

The data in table 2 shows that the monitoring and respectively the exact forecasting of the electricity consumption on ETSs can decrease the cost of electricity for the railway transport.

One solution to eliminate these additional costs, given SCADA's ability to perform remote energy management, is the dynamic modification of the supply scheme of the railway sections based on real-time consumption [11]. This opens up the possibility of simply redirecting energy from an ETS that exceeds forecasted consumption to another ETS, decreasing the likelihood of forecast errors and increasing cost-effectiveness.

Figure 3 shows how the ETSs that provide power to railway sections can be integrated in a SCADA system [12].

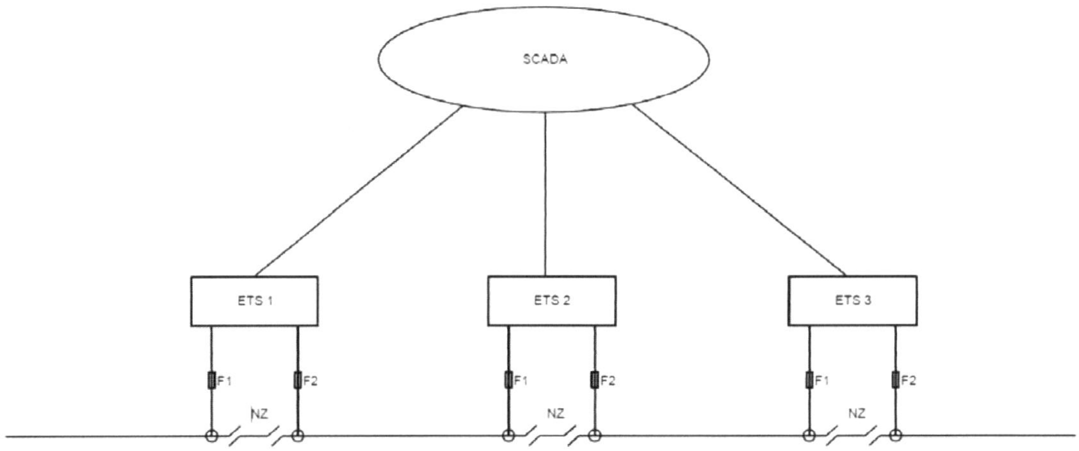


**Figure 3.** Block diagram of SCADA remote management of the electricity supply of the railway sections.

The process of streamlining the electricity consumption necessary for the railway electric traction through the SCADA system includes permanent monitoring on each ETS of the forecasted / realized consumption by hours / days / months, creating and sending as accurate consumption forecasts as possible to the supplier for the following periods, based on the analysis of the real consumption values previously registered.

## 5. Conclusion

The calculation algorithm for determining the electricity demand for railway section can be used as a means of reducing electricity consumption and cost in the railway traction system. The proposed optimization solution accomplishes a minimization of the electricity consumption of individual railway sections. The efficiency of energy consumption by implementing remote management in the SCADA system provides a reduction or even avoidance of any penalties charged by the electricity supplier for not being included in the consumption forecast for an ETS for a certain time interval. The optimization of the calculation algorithm for electricity demand of railway sections, using MATLAB software together with the application of the remote control and remote management system (SCADA) are necessary and viable measures for increasing the electrical efficiency and cost-effectiveness of the railway electricity transport.

## References

[1] Gupta S, Mahajan P and Garg R 2016 Tractive energy optimization in railway electric traction system *1st IEEE Int. Conf. on Power Electronics, Intelligent Control and Energy Systems ICPEICES 2016* (Delhi) p 1-5

[2] Dongre N R and Sindekar A S 2018 Optimization of Energy Consumption In Electric Traction System By Using Interior Point Method *J. Electr. Electron. Eng.* **13** 9-15

[3] Van der Kooij RBK, Landmark AD, Seim AA and Olsson NOE 2017 The effect of temporary speed restrictions, analyzed by using real train traffic data *Transp. Res. Proc* **22** 580-587

[4] Huang X Liao Q Li Q. et al. 2020 Power management in co-phase traction power supply system with super capacitor energy storage for electrified railways *Rail. Eng. Science* **28** 85–96 (2020)
[5] Calderaro V; Galdi V; Graber G and Piccolo A 2014 An algorithm to optimize speed profiles of the metro vehicles for minimizing energy consumption *Int. Symp. on Power Electronics, Electrical Drives, Automation and Motion (SPEEDAM)* (Sorrento) pp 813-819
[6] Song Y and Song W 2016 A novel dual speed-curve optimization based approach for energy-saving operation of high-speed trains *IEEE trans. Intell. Transp. Syst.* **17** 1-12
[7] Su S, Tang T, Li X and Gao Z 2013 Optimization of multi train operations in a subway system *IEEE trans. Intell. Transp. Syst.* **15** 673-684
[8] Yadav G and Paul K 2021 Architecture and security of SCADA systems: A review. *Int. J. Crit. Infrastruct. Prot.* **34**
[9] Pan Z Chen M Lu S Tian Z and Liu Y 2020 Integrated Timetable Optimization for Minimum Total Electrical Energy Consumption of an AC Railway System *IEEE Transactions on Vehicular Technology* **69** 3641-3653
[10] Romanian Ministry of Transport 2006 Order nr. 34/2006
[11] Uglešić I, Pavić I, Milardić V and Mandić M 2004 *Calculation of Electric Power Supply for the First Phase Electric Traction Substitution on Railway Track Zagreb - Rijeka and on Connected Tracks* (Zagreb: Faculty of Electrical Engineering and Computing)
[12] Ke BR Lin C and Chien H 2012 Improvement of Freight Train Timetable for Single-Track Railway System *Int. Symp. on Computer, Consumer and Control (IS3C 2012)* (Taichung) pp 456-459

# Optimal simulation and experimental research on the operation system of dumper of the stemming production line

**F L Qi[1*], Y Q Zhao[1], G Y Zhang[1] and J X Tian[1]**

[1] School of Mechanical Engineering, Shenyang Jianzhu University, China
[2] Inner Mongolia North Heavy Industries Group Corp ltd, Baotou, Neimenggu, China

* alianqi@163.com

**Abstract.** In order to further improve the running speed and stability of the feed truck in the mud production line, optimization simulation and experimental research on the running system of the feed truck were carried out. First, based on the analysis of the operating system of the material truck, Then, according to the experimental plan, an optimization experiment on the project site was carried out, and it was confirmed in a quantitative form that the operating efficiency and stability of the optimized truck operating system have been significantly improved. Finally, the experimental results are applied and analyzed, and a mathematical model that can be used in actual engineering is obtained, so as to reflect the actual engineering application value.

## 1. Introduction

As a kind of refractory material for blocking the iron outlet, stemming is increasing with the increase of iron and steel demand, at the same time, it also puts forward higher requirements for the performance of the stemming production line. During the operation of stemming production, the operation system of the hopper car has a great influence on the operation of the whole production line, while in the actual operation of the hopper car [1,2], there are always poor operation efficiency and smoothness. At the same time, in the research of the batching control system and its related aspects, the focus is often on the weighing and path planning of the hopper car, while the research on the running speed and smoothness of the dumper is very little.

In conclusion, this study will gradually carry out relevant research on the operational efficiency (speed) and smoothness of the hopper car from three aspects: simulation, experiment and application analysis, taking into account the actual conditions of the hopper car operation and the current relevant research background. Fill the vacancy of relevant research on the running speed and stability of batching truck.

## 2. Simulation analysis of system optimization

### 2.1. Analysis on operation system of hopper car

Comprehensively considering the actual operation of the stemming production line, the hopper car part is analyzed. The operation of the hopper car is shown in figure 1. Through the analysis of the hopper car operation system, the operation system control model shown in figure 2 can be obtained.

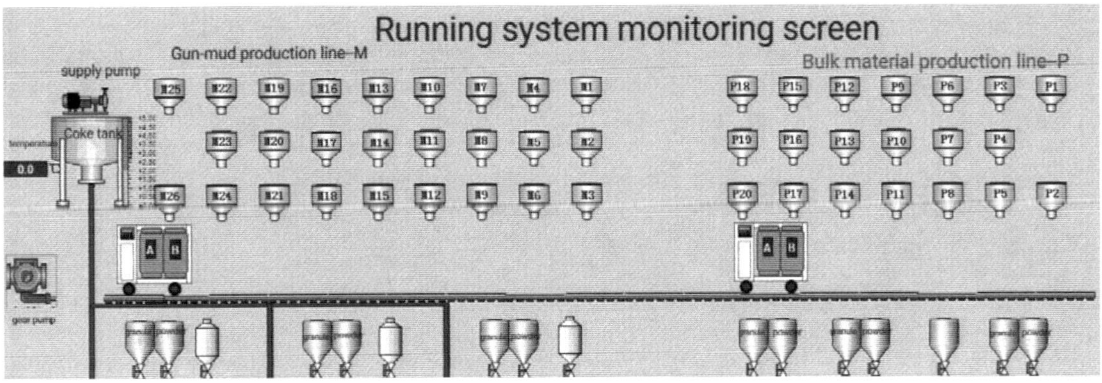


**Figure 1.** Schematic diagram of hopper car operation system.

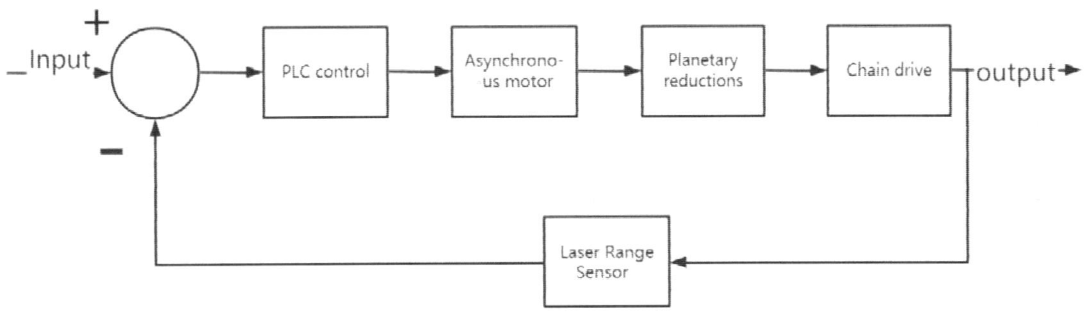


**Figure 2.** Control model of hopper car operation system.

In the control model of the running system, the theoretical speed value of the asynchronous motor is:

$$n = \frac{60f}{P(1-S)} \tag{1}$$

$$S = \frac{(n_1 - n_0)}{n_1} \times 100\%, \quad n_1 = \frac{60f}{P} \tag{2}$$

Thus, by combining the control model of the operating system, the mathematical relationship between the operating speed V and the frequency f of the hopper car is obtained as follows:

$$v = \frac{60\pi d f^2}{K_1 K_2 n_0 P^2} \tag{3}$$

$F$ is the frequency; $P$ is the extreme logarithm; $S$ is slip rate and $n_0$ is rated speed, $n_1$ is synchronous speed; $d$ is hub diameter; $K_1$ is Planetary Reducer Reduction Ratiohub diameter; $K_2$ is the sprocket reduction ratio. Some of the selected parameters are shown in table 1.

**Table 1.** Control model partial parameter table.

| Pole logarithm | Rated speed $(r/min)$ | Hub diameter(mm) | Reduction ratio K1 | Reduction ratio K2 |
|---|---|---|---|---|
| 2 | 1420 | 320 | 29 | 310/230 |

### *2.2. Establishment and simulation analysis of system optimization scheme*

As the running environment of stemming production line is complicated, it is difficult to describe the running system of the truck in detail with accurate mathematical model. Therefore, the optimization will be carried out with the help of fuzzy logic control algorithm to better describe the actual running situation of the truck.

According to the actual operation, by adding the designed fuzzy controller [3-5], an optimal simulation control model can be established as shown in figure 3, and then the control model established can be used for optimization simulation analysis.

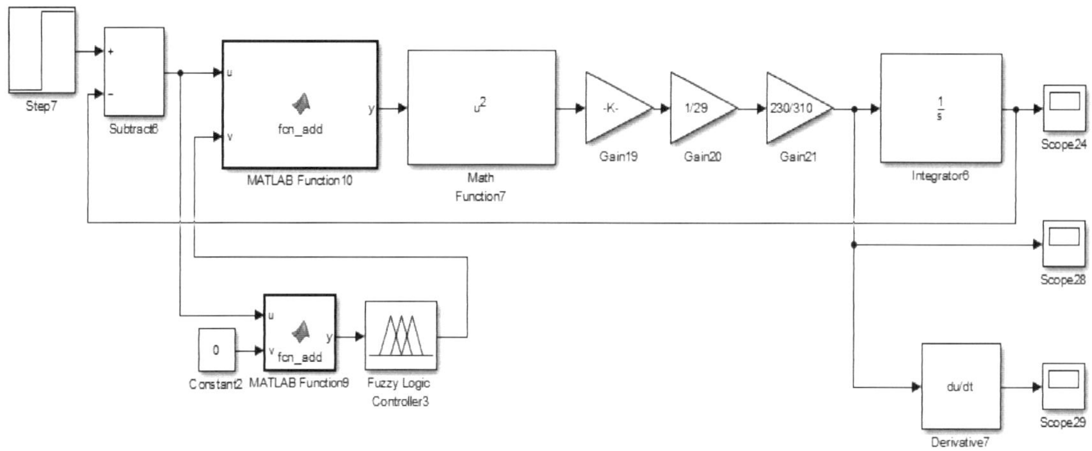


**Figure 3.** Optimized simulation model of hopper car system.

Speed optimization aspect: By comparing the simulation results before and after optimization, the running curve of the hopper car shown in figure 4 can be obtained, in which B and A represent the displacement curves of the hopper car at different target positions before and after optimization respectively. According to the curve analysis in the diagram, it can be seen that under the same circumstances, the time required for the optimized hopper car running system to reach the target position is less than before optimization. It further highlights the better speed optimization effect.

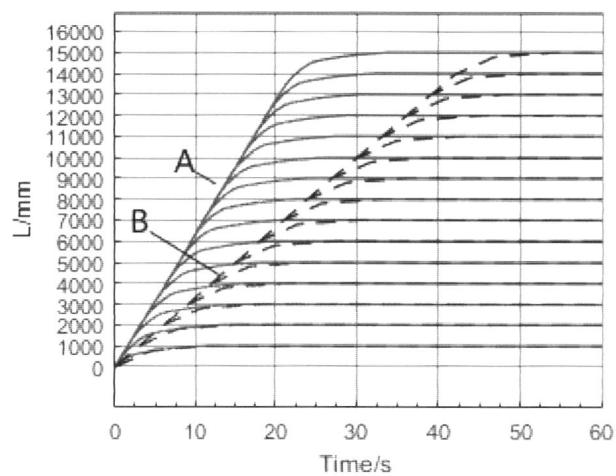


**Figure 4.** Operating curve before and after optimization of truck.

In terms of smoothness optimization: the change of acceleration before and after optimization is shown in figure 5, in which B and A represent the situation before and after optimization respectively.

Through analysis, it can be seen that the optimized system avoids the great impact caused by sudden and large change of speed, thus the truck can run more smoothly and improve the smoothness of system operation.

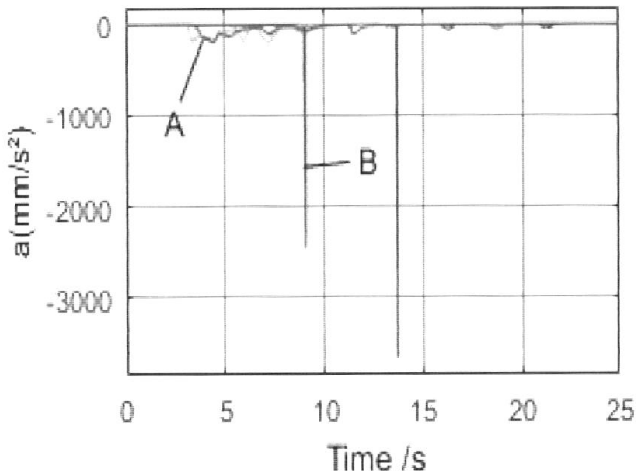


**Figure 5.** Optimized comparison diagram of acceleration of front and rear trucks.

## 3. Joint simulation analysis

After completing the optimization simulation of control system simulation with MATLAB-Simulink, the simulation experiment of engineering application process can be carried out by establishing the connection between WinCC and MATLAB to verify the optimization effect. At the same time, the feasibility of field experiment in the later stage can be verified according to the results of simulation experiment.

In practical engineering application, the operation of fuzzy controller is relatively complex. Neither WinCC nor S7-300PLC can build the fuzzy controller module alone. Therefore, the operation of control algorithm can be realized with the help of the powerful operation function of MATLAB, and the communication between WinCC and MATLAB can be realized by OPC technology to transmit relevant data [6-7].

The structure of practical application process is shown in figure 6. Among them, the controller model shown in figure 7 will be used in MATLAB, and the joint simulation experiment will be carried out according to the designed application process.

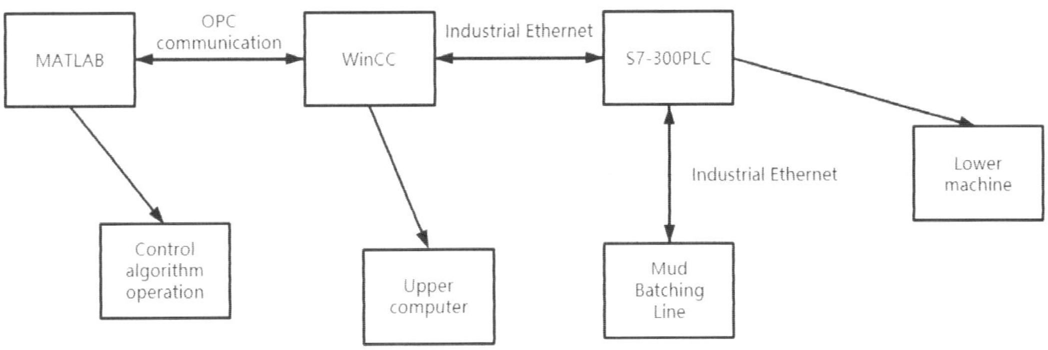


**Figure 6.** Structure diagram of engineering application process.

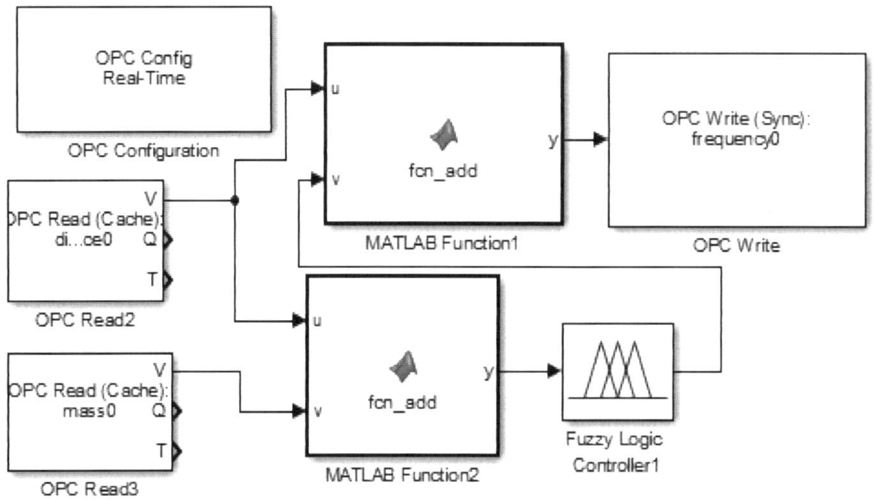


**Figure 7.** Practical application model diagram of fuzzy controller.

The simulation experiment results are shown in figure 8, in which application represents the operation curve of joint simulation and simulation represents the operation results of system simulation. Through comparative analysis, it can be seen that there is little difference between the joint simulation and system simulation results, and the operation track is almost the same, which well verifies the reliability of system simulation and shows the practical operability of the next field experiment.

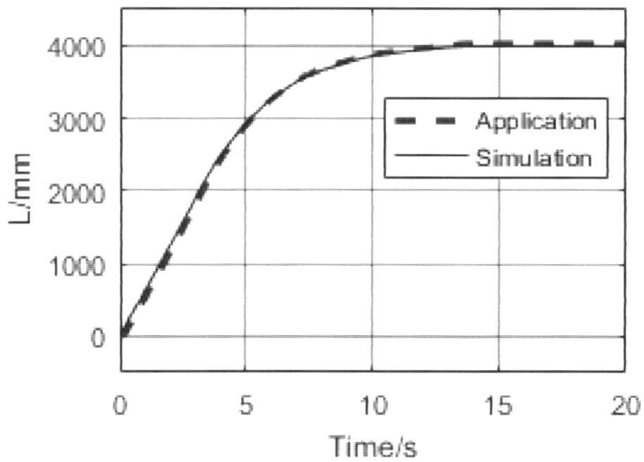


**Figure 8.** Comparison diagram of hopper car simulation and joint simulation.

## 4. Optimization experiment and analysis

The construction of the optimization experimental platform for the operation control system of the hopper car is based on the mechanical equipment of the existing gun mud batching control system project. The experimental site is shown in figure 9. Based on the existing equipment, the software part is designed and built, mainly including three parts: PLC programming, WinCC control picture building and MATLAB module building.

**Figure 9.** Picture of experimental site.

Among them, the PLC programming part mainly completes the programming of logic control program; The construction of WinCC control screen mainly reflects the functional parts to be controlled in the program in WinCC according to the actual operation process of hopper car and the situation of PLC logic control program. The specific control screen diagram of hopper car operation system optimization experimental platform is shown in figure 10 The module construction of Matlab can fully refer to the design of joint simulation.

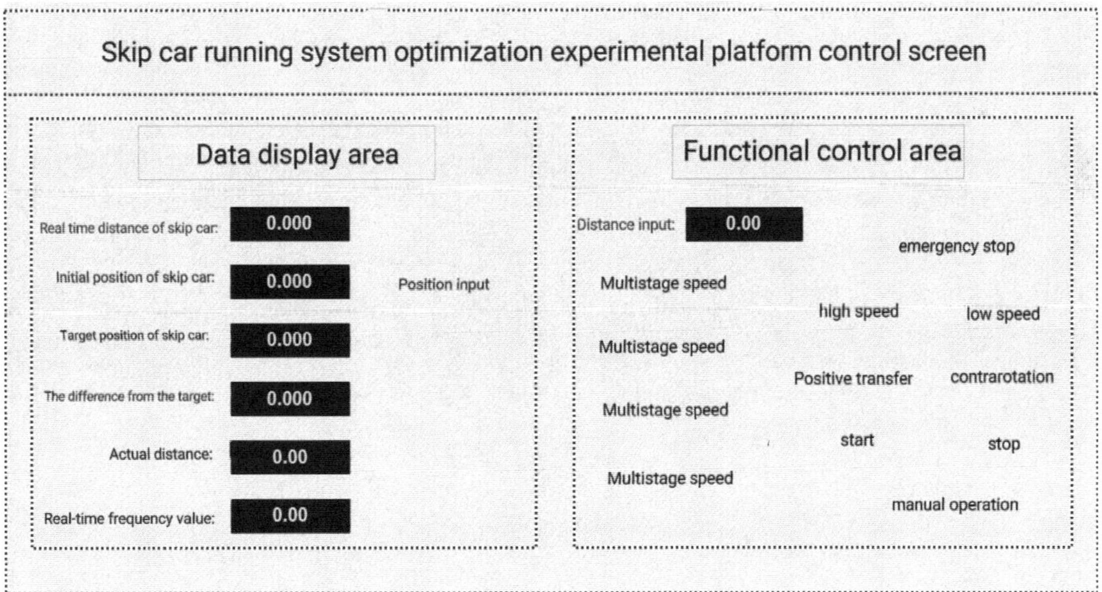


**Figure 10.** Control picture of optimization experiment platform of charging car operation system.

*4.1. Analysis of experimental results of speed optimization*

By sorting out and comparing the time data required for the operation of the batching truck at different target positions before and after optimization, the comparison diagram of the operation time of the batching truck as shown in figure 11 can be obtained. Through the comparative analysis in the figure, it can be seen that with the increase of the running distance, the time difference between the operation before and after optimization increases at the same target position, that is, the optimization effect is becoming more and more obvious.

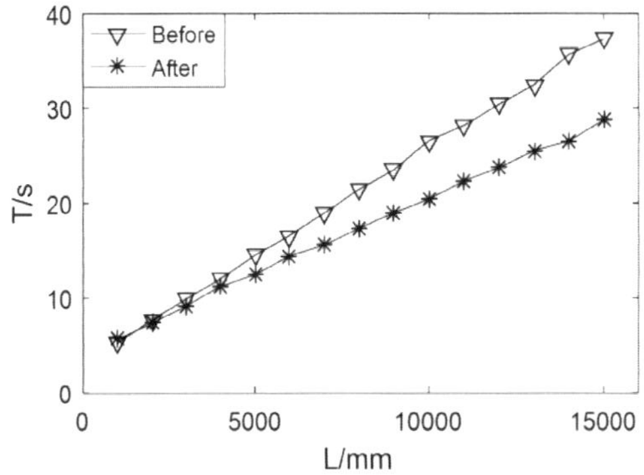


**Figure 11.** Comparison diagram of operation time of hopper car.

According to the data in the figure, the total operation time of the hopper car after optimization is $T_A$ = 259.39 s, and the total operation time before optimization is $T_B$ = 320.29 s. Therefore, the optimization efficiency of the operation speed can be estimated, and the optimization efficiency is :

$$\eta = \frac{T_B - T_A}{T_B} \times 100\% = 19.02\% \tag{4}$$

That is to say, the running speed of the optimized charging car running system is increased by about 19.02% compared with that before optimization, which further verifies the reliability of the hopper car simulation and optimization process.

### *4.2. Analysis of experimental results of stationarity optimization of stationarity optimization*

During the experiment, in terms of the running stability of the hopper car, we can get very intuitive experimental results by observing the running stability of the hopper car. However, in order to compare the effect of stability optimization more clearly and quantitatively, a simple single pendulum motion device as shown in figure 12 is adopted in the experiment. The single pendulum motion device in the figure is to hang a small ball on the support through a thin rope, fix the device on the hopper car during the experiment, and ensure that the motion direction of the small ball is consistent with the running direction of the hopper car. The maximum angle of single pendulum motion shall be recorded during each experiment $\theta$, through the comparison of maximum angle $\theta$, the optimization effect can be seen more clearly and intuitively.

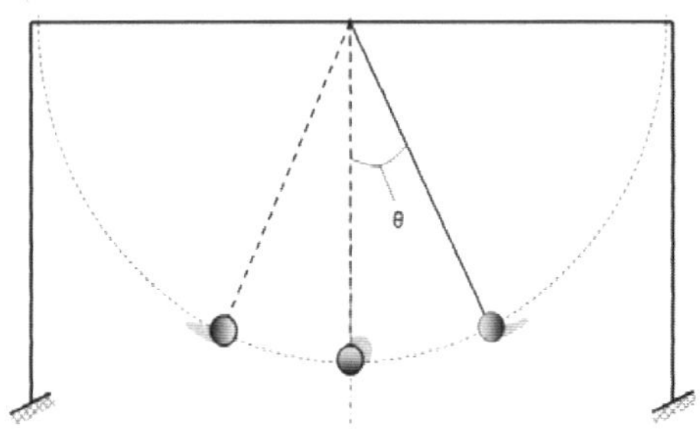


**Figure 12.** Simple pendulum motion device.

By making use of the recorded maximum motion swing angle theta data at different target positions before and after optimization, the corresponding comparative analysis can be made, and the comparative graph of running smoothness shown in figure 13 can be obtained. From the analysis in the graph, it can be seen that before optimization, the maximum motion swing angle $\theta$ of a single pendulum, except $\theta$ = 6.3 degrees at the condition of L = 1000 mm, is obtained, (because of the small distance and the low starting speed at three-stage speed, theta is relatively small). The maximum swing angle Theta of the single pendulum after optimization is maintained at about 9.4 degrees, and the maximum swing angle Theta of the single pendulum after optimization is maintained at about 3.3 degrees. The maximum angle Theta of the single pendulum motion during each experiment after optimization is significantly reduced compared with that before optimization, which means that the smoothness of the barrel after optimization has been significantly improved.

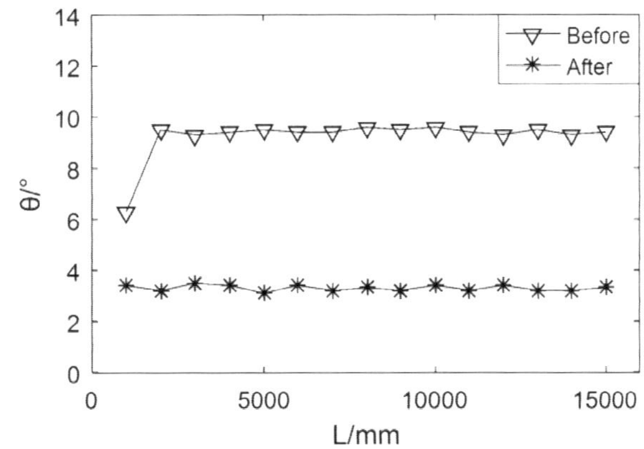


**Figure 13.** Comparison diagram of operation stability of hopper car.

Through the data in the figure, the optimization of stationarity can be estimated. From the known data, the average value of the maximum motion swing angle of the simple pendulum before optimization can be obtained $\theta_B = 9.436°$ ; The average value of the maximum motion swing angle of the optimized simple pendulum $\theta_A = 3.287$ °, from which the stationarity optimization efficiency is:

$$\eta = \frac{\theta_B - \theta_A}{\theta_B} \times 100\% = 65.16\% \tag{5}$$

The stability of the operation system of the batching truck after optimization is about 65.16% higher than that before optimization, and then the optimization degree of the stability of the operation system of the batching truck is shown in quantitative form.

## 5. Application analysis of optimization results

In order to more conveniently use the results of optimization research in engineering projects, the data results obtained in the experimental process are processed. The obtained experimental data are shown in figure 14. At the same time, the corresponding curve fitting is carried out on the obtained data diagram by using the tools in MATLAB. In the process of curve fitting, the method of n-order polynomial can be selected for curve fitting, in which the greater the value of N.

The better the fitting effect of the curve, because the curve fitting effect is too poor when $n < 2$. When n is greater than 8, although the curve fitting effect is excellent, it will lead to too high order of the fitting curve, resulting in difficulties in practical application. Therefore, an analysis and research are carried out for the case that n is 2-8 in turn.

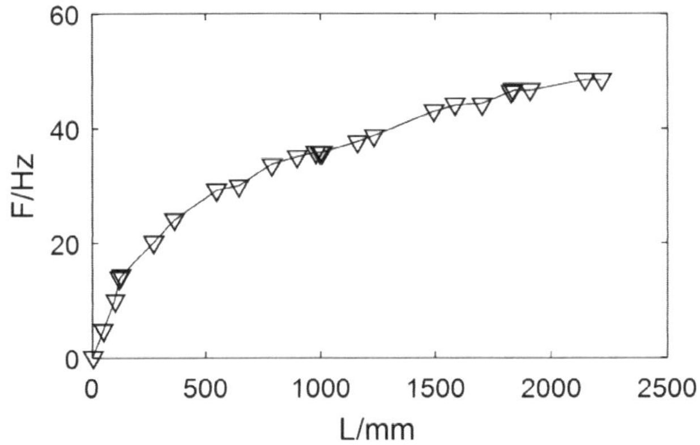


**Figure 14.** Experimental data diagram.

In order to better compare the curve fitting effects under different n, the residuals in curve fitting of different orders can be analyzed. The obtained residuals are shown in figure 15. The smaller the absolute value of residuals R, the better the curve fitting effect. From the variation of residuals in the figure, it can be seen that when n = 2,3 (2,3 power), the residuals are larger than n = 4 – 8 (4 - 8 power).

The absolute value of residual error is the largest when n = 2. It can be seen that the curve fitting effect when n = 2,3 is not as good as that when n = 4 - 8, and the curve fitting effect when n = 2 is relatively the worst; when n = 4 - 8, the residual has the same change trend and the difference of residual values is small. It can be seen that the difference of curve fitting effect is small when n = 4 - 8.

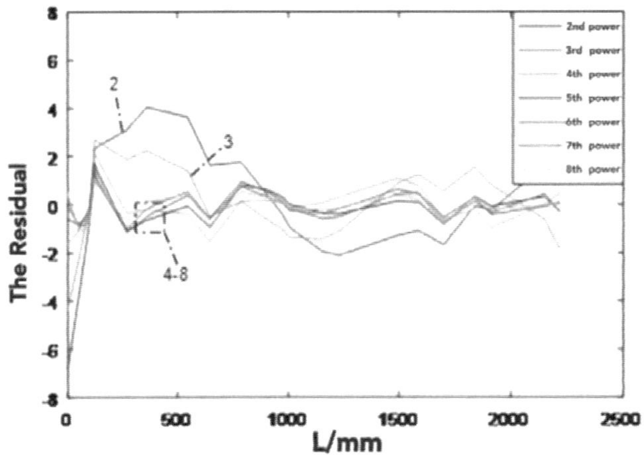


**Figure 15.** Residual diagram.

When the residual is obtained, the residual modulus corresponding to different values of N can also be obtained, so as to obtain the variation diagram of residual modulus value as shown in figure 16. The smaller the value of residual modulus (the value of residual modulus ≥ 0), the better the effect of curve fitting. It can be seen from figure 16 that the value of residual modulus decreases with the increase of N, and the change of modulus value is small from n = 4, and its values are relatively low. Therefore, on the premise of pursuing the same curve fitting effect and considering the actual application of the project, n = 4 can be selected for curve fitting. The quartic curve fitting diagram shown in figure 17 can be obtained, and the polynomial function under its conditions can be obtained. The function is shown in formula:

$$F = -1.1\times10^{-11}\times L^4 + 5.7\times10^{-8}\times L^3 \\ -0.0001\times L^2 + 0.093\times L + 1.6368 \tag{6}$$

In the formula: F is the motor input frequency; L is the target distance.

The obtained polynomial function can be used in the operation of subsequent batching vehicles, which has certain practical engineering application significance.

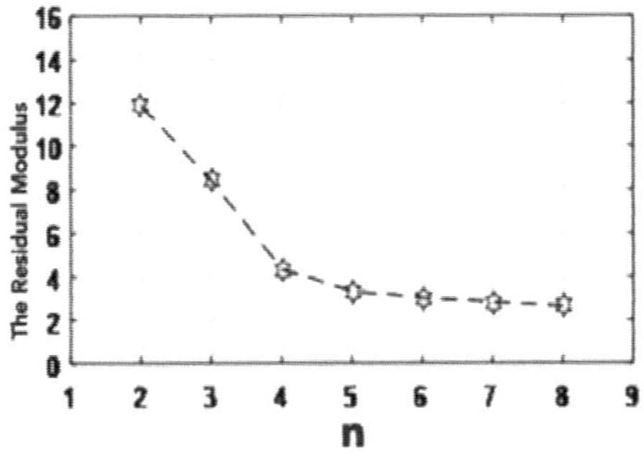


**Figure 16.** Residual modulus variation diagram.

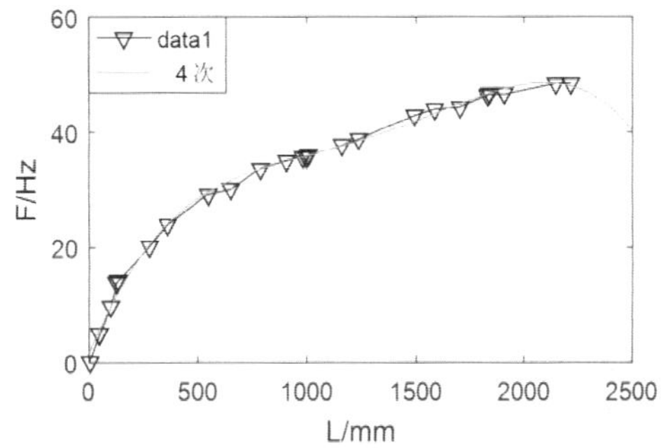


**Figure 17.** Quartic curve fitting diagram.

## 6. Conclusion

Aiming at the running speed (efficiency) and stability of the hopper car running system of the stemming production line, the optimization simulation of the system is analyzed based on the fuzzy logic control algorithm, and the results show the optimization effect of the optimization system.

Through the joint simulation analysis, the reliability of the system simulation is verified, and the practical operability of the field experiment is confirmed. Then, relevant field experiments are carried out on the optimization problem, which shows the optimization effect of the hopper car operation system in the form of quantification. At the same time, the application of the experimental results is analyzed, and the specific mathematical model is obtained by using the experimental data, which provides a data basis for the later practical engineering application.

The optimization simulation and Experimental Research on the hopper car operation system of stemming production line has strong practical engineering significance. At the same time, it also provides ideas for the optimization control of relevant aspects in other industries.

## References

[1] LI J J 2013 Research on optimal control method of batching system *D. NEU.* **43**

[2] Dong T 2017 Design of an automatic weighing and packaging control system *J. Control Engineering of China.* **24** 2

[3] Song S L and Yao J Q 2017 Research on Optimization of mine drainage control system based on fuzzy control *J.Control Engineering of China.* **24** 10

[4] Wen Z and Sun H K 2017 Matlab intelligent algorithm *M. Tsinghua University.* **197**

[5] Zhao Y Q 2020 Research on Optimization of gun mud batching system based on fuzzy control *J. Development and Innovation of Machinery and Electrical Products.* **33** 4

[6] Li Z Y and Peng X M 2011 Research on software interface method between WinCC and MATLAB *J. Journal of Logistical Engineering University.* **27** 6

[7] Li X and Lei J Y 2018 Real time monitoring system of water tank based on OPC Technology *J. Process Automation Instrumentation.* **39** 12

# Verification of elastic wave velocity tomography based on compressive sampling

**X Qiao[1*], Y Kobayashi[1], K Oda[1] and K Nakamura[1]**

[1] Department of Civil Engineering, College of Science and Technology, Nihon University, Tokyo, Japan

* cssi20001@g.nihon-u.ac.jp

**Abstract**. This paper verifies the compressive sampling(CS)-based elastic wave velocity tomography(EWT) technique. Conventional elastic wave velocity tomography necessitates the quantity of effort and time. Nevertheless, because the central premise of CS algorithm is to recover the original velocity distribution from random measurement subsampling, compressive sampling is adopted at overcoming this challenge. CS algorithm chose fewer paths from entire measurements during the measurement stage. After that, the L1-minimization optimization method adopts to recover the damage area and magnitude in the visualization stage. Because of hugely reduces the measurement workload and a high level of accuracy reconstruction, CS algorithm is more effective in comparison with the conventional algorithms of elastic wave velocity tomography.

## 1. Introduction

Numerous infrastructures were constructed by using concrete globally, such as skyscrapers, factory buildings, bridges, dams, and ports. These buildings gradually age over time. Because the damage of concrete exists inside and is unclear to be found although it has a significant impact on the safety of the structure, it is essential to utilize non-destructive testing to check whether the structure is damaged and whether it needs repair to make it safe. Therefore, it is essential to identify the damage by using non-destructive testing for these aging structures efficiently.

Under the premise of damaging the structures, the transmitter launches the elastic wave and receives the elastic wave by sensors to visualize the internal condition of the structures. Events in elastic wave tomography release elastic waves, and placed receivers subsequently collect the elastic waves. Based on the grid division in advance, the travel time of each event is calculated to obtain the velocity field distribution to determine the extent and area of damage.

The existing inversion algorithms such as SIRT and ART are widely applied after the measurement phase. In this phase, visualize the internal damage of the concrete could be obtained as a velocity distribution. However, these algorithms have difficulties to obtain the distribution accurately if the number of measurement data is insufficient. Generally large number of measurement data is needed and it requires a lot of costs. Because of the sparseness of concrete damage, CS theory could overcome these difficulties while obtaining accurate results requires a small measurement workload.

In this paper, the second part introduces the basic theory of compressive sampling, theoretical calculation process, and conditions. The third part introduces the theory of compressive sampling combined with elastic wave tomography and determines the relevant parameters to ensure the feasibility of algorithm. In the fourth part, results of numerical simulation is introduced and the results in comparison with the results of SIRT to reveal the capability of CS algorithm.

## 2. Compressive sampling algorithm

E. Candes, D. Donoho, and T. Tao introduced the compressive sampling theory in 2006 as a novel view. [1-4]. For a significant quantity of data acquisition and compression, CS theory suggests obtaining the data in a compressed format rather than obtaining the total data and then compressing it. To retain the integrity of the original signal, the samples frequency has to be more than twice the highest frequency during sampling, according to classic Nyquist sampling theory. The traditional theory has some defects; an enormous volume of sampling data is required because the signal frequency is reasonably high. However, the sparseness of the signal, rather than the frequency of sampling, is the feature from a different perspective of compressed sensing that must be considered, and the original signal can be reconstructed if the signal form is sparse enough [5]. Figure 1 depicts the basic flow chart:

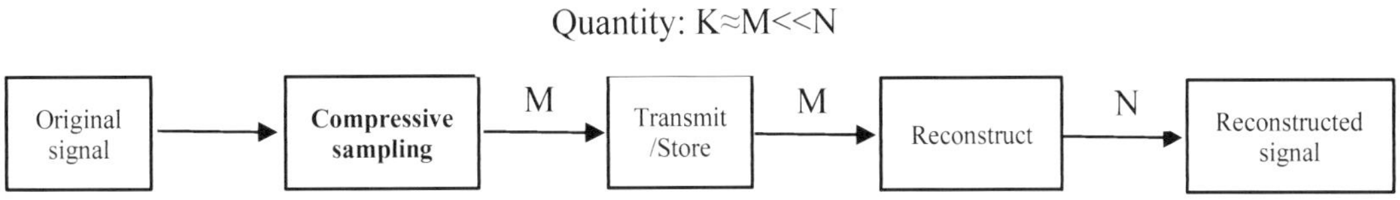


**Figure 1.** Flow chart of CS theory.

Compressive sampling has the main characteristic that the signal could be recovered from a much fewer measurements samples than traditional methods. Typically, one-dimensional signal x is regarded as a $n \times 1$ row vector, which elements are x [n], n = 1,2,...,n belong to $R^N$. Simultaneously, the signal in the high-dimensional form can be vectorized into a lengthy one-dimensional vector. The signal x is able to sparse by $n \times n$ basis matrix $\psi$ and expressed as equation (1). It is important that the signal x would be more compressible if it is sparse sufficiently.

$$x = \Psi\alpha \tag{1}$$

In this formula, $\alpha \in R^N$ denotes a vector with k non-zero elements (k is more far smaller than n), and $\psi \in R^{N\times N}$ represents a basis matrix, $\Phi$ named measurement matrix as a random matrix(normally is Gaussian or Bernoulli matrix), and y is described as:

$$y = \Phi\psi\alpha + n = \Theta\alpha + n \tag{2}$$

The transfer matrix $\Theta$ has a dimension of $m \times N$, and the number of rows is substantially lower than the number of columns. $\Psi$ describes the sparse matrix. The noise should not be ignored throughout the measuring procedure and represented as n in the formula. Generally, recovering $\alpha$ from the measured value y is an ill-posed problem. Convex optimization algorithms can solve this mathematical problem. In order to employ this theory, there is a condition named restricted isometry property(RIP) that the signal x and the transition matrix $\Theta$ must satisfy [6], as shown below:

$$1-\delta \leq \frac{\|\Theta v\|_2}{\|v\|_2} \leq 1+\delta, \delta>0 \tag{3}$$

The restricted isometry property (RIP) states that there exists an isometric constant $\delta$ that fulfills the above connection for each integer k. In addition, $v \in R^N$ is all K-spare(the degree of sparsity is k) vector, and $\delta$ is minimum value that fulfills the vector v with any sparsity. Both the conditions of each row of the matrix $\Phi$ have to be orthogonal, and the value must be much less than k. The following equation could decide the measurement quantity:

$$m > \mu \cdot K \cdot \log(N/K) \tag{4}$$

In the above function, μ is consistent that has a fixed value; this is 4 in this study. Assuming that all the above requirements are satisfied, from the vector y, vector x can be accurately rebuilt. Then, the following convex optimization method equation is applied to crack this ill-posed issue.

$$\min \|\alpha\|_1 \text{ subject to } \|\Theta\alpha - y\|_2 < \varepsilon \tag{5}$$

In the above formula, $\|\cdot\|_1$ and $\|\cdot\|_2$ correspondingly, the l1-norm and l2-norm, and the noise boundary is symbolically represented as ε.

## 3. Implementation of CS algorithm in elastic wave velocity tomography

### *3.1. EWT theoretical elaboration*

Elastic wave tomography is used in the field of civil engineering to identify and detect internal damage of the structure and aged concrete components such as pier and wall [7]. The basic principle is that inhomogeneity of media alter the propagation speed of the elastic wave, which means that the propagation speed in undamaged concrete is fixed, although the propagation speed decreases if elastic wave encounters holes or cracks in the structures. Thus, the graphics of velocity distribution is exhibited to judge the inside of concrete damage condition. The travel time on a measurement path could exhibit as follow:

$$T_i = \int_{L_i} \frac{1}{V(x,y)} dL = \int_{L_i} s(x,y)\, dL \tag{6}$$

$T_i$ is the travel time in $i^{th}$ path, and $L_i$ is the $i^{th}$ measurement path length from the source to sensor, V(x, y) is velocity function, and s(x, y) is slowness. The reciprocal of the velocity is titled slowness [8]. In order to locate the defect degree and position in the specimen more precisely, the whole inspection area is divided into subdivision units. The sum of the travel time of each subdivision unit can be expressed as follow:

$$T_i = \sum_{j=1}^{w} \frac{a_{ij}}{V_j} = \sum_{j=1}^{w} a_{ij}\, s_j \tag{7}$$

where $a_{ij}$ is the travel length of $j^{th}$ grid in $i^{th}$ path, $V_j$ is velocity of $j^{th}$, and slowness in path of $j^{th}$ is display as $S_j$. The travel time in each path is described above, and following formula can be expressed all the measurement paths.

$$T = A \cdot \frac{1}{V} = A \cdot s = A \cdot (s_0 + \Delta s) \tag{8}$$

With T is the vector of measurement travel time, and A is travel length matrix, V is velocity, S is the vector of the slowness, $S_0$ is vector of the standard slowness, and Δs is the vector of the difference of the slowness.

Through the above calculations, the equation describing the internal velocity of concrete can be achieved. Traditional algorithms(such as ART, SIRT, etc.) iterate on the relative measurement data to construct a velocity distribution image of the internal concrete structure. However, these algorithms have their own limitations. For example, these algorithms need a high ratio of measurement paths, a lot of data, and colossal labor costs. This paper applied the algorithm to elastic wave tomography named compressive sampling, for eliminating these shortcomings.

*3.2. CS applying in EWT*

If the damage of a concrete structure is sparse, a few local parts are damaged and most parts of the structure are healthy. This feature is the basic principle that CS algorithm could apply in elastic wave tomography. In general, the propagation speed of elastic wave in typical concrete structures is around 4000m/s. Therefore, the situation of the inside concrete damage could be judged by determining the slowness in each area of the concrete structures.

The theory of elastic wave tomography combined with compressive sampling is shown as follows:

$$T = A\cdot(s_0+\Delta s) \tag{9}$$

Where, T is the measurement travel time vector, and $S_0$ is the standard slowness in totally healthy condition, $\Delta s$ is the slowness difference vector that is obtained as difference between actual slowness and standard slowness. A is a coefficient matrix of the travel lengths that is generated from all of the ray-path matrices. If $T_0$ is the travel time vector if the specimen is healthy, and $T_0$ is described as

$$T_0 = A * s_0 \tag{10}$$

Where y is the random sampling results of the difference between the measured travel time and the standard travel time. $\Phi$ is the measurement matrix obeys random Bernoulli distribution and contains only one "1" emerge in each row and column. 1 means that the travel time data are selected, and 0 means these data are omitted on the measurement stage. $\Phi$ is shown as follows:

$$\Phi=\begin{bmatrix} 1 & 0 & 0 & \cdots & 0 \\ 0 & 0 & 1 & \cdots & 0 \\ 0 & 1 & 0 & \cdots & 0 \\ \vdots & \vdots & \vdots & \ddots & \vdots \\ 0 & 0 & 0 & \cdots & 1 \end{bmatrix}_{m\times N} \tag{11}$$

After the sampling stage, the image of elastic wave tomography is reconstructed by using equation (12), which employed L1- minimization algorithm [8].

$$\Delta\hat{s} = \text{argmin}\ (\|\Theta\Delta s-y\|_2 + \lambda\|\Delta s\|_1),\ \text{ where } \Theta = \Phi \cdot A \tag{12}$$

With $\Delta\hat{s}$ is a vector that shows the difference of the slowness from the standard slowness and $\lambda$ is the regularization parameter, that is 0.083 in this study. $\|\cdot\|_1$ and $\|\cdot\|_2$ are l1-norm and l2-norm correspondingly [9]. Both the RIP and incoherence criteria are satisfied by the transfer matrix $\Theta = \Phi\cdot A$.

$$y = \Phi\cdot(T - T_0)=\Phi \cdot A \cdot\Delta s \tag{13}$$

## 4. Numerical simulation results and analysis

*4.1. Model and conditions for the numerical analysis*

The findings of the CS algorithm and the results of SIRT algorithm are compared on a series of numerical investigations in order to verify CS algorithm on the elastic wave velocity tomography. In SIRT algorithm, the travel time residuals are proportionately distributed in each pathed cell, besides correcting the slowness factor from ray lengths and travel time residuals in a selected cell [10]. The following is the basic information for the numerical simulation model: the concrete slab model dimension is 10 m × 10 m, there are 100 random events and 20 receivers. Original velocity distribution is shown as figure 2. The blue part is the concrete in healthy condition, velocity is 4000 (m/s), and the red part is the concrete in damage condiction, velocity is 3000 (m/s). The twenty receivers are installed on the edges of the model with every two cells interval, the mesh information and receivers location exhibited in figure 2. From the previous section, a total of 2000 samples, matrix Φ randomly selected 50 measurement paths. In this study, the chosen quantity of paths satisfies the condition of accurately reconstructing the outcomes. Therefore, the dimension of the transfer matrix $\Theta = \Phi * A$ is 50×100. Then use the L1 minimization algorithm to restore the reconstruction image.

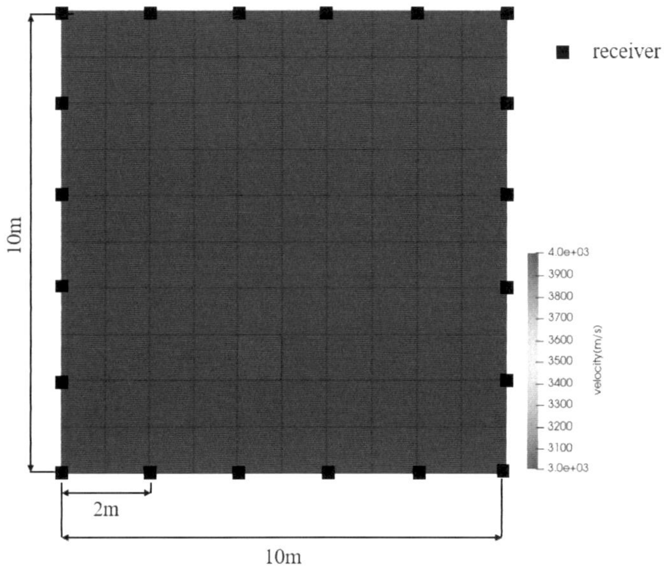


**Figure 2.** Original velocity distribution.

*4.2. Results and discussion*

Figures 3(a) and 3(b) illustrate the reconstructed elastic wave velocity distribution based on CS and SIRT, respectively. The reconstructed results show that the red part is the damaged area with lower velocity distribution, and higher speed distributions are healthy condiction displays as navy blue area. The CS algorithm, in comparison with SIRT algorithm, uses 50 measurement paths instead of 2000. It is impressive that these two algorithms could recover the damage of target quite clearly although the number of measurement data is largely different.

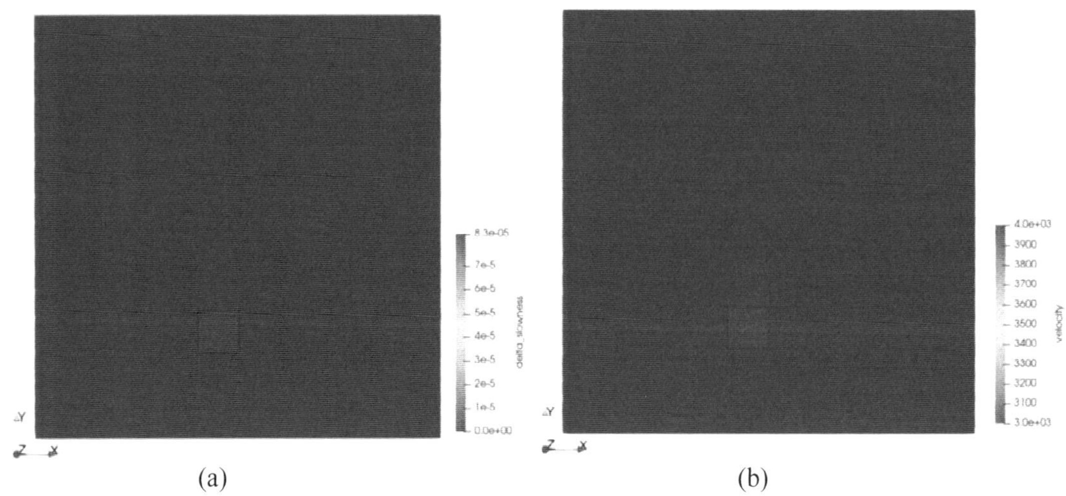


**Figure 3.** Reconstructed velocity distribution (a) Results on CS Algorithm. (b) Results on SIRT Algorithm.

## 5. Conclusion

This study verified elastic wave velocity tomography based on CS algorithm. Through the series of numerical investigation, the effectiveness of the CS algorithm applied on elastic wave tomography was confirmed. Considering the sparseness of concrete failure under normal circumstances, this is the prerequisite for utilizing CS theory. After randomly sampling the elastic wave testing in the measurement stage and restoring the reconstructed calculation, the accurate results that are comparable with SIRT are obtained.

The CS algorithm requires fewer measurement paths than the classic SIRT algorithm, which greatly reduces the amount of calculation while maintaining the same accuracy. The calculation results show that this method can accurately locate the position of concrete injury and detect the degree of damage with extraordinary efficiency.

## References

[1] Donoho D L 2006 Compressed sensing *IEEE Trans. Inf. Theory* **52** 1289–306

[2] Emmanuel J. Candès and Michael B. Wakin 2008 An Introduction To Compressive Sampling *IEEE Signal Process. Mag.* **77** 6897–912

[3] Candès E J, Romberg J and Tao T 2006 Robust uncertainty principles: Exact signal reconstruction from highly incomplete frequency information *IEEE Trans. Inf. Theory* **52** 489–509

[4] Candès E J, Romberg J K and Tao T 2006 Stable signal recovery from incomplete and inaccurate measurements *Commun. Pure Appl. Math.* **59** 1207–23

[5] Xin, Q., Yoshikazu, K., Kenichi, O., & Katsuya, N The application of compressive sampling in elastic wave velocity tomography.

[6] Wang, W.; Wang, C.; Bao, Y.; Li, H The study of compressive sampling in ultrasonic computerized tomography. In Proceedings of the 22nd Annual SPIE Smart Structures and Materials + Nondestructive Evaluation and Health Monitoring, San Diego, CA, USA, 8–12 March 2015.

[7] Hashimoto K, Shiotani T, Nishida T and Miyagawa T 2017 Application of elastic-wave tomography to repair inspection in deteriorated concrete structures *J. Disaster Res.* **12** 496–505

[8] Jiang B, Jia P, Zhao W, Wang W The application of compressive sampling in rapid ultrasonic computerized tomography (UCT) technique of steel tube slab (STS). *PloS one*. **13** 1-15

[9] Jiang B, Zhao W and Wang W 2017 Improved ultrasonic computerized tomography method for STS (steel tube slab) structure based on compressive sampling algorithm *Appl. Sci.* **7**

[10] Sassa K 1988 Suggested methods for seismic testing within and between boreholes *Int. J. Rock Mech. Min. Sci.* **25** 449–72

# Using natural stone for buildings – an analysis of educational programs in Romania

**D Rusu[1] and R Muntean[1*]**

[1] Transilvania University of Brasov, Faculty of Civil Engineering, Romania

* radu.m@unitbv.ro

**Abstract**. The topic of sustainability is of utmost importance today, especially referring to construction industry – a sector that contributes with almost 40% to the global CO2 emissions. This subject comprises various areas of research and knowledge and one study cannot cover all of them; the studies are usually focusing on a restrained field starting from one of the sustainability characteristics. Some principles of sustainable development are referring to the use of natural products, using local raw materials, using traditional construction methods, and respecting the cultural heritage. Therefore, in this context, the use of natural stone could be a sustainable solution especially in communities that historically used stone as a building material. Considering the value which the use of natural stone in new and existing buildings could bring for a more sustainable society, the current study aims to explore how topics like use of natural stone, stone processing, stone restauration and others related to them are included in the educational programs and curricula of higher education and vocational education and training in Romania. The results are compared with the situation from other European countries in order to formulate some conclusions and recommendations.

## 1. Introduction

The topic of sustainability is of utmost importance today, especially referring to construction industry – a sector that contributes with almost 40% to the global CO2 emissions [1,2]. Dealing with sustainability in relation to buildings is an extensive task, which cannot be easily solved with one unique solution, but with interventions in different directions.

The governments, the companies, the society and the people should change the way they act, the way they make business, their approaches and more important their mindset (their beliefs). It is well known that education is one of the most powerful tools to contribute to these changes, if the appropriate information is integrated into educational programs and contents.

One principle of sustainability is the sustainable use and management of natural resources to which it can be added the use of local raw materials, the use of traditional construction methods, and respecting the cultural heritage [3], which all leads to the use of natural stone as a sustainable solution for building materials. The natural stone is one of the greenest materials: it is not produced in a factory, it is not processed with other chemicals, has a long lifetime, is 100% recyclable and causes no emissions [4]. Moreover, the extraction of stone requires less energy than other building materials [5]. And although until now, the industry was reluctant in using natural stone for construction, mainly due to its higher initial cost, the lead time and the necessary skills to put it into practice, with the right measures these factors can become arguments for using natural stone. Measured on an entire life cycle basis, the natural stone is also financially competitive [6]. In accordance with the circular economy model, the raw materials should be kept in use as long as possible, so the natural stone in buildings is highly sustainable. Further, the scientists and stakeholders should approach the technical challenges in the stone sector and find solutions on how this sector can contribute effectively to economic development with a more responsible and sustainable approach, by following best practices towards a circular economy [7].

Besides respecting the environmental principle of a very short transport chain, using the natural stone quarried from the region or from the closest stone mine, which in the present study refers to Europe, means positive aspects for the social responsibility, as this is certainly made without child labor [5].

Natural stone was historically used for buildings (castles, palaces, churches and other impressive structures) both for interior and exterior decoration and for the structures, but also today, the modern architecture uses natural stone as a building material. This raw material has a broad range of uses in the preservation of monuments and renovations and also in modern architecture projects.

In order to enhance its utilization, the people should firstly learn about its properties and how to use it properly. For example, it is important to know how and where different types of natural stone can be use – the suitability depends on the stone composition and the environmental requirements (loads and weather conditions). Besides that, it is also equally important to know the techniques to process the stone. By improving and developing this ability the stone can be valued in various ways, showing its unique and versatile character.

Taking these aspects into account, this study aims to explore how topics like use of natural stone, stone processing, stone restauration and others related to them are included in the educational programs and curricula of higher education and vocational education and training in Romania.

The research was part of the BKStone project co-funded by the European Union – project number: 610483-EPP-1-2019-1-AL-EPPKA2-CBHE-JP (figure 1), whose main objective is fostering the modernization and development of the stone sector in Western Balkans partner countries through strengthening the relationship between the higher education institutions and the industry, and internationalizing curriculum and methods. In order to develop the training strategies to meet the challenges in terms of quality, delivery and relevance, an analysis of the current situation should firstly be performed [8].

**Figure 1**. Logo of the project BKStone - Higher Education–Enterprise platform for fostering, modernisation and sustainable growth in natural stone industry in Western Balkans.

The present study is structured in two main parts. In the first part the research methodology and the steps followed throughout the research are described, displaying afterwards a summary of the results. The second part presents some conclusions and future research directions intending to support the formulation of new strategies and training developments in terms of education related to natural stone.

## 2. Research

The research presented in this paper represents the first step of a more extensive research within the project which aims to contribute to the elaboration of strategies for specialization in the natural stone sector and development of training initiatives and educational materials in accordance with the project objectives [8].

The main objective of this research was to determine to which extent the study of natural stone is present in the educational programs within higher education (HE) and vocational education and training (VET) in Romania, as basis for identifying the needs for further training initiatives, for new educational materials and for new courses and contents on this topic. The determined situation is afterwards compared with the one in two other European countries – Greece and Spain.

### *2.1. Research methodology*

The research was conducted between January and March 2021 by the authors as part of the above-mentioned project. The first activity before starting was to define the research methodology and tasks for each phase of the research. The steps are summarized in the figure 2 below and detailed in the next section.

**Figure 2**. Research steps.

The first step consisted of establishing the secondary objectives of the research, namely:

a) identifying the educational programs which integrate the study of natural stone as a building material;
b) identifying the courses or modules within a course which integrate topics related to natural stone;
c) compiling the data gathered and summarizing it;
d) comparing the results with the findings from other European countries.

In order to collect relevant data for the present study, the research team performed the following tasks:

a) identifying the he institutions and vet organizations offering education and training in the field of construction and related areas;
b) collection of publicly available information from internet websites, publications, official reports, etc.;
c) contacting directly the identified institutions through calls or e-mails;
d) discussions (semi-structured interviews) with professors and trainers from the identified educational institutions.

The team searched and organized the data from two perspectives: independent courses on natural stone and courses which include references on natural stone, both grouped on higher education and on vocational education level. The search tried also to identify if the courses approach the learning in a theoretical and practical way.

The raw data gathered was put together and analysed within several focus groups, based on the previously established criteria. The methods used for analyzing the information were documents study, grouping of data, categorizing of data, analyzing of curriculum and semi-structured interviews for clarifying the unclear aspects. Although the data seemed to be quantitative, it was in fact of qualitative type, therefore open to more subjective interpretation. To avoid misinterpretation, the team tried to always follow the rules previously established and to discuss together all the elements that were not unanimously accepted.

The authors tried to identify all the study programs from the Romanian universities and the courses offered by private organizations in this field, however the present research is not a quantitative one, but a qualitative one, as the main scope was to assess the content of the existing courses.

The results were compiled into a final report, including tables and graphics where possible, in order to facilitate the understanding. Finally, these findings were compared with the results obtained from a similar study conducted in Spain and Greece and some conclusions and recommendations were formulated.

### *2.2. Research results*

A total of 48 public universities of Romania were included in the analysis, of which after a first sorting it remained a number of 9 universities to be included in the further analysis. The educational institutions cover all the regions of Romania and are mainly based in the biggest cities of the country, like Bucharest,

Cluj-Napoca, Iași, Timișoara, Brașov. The researchers have explored the universities offerings at both bachelor and master level.

In Romania, the high school education includes the following profiles: theoretical branch (humanistic and real profiles), technological branch (technical profiles, services, natural resources and environmental protection) and vocational branch (military, theological, sports, artistic and pedagogical profiles). The curricula for the state educational institutions from the technological branch is established by the Ministry of National Education and Scientific Research.

Regarding the training organizations subject to the analysis, their identification could not be exhaustive, as there is no central database with all the companies, but the aim was to identify providers of courses in each region of Romania, especially in the big cities, but also in smaller ones.

*2.2.1. Higher Education.* In Romania there are no university programs entirely dedicated to the study of stone. However, there are several courses and teaching modules focusing on related aspects, mainly in the following areas: civil engineering, architecture, geology, landscaping. These are present in the study programs from the most important universities in Romania and are summarized in the table 1.

Mainly, the undergraduate courses like Building materials and Construction materials which ultimately refers to the same thing, just the formulation is slightly different, are the ones which offers the widest content related to natural stone, although not in a very thorough manner. These courses provide an introductory overview of the various building materials used in construction: the physical and mechanical properties, the fundamental principles of manufacturing techniques, the practical use in civil engineering domain, the selection of construction materials according to application, in terms of security, durability and economy. The classes include references to all the building materials not only stone.

**Table 1**. Courses and study programs in universities of Romania covering topics related to natural stone.

| Field of study | Bachelor study programs | Master study programs | Courses |
|---|---|---|---|
| **CIVIL ENGINEERING** | *Civil, Industrial and Agricultural Constructions* | Materials and Performance Products for Constructions, Geotechnical Engineering, Rehabilitation and Increasing Construction Safety | Building Materials, Engineering Geology, Geotechnical Engineering, Geotechnics, Foundations, Maintenance and rehabilitation of roads and railways, Tunnels and metropolitan underground construction, Support structures in Geotechnical Engineering |
| **ARCHITECTURE** | *Architecture* | Restoration and patrimonial regeneration | Architectural Construction in Masonry and Concrete, Heritage Protection, Construction Materials, Restoration of monuments |
| **GEOLOGY** | *Geological Engineering* | | Construction rocks, Engineering Geology, Non-metallic mineral resources, Non-metalliferous deposits, Rock Mechanics |
| **LANDSCAPING** | *Landscaping* | | Restoration of parks and gardens, Roads and embankments |

At master level, the courses include more detailed information, in completion of the specialization in the field of constructions. For example, the master program Materials and Performance Products for Constructions within the Technical University „Gheorghe Asachi" Iași proposes a theoretical and practical approach to the problem of the construction industry and the materials to be used for new, high-performance construction structures.

It cannot be excluded the possibility of existing other courses and classes where information about the natural stone is present, but when the course name or the course's content were not explicit, they could not be identified. For this study we assume that these cases are rather singular and they would not considerably influence the results and hence the conclusions formulated. This situation could have been avoided in the case of an exhaustive research where surveys would have been sent to all the professors from all specialization throughout the country's universities.

*2.2.2. Vocational education and training.* The technological branch which has classes related to stone in the curricula, includes the fields: construction, installations and public works; construction materials; mining, oil and gas. In the vocational branch only, the artistic profile includes classes related to stone at the specialization architecture, plastic and decorative arts. The study programs from the vocational education include a practical component where the goal is for students to acquire practical abilities and hands-on experience.

From the existing teaching modules, the following includes some references to natural stone: Material processing equipment and machinery; Construction work for structures; Finishing and insulation works for constructions; Raw materials; Technology for the manufacture of concrete, cement and plaster products; Manufacturing control in the construction materials industry; Construction materials; The scientific basis of conservation and restoration; Materials and techniques used in conservation and restoration.

Besides the state educational institutions, there are organizations (mostly particular companies) that offers training courses and certification exams related to stone, mainly for the qualification of: Mason, stonemason, plasterer; but also, for the qualifications: Paver; Roads and bridges technician; and Worker and finisher in construction. The focus here is more on the practical formation and the graduates should pass a both theoretical and practical exam in order to obtain a certificate of professional competence recognized at national and international level.

## 3. Comparison with other countries

A similar study conducted for Spain and Greece revealed partly similar results. In absolute numbers, the number of universities offering courses on natural stone related topics in the 3 countries is approximately the same: 9, 7 and respectively 11, but it can be observed that 44% of the Greek universities have dedicated content.

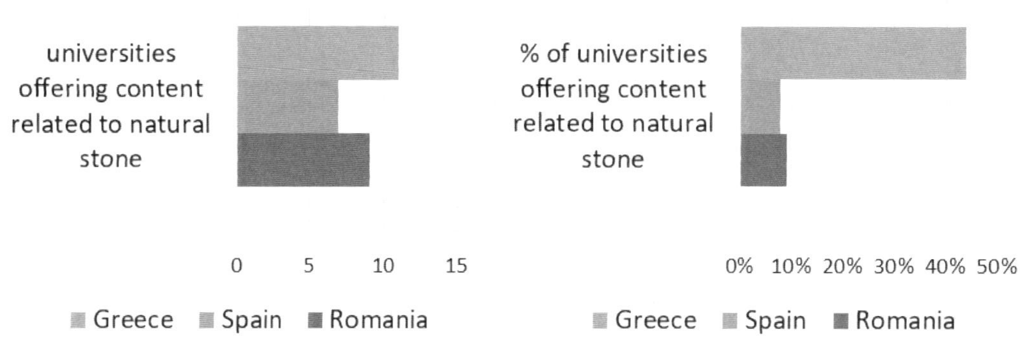


**Figure 3**. Comparison of educational programs offering content related to natural stone in Greece, Spain and Romania.

In Greece and Spain are no higher education programs focused completely on teaching aspects related to stone. However, there are several courses which include such references. In Spain, within the programs for a degree in technical architecture, are a block of contents that are exclusively dedicated to stone: Structures with stone materials (University of Burgos), Stone materials (University of La Coruña).

An interesting finding is that the Polytechnic University of Madrid offers a master program in building rehabilitation which proposes a Stonework Workshop. It is a unique project within the training offer of Spanish architecture universities and constitutes one of the optional subjects. Most of the time the students are carving the stone by entirely manual procedures.

A summary of the courses from universities of Greece and Spain with content related to stone is presented in the table 2, organized on the two fields of study: engineering and architecture.

Besides the courses included in the below table, the natural stone is presented as a transversal topic in more programs, like for example the master's program in Building rehabilitation from the University of La Coruña.

In conclusion, in Spain exists more courses and in Greece are more universities offering education related to natural stone. This might be due to their historical heritage and the traditional construction methods and the current need to conserve the existing stone buildings. Also, due to the warmer weather, the buildings were traditionally made of stone, in order to avoid the heating during summer, so the heritage built of stone is much more extensive than in Romania.

**Table 2**. Courses and study programs in universities of Greece and Spain covering topics related to natural stone.

| Field of study | Bachelor study programs | Master study programs | Courses |
|---|---|---|---|
| | | ***GREECE*** | |
| **ENGINEERING** | *Civil Engineering, Mining and Metallurgical Engineering, Chemical Engineering, Mining and Geotechnical Engineering, Mineral Resources Engineering* | *Protection, Conservation and Restoration of Architectural Monuments, Protection, Conservation and Restoration of Works of Art and Mechanisms* | Construction Materials, Structural Materials, Marbles and Industrial Minerals, Building and Ceramic Materials, Marbles and Aggregate Materials, Building Materials, Marble and Quarry Materials Exploitation and Processing, Technology of Structural Materials – Composite Materials, Technology of Building Materials, Building and aggregate materials, Historical Construction Materials Pathology & Technology of Conservation & Restoration, Construction technology of historical buildings, Masonry Engineering |
| **ARCHITECTURE** | *Architecture, Conservation of Cultural Heritage, Interior Architecture* | *Conservation and Restoration of Historic Buildings, Conservation of Building Materials* | Technology – Building Construction and Materials , Building Material Technology, Stone Conservation, Structural Art, The art of construction, Technology of Materials, Decay and conservation of construction materials of monuments and architectural surfaces, Conservation of materials and structural systems of monuments and archaeological sites, Construction Materials, Repair & Strengthening of Structures, Introduction to the pathology & restoration of monuments & building materials, Science and technique of building materials and materials of architectural surfaces |

| | | | |
|---|---|---|---|
| | | ***SPAIN*** | |
| **ENGINEERING** | *Civil Engineering, Building, Engineering Technology, Civil and Territorial Engineering, Mining and Energy Resources Engineering* | *Building Rehabilitation* | Materials Fundamentals. Chemistry and Geology, Building Materials, Construction Materials, Sustainable exploitation of mining resources, Stonework workshop |
| **ARCHITECTURE** | *Technical Architecture, Architecture Studies, Fundamentals of Architecture* | | Structures with stone materials, Stone materials, Stonework works, Building materials, Construction materials for construction systems |

In Spain exists five study programs of VET related to stone for the following specializations: Technician in Natural Stone; Technician in Construction; Diploma of Education in Plastic Arts and Design in Artistic Reproductions in Stone; Diploma of Higher Education in Plastic Arts and Design in Art Applied to Stone; Diploma of Higher Education in Plastic Arts and Design in Stone Sculpture Techniques. As a continuing professional training exists the following courses: Natural stone production; Natural stone handcraft and restoration works, Natural stone treatment and placement; Recovery of cultural landscapes. Dry stone technique. The authors could not identify study programs in Greece related to natural stone, only some courses which included topics on it. As it can be observed, in Spain there are more courses and professional trainings dedicated to stone processing and production as in Romania and Greece.

## 4. Conclusions

The natural stone is a sustainable building material having the following characteristic in accordance with the principles of sustainability: its production saves energy, it is environmentally friendly, it is durable and recyclable and its quarrying from local mines saves CO2. To all these advantages there can be added its unique character which offers numerous design options for architects and builders.

So, for enhancing the use of natural stone for buildings, firstly the people should learn about its properties and the technologies and techniques used for its effective utilization. This was the rationale behind the present study, which aimed to determine the current situation of the educational programs on natural stone in Romania and compared to other European countries.

The study revealed that there are no dedicated university programs in any of the three countries, which in fact may not be needed, but it can be stated that there is a need of integrating the topics related to natural stone as a transversal topic in more courses and especially from a sustainable development perspective.

Although the situation in the three countries seems similar, the difference consists in the qualitative aspects of the identified courses. The presence of theoretical and practical topics related to natural stone as a building material in Romania is very limited and rather impromptu. It can be said that here exists a need of extending the content by including more references on natural stone in the identified classes but also in other disciplines. The scope is to make the students aware of the advantages of using natural stone as a building material and to familiarize them with the possibilities of integrate it into modern projects but also to learn about the methods to conserve the stone from existing buildings.

In the higher education programs, it was identified a very low presence of laboratory hours for tests and experiments on rocks and building materials. Within VET, it was identified a more practical approach to the study of natural stone, existing also dedicated courses or training modules which deal with the processing and use of natural stone within buildings.

But in both cases, Higher Education and Vocational Training and Education, the research has revealed that the topic *using natural stone for buildings* is not very present in the educational content and learning material. Moreover, even if present, it is not from the perspective of promoting sustainability and the use of stone as a sustainable material.

Depending on the objectives set, future research could focus on identifying the best approaches in teaching about natural stone use in construction sector or on analyzing the effectiveness of a new course implementation in relation to the use of stone as a building material. Another research direction could be the exploration of the factors which can contribute to increase the use of stone in construction besides education.

In conclusion, a new approach in developing the curricula and the educational materials would be needed, generally more oriented towards sustainable solutions. The main purpose is to make students aware of the importance of sustainability, to teach them the basic principles of sustainability and to develop the fundamental competences in order for them to become promoters of sustainability in the future.

Based on the conclusions formulated from this research, the project team is aiming to further develop new curricula for courses and possible new study programs in the higher education as well in the vocational and training education, in the following fields: quarrying, manufacturing, marketing, heritage and restoration.

**References**

[1] Kibert C J 1994 Establishing Principles and a Model for Sustainable Construction *Proc. First Int. Conf. CIB TG 16 Sustain. Constr.* 3–12
[2] Neill P 2020 Environment Journal, https://environmentjournal.online/articles/emissions-from-the-construction-industry-reach-highest-levels/, accessed on 20.09.2021
[3] Rusu D 2021 *Aspecte privind managementul sustenabil în construcții* (Aspects of sustainable construction management) Brasov: Ed. Univ. Transilvania
[4] Natural Stone Institute, https://www.naturalstoneinstitute.org/consumers/sustainable-natural-stone/, accessed on 15.09.2021
[5] Deutscher Naturwerkstein-Verband e.V. 2015 Nachhaltigkeitsaspekte bei Naturwerksteinen, https://www.natursteinverband.de/fileadmin/user_upload/Nachhaltige_Natursteine_2015-07.pdf
[6] Klemm A and Wiggins D 2016 Sustainability of natural stone as a construction material *Sustainability of Construction Materials* pp 283–308
[7] Careddu N, Di Capua G and Siotto G 2019 Dimension stone industry should meet the fundamental values of geoethics *Resour. Policy* **63**
[8] BKStone, www.bkstoneproject.com, accessed on 15.09.2021

# Cement replacement by used paints in plasters mortars mixtures

**AC Siomin[1*], ML Ținținșan[1], A Zaharie[1] and DL Manea[1]**

[1] Technical University of Cluj-Napoca, Faculty of Civil Engineering, Romania

* cristian.siomin@googlemail.com

**Abstract**. This research started from the fact that humanity exploits on an industrial scale and chaotically consumes the raw materials that this planet offers, without realizing the importance of reusing waste and the possibility of introducing them in the production of plaster mortars. Thus, an attempt is made to create a balance between raw materials and the exploitation of new resources, so that future generations can have access to raw materials. Plaster mortar mixtures were made by using waste from construction sites, respectively certain percentages of the amount of cement were replaced with used paints. In order to reduce the costs of transporting and transferring waste to the landfill, most of the executors (builders) at the stage of systematization of the land, bury all the waste, not respecting the conditions given by the environmental agreement. Through a waste management, construction materials are recovered, in this case used paint and reused in the preparation of plaster mortar. Starting from a plaster mortar mixture, the cement was replaced in a percentage of 10% and 15% with used paint, resulting in significantly higher bending strengths. We must be responsible and contribute to the protection of the environment by recycling/reusing waste from construction materials.

## 1. Introduction

In 2015, the United Nations set through the 2030 Agenda, 17 sustainable development goals to be implemented by all member countries by 2030.

The 12th goal needed for sustainable development is the production and consumption of products/materials in a responsible manner [1].

Starting from the importance of sustainable development, as well as the exponential growth regarding the manufacture of construction materials, in the experimental program, used paint was recovered from construction sites to create new recipes for plaster mortars, by replacing cement in different percentages [2].

If we add the population growth by 2050 to 9.6 billion inhabitants, it is imperative to find the best waste management solutions, so as to reduce the use of resources and raw materials used in the construction of buildings.

## 2. Experimental Program

The experimental program developed within the laboratory of the Faculty of Constructions of the Technical University of Cluj-Napoca started from a standard recipe of plaster mortar, called CSIV (composition: Portland cement CEM I 52.5R without addition [3, 4], sand [5] and water [6]), in which the cement was gradually replaced with water-based used paint, according to the table below (table 1):

**Table 1.** Own recipes.

| | Portland Cement Quantity CEM I without addition [kg] | Quantity of Used paint [l] | Quantity of sand [kg] | Quantity of the water [l] |
|---|---|---|---|---|
| **CSIV** | 2.695 | - | 6.960 | 1.500 |
| **CSIV 10C** | 2.426 | 0.269 | 6.960 | 1.100 |
| **CSIV 15C** | 2.291 | 0.404 | 6.960 | 0.905 |

where:
CSIV -Plaster mortar (standard recipe);
CSIV 10C -Plaster mortar in which the cement was replaced with 10% used paint;
CSIV 15C -Plaster mortar in which the cement was replaced with 15% used paint.

## 3. Results

The experimental program included 2 stages: determinations on fresh mortar and determinations on hardened mortar [7].

### *3.1. Determinations on fresh mortar*

The determination of the density in the fresh state was performed by introducing the freshly homogenized mortar in a cylindrical vessel with known characteristics (vessel volume V = 1 liter).

The mortar was compacted according to its consistency, so for the consistency greater than 6 cm the mortar in the vessel was imposed in 25 different places along the entire height with a metal rod with a diameter of 10–12 mm.

With the help of a balance, the vessel was weighed with fresh mortar, obtaining the mass of fresh mortar.

In the fresh state, for all the 3 recipes of plaster mortars, determinations on the density were performed, and the results are presented in table 2.

**Table 2.** Density of plaster mortars.

| | Apparent Density Pa [kg/m³] |
|---|---|
| **CSIV** | 2.695 |
| **CSIV 10C** | 2.426 |
| **CSIV 15C** | 2.291 |

### *3.2. Determinations on hardened mortar*

*3.2.1. Bending tensile strength.* To test the tensile strength by bending, the plaster mortar was poured into molds (figure 1), after which it was compacted on the vibrating table (figure 2), obtaining 9 specimens with the size of 40x40x160 mm. These were tested at intervals of 3 days, 7 days and 28 days, and the results are presented in figure 4.

**Figure 1.** Plaster mortar poured in molds.

**Figure 2.** Plaster mortar compacted on the vibrating table at 60 shocks.

The determination of the tensile strength by bending was performed with the help of the automatic bending test machine - Fruhling-Michaels equipment, in which the obtained specimens were introduced (figure 3). After fixing a specimen, the equipment was started, which locks automatically when the prism is broken, and the value of the breaking force was read.

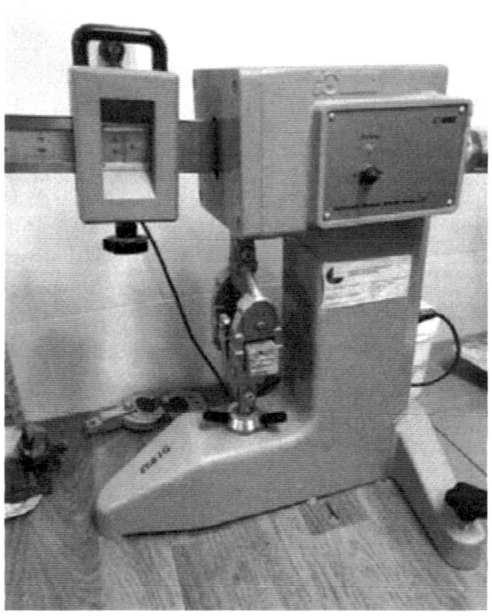

**Figure 3.** Test specimens to determine the tensile strength by bending.

The bending tensile strength represents the unit effort by reporting the maximum bending moment to the resistance modulus of the section.

Thus, the tensile strength by bending was calculated with the equation (1):

$$Rti = \frac{3}{2}\frac{Pl}{a^3} \quad \left[\frac{N}{mm^2}\right] \tag{1}$$

where:

P - breaking force [N];
l - distance between supports [cm];
a - side of section [cm].

The obtained results are presented in the figure below, respectively figure 4:

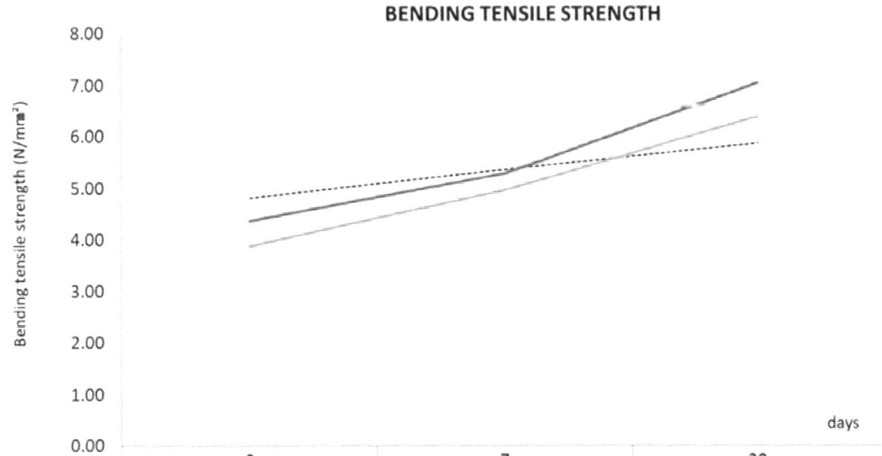


| | 3 | 7 | 28 |
|---|---|---|---|
| CS IV | 4.80 | 5.36 | 5.86 |
| CSIV 10C | 4.37 | 5.29 | 7.04 |
| CSIV 15C | 3.88 | 4.96 | 6.38 |

**Figure 4.** Bending tensile strength for CSIV 10C - CSIV15C plaster mortar recipes compared to CS IV.

*3.2.2. Compressive Strength.* Using the hydraulic press provided by the laboratory of the Faculty of Constructions, the test specimen heads were tested for compression at the same time intervals as for determining the tensile strength by bending, respectively 3 days, 7 days and 28 days (figure 5 and figure 6).

After centering the specimen in the hydraulic press on the lower plate, bring the fixed plate as close as possible to the upper surface of the sample. The hydraulic press is started, and when the sample is destroyed the manometer locks on the value of the breaking resistance.

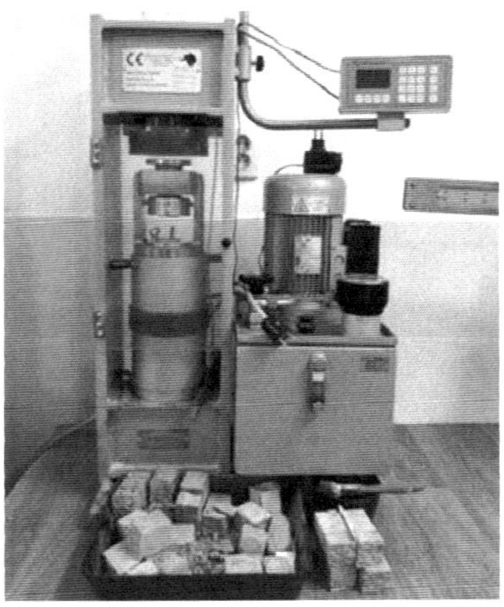

**Figure 5.** Test specimens to determine the compressive strength.

**Figure 6.** Test heads tested at the hydraulic press.

Compressive strength is calculated by the equation (2):

$$Rc = \frac{N}{A} \quad \left[\frac{N}{mm^2}\right] \tag{2}$$

where:

N - compression force;

A - the area of the sample required for compression.

The results regarding the compressive strength are presented in the figure below, respectively figure 7:

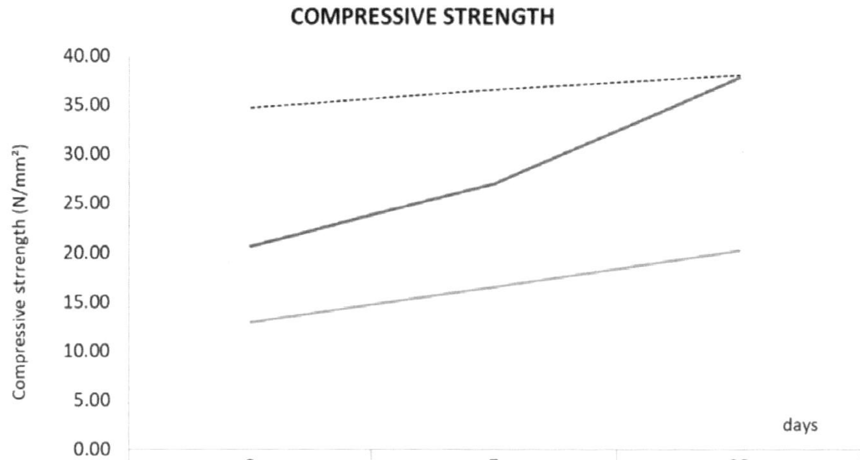


| | 3 | 7 | 28 |
|---|---|---|---|
| CS IV | 34.65 | 36.52 | 38.03 |
| CSIV 10C | 20.64 | 26.99 | 37.72 |
| CSIV 15C | 12.95 | 16.48 | 20.13 |

**Figure 7.** Compressive strength for CSIV 10C - CSIV 15C plaster mortar recipes compared to CS IV.

## 4. Interpretation of results

### *4.1. Contributions*

The personal contribution of the experimental program is the replacement of a percentage of 10% and 15% of the amount of cement with used paint for the preparation of plaster mortars.

The results obtained by recycling used paint as a cement substitute in different percentages, respectively 10% and 15% of the amount of cement used for the preparation of plaster mortar, highlight the contribution to the reduction of raw materials, respectively the recycling of construction waste and environmental protection.

### 4.2. *Bending strength*

By replacing the cement with 10% used paint, according to the CSIV 10C plaster mortar recipe, you can see the increase in bending strength at 28 days compared to the CSIV standard recipe by 20.13% (See figure 8).

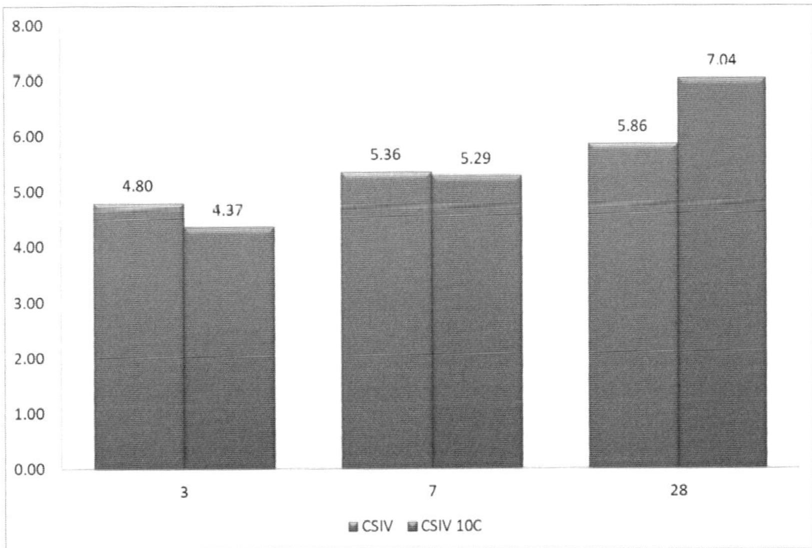


**Figure 8.** CSIV bending strength - CSIV 10C [N/mm²] (Rti at 3 days, 7 days, 28 days).

By replacing the cement with 15% used paint, according to the CSIV 15C plaster mortar recipe, you can see the increase in bending strength at 28 days compared to the CSIV standard recipe by 8.87% (See figure 9).

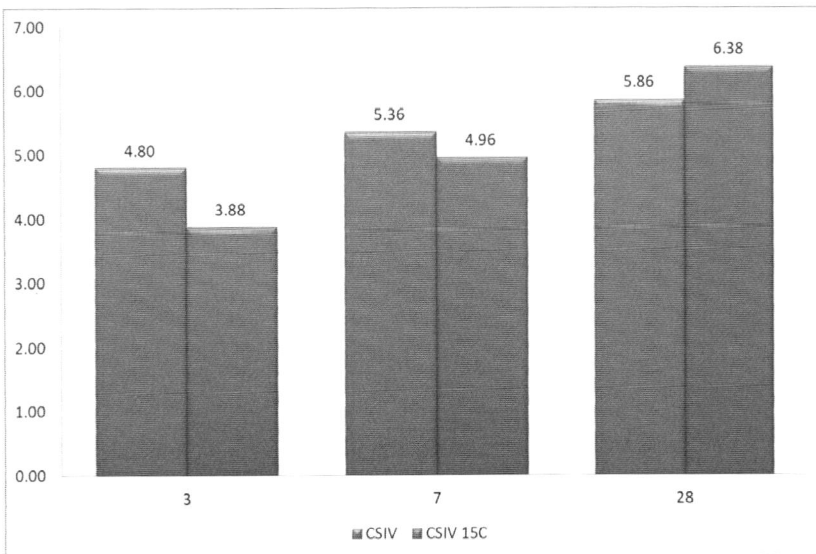


**Figure 9.** CSIV bending strength - CSIV 10C [N/mm²] (Rti at 3 days, 7 days, 28 days).

4.3. *Compressive strength*

By replacing the cement with 10% used paint, an insignificant decrease of the compressive strength in the CSIV 10C recipe can be observed at 28 days compared to the mortar in the CSIV standard recipe, by only 0.82% (See figure 10).

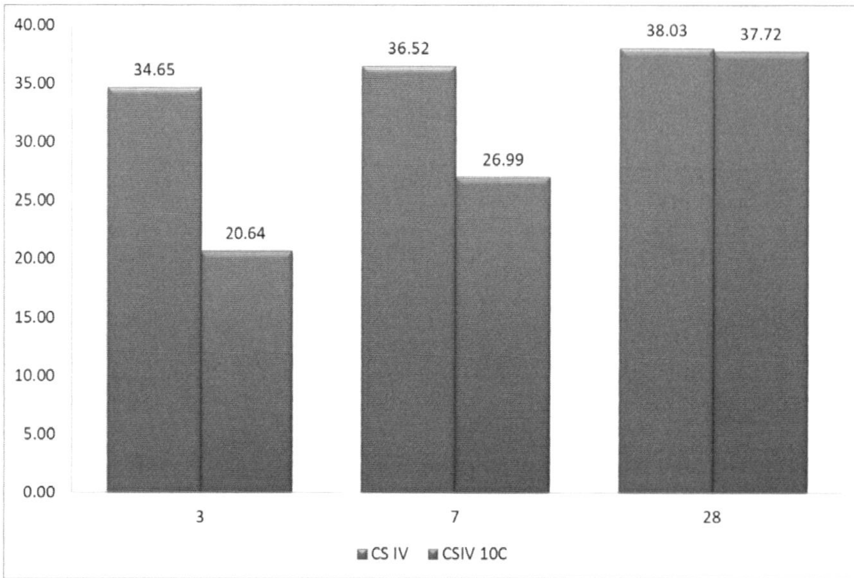


**Figure 10.** CSIV - CSIV 10C compressive strength [N/mm²] (Rc at 3 days, 7 days, 28 days).

In the case of the CSIV 15C mortar recipe, by replacing the cement with 15% used paint, a decrease of the compressive strength compared to the mortar from the CSIV standard recipe by 47.07% can be observed at 28 days (figure 11).

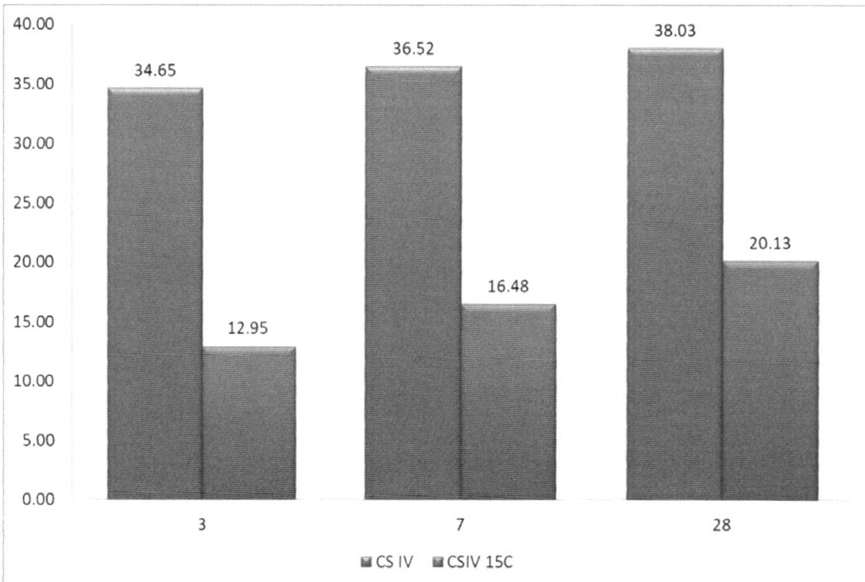


**Figure 11.** CSIV compressive strength - CSIV 15C [N/mm²] (Rc at 3 days, 7 days, 28 days).

## 5. Conclusions

The results obtained from the experimental program, highlight the fact that the most efficient recipe, both fresh and hardened is CSIV 10C, by replacing 10% of the amount of cement with used paint for the preparation of plaster mortars (figure 12).

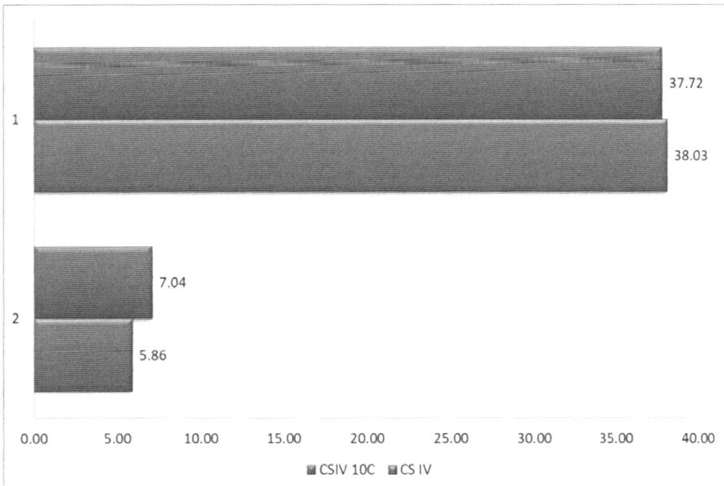


**Figure 12.** CSIV 10C recipe strengths compared to CS IV [N/mm$^2$] (1 - Compressive strength, 2- Bending tensile strength).

The explanation about the resistance variation will be the objective for the next step of the experimental program, using MRI and SEM type investigations. These results will be presented in the following papers.

The results obtained show that used paints can be used in the preparation of plaster mortars. This aspect will be an important factor for the protection of the environment.

**Acknowledgement**

This paper was financially supported by the Project "Entrepreneurial competences and excellence research in doctoral and postdoctoral programs - ANTREDOC", project co-funded by the European Social Fund financing agreement no. 56437/24.07.2019.

## References

[1] United Nations 2015 The 2030 Agenda for Sustainable Development
[2] Feng E, Sun J and Feng L, 2018 Regeneration of paint sludge and reuse in cement concrete *E3S Web Conf.* vol 38
[3] European Committee for Standardization 2011 SR EN 197-1 Cement Part 1: Composition, specifications and conformity criteria for common cements *ASRO*
[4] European Committee for Standardization 2011 SR EN 413-1 Cement for masonry. Part 1: Composition, specifications and conformity criteria *ASRO*
[5] European Committee for Standardization 2009 SR EN 13139/C91 Aggregates for mortars *ASRO*
[6] European Committee for Standardization 2003 SR EN 1008:2003 Preparation water for concrete *ASRO*
[7] European Committee for Standardization 2020 SR EN 1015 1 ÷ 21/2020 on Methods for testing mortars *ASRO*

# Structural analysis and optimal retrofitting solutions for historical monument with masonry walls

**I Sococol[1*], P Mihai[1], T C Petrescu[1] and D T Babor[1]**

[1] "Gheorghe Asachi" Technical University of Iasi, Faculty of Civil Engineering and Building Services, Iasi, Romania

* ion.sococol@student.tuiasi.ro

**Abstract.** "Braila Emergency County Hospital (ECH)" historical monument represents one of the main objectives of structural rehabilitation in both Braila county and Braila city. Thus, a structural strength verification for Pavilion B, Section 2, using ETABS software was performed, in order to find optimal retrofitting (rehabilitation) solutions. This analytical procedure proved to be beneficial and absolutely necessary, due to observations regarding the spatial contribution of masonry shear walls and finally the proper establishment of capable/design efforts. Following the structural calculation, it was concluded that the best solution would be to use composite materials for retrofitting a certain number of masonry shear walls, as classical retrofitting methods proved to be more expensive and structurally inefficient. Furthermore, in the context of the COVID-19 pandemic, the solution to use composite materials is beneficial, as it would entail no personnel or patients to be evacuated during the rehabilitation works, i.e. the hospital being able to continue to operate.

## 1. Introduction

The COVID-19 pandemic, which affected the entire humanity in more than one way, has highlighted the lack of emergency hospitals which could be used in the fight against the disease. In these circumstances, many countries seek to solve this pressing problem. So far, several solutions were put forward:

- the full capacity use of available spaces in existing hospitals;
- the rehabilitation of existing hospitals which require such works;
- the development of new constructions, which can be "traditional" hospitals or innovative buildings:
  - „modular composite building in urgent emergency engineering projects (Wuhan Thunder God Mountain/ Leishenshan hospital)" [1];
  - „off-site construction for emergencies" [2].

As such, the current article presents a study concerning the checking of lateral structure from a seismic point of view regarding the "Braila Emergency County Hospital (ECH)" (see figures 1 - 8), which is a "historical monument" [3-8].

IOP Conf. Series: Materials Science and Engineering 1242 (2022) 012035 doi:10.1088/1757-899X/1242/1/012035

**Figure 1.** Main entrance of the ECH Hospital.

**Figure 2.** Lateral left façade of the ECH Hospital.

**Figure 3.** Lateral right façade of the ECH Hospital.

**Figure 4.** Continuation of lateral left façade of the ECH Hospital.

**Figure 5.** Inner courtyard of the ECH Hospital.

**Figure 6.** Inner courtyard of the ECH Hospital.

**Figure 7.** Part of back façade of the ECH Hospital.

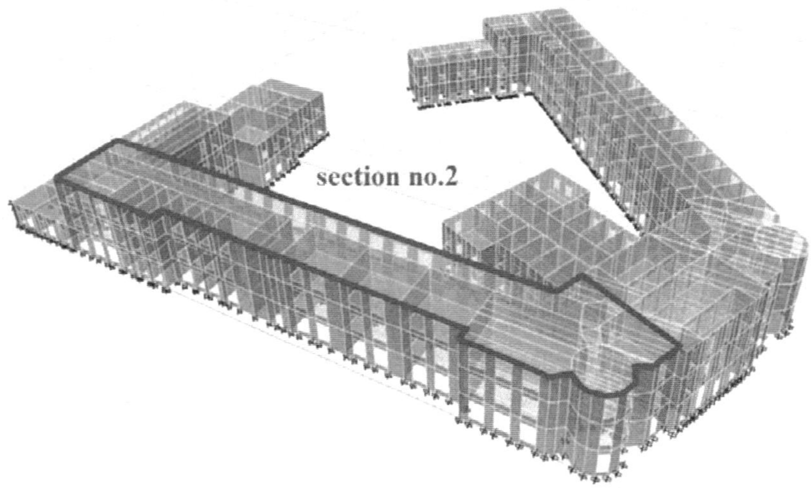


**Figure 8.** 3D representation of the lateral resistance structure for ECH Hospital – Pavilion B and outline of the analysed section of the structure (with ETABS software).

For proper knowledge of the building, some data and general aspects regarding the physical condition of the historical monument and data regarding the geometry of the masonry structure was specified.

General construction data (according to P100-3 norm [9]) (figures 1 - 8):

- Date (period) of execution: 1923 – 1928;
- Number of stories: partial – 1st storey; partial – 2nd storey; partial – 3rd storey;
- Plan form: irregular form;
- Vertical continuity, elevation form and dimensions: vertical discontinuity, regular form with 4.85 m height storey;
- Masonry type: unreinforced;
- Nature of masonry elements and finishing works: degradable masonry elements and mortar finishing;
- Type and materials of the floors: reinforced concrete slabs;

- Type and materials of the roof: roof with a special architectural configuration with structural wood frame structure and ceramic elements;
- Nature of the soil: foundation soil sensitive to moisture;
- Type and materials of ground foundations: continuous base masonry and stone foundations.

Relevant aspects related to the physical condition of the masonry construction „Braila Emergency County Hospital – Pavilion B, section 2" for the assessment of earthquake safety (according to P100-3 norm [9]) (see figure 1 - figure 8):

- Degradation of masonry elements through: *capillary rise of water, freeze-thaw effects, mortar degradation;*
- Foundation soil collapse: *in the central area, adjacent to section 2, due to storey difference of the building, respectively due to structural mass;*
- Deterioration of RC slabs from vertical loads (local cracks, excessive deformations, vibrations): *there is no damage (concrete cracks, deformations etc.);*
- Seismic damage of the structure: *no significant seismic damage to the masonry shear walls.*

Information on construction specific details to the historical monument Braila (ECH) – Pavilion B, section 2 (according to P100-3 norm [9]):

- Type and quality of corner connections between masonry shear walls, ramifications and intersections: *proper weaving of brick masonry and proper quality of masonry shear walls corner intersections;*
- Type and quality of connections between RC slabs and masonry shear walls: *there exist RC belts that ensure the transfer of mass and inertial forces;*
- Type of lintels: *rectangular reinforced concrete;*
- Details of time interventions on construction: *such details are not available;*
- The creation of new door and window openings: *such openings do not exist;*
- Organization of structural elements with high vulnerability: *the existence of degraded ceramic elements due to natural weather and rows with expelled mortar on the outside wall faces;*
- Existence of basement and type of basement: *partial basement;*
- Basement wall materials: *solid brick masonry;*
- Foundation height: *4 m;*
- Existence of waterproofing: *no waterproofing exists;*
- Site topography: *flat site;*
- Nature of foundation soil: *soil sensitive to moisture.*

According to P100-3 [9] norm, the level 2 methodology was used for detailed seismic evaluation through response spectra method. This methodology was chosen due to the complexity of the lateral resistance structure of Braila Emergency County Hospital – Pavilion B, section 2.

## 2. Lateral resistance capacity determination of masonry shear walls loaded with horizontal forces for section no. 2

A "global analysis regarding the seismic response" [10,11] of the „masonry structure" [12] was performed in the first phase and the partition into five modules was subsequently established. Thus, it was possible to perform separately the seismic analysis for each section. Due to the complexity of the seismic verification, the calculation steps only for one module are represented (see figure 8).

Thus, a "modal analysis of the masonry structure" [13] was carried out. For the first three modes of vibration, a local translation effect of the marginal walls was observed (see figure 9 – figure 14). Modal periods and frequency results can be consulted in table 1, the structure presenting a fundamental vibration period of 0.292 s.

Furthermore, it is possible to observe that the modal load participation ratio is consistent on the global Y direction – the "short" direction of the masonry structure (see table 2). Corresponding to these analytical results, in table 3, the maximum drift and the Avg (average) drift for the historical masonry structure on the two global directions of the structural spectrum are presented; this holds for the design spectrum method of the seismic loading.

Another important effect which can be observed in the dynamical analysis is the important lateral drift of the structural areas with large lateral rigidity, as well as the ones which are located in the rectangular intersection area of the structural walls (see figure 15 and figure 16).

The relative virtual work/volume, which is graphically depicted in figure 17 and figure 18, presents data regarding the main masonry mullions, which are the first ones to become "active" in the dissipation of the seismic energy induced by an earthquake.

The results obtained following the calculus to seismic action design are expressed as maximums: storey displacements, storey shears, storey overturning moments, storey stiffness (see table 4, figure 19 – figure 22). These results locate the existence of maximum values between storey 1 and storey 2.

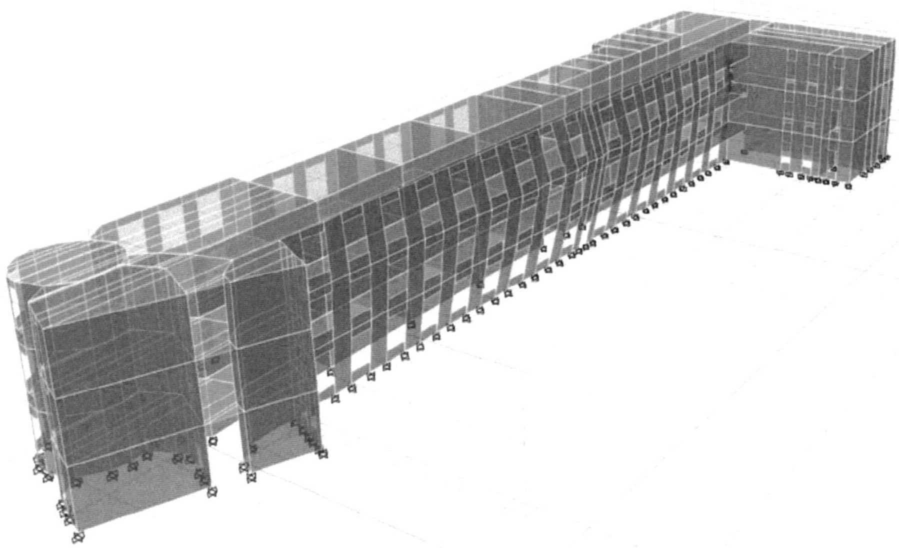

**Figure 9.** The first eigenmode of the vibration of the structure – 3D view.

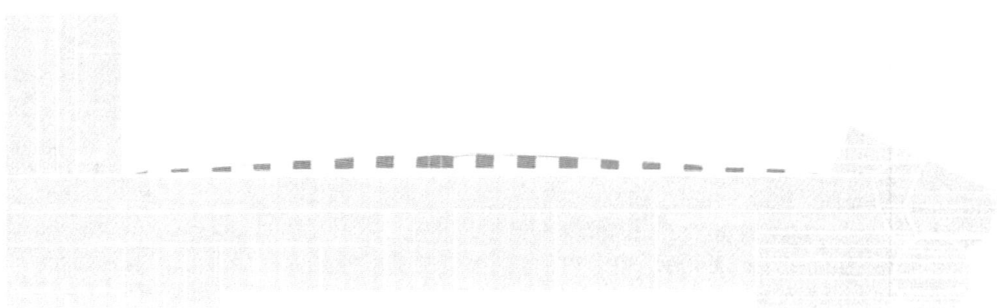

**Figure 10.** The first eigenmode of the vibration of the structure – top view.

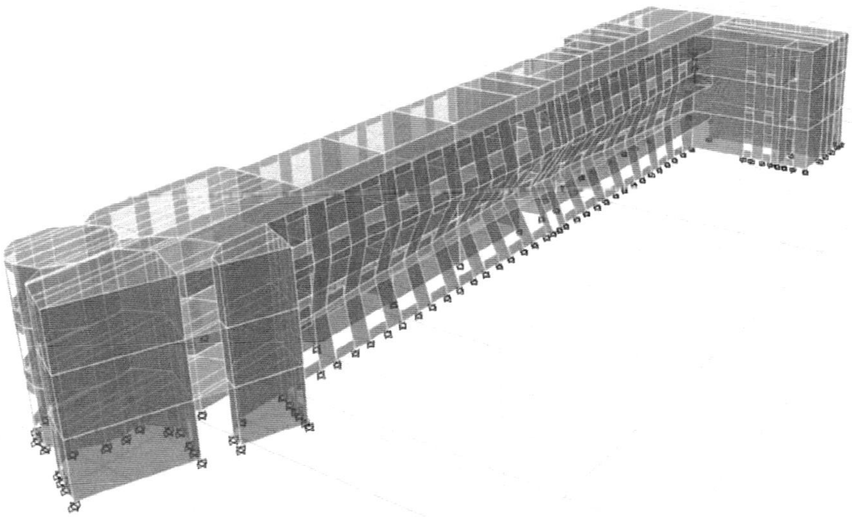

**Figure 11.** The second eigenmode of the vibration of the structure – 3D view.

**Figure 12.** The second eigenmode of the vibration of the structure – top view.

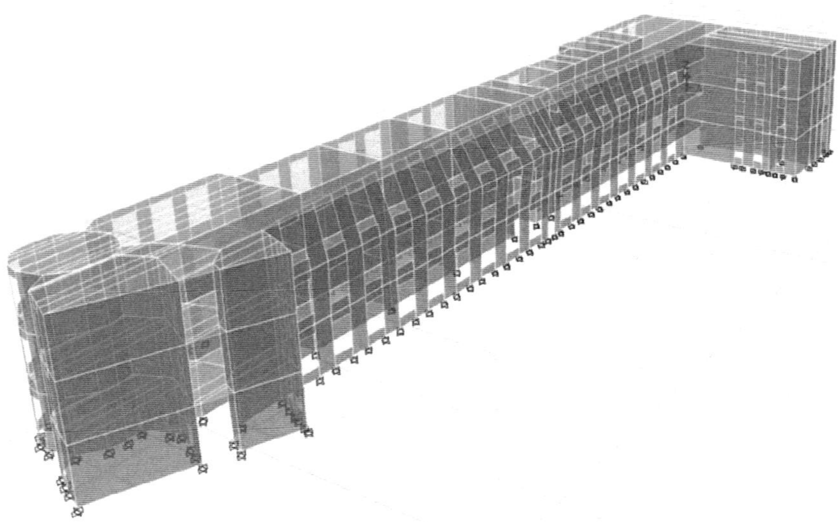

**Figure 13.** The third eigenmode of the vibration of the structure – 3D view.

**Figure 14.** The third eigenmode of the vibration of the structure – top view.

**Table 1.** Modal periods and frequency results for historical masonry structure.

| Case | Mode | Period (s) | Frequency (cyc/s) | Circular Frequency (rad/s) | Eigenvalue ($rad^2/s^2$) |
|---|---|---|---|---|---|
| modal | 1 | 0.292 | 3.424 | 21.5162 | 462.9464 |
| | 2 | 0.263 | 3.805 | 23.9048 | 571.4375 |
| | 3 | 0.21 | 4.762 | 29.9192 | 895.1598 |
| | 4 | 0.186 | 5.391 | 33.871 | 1147.248 |
| | 5 | 0.185 | 5.397 | 33.91 | 1149.885 |
| | 6 | 0.184 | 5.428 | 34.1048 | 1163.1386 |
| | 7 | 0.184 | 5.431 | 34.1252 | 1164.5295 |
| | 8 | 0.183 | 5.462 | 34.3208 | 1177.9175 |
| | 9 | 0.183 | 5.478 | 34.4182 | 1184.6097 |
| | 10 | 0.181 | 5.517 | 34.6648 | 1201.6484 |
| | 11 | 0.181 | 5.529 | 34.7391 | 1206.8055 |
| | 12 | 0.18 | 5.543 | 34.8269 | 1212.9147 |

**Table 2.** Modal load participation ratios for historical masonry structure.

| Case | Type of lateral load consideration | Global direction | Static (%) | Dynamic (%) |
|---|---|---|---|---|
| modal | acceleration | X | 2.55 | 0.96 |
| | | Y | 77.24 | 59.68 |
| | | Z | 0 | 0 |

**Table 3.** Maximum drift and Avg drift for historical masonry structure.

| Storey | Seismic load combination | Global direction | Max drift (mm) | Avg drift (mm) | Ratio (Max drift/ Avg drift) |
|---|---|---|---|---|---|
| 1 | design spectrum | X | 2.796 | 1.397 | 2.002 |
| 2 | | | 22.482 | 11.485 | 1.957 |
| 3 | | | 41.883 | 21.091 | 1.986 |
| 1 | | Y | 26.623 | 13.576 | 1.961 |
| 2 | | | 22.559 | 11.603 | 1.944 |
| 3 | | | 42.279 | 21.402 | 1.975 |

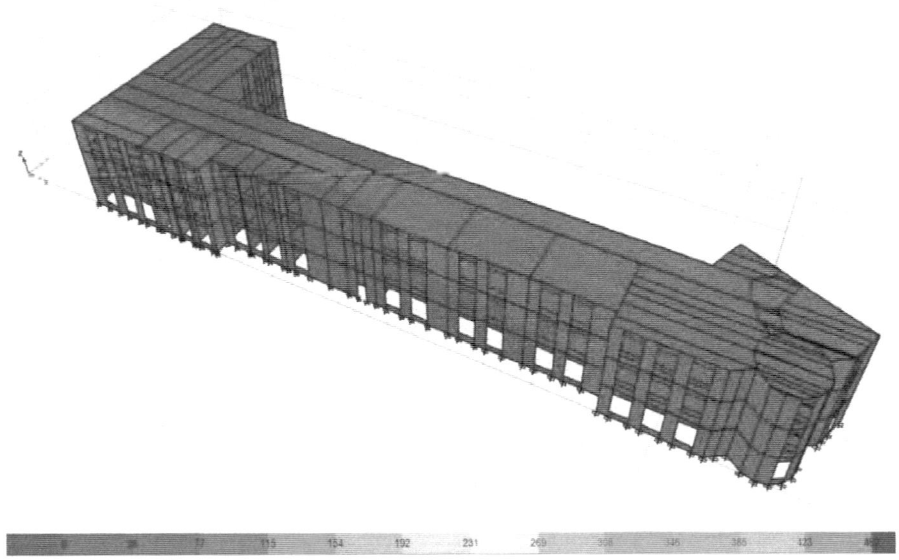


**Figure 15.** Graphical representation of the "lateral displacements" for structural masonry system in the global X direction of the structure.

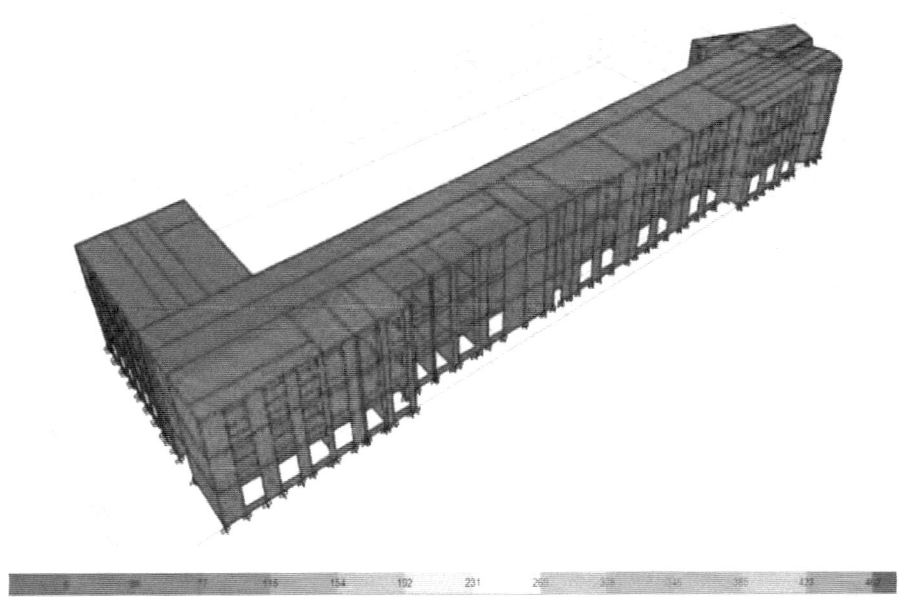


**Figure 16.** Graphical representation of the "lateral displacements" for structural masonry system in the global Y direction of the structure.

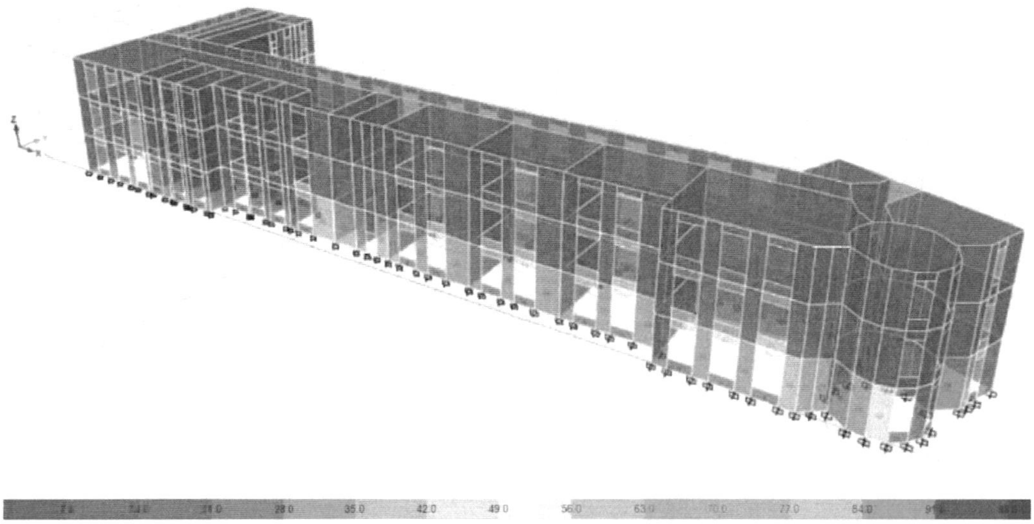

**Figure 17.** Graphical representation of the "relative work/volume" for structural masonry systems in the global X direction of the structure.

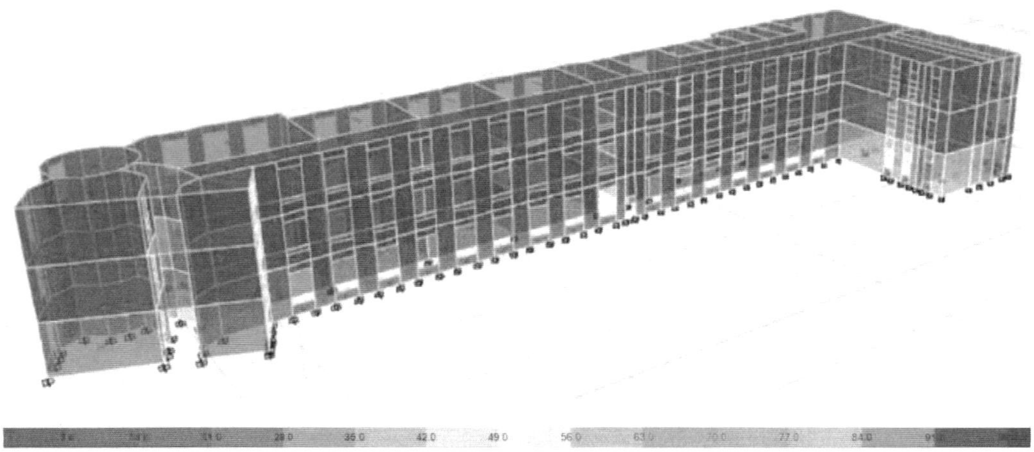

**Figure 18.** Graphical representation of the "relative work/volume" for structural masonry systems in the global Y direction of the structure.

**Table 4.** Information data related to figure 19 - figure 22.

| Results type | Seismic load combination | Storey range | Global direction | Graphical representation |
|---|---|---|---|---|
| Maximum storey displacements | design spectrum | all stories | X – blue color<br>Y – red color | figure 19 |
| Maximum storey shears | | | | figure 20 |
| Maximum storey overturning moments | | | | figure 21 |
| Maximum storey stiffness | | | | figure 22 |

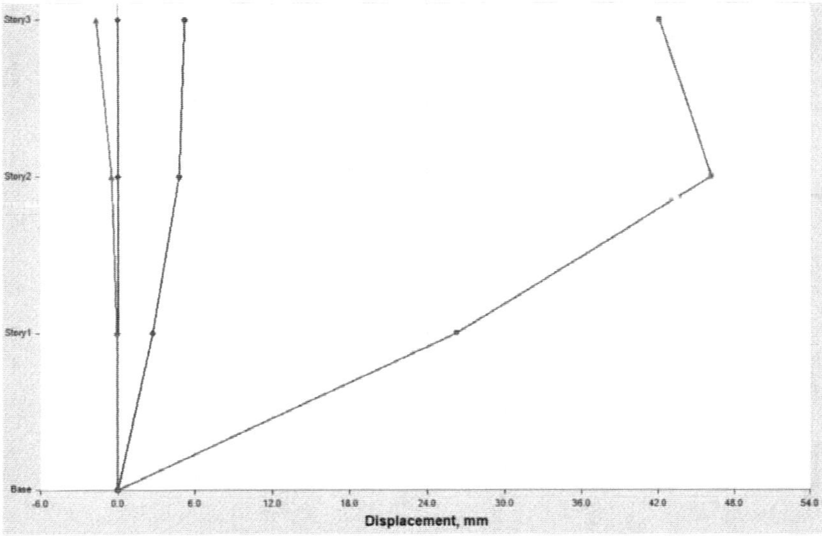


**Figure 19.** Representation of maximum storey displacements.

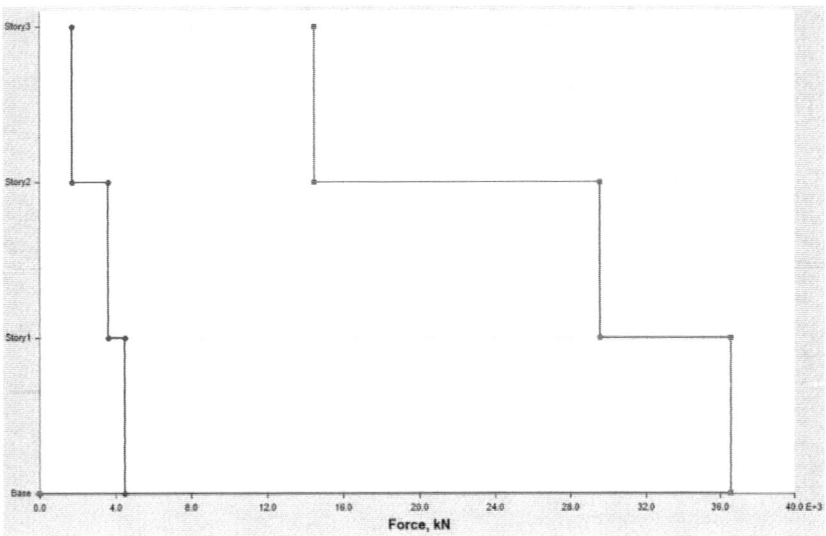


**Figure 20.** Representation of maximum storey shears.

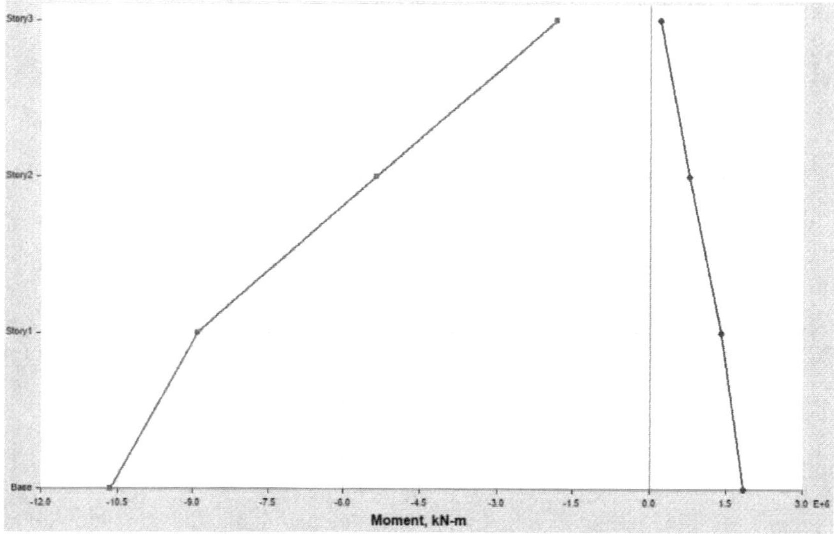


**Figure 21.** Representation of maximum storey overturning moments.

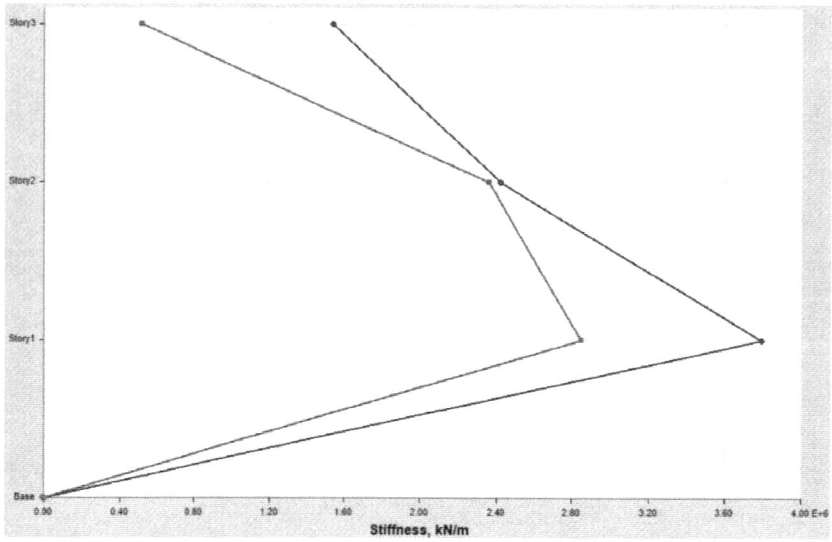


**Figure 22.** Representation of maximum storey stiffness.

In the second phase, the calculation regarding the strength capacity determination of the masonry shear walls was performed, depending on the failure mechanism associated with the capable efforts (see table 5).

According to P100-3 standard [9], three possible failure mechanisms were considered: eccentric compression failure (see equation (1)), horizontal sliding failure (see equation (2)) and diagonal cracking mechanism (see equation (3)):

- with the relation from the P100-3 norm [9], the design value of the shear force associated with the eccentric compression failure mechanism of an unreinforced masonry shear wall loaded by the „$N_d$" axial design force was calculated:

$$V_{f1} = \frac{N_d}{c_p \lambda_p} \cdot (1 - 1.15 \cdot \nu_d) \tag{1}$$

where: $\lambda_p = H_p/l_w$ – the form factor of the masonry wall; $H_p$ – masonry wall height; $l_w$ – masonry wall length; $c_p$ – coefficient that depends on the fixing conditions at the end regions of the masonry wall ($c_p$ = 2.0 for the console masonry wall, $c_p$ = 1.0 for the double fixed masonry wall at the end regions); $\sigma_0 = N_d/tl_w$ – stress corresponding to $N_d$ axial design force; $t$ – masonry wall thickness; $\upsilon_d = \sigma_0/f_d$ – normalized axial stress; $f_d$ – design value of the masonry compressive strength.

- the design value of the shear force associated with the horizontal sliding failure mechanism was calculated, using the formula:

$$V_{f21} = \frac{1.33}{CF\gamma_M} \cdot (f_{vko} \cdot \frac{l_{ad}}{l_c} + 0.4 \cdot \sigma_d) \cdot t \cdot l_c \tag{2}$$

where: $l_c = 1.5l_w - 3(M_d/N_d)$ – the length of the compressed area of the masonry wall cross section, taking into account the alternating seismic force effect; $M_d$ – design bending moment effort; $N_d$ – axial design force; $l_{ad} = 2l_c - l_w$ – the length of the masonry wall cross section on which the adhesion is active.

- the design value of the shear force associated with diagonal cracking mechanism was calculated, with the formula:

$$V_{f22} = \frac{t \cdot l_w \cdot f_{td}}{b} \cdot \sqrt{1 + \frac{\sigma_0}{f_{td}}} \tag{3}$$

where: $b$ – coefficient with $1.0 \le b = \lambda_p \le 1.5$ values; $f_{td}$ – masonry design strength to principal tension efforts.

In these conditions, the minimum value of the shear force associated with these three masonry failure mechanisms represents the capable strength value to this sectional effort (see equation (4) and equation (5)):

$$V_{f2} = \min(V_{f21}, V_{f22}) \tag{4}$$

$$V_{cap} = \min(V_{f1}, V_{f2}) \tag{5}$$

For the case: $V_{f1} < V_{f2}$ - unreinforced masonry shear walls with ductile behaviour was considered.
For the case: $V_{f2} < V_{f1}$ - unreinforced masonry shear walls with non-ductile (fragile) behaviour was considered.
All estimations in table 5 it were performed with ETABS software, respecting the provisions of the P100-3 normative [9].

**Table 5.** Design values of the capable shear force for masonry shear walls (section no.2).

| PN[a] | $V_{f1}$ (kN) | $V_{f21}$ (kN) | $V_{f22}$ (kN) | $V_{f2}$ (kN) | $V_{cap}$[b] (kN) | $V_{Ed}$ (kN) | Conclusion (RSW/NRSW)[c] |
|---|---|---|---|---|---|---|---|
| P1 | 1514.5 | 220.1 | 429.7 | 220.1 | 220.1 | 374.44 | RSW |
| P2 | 256.92 | 119.9 | 204.04 | 119.9 | 119.9 | 144.78 | RSW |
| P3 | 216.1 | 147 | 244.3 | 147 | 147 | 23.51 | NRSW |
| P4 | 106.7 | 110.80 | 170.57 | 110.80 | 106.7 | 46.92 | NRSW |
| P5 | 42.89 | 85.2 | 108.30 | 85.2 | 42.89 | 14.05 | NRSW |
| P6 | 42.74 | 85.2 | 108.11 | 85.2 | 42.74 | 10.15 | NRSW |
| P7 | 42.55 | 85.2 | 107.90 | 85.2 | 42.55 | 8.32 | NRSW |
| P8 | 42.52 | 85.2 | 107.90 | 85.2 | 42.52 | 6.60 | NRSW |
| P9 | 42.57 | 85.2 | 107.92 | 85.2 | 42.57 | 4.91 | NRSW |
| P10 | 43.44 | 85.2 | 108.94 | 85.2 | 43.44 | 7.98 | NRSW |
| P11 | 28.50 | 55.38 | 88.37 | 55.38 | 28.50 | 2.05 | NRSW |
| P12 | 28.53 | 55.38 | 88.41 | 55.38 | 28.53 | 2.58 | NRSW |
| P13 | 43.45 | 85.2 | 108.95 | 85.2 | 43.45 | 7.70 | NRSW |
| P14 | 42.56 | 85.2 | 107.90 | 85.2 | 42.56 | 4.58 | NRSW |
| P15 | 42.48 | 85.2 | 107.80 | 85.2 | 42.48 | 6.24 | NRSW |
| P16 | 42.43 | 85.2 | 107.75 | 85.2 | 42.43 | 7.96 | NRSW |
| P17 | 42.40 | 85.2 | 107.71 | 85.2 | 42.40 | 9.91 | NRSW |
| P18 | 42.35 | 85.2 | 107.65 | 85.2 | 42.35 | 12.20 | NRSW |

*Table 5 (continue)*

| | | | | | | | |
|---|---|---|---|---|---|---|---|
| P19 | 42.23 | 85.2 | 107.52 | 85.2 | 42.23 | 15.125 | NRSW |
| P20 | 41.97 | 85.2 | 107.21 | 85.2 | 41.97 | 20.44 | NRSW |
| P21 | 58.10 | 85.2 | 125.61 | 85.2 | 58.10 | 45.82 | NRSW |
| P22 | 38.68 | 63.9 | 103.05 | 63.9 | 38.68 | 81.83 | RSW |
| P23 | 8.59 | 31.95 | 48.34 | 31.95 | 8.59 | 49.54 | RSW |
| P24 | 11.32 | 34.08 | 55.83 | 34.08 | 11.32 | 53.98 | RSW |
| P25 | 515.88 | 234.3 | 376.16 | 234.3 | 234.3 | 78.68 | NRSW |
| P26 | 1663.91 | 387.66 | 683.34 | 387.66 | 387.66 | 75.34 | NRSW |
| P27 | 116.01 | 104.37 | 179.82 | 104.37 | 104.37 | 127.06 | RSW |
| P28 | 18.56 | 46.86 | 71.07 | 46.86 | 18.56 | 59.40 | RSW |
| P29 | 264.79 | 235.19 | 390.26 | 235.19 | 235.19 | 98.72 | NRSW |
| P30 | 104.51 | 103.52 | 169.64 | 103.52 | 103.52 | 48.85 | NRSW |
| P31 | 32.39 | 68.16 | 93.76 | 68.16 | 32.39 | 8.92 | NRSW |
| P32 | 796.74 | 234.3 | 493.73 | 234.3 | 234.3 | 297.69 | RSW |
| P33 | 418.19 | 170.4 | 357.01 | 170.4 | 170.4 | 126.55 | NRSW |
| P34 | 27.45 | 63.9 | 86.36 | 63.9 | 27.45 | 2.90 | NRSW |
| P35 | 30.11 | 63.9 | 90.38 | 63.9 | 30.11 | 45.96 | RSW |
| P36 | 59.30 | 63.9 | 134.71 | 63.9 | 59.30 | 16.79 | NRSW |
| P37 | 234.28 | 117.15 | 285.71 | 117.15 | 117.15 | 118.94 | RSW |
| P38 | 32.02 | 63.9 | 93.23 | 63.9 | 32.02 | 35.35 | RSW |
| P39 | 333.178 | 178.92 | 304.19 | 178.92 | 178.92 | 42.07 | NRSW |
| P40 | 24.922 | 63.9 | 82.46 | 63.9 | 24.922 | 7.26 | NRSW |
| P41 | 342.6 | 187.44 | 307.14 | 187.44 | 187.44 | 10.58 | NRSW |
| P42 | 24.02 | 63.9 | 81.05 | 63.9 | 24.02 | 5.013 | NRSW |
| P43 | 176.64 | 127.8 | 222.17 | 127.8 | 127.8 | 21.46 | NRSW |
| P44 | 1194.72 | 334.41 | 577.20 | 334.41 | 334.41 | 54.38 | NRSW |
| P45 | 61.15 | 87.33 | 128.87 | 87.33 | 61.15 | 5.75 | NRSW |
| P46 | 8.56 | 36.21 | 48.26 | 36.21 | 8.56 | 7.86 | NRSW |
| P47 | 0.31 | 12.8 | 10.43 | 10.43 | 0.31 | 2.64 | RSW |
| P48 | 41.52 | 72.42 | 106.18 | 72.42 | 41.52 | 32.05 | NRSW |
| P49 | 7.41 | 31.95 | 44.84 | 31.95 | 7.41 | 18.7 | RSW |
| P50 | 24.19 | 55.38 | 81.05 | 55.38 | 24.19 | 13.39 | NRSW |
| P51 | 39.17 | 72.42 | 103.10 | 72.42 | 39.17 | 14.18 | NRSW |
| P52 | 7.018 | 31.95 | 43.65 | 31.95 | 7.018 | 10.96 | RSW |
| P53 | 16.86 | 55.38 | 68.105 | 55.38 | 16.86 | 13.42 | NRSW |
| P54 | 37.42 | 76.68 | 100.94 | 76.68 | 37.42 | 17.19 | NRSW |
| P55 | 81.5 | 95.85 | 149.16 | 95.85 | 81.5 | 58.31 | NRSW |
| P56 | 700.6 | 195.96 | 365.90 | 195.96 | 195.96 | 280.05 | RSW |
| P57 | 47.76 | 63.9 | 116.44 | 63.9 | 47.76 | 63.59 | RSW |
| P58 | 34.29 | 57.51 | 97.56 | 57.51 | 34.29 | 48.87 | RSW |
| P59 | 4.29 | 27.69 | 34.315 | 27.69 | 4.29 | 19.95 | RSW |
| P60 | 4.37 | 27.69 | 34.63 | 27.69 | 4.37 | 16.72 | RSW |
| P61 | 4.41 | 27.69 | 34.74 | 27.69 | 4.41 | 17.54 | RSW |
| P62 | 4.38 | 27.69 | 34.63 | 27.69 | 4.38 | 18.61 | RSW |
| P63 | 4.36 | 27.69 | 34.57 | 27.69 | 4.36 | 20.53 | RSW |
| P64 | 4.35 | 27.69 | 34.48 | 27.69 | 4.35 | 24.81 | RSW |
| P65 | 6.54 | 27.69 | 42.20 | 27.69 | 6.54 | 28.32 | RSW |
| P66 | 2472.85 | 291.1 | 543.41 | 291.1 | 291.10 | 202.01 | NRSW |
| P67 | 2430.19 | 291.1 | 537.51 | 291.1 | 291.10 | 147.05 | NRSW |
| P68 | 125.47 | 69.22 | 120.96 | 69.22 | 69.22 | 41.95 | NRSW |
| P69 | 1634.78 | 221.88 | 451.49 | 221.88 | 221.88 | 144.50 | NRSW |
| P70 | 30522.7 | 1699.7 | 2914.84 | 1699.7 | 1699.7 | 220.88 | NRSW |
| P71 | 183.49 | 127.8 | 227.24 | 127.8 | 127.8 | 30.76 | NRSW |
| P72 | 2331.7 | 447.3 | 813.23 | 447.3 | 447.3 | 49.54 | NRSW |
| P73 | 2226.6 | 438.78 | 794.01 | 438.78 | 438.78 | 84.19 | NRSW |
| P74 | 2693.29 | 440.91 | 898.02 | 440.91 | 440.91 | 231.52 | NRSW |
| P75 | 4524.15 | 545.28 | 1192.34 | 545.28 | 545.28 | 209.15 | NRSW |
| P76 | 1192.04 | 399.38 | 890.19 | 399.38 | 399.38 | 523.13 | RSW |
| P77 | 1868.73 | 488.12 | 1136.64 | 488.12 | 488.12 | 576.86 | RSW |
| P78 | 1119.66 | 492.56 | 800.80 | 492.56 | 492.56 | 190.39 | NRSW |
| P79 | 334.41 | 306.19 | 434.75 | 306.19 | 306.19 | 196.56 | NRSW |

IOP Conf. Series: Materials Science and Engineering 1242 (2022) 012035 doi:10.1088/1757-899X/1242/1/012035

*Table 5 (continue)*

| | | | | | | | |
|---|---|---|---|---|---|---|---|
| P80 | 100.39 | 95.85 | 167.64 | 95.85 | 95.85 | 114.03 | RSW |
| P81 | 71.93 | 76.68 | 143.70 | 76.68 | 71.93 | 122.20 | RSW |
| P82 | 293.98 | 176.79 | 283.97 | 176.79 | 176.79 | 149.70 | NRSW |
| P83 | 351.387 | 189.57 | 311.10 | 189.57 | 189.57 | 60.513 | NRSW |
| P84 | 150.47 | 91.59 | 240.09 | 91.59 | 91.59 | 356.53 | RSW |
| P85 | 2054.97 | 345.06 | 852.81 | 345.06 | 345.06 | 776.84 | RSW |
| P86 | 1032.55 | 258.80 | 468.26 | 258.80 | 258.80 | 338.72 | RSW |
| P87 | 962.144 | 258.80 | 448.91 | 258.80 | 258.80 | 358.02 | RSW |
| P88 | 942.26 | 258.80 | 443.48 | 258.80 | 258.80 | 353.88 | RSW |
| P89 | 891.88 | 258.80 | 429.78 | 258.80 | 258.80 | 335.67 | RSW |
| P90 | 599.231 | 172.53 | 287.78 | 172.53 | 172.53 | 262.90 | RSW |
| P91 | 351.03 | 151.76 | 272.69 | 151.76 | 151.76 | 160.64 | RSW |
| P92 | 145.36 | 67.45 | 116.21 | 67.45 | 67.45 | 53.21 | NRSW |
| P93 | 137.84 | 67.45 | 112.72 | 67.45 | 67.45 | 52.10 | NRSW |
| P94 | 879.82 | 293.94 | 493.54 | 293.94 | 293.94 | 230.64 | NRSW |
| P95 | 334.94 | 97.98 | 177.91 | 97.98 | 97.98 | 84.71 | NRSW |
| P96 | 355.01 | 97.98 | 184.50 | 97.98 | 97.98 | 96.50 | NRSW |
| P97 | 790.41 | 220.45 | 412.36 | 220.45 | 220.45 | 269.28 | RSW |
| P98 | 164.015 | 74.55 | 159.10 | 74.55 | 74.55 | 30.11 | NRSW |
| P99 | 106.53 | 61.77 | 126.50 | 61.77 | 61.77 | 17.08 | NRSW |
| P100 | 161.71 | 77.745 | 154.52 | 77.745 | 77.745 | 20.14 | NRSW |
| P101 | 190.56 | 85.20 | 167.17 | 85.20 | 85.20 | 19.08 | NRSW |
| P102 | 81.37 | 55.38 | 109.44 | 55.38 | 55.38 | 6.73 | NRSW |
| P103 | 196.9 | 85.20 | 170.95 | 85.20 | 85.20 | 10.74 | NRSW |
| P104 | 983.4 | 319.5 | 520.00 | 319.5 | 319.5 | 202.08 | NRSW |
| P105 | 34.70 | 68.16 | 97.04 | 68.16 | 34.70 | 37.54 | RSW |
| P106 | 61.90 | 85.20 | 129.82 | 85.20 | 61.90 | 38.81 | NRSW |
| P107 | 47.10 | 72.42 | 113.42 | 72.42 | 47.10 | 52.33 | RSW |

[a] PN – **P**ier **N**ame (see figure 23).
[b] $V_{cap} = min(V_{f1}, V_{f2})$ – capable shear force.
[c] (RSW/NRSW) – **R**etrofitting **S**hear **W**alls/ **N**ot **R**etrofitting **S**hear **W**alls.

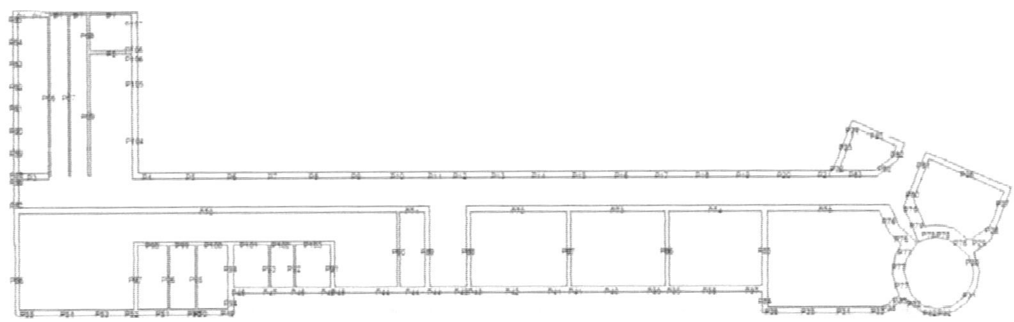

**Figure 23.** Numbering of masonry piers for section no. 2 of the structural masonry system.

**3. Lateral strength capacity verification of the structural masonry shear walls with rigid slabs in horizontal plane for Braila ECH – Pavilion B, section no.2**

The level 2 methodology was used to verify the lateral strength capacity of the masonry shear walls. Thus, the values of the sectional design efforts with strength capacity calculated for each masonry shear wall, were compared for ULS. The results are summarized in table 5.

Furthermore, the value of the „$R_3$" indicator was established with the necessity to confirm the seismic risk class of the structure (see equation (6)) (mentioning the existence of "$R_1$" and "$R_2$" indicators for a complete technical expertise). The "$R_1$" and "$R_2$" indicators have not been established because the complete technical expertise of the masonry structure is not an objective of this paper (article).

The determination of the „$R_3$" indicator for the whole building was calculated with the relation (according to P100-3 norm [9]):

$$R_3 = \frac{\sum_{jd} V_{fd} + \sum_{kf} V_{ff}}{\sum F_{bi}} \tag{6}$$

where: $V_{fd}$ – design value of the shear force capacity for masonry shear walls with ductile failure; $V_{ff}$ – design value of the shear force capacity for masonry shear walls with fragile failure; $F_{bi} = V_{Ed}$.

Applying equation (6), the „$R_3$" indicator for X global structure direction is equal to:

$$R_3^X = \frac{8622.465}{3090.6} = 2.789 \tag{7}$$

Applying equation (6), the „$R_3$" indicator for Y global structure direction is equal to:

$$R_3^Y = \frac{6379.48}{7345.8} = 0.868 \tag{8}$$

In these conditions, the seismic response of the structure is favorable and the lateral structure does not present major deficiencies. However, it can be observed that, for each principal direction of the masonry structure, a different classification in the seismic risk class is made.
Thus, according to P100-3 norm [9], the risk class associated with „$R_3$" indicator for X direction is IV and for Y direction is III.
Finally, the representation of shear masonry walls that need retrofitting intervention can be studied in figure 24.

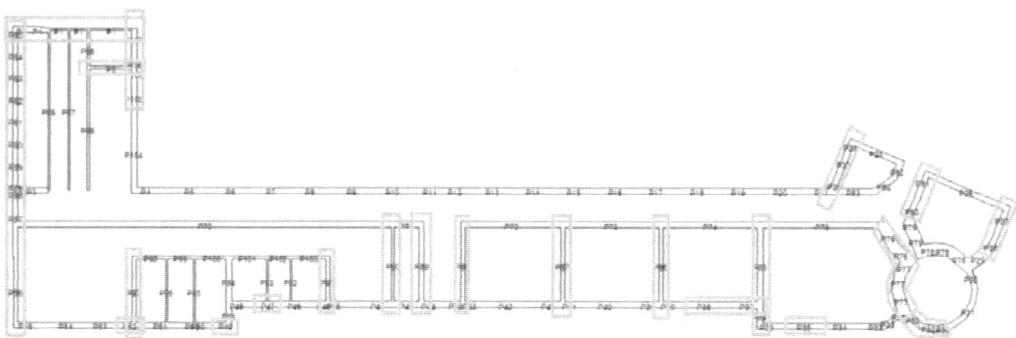

**Figure 24.** Representation of masonry shear walls with exceeding strength capacity and required retrofitting intervention.

## 4. Structural retrofitting intervention suggestions for Braila ECH – Pavilion B, section no.2

Due to the structural complexity and special regime of use, the "classic retrofitting interventions" [14] of the masonry shear walls with exceeding lateral strength capacity cannot be considered valid.

Moreover, the moisture-sensitive soil has produced differentiated settlements, these being associated with stone and brick foundations and being compounded by the presence of the basement only in certain areas of the building (partial basement).

Another aspect to be taken into account in this consideration (non-utilization the classic retrofitting interventions) is the increase of the lateral structure mass associated to architectural modifications (which are not valid for historical monuments).

In these conditions, it is considered useful to retrofit the masonry shear walls with "composite materials" [15-26] (ex.: externally bonded Fiber Reinforced Polymer (FRP) reinforcement – see figure 25).

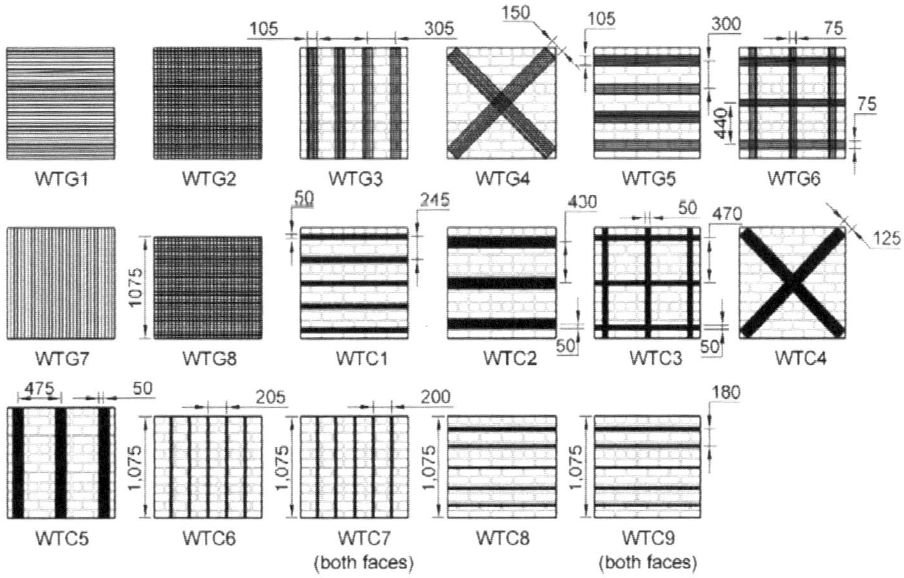


**Figure 25.** Retrofitting methods with FRP (**F**iber **R**einforced **P**olymer) reinforcement [27].

This methodology positively influences the seismic response of the structure by increasing the lateral strength capacity of the masonry shear walls, without negative implications regarding the foundations or the increase of the mass of the building.

Moreover, this retrofitting method would not influence the architectural characteristics of the historical monument Braila ECH – Pavilion B.

## 5. Conclusions

Historical monuments require a special volume of analytical calculation and increased attention to the establishment of retrofitting interventions. In the case of Braila ECH – Pavilion B, it the masonry shear walls with a reduced lateral strength capacity to the action of horizontal seismic force were analysed and a retrofitting method of structural intervention for this type of walls that would not affect the infrastructure construction was proposed.

Moreover, by choosing composite materials to rehabilitate masonry walls susceptible to horizontal seismic action, further advantages can be observed. Besides the fact that composites are generally easier to apply and have a lower workload than "classic" materials, it is not necessary to evacuate hospital personnel or patients. In the current context of the COVID-19 pandemic, this may prove a crucial aspect.

The analytical calculation proved to be efficient and very important, such that the failure mechanisms of the masonry shear walls were correctly determined.

In these conditions, the lateral structure registered in some masonry shear walls all three failure mechanisms: eccentric compression failure, horizontal sliding failure and diagonal cracking mechanism.

## Acknowledgments

This research paper is sponsored by "Gheorghe Asachi" Technical University of Iasi with grant number GI/R16_Drd/2021.

## References

[1] Chen L K *et al.* 2021 *Automation in Construction* **124** 103555
[2] Gbadamosi A Q, Oyedele L, Olawale O and Abioye S 2020 *Progress in Disaster Science* **8** 100130
[3] Aghabeigi P and Farahmand-Tabar S 2021 *Engineering Structures* **240** 112418
[4] Bertolesi E, Adam J M, Rinaudo P and Calderon P A 2019 *Engineering Structures* **180** 67-88

[5] Milani G 2013 *Engineering Failure Analysis* **34** 761-778
[6] Foraboschi P 2013 *Engineering Failure Analysis* **33** 281-314
[7] Gunes B, Cosgun T, Sayin B and Ceylan O 2019 *Engineering Failure Analysis* **105** 527-544
[8] Ponte M, Bento R, Costa A A, Quelhas B, Guedes J M, Ilharco T and Lopes V 2021 *International Journal of Disaster Risk Reduction* **60** 102172
[9] MDRAPFE 2019 *Cod de proiectare seismică - Partea a III-a - Prevederi pentru evaluarea seismică a clădirilor existente, indicativ P100-3/2019* (București: Monitorul Oficial al României)
[10] Budescu M and Ciongradi I 2014 *Inginerie Seismică* (Iași: Politehnium)
[11] Stratan A 2014 *Dinamica structurilor și inginerie seismică. Note de curs* (Timișoara: Universitatea Politehnica Timișoara)
[12] Paulay T and Priestley M J N 1992 *Seismic design of reinforced concrete and masonry buildings* (USA: John Wiley & Sons)
[13] MDRAP 2013 *Cod de proiectare seismică - Partea I - Prevederi de proiectare pentru clădiri, indicativ P100-1/2013* (București: Monitorul Oficial al României)
[14] Arsenie G *et al.* 1997 *Soluții de Consolidare a Construcțiilor Avariate de Cutremure* (București: Editura Tehnică)
[15] Budescu M *et al.* 2001 *Reabilitarea construcțiilor* (Iași: Editura Vesper)
[16] Gattulli V, Lampis G, Marcari G and Paolone A 2014 *Engineering Structures* **75** 604-618
[17] Wang X, Ghiassi B, Oliveira D V and Lam C C 2017 *Engineering Structures* **134** 11-24
[18] Carozzi F G, Milani G and Poggi C 2014 *Composite Structures* **107** 711-725
[19] D'Ambrisi A, Mezzi M, Feo L and Berardi V P 2014 *Composite Structures* **113** 264-271
[20] Carozzi F G, Poggi C, Bertolesi E and Milani G 2018 *Composite Structures* **187** 466-480
[21] Capozucca R and Magagnini E 2020 *Composite Structures* **235** 111735
[22] Milani G, Fagone M, Rotunno T, Grande E and Bertolesi E 2020 *Composite Structures* **241** 112074
[23] Santis S and Felice G 2021 *Composite Structures* **275** 114508
[24] Razavizadeh A, Ghiassi B and Oliveira D V 2014 *Construction and Building Materials* **64** 387-397
[25] Bernat-Maso E, Gil L and Escrig C 2015 *Construction and Building Materials* **84** 169-183
[26] Santis S, AlShawa O, Felice G, Gobbin F, Roselli I, Sangirardi M, Sorrentino L and Liberatore D 2021 *Construction and Building Materials* **307** 124962
[27] Mahmood H and Ingham J M 2011 *Journal of Composites for Construction* **15** (5)

# Optimization method based on genetic algorithm for the mass compensation of the rotor on-line dynamic balance system

**Q L Song[1], Z Wang[1]* and L Liang[1]**

[1] School of Mechanical Engineering, Shenyang Jianzhu University, Shenyang, China

* juven1126@163.com

**Abstract.** The rotor is the key component of the rotary drilling rig, it can realize the rotation and feed movement of the drill. the rotor unit of the rotary drilling rig should take on-line dynamic. Because unbalanced faults would lead to vibration. In order to further improve the pricision of on-line dynamic balance, an optimization strategy based on Genetic Algorithm for the mass compensation of the rotor on-line dynamic balancing system is proposed. A mass compensation optimization model is established. By the way of taking the phase of counterweight movement as the optimization variable. At the same time,taking the residual stress under dynamic balance as the optimization objective. According to the principle of Genetic Algorithm, the movement phase of the counterweight can be calculated when the residual force reaches the minimum value. By performing simulation analysis, the residual force can be calculated under different rotating speeds and different amount of unbalance, the residual force after optimizing balance can be reduced greatly. The simulation results showed that the residual force of the rotor after optimizing balance can decrease by 95.74% on average. Therefore, an optimization strategy based on Genetic Algorithm for the mass compensation of the rotor can effectively improve dynamic balance precision and operating performance of rotary drilling rig.

## 1. Introduction.

Rotary drilling rig, as a kind of construction machinery, is the key equipment for construction and underground exploration in construction industry. It takes advantage of built-in rotating device. During operation, fast drilling can be achieved by driving the auger stem [1]. To realize the rotation and feed movement of the drill, rotor unit is the key component of the rotary drilling rig. The balance condition of the rotor directly affects the normal operation of the rotary drilling rig. At present, rotor unbalance vibration is still an important cause of rotating machinery failure, and the loss caused by it occupies a large proportion of the total loss in machinery production. In order to ensure production safety and economic efficiency, there is an urgent need to conduct dynamic balancing studies on rotors of rotary drilling rigs. The vibration of the rotor caused by external force will affect the stability and the accuracy of the drilling rig, even the drill itself. The rotor unit should take on-line dynamic balance. Because unbalance fault would lead to vibration problem. On-line dynamic balancing means that the rotor can be balanced without operating the unit under the normal working state of the rotating machine. In order to realize the high-precision balance quickly and efficiently and ensure the real-time monitoring of rotor vibration during operation without shutdown. Therefore, on-line dynamic balance is the efficient solution for the unbalanced of drilling rig.

Recently, the on-line dynamic balance technology for rotor has become a research hotspot in the world. Fang Jiancheng [2] and others proposed a high-precision number online dynamic balance inspection device based on FPGA. It can carry out high-precision online inspection of rotor unbalance and provide data analysis for rotor dynamic balance. Gao Jinji [3] and others designed an on-line automatic balance controller for rotating machinery. It has the advantages of strong anti-interference ability, high accuracy and fast response. Jiang Hongyan [4] proposed a new type of high-speed rotor dynamic balancing head, based on the coupling effect of electromagnetic field. It can obtain the radial electromagnetic force equivalent to the centrifugal force of the rotor at the plane of weighting and achieve high precision rotor balancing efficiently. Sun Changjing and Bai Caibo [5] jointly proposed the liquid type high-speed rotor on-line dynamic balance head. It controls the distribution of plane mass with the help of the flow of trace liquid in the balance head to achieve on-line dynamic balance. Mei Xuesong [6] proposed online balancing device for rotor of high-speed rotating machinery. It can reduce the spindle vibration and ensure the stable operation of high-speed rotating machinery. He Chengshan [7]and others proposed an on-line dynamic balance method of based on the principle of parallel decomposition of forces. Zhang Ke [8] and others tested the balance characteristics of an electromagnetic slip ring type on-line dynamic balance device, and carried out on-line dynamic balance experimental research based on the influence coefficient method. It laid a foundation for the research of double-sided dynamic balance of high-speed spindle. Tu Wei [9] proposed an optimization method for the Mass Compensation through the optimization study of online dynamic balancing. It achieves higher accuracy through the movement of counterweights.

In summary, the research on on-line dynamic balance technology has gradually matured, but for the mechanical dynamic balance device, how to improve its accuracy and efficiency when the device is balanced is still an urgent problem to be solved. To achieve higher accuracy and efficiency of on-line dynamic balance, The paper will take the rotor unit as the research object, establish the optimization model for the mass compensation of the rotor on-line dynamic balancing system based on the principle of force balance model, and solve the optimal movement strategy of counterweight based on Genetic Algorithm. Consequently, it is of far-reaching significance to carry out research on rotor online dynamic balancing quality compensation optimization model to improve production efficiency, avoid production accidents and promote the development of rotary drilling rigs in the direction of intelligence and high speed.

## 2. Optimization method based on Genetic Algorithm for the Mass Compensation

### *2.1. Optimization method for the Mass Compensation*

Balance head method is one of the important methods for on-line dynamic balance of rotor. Its core content is to monitor the vibration of the rotor in real time through the sensor when the rotor is running continuously. By adjusting the size and phase of the two counterweights in the balance head when the sensor detects that the unbalanced vibration reaches the set threshold, the purpose is to make the correction balance equal to the original unbalance and the two are in the opposite direction [9].

An optimization model for the mass compensation was established by taking the phase of counterweight movement as the optimization variable and the residual force under dynamic balance as the optimization objective [10]:

$$\min f(\varphi_A,\varphi_B)=\left(\begin{array}{l}\left(2F'\cos\left(\dfrac{\pi(\varphi_B-\varphi_A)}{360}-\dfrac{\theta_B-\theta_A}{2}\right)\right)^2+W^2+2F' \\ \cos\left(\dfrac{\pi(\varphi_B-\varphi_A)}{360}-\dfrac{\theta_B-\theta_A}{2}\right)W\cos\left(\dfrac{\pi(\varphi_B-\varphi_A)}{360}-\dfrac{\theta_B-\theta_A}{2}-\eta\right)\end{array}\right)^{1/2} \tag{1}$$

Where $F'$ is the centrifugal force that is produced by counterweight A and counterweight B.( $\varphi$ is between A and B)

$$F_A=m_A\omega^2 r_A \tag{2}$$

$$F_B = m_B \omega^2 r_B \tag{3}$$

W is the inertia force of the unbalanced mass, $\theta_A$ and $\theta_B$ are represent the initial phase of counterweight A and B in the mechanical balance head, $\varphi_A$ and $\varphi_B$ are the movement angle value of counterweight A and B in the mechanical balance head in poplar coordinates, $\eta$ is the phase of the inherent unbalance of the rotor.

### 2.2. *Optimization solution based on Genetic Algorithm.*

Genetic Algorithm is to find the optimal solution by simulating the biological evolution process of natural selection and genetic principles of animals and plants in nature [11]. As a random search algorithm, when solving optimization problems, group search can be carried out without auxiliary information. In the face of complex phenomena, it won't easily fall into local optimization and can find the global optimal solution [12].

Here defining the mathematical model of Genetic Algorithm as a 7-tuple $GA = (C, N_{pop}, N_{gen}, f, \Omega, \psi, \Gamma)$. It respectively represents the core contents of Genetic Algorithm: coding method, initial population, the number of iterations, individual fitness, selection, crossover and variation.

For the binary single objective optimization problem studied in this paper, Genetic Algorithm can search the best position of two counterweights from the global point of view, so as to minimize the residual unbalance force and further optimize the dynamic balance of the rotor. The specific steps are as follows:

(1) Set the initial population, include population size is N, initial iteration m=1 and total number of iterations; maximum evolution algebra is M, probability of crossover and variation.

(2) After binary coding, N sets of initial populations are randomly generated $(\varphi_A, \varphi_B)$.

(3) Calculate fitness of individuals in populations according to the fitness function: $f(\varphi_A, \varphi_B)_{n,m}$ (n is the number of individuals, where n=1,2,3,······N; m is current iterations, where m=1,2,3······M); Thus the probability of individual being selected can be obtained:

$$p_{n,m} = f(\varphi_A, \varphi_B)_{n,m} / \sum_{n=1}^{N} f(\varphi_A, \varphi_B)_{n,m} \tag{4}$$

and cumulative probability can be obtained:

$$P_{n,m} = \sum_{n=1}^{N} p_{n,m} \tag{5}$$

Then r is the random number, which is taken in the range [0,1], if r < $P_{n,m}$, the n-th individual will be passed on to the next generation at the m-th iteration—otherwise, the n-th individual will be eliminated.

(4) Select some individuals according to the probability of crossover and variation from the population, then achieve crossover and variation according to the prescribed principles and produce new populations.

(5) Let m=m+1, if m=M, algorithm end—otherwise turn to step 2 the algorithm continues to run.

(6) Output the best solution after optimization as the minimum value of the residual unbalance force and output the movement of counterweight $(\varphi_A, \varphi_B)$.

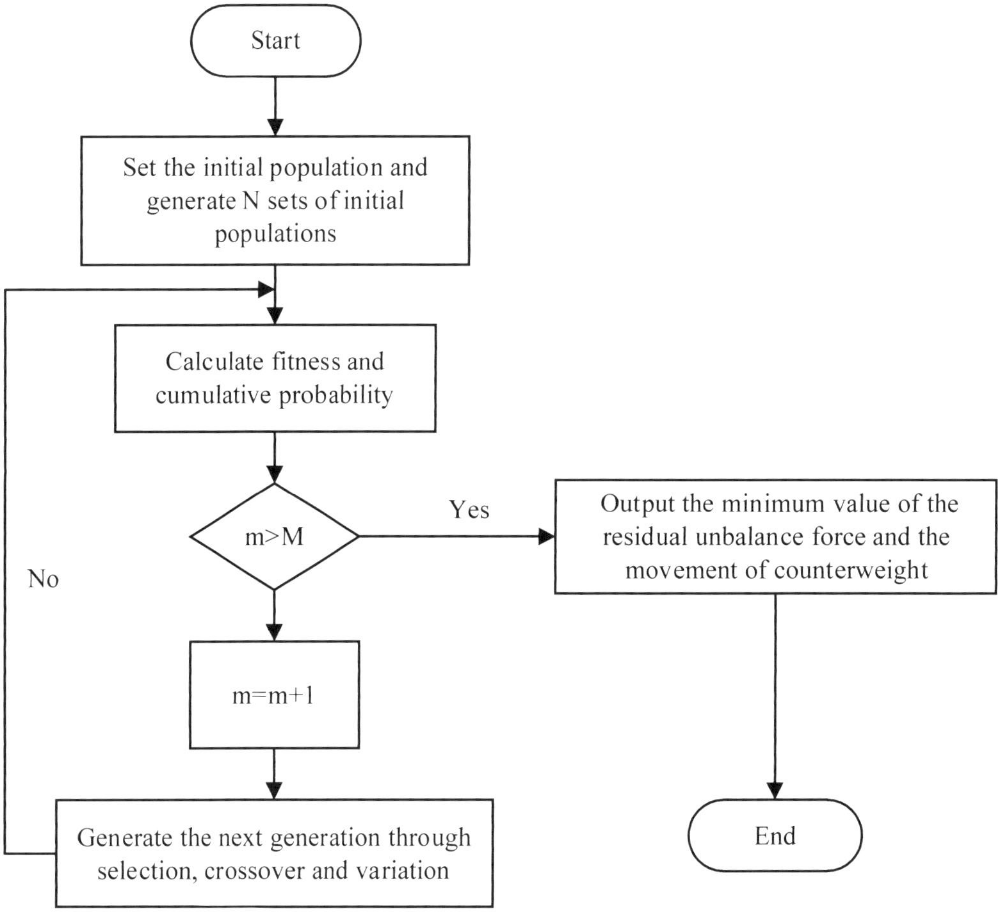


**Figure 1.** Genetic Algorithm flow chart.

## 3. Optimization simulation of rotor online dynamic balance

An online dynamic balance optimization simulation experiment was established. In order to solve the moving phase of the counterweight at different rotate speeds and analyze the optimization results of the original unbalance of the rotor. This simulation research is mainly aimed at the mechanical dynamic balance head. The way to achieve the rotar dynamic balance is the force balance method. Since the initial unbalance of the balance head is zero, under the premise of no load, the inherent balance of the rotor is simulated by the external unbalance. The moving strategy of the counterweight is solved by the optimization model and applied to the simulation, and the feasibility of the scheme is verified by comparing the unbalanced force of the rotor before and after the optimization balance.

Suppose the unbalances of 8.5 g·mm and 22.0 g·mm are added and placed at the positions of 4 mm and 1 mm respectively. Six different rotate speeds of 1000 r/min, 1500 r/min, 2000 r/min, 2500 r/min, 3000 r/min and 3500 r/min are selected to simulate the inherent unbalance of the rotor. In order to avoid excessive rotor vibration in the initial conditions, the initial phase of the counterweight is set to 0° and 180°. The optimal moving phase of the two counterweights is calculated based on the mass compensation optimization model of the genetic algorithm as shown in table 1.

**Table 1.** Phase of the counterweight after optimization.

| Speed/(r/min) | Phase of 8.5 *g·mm* Mass/° | | Phase of 22.0 *g·mm* Mass/° | |
|---|---|---|---|---|
| | A | B | A | B |
| 1000 | 98 | 262 | 285 | 155 |
| 1500 | 102 | 265 | 40 | 270 |
| 2000 | 98 | 260 | 65 | 180 |
| 2500 | 99 | 262 | 220 | 335 |
| 3000 | 98 | 261 | 120 | 230 |
| 3500 | 99 | 261 | 160 | 280 |

Table 1 can get the balance phases that the counterweights A and B should reach under different rotate speeds and different unbalanced amounts. Since the mass of the counterweight is much greater than the external unbalanced mass, under the same unbalanced amount, even if the rotate speed is different, the balance phase of the counterweight is almost the same, which means that the correction mass can move in a small range to realize the rotor movement balance. The phase calculated by the optimization model is the theoretically optimal position of the calibration mass. However, due to the complexity of the actual situation, it is difficult for the dynamic balancing device to achieve the balance accuracy of the calibration mass under automatic balance, and manual adjustment is required to make the counterweight reach the above optimal position.

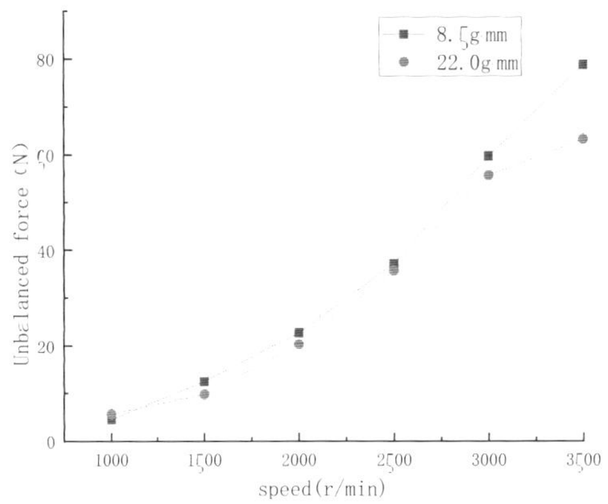


**Figure 2.** The unbalance force before the optimized balance.

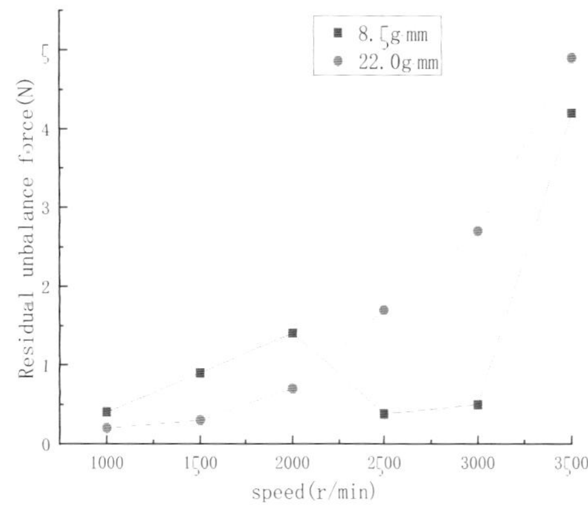


**Figure 3.** The residual unbalance force after the optimized balance.

Figure 2 shows the unbalanced force of the rotor before optimizing the balance. When the rotate speed is low, the difference in centrifugal force produced by different unbalanced masses is small. As the speed increases, the unbalanced force begins to increase, and the centrifugal force generated by a larger unbalanced amount increases more obviously. Due to the different positions of unbalanced masses during the simulation study, the rotation radius of 8.5 g·mm unbalanced placement is four times that of 22.0 g·mm, so the unbalanced force generated by the small unbalanced amount is greater. When the rotary drilling rig is running at high speed, if the rotor is not dynamically balanced on-line, a great unbalanced force will be generated, causing the drilling rig to vibrate, thereby affecting the working accuracy and causing great harm to the drilling rig.

Figure 3 shows the residual unbalanced force after optimized balance. The minimum residual force is 0.07 N, which is caused by an unbalance of 22 g·mm externally. The speed at this time is 1000 r/min. The maximum residual force is 4.9 N, which is caused by an unbalance of 22 g·mm, and the speed at this time is 3500 r/min. Comparing figure 2 with figure 3, the residual unbalanced force is compared with the unbalanced force before the optimized balance, the unbalanced force has been greatly reduced, and the rotor unbalanced force has been reduced by an average of 95.74% after the optimization of the mass compensation. However, as the speed increases, the residual unbalanced force increases significantly, indicating that the speed has an effect on the balance ability of the balancing device. The higher the speed, the greater the decrease in the balance ability. It can be seen from the overall downward trend that the phase of the corrected mass obtained by this optimization method is correct, and the rotor can be effectively balanced.

## 4. Conclusions.

An optimization method based on Genetic Algorithm for the mass compensation of the rotor on-line dynamic balance system is proposed. It targets at the research on the dynamic balance of rotor in rotary drilling rig. It bases on fully learning the on-line dynamic balance theory proposed by domestic and foreign scholars. It will achieve the goal of making the balance effect better when the rotor rotates. A mass compensation optimization model was established by taking the phase of counterweight movement as the optimization variable and the residual force under dynamic balance as the optimization objective.

Through adding unbalanced quantity of 8.5 g·mm and 22.0 g·mm and selecting 6 different speed to simulate the inherent unbalance quantity of rotor. Taking advantage of genetic algorithm in finding the global optimal solution, the minimum value of residual unbalance force and the movement value of counterweight are obtained. Through the analysis of the unbalanced force image before balance optimization, it is found that the speed has a great influence on the unbalanced force of the rotor. The higher the speed, the greater the unbalanced force. The residual unbalanced force is compared with the unbalanced force before the optimized balance, the unbalanced force has been greatly reduced, and the rotor unbalanced force has been reduced by an average of 95.74% after optimizing balance.

Therefore, the optimal phase of balanced rotor can be obtained by mass compensation optimization model, the unbalanced force during the rotation of the rotor can be greatly reduced, improve the operation accuracy of rotary drilling rig effectively. It lays a foundation for reducing the vibration, loss and service life of rotary drilling rig, and also provides support for the working quality. It has certain application value in the construction industry. In the future, the model will be further optimized to achieve more accurate dynamic balance.

## Acknowledgements.

This work was supported by National Science Foundation of China: 51805337, National Science Foundation of China: 52175107, Key Laboratory of Vibration and Control of Aero-Propulsion System, Ministry of Education, Northeastern University: VCAME202008 and Scientific Research Project of Liaoning Provincial Education Department: lnjc202010.

## References.

[1] Tang Y Q and Shu D W 2010 Optimum Design of Spiral Drill Pipe on Rotary Drill *Tool Engineering* **44** 76-80
[2] Fang J C, Han F J, Liu H, Liu G and Li J K 2007-09-19 High accuracy number online dynamic balance detecting device based on FPGA Bei Jing CN101038228A
[3] Gao J J, Zhao Q L, Jiang Z N, Wang H Q, Yao J F and Xu L 2012-06-27 On-line automatic balancing controller of rotating machinery Bei Jing CN102520631A
[4] Jiang H Y and Wang J Y 2007 Novel on-line dynamic balance head of high-speed rotor *New Technology & New Process* **12** 40-43
[5] Sun C J and Bai C B 2011-12-21 Liquid type on-line dynamic balance head of high-speed rotator Zhe Jiang CN202083521U

[6] Mei X S, Du Z, Jiang G D, Tao T, Dong J X and Zhang Y Su 2010-03-17 Rotor online automatic balancing device of high-speed rotating machine Shan Xi CN101672342
[7] He C S, Song J Y and Chen Y W 2012-07-04 Fan rotor on-line dynamic balance method based on force parallel decomposition principle Gan Su CN102539071A
[8] Zhang K, Zhang C Y and Zhang L X 2018 Performance Analysis and Experiment of Electromagnetic Ring Balancer During Operation *Journal of Vibration,Measurement & Diagnosis* **38** 34-38
[9] Tu W 2019 Study on the High Speed Spindle Online Dynamic Balance Vibration Signal and Optimization of Dynamic Balance Regulating Strategy *Shen Yang Jian Zhu University*
[10] Wang Z, Zhang B and Zhang K 2021 Optimization of Quality Compensation Strategy fro On-Line Dynamic Balance of Spindle System *Journal of Vibration,Measurement & Diagnosis* **41** 164-169+206
[11] Li Y M 2012 An Improved Genetic Algorithm and Its Application *South China University of Technology*
[12] Mi B F 2014 The improvement research on evolution strategy of genetic algorithm *Northeast Agricultural University*

# Improving the properties of mortars by using recycled paper ash flakes

**M L Țințișan[1*], A C Siomin[1], A Zaharie[1] and D L Manea[1]**

[1] Technical University of Cluj-Napoca, Faculty of Civil engineering, Romania

* coniac.loredana@gmail.com

**Abstract**. This study addresses the need to improve life on earth by finding a way to reintroduce ubiquitous waste back into the system. On a hand, there is the amount of cellulosic waste that can no longer be reintroduced into the paper recycling process. On the other hand, the amount of cement used globally, is growing alarmingly. My study seeks the balance of a mix of these two materials, which would generate desirable effects both from an engineering and ecological point of view. Paper ash, is a material made of particles of various sizes and can play various roles in the composition of mortars. The efficacy of particles up to 0.025 mm in size (powder ash) as a cement substitute and the efficiency of particles in size between 0.025 and 8 mm (flakes) as a pozzolanic additive are demonstrated. The results of the mechanical tests to which the samples were subjected show the possibility of reducing the amount of cement used in standard recipes by adding ash from recycled paper, without compromising the physical and mechanical properties of the mortars. The cement economy can rise up to 20 percent while the mechanical strength increases by 5-10 percent compared to the basic recipe.

## 1. Introduction

Studies regarding the integration of different types of waste into construction materials have been carried out since ancient times, as construction has always been one that encompasses enormous amounts of energy and materials. In recent history, due to the evolution of mankind, the diversity of waste embedded in construction works is beginning to grow. With the Industrial Revolution of the 18th century, we began to use not only organic waste, but also residues from new sectors of industry, large producers of waste.

A new trend in the residential sector is the return to the origins. The construction industry is increasingly experiencing this return to original, organic, eco-friendly materials, in response to the ecological problems that the planet is facing. This fact emphasizes the need to study the characteristics of these materials and techniques, in order to be able to be revived and adapted to the current needs.

The idea of using ash in construction materials is not new - ash from various materials has been studied over the years in laboratories, but the idea of ash from paper and other cellulosic waste is a relatively new, studied by other researchers, but with immense potential and with mysteries still undiscovered [1-3]. The main motivation of the study is therefore based on the personal contribution to improving life on earth, by identifying a method to reintroduce into the system a ubiquitous waste. Firstly, there is the amount of pulp that can no longer be reintroduced into the paper recycling process. Secondly, there is the amount of cement used globally, which is alarmingly and threateningly growing. The study seeks the balance of a mix of these two materials, which will generate desirable effects both from an engineering and environmental point of view.

## 2. Materials used and studied recipes

### *2.1. Basic recipe*

The basic recipe from which the recipes with the addition of RPA and AF were derived, is a recipe in the M25 class [4], with an average compressive strength at 28 days of 39 N/mm$^2$. A CS4 type cement-based rendering mortar is analyzed [5]. The volumetric proportion of the basic ingredients is 28% binder and 72% sand, with special attention to the water/cement ratio, which falls within the limits provided in standards.

#### *2.1.1. Component materials used in the experimental program.*

The binder used is a high quality non-additive cement, type CEM I 52.5R. In the laboratory, determinations were made on the cement powder in order to establish its properties. The specific density of the cement powder is 2.94 g/cm$^3$, obtained by the pycnometer method [6]. The initial setting time was set with the Vicat device at 45 minutes [7-8].

The aggregate used is a river aggregate, from the same batch throughout the experimental program. The 0-4 mm raw aggregate was screened under laboratory conditions. They were used as a percentage according to the particle size curve in figure 1, species 0-0.25, 0.25-0.5 mm, 0.5-1 mm, 1-2 mm and 2-4 mm [9].

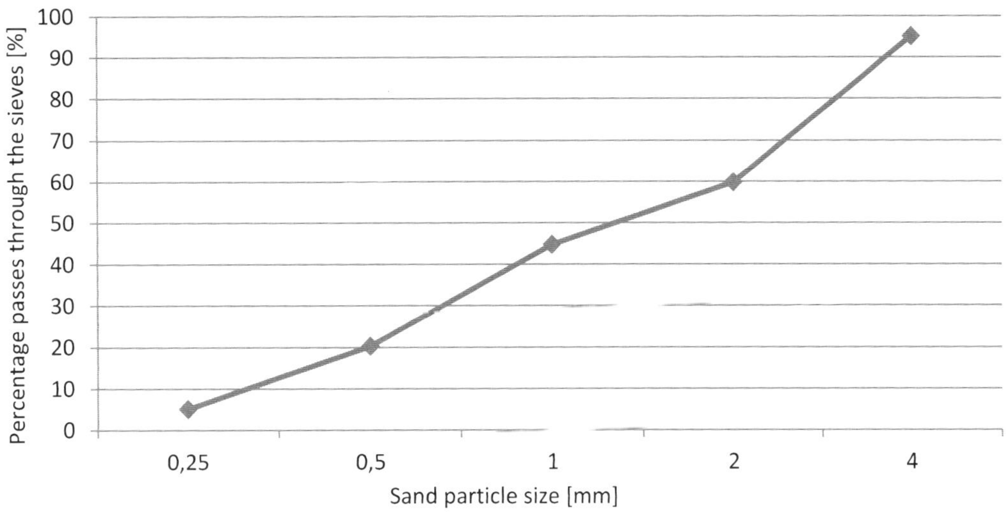


**Figure 1.** Granulometric curve of the aggregate used in the recipes included in the experimental program. Percentage passages on 0.25, 0.5, 1, 2 and 4 mm square mesh sizes are analyzed.

The waste used in the experimental study is ash paper. The raw waste of paper and cardboard was incinerated, the resulting ash being sifted in order to separate the particles by size. Two categories have been established. Recycled paper ash - RPA which has a particle size equal to that of cement particles, 0.063 mm (figure 2). The second category is ash flakes (see figure 3), with specimens 0.063-0.25 mm, 0.25-0.5 mm, 0.5-1 mm, 1-2 mm and 2-4 mm.

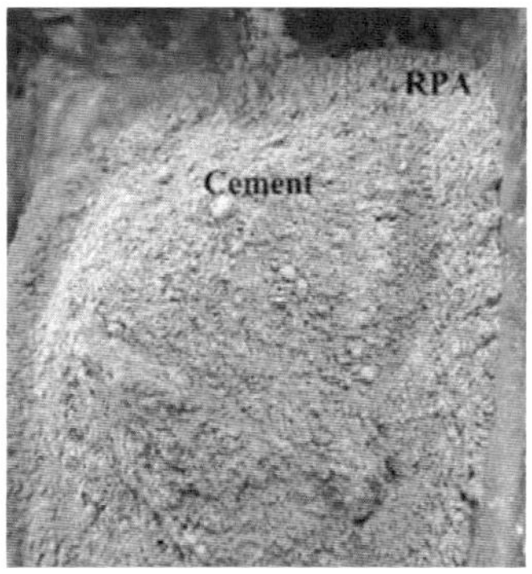


**Figure 2.** Cement powder compared to ash – RPA.

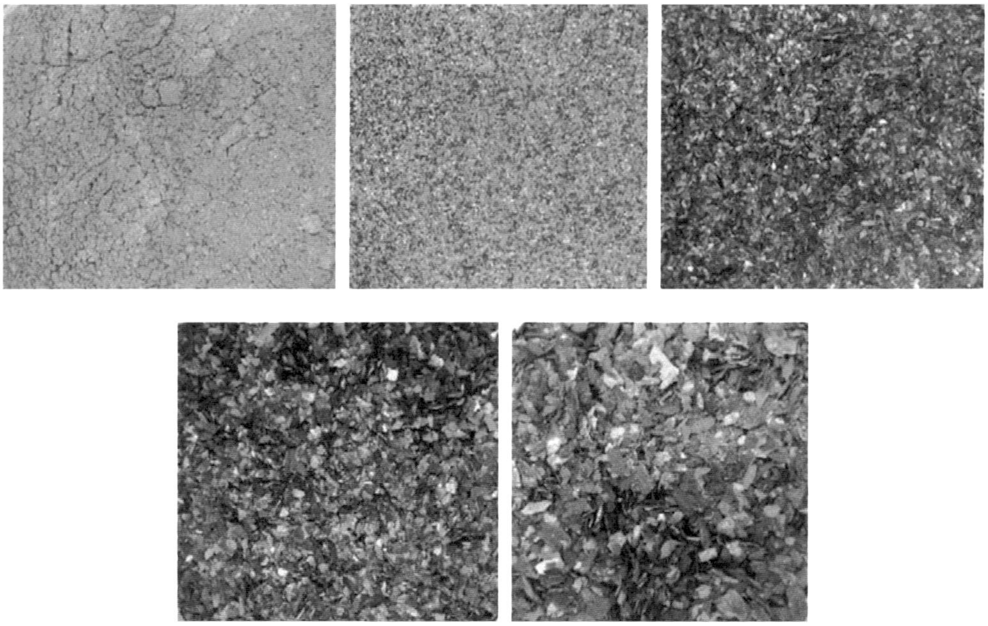

**Figure 3.** Ash flakes –AF samples of each.

### *2.2. Studied recipes*

This paper presents the evolution of physical-mechanical characteristics of hybrid recipes, which combined the best results from previous research, namely the replacement of 10% of the amount of cement with RPA [10], with the addition of coarse ash flakes in the recipe, under form the of additive, calculated as a percentage by mass of binder. The aim of this research is to integrate as much paper waste as possible in the composition of the mortars without compromising the physical properties of the mortar. The behavior of the fly ash added to the mass of the mortar is expected to be similar to a dispersed reinforcement.

The short name, the meaning of the name and also the composition of each studied recipe are presented in table 1.

For each of the studied recipes, 9 parallelepiped specimens with dimensions of 4x4x16 cm and 2 cubic samples with a side of 10 cm were cast (see figure 4). The parallelepiped samples have been tested for bending, the ends resulting from their rupture, being tested for compression, at the time intervals established according to SR EN 196-1: 2020 [11], namely 3, 7, and 28 days. The samples were kept for testing at the interval established under normal conditions in accordance with SR EN 1015-2: 2001 / A1: 2007 [12].

**Table 1.** Studied recipes.

| Short name | Composition |
|---|---|
| CS IV | Standard plaster mortar |
| CS IV+10C | Standard CS IV mortar, with 10% cement replaced with recycled paper ash |
| CS IV+10F | Standard CS IV mortar, with 5%* addition of ash flakes |
| CS IV+10C+1F | Standard CS IV mortar, with 10% cement replaced with recycled paper ash and with 1%* addition of ash flakes |
| CS IV+10C+5F | Standard CS IV mortar, with 10% cement replaced with recycled paper ash and with 5%* addition of ash flakes |
| CS IV+10C+10F | Standard CS IV mortar, with 10% cement replaced with recycled paper ash and with 10%* addition of ash flakes |
| CS IV+10C+15F | Standard CS IV mortar, with 10% cement replaced with recycled paper ash and with 15%* addition of ash flakes |

*The percentage of ash flakes is calculated according to the amount of cement

**Figure 4.** Cubic samples cast in the experimental program.

## 3. Experimental conditions

### *3.1. Characteristics of mortars in fresh state*

The bulk density of the fresh mortar [13] is an indicator of the amount of water needed to hydrate the binders. Table 4 shows the bulk density of the fresh mortar values of all recipes. Figure 5 shows the density varies depending on AF added quantity.

**Table 2.** The bulk density of fresh mortar.

| Recipe | Density [kg/m$^3$] |
|---|---|
| CS4 | 2242 |
| CS4+10% | 2238 |
| CS4+5%F | 2205 |
| CS4+10C+1F | 2250 |
| CS4+10C+5F | 2195 |
| CS4+10C+10F | 2117 |
| CS4+10C+15F | 2077 |
| CS4+10C+20F | 1992 |

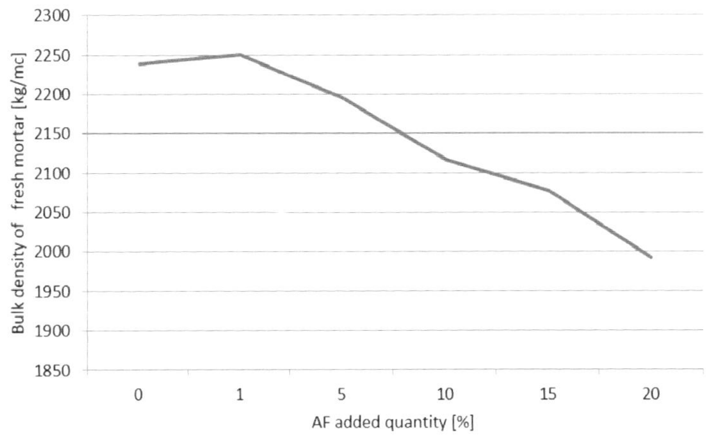


**Figure 5.** Bulk density of fresh mortar variation.

The water-cement ratio but also the water-cementitious materials ratio is a measurement that directly influences the mechanical strength and durability of the mortar. The water-cementitious ratio, unlike the water-cement ratio, also takes into account the amount of additive used in the addition of mortar, in the present study, recycled paper ash (w/cm= amount of water in relation to the amount of cement + RPA). A small w/c ratio describes a stronger and more durable mortar [14]. Table 3 contains the amount of water needed to reach the constant consistency, proposed at the beginning, for each recipe studied, the w/c ratio but also the w/cm ratio.

**Table 3.** Added water quantity, w/c ratio and w/cm ratio.

| Recipe | water [mL] | cement [kg] | AF [kg] | w/c ratio | w/cm ratio |
|---|---|---|---|---|---|
| CS4 | 1447 | 2.695 | 0 | 5.37 | |
| CS4+10% | 1560 | 2.426 | 0.269 | | 5.79 |
| CS4+5%F | 1610 | 2.695 | 0 | 5.97 | |
| CS4+10C+1F | 1500 | 2.426 | 0.269 | | 5.57 |
| CS4+10C+5F | 1540 | 2.426 | 0.269 | | 5.71 |
| CS4+10C+10F | 2000 | 2.426 | 0.269 | | 7.42 |
| CS4+10C+15F | 2200 | 2.426 | 0.269 | | 8.16 |

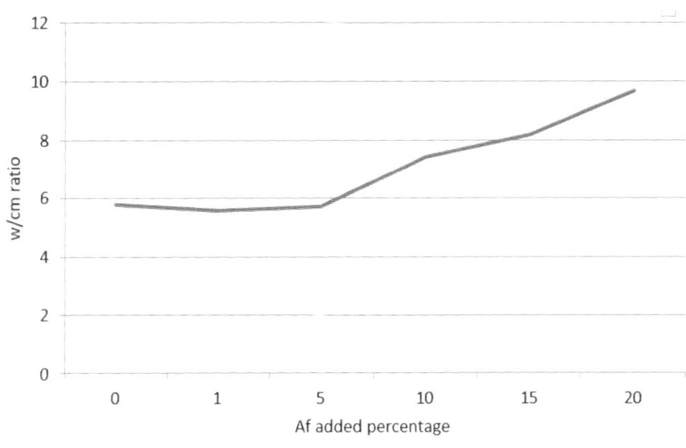


**Figure 6.** w/c ratio variation.

### *3.2. Test results for mechanical properties of hardened mortars*

Table 4 shows that the average values of the compressive strength obtained at the time intervals set for each recipe in which a percentage of the amount of cement was replaced with RPA and the mixture was added with AF.

**Table 4.** Compression strength values.

| Recipe | Load [N] | Compression strength [N/mm$^3$] | Age |
|---|---|---|---|
| CS IV | 55.82 | 34.85 | |
| CS IV+10C | 41.83 | 26.08 | |
| CS IV+10C+1F | 27.75 | 17.34 | 3 days |
| CS IV+10C+5F | 34.91 | 21.81 | |
| CS IV+10C+10F | 13.60 | 8.50 | |
| CS IV | 58.71 | 36.69 | |
| CS IV+10C | 50.44 | 31.52 | |
| CS IV+10C+1F | 47.95 | 29.97 | 7 days |
| CS IV+10C+5F | 49.60 | 31.10 | |
| CS IV+10C+10F | 31.60 | 19.76 | |
| CS IV | 62.57 | 39.11 | |
| CS IV+10C | 66.80 | 41.74 | |
| CS IV+10C+1F | 56.40 | 35.25 | 28 days |
| CS IV+10C+5F | 56.56 | 35.35 | |
| CS IV+10C+10F | 34.55 | 21.59 | |

The results of compressive strength tests are presented in the form of a graph of their evolution at intervals of 3, 7 and 28 days (figure 7). These values represent the average results of 3 samples tested for each recipe.

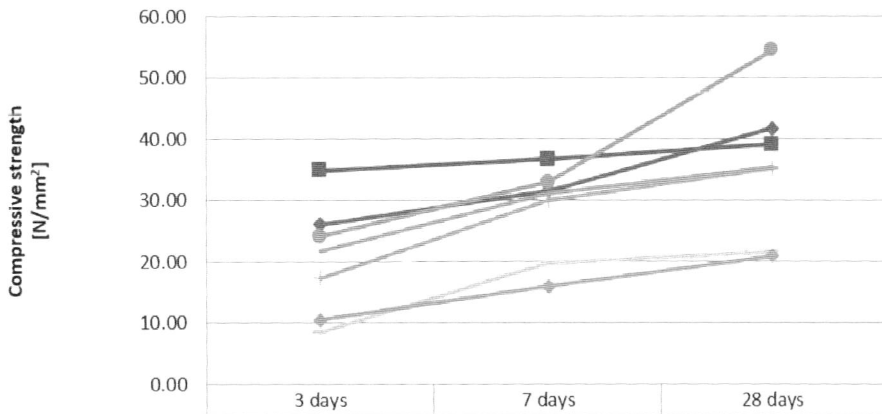


| | 3 days | 7 days | 28 days |
|---|---|---|---|
| CS4 | 34.85 | 36.69 | 39.11 |
| CS4+ 10% | 26.08 | 31.52 | 41.74 |
| CS4+5% FULGI | 24.09 | 32.88 | 54.54 |
| CS4+10C+1F | 17.34 | 29.97 | 35.25 |
| CS4+10C+5F | 21.81 | 31.10 | 35.35 |
| CS4+10C+10F | 8.49 | 19.76 | 21.59 |
| CS4+10C+15F | 10.47 | 15.86 | 20.90 |

**Figure 7.** Variation of compressive strength at 3,7 and 28 days in N/mm$^2$.

Table 5 shows the average values of bending strength obtained at the time intervals set for each recipe in which a percentage of the amount of cement has been replaced with RPA and the mixture has been added with AF.

**Table 5.** Bending strength values.

| Recipe | Load [N] | Bending strength [N/mm$^3$] | Age |
|---|---|---|---|
| CS IV | 1906.67 | 4.47 | |
| CS IV+10C | 1803.33 | 4.23 | |
| CS IV+10C+1F | 1746.67 | 4.09 | 3 days |
| CS IV+10C+5F | 1700.00 | 3.98 | |
| CS IV+10C+10F | 980.00 | 2.30 | |
| CS IV | 2226.67 | 5.22 | |
| CS IV+10C | 2270.00 | 5.32 | |
| CS IV+10C+1F | 2600.00 | 6.09 | 7 days |
| CS IV+10C+5F | 2286.67 | 5.36 | |
| CS IV+10C+10F | 1416.67 | 3.32 | |
| CS IV | 2593.33 | 6.08 | |
| CS IV+10C | 2710.00 | 6.35 | |
| CS IV+10C+1F | 2596.67 | 6.09 | 28 days |
| CS IV+10C+5F | 2906.67 | 6.81 | |
| CS IV+10C+10F | 2183.33 | 5.12 | |

The results of bending strength tests are presented in the form of a graph of their evolution at intervals of 3, 7, and 28 days (figure 8). These values represent the average results of 3 samples tested for each recipe.

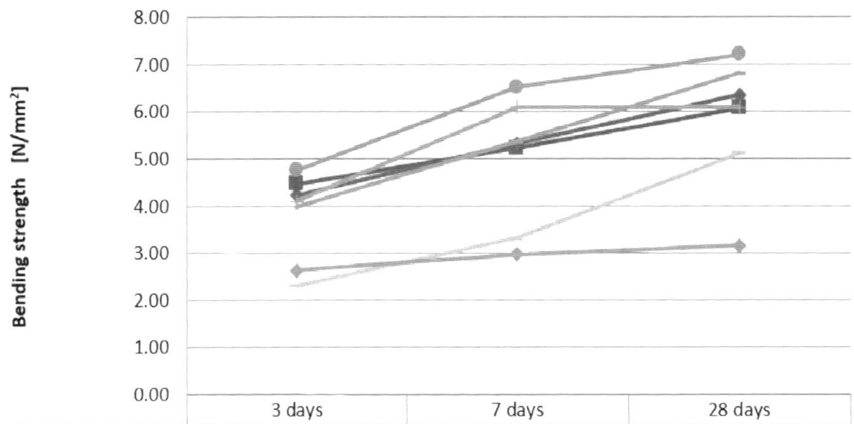


| | 3 days | 7 days | 28 days |
|---|---|---|---|
| CS4 | 4.47 | 5.22 | 6.08 |
| CS4+ 10% | 4.23 | 5.32 | 6.35 |
| CS4+5% FULGI | 4.75 | 6.52 | 7.20 |
| CS4+10C+1F | 4.09 | 6.09 | 6.09 |
| CS4+10C+5F | 3.98 | 5.36 | 6.81 |
| CS4+10C+10F | 2.30 | 3.32 | 5.12 |
| CS4+10C+15F | 2.63 | 2.97 | 3.15 |

**Figure 8.** Variation of bending strength at 3,7 and 28 days in N/mm$^2$

## 4. Discussions and conclusions

Bulk density of fresh mortar tends to decrease with the increase of the percentage of cement replaced with ash, also together with the increase in the percentage of AF (figure 5), a not at all surprising effect, given the difference in density between cement powder and RPA powder.

Regarding the graph of the water- cementitious materials ratio (figure 6), a significant increase can be observed. Taking into account the limit of 6 imposed by norms, it is observed the maintenance under this limit of the recipes with AF goes up to 5%. Above this percentage the values increase exponentially. This can be interpreted as a direct reaction to the need for more water to moisturize the RPA and AF.

Although the evolution of the results of compressive and bending strengths is important during the 28 days of maturation (see figure 7 and 8), the best conclusions can be drawn from the analysis of the compressive and tensile strength graphs at 28 days (see figure 9 and 10). The evolution over time of bending strength is linear, it fits in approximately the same pattern for all the recipes in the study. Ignoring the special results obtained by the recipe CS IV + 5F because the purpose of the research is to introduce a large amount of ash in the composition of the mortar without compromising its physical and mechanical properties, the best result from this point of view is CS VI+5C+5F. At 3 days it starts with a deficit of 10% but at 28 days it exceeds the results of the standard recipe by 12%. The conclusion of the results obtained for these mixtures in terms of their bending behavior is that, in the first days of application, the mortar does not excel, but during maturation, the evaporation of the mixing water leaves behind a homogenous and compact structure, which leads to very good results 28 days after casting (see figure 8).

Compressive strength registers various trends of evolution over time (see figure 7). The results are unsatisfactory in the case of hybrid recipes. Almost all the results are below the expectations set by the basic recipe. However, we can say that the previously chosen recipe, CS IV + 10C + 5F has a decrease in compressive strength at 28 days of only 9% (see figure 10), which leads to the conclusion that it is a result that can be accepted as positive.

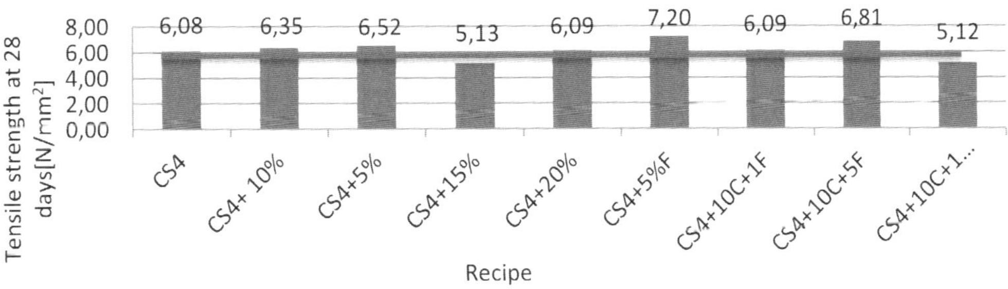


**Figure 9.** Variation of bending strength at 28 days in N/mm$^2$.

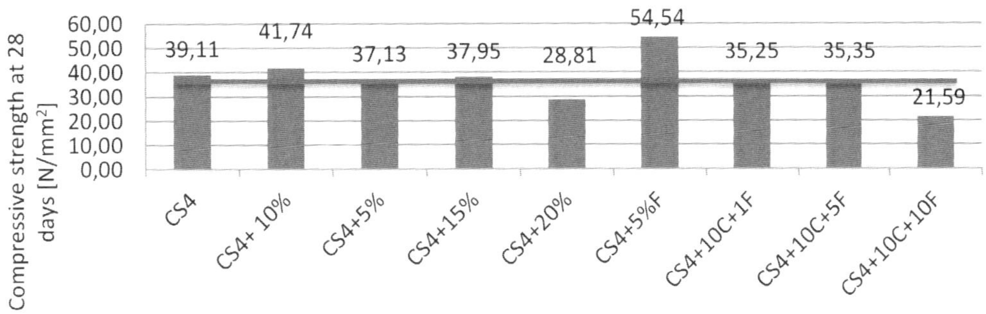


**Figure 10.** Variation of compressive strength at 28 days in N/mm$^2$.

At this stage of the research, we can say that ash can replace up to 10% of the amount of cement needed in the mortar recipe and ash flakes can be added in up to 5% of the binder quantity (recipe CS IV+10C+5F), without significantly affecting the mechanical properties.

Paper ash is a waste obtained by processing recycled paper, so it is a material that has zero costs, a plus from an economic point of view being the fact that in the combustion process is obtained caloric energy and biogas, which can be transformed into alternative energy sources. The ash, which until now was also a waste, is collected and used at no additional cost, to bring a significant benefit to the cement both from an economic point of view and from the protection of the environment. The study showed that ash can be present in the mortar composition in significant quantities, up to 15% of binder quantity. Relating this percentage to the amount of cement used globally annually, 375 million tons of recycled paper ash can be reintroduced into the system. The contribution of ash is not only non-destructive on the properties of cement-based mortars, but we can say that it significantly improves its behavior in tensile tests. In-depth behavioral studies at the molecular level will be performed as a result of this research, in order to formulate a concrete conclusion regarding the crack resistance and ductility of new mixtures.

**Acknowledgment**
This paper was financially supported by the Project "Entrepreneurial competences and excellence research in doctoral and postdoctoral programs - ANTREDOC", project co-funded by the European Social Fund financing agreement no. 56437/24.07.2019

**References**

[1] Ahmad S 2013 Study of Concrete Involving Use of Waste Paper Sludge Ash as Partial Replacement of Cement **3** *IOSR Journal of Engineering (IOSR-JEN)* 2278-8719
[2] Raut S, Ralegaonkar R, Mandavgane S 2013 Utilization of recycle paper mill residue and rice husk ash in production of light weight bricks **13** *Archives of Civil and Mechanical Engineering* p 269-275
[3] Wong H S, Barakat R, Alhilali A, Saleh M and Cheeseman C 2013 Hydrophobic concrete using waste paper sludge ash **70** *Cement and Concrete Research*
[4] SR EN 1996-1-1 Eurocode 6: Design of masonry structures
[5] SR EN 998-1:2016 - Specification for mortar for masonry
[6] SR EN 196-4:2008 - Methods of testing cement - Part 4: Quantitative determination of constituents
[7] SR EN 196-1:2016 Methods of testing cement-Determination of strength
[8] SR EN 196-3:2016Methods of testing cement-Determination of setting times and soundness
[9] SR EN 13139:2003 / C91:2009 Aggregates for mortar
[10] Tintisan M L, Siomin A C and Zaharie A 2021 Paper ash used as substitute of cement, in cement mixtures
[11] EN 196-1:2016 Methods of testing cement - Part 1: Determination of strength
[12] EN 1015-2:1998/A1:2006 Methods of test for mortar for masonry - Part 2: Bulk sampling of mortars and preparation of test mortars
[13] EN 1015-6:2001/A1:2007 Methods of test for mortar for masonry - Part 6: Determination of bulk density of fresh mortar
[14] Levy S M. 2012 *Construction Calculation Manual*

# The analysis of TBM cutter stress under different construction environment

**Z N Wang[1], Q Y Wang[1]*, B B Hu[1], C Wei[1] and Q N Wu[1]**

[1] School of Mechanical Engineering, Shenyang Jianzhu University, Shenyang, China

* 1404589743@qq.com

**Abstract**. As the country's underground tunnels continue to expand, underground construction technology has been developed rapidly. In order to meet the development needs of construction technology in China, Tunnel Boring Machine (TBM) has made great progress due to its own advantages. In this paper, feed amount of disc cutter on rock crushing under complex working conditions is considered. The mechanism of rock fragmentation is analyzed. The force between disc cutter and the rock is studied. The accuracy of the model is verified by setting up a TBM rock fragmentation experiment platform. This study provides technical support for TBM construction parameters under different working conditions.

## 1. Introduction

TBM is an interdisciplinary collection of machinery, automation, information, geotechnical and other disciplines. After a period of development, TBM has achieved considerable development and is now widely used in tunnel boring in China [1]. TBM belongs to high-end technical equipment. And a few developed countries are in a leading position in equipment production, equipment assembly and tunneling technology due to their mastery of core technology. The development of TBM also reflects the national level of science and technology. The cutter is arranged on the cutter head which is almost like in the disc form. The cutting wheel rotates to drive the cutters rotate. At the same time, the cutter head is squeezed by the hydraulic cylinder so that the cutter can squeeze the rock, which can achieve the rock breaking in the rotation process.

Scholars have carried out some researches on TBM cutter head under different working conditions. Guoqiang Liu et al. had carried on the optimization design and analyze the tool between the rock breaking mechanism of the cutter [2]. Yimin Xia et al. established the rock breaking load model of the cutter and analyzed the change of the lateral force of the cutter under different stresses [3]. Min Zhang et al. studied the relationship between cutter failure form and geology, and analyzed the distribution of cutter ring fracture and microstructure [4]. Qing Tan et al. studied the development of crack and the variation of rock breakage state by considering the influence of friction, and experimental verification is carried out [5]. Lu Feng used finite element software to establish the rock breaking model of cutter, and analyzed the influence relationship between the structure of cutter and rock breaking effect [6]. Xuhui Zhang et al analyzed the dynamic process of rock breaking under the influence of different tool spacing [7]. Although some work has been done, the mechanism of rock breaking under different surrounding rock conditions has not been thoroughly studied.

Based on the working principle of TBM, this paper analyzes the force of the cutter, the feed of the cutter and the rock breakage under different conditions of surrounding rock. The accuracy of the simulation model is verified by rock crusher experiment, which provides technical guidance for the structural design of cutter, and has important engineering significance.

## 2. The force analysis of TBM cutter

Surrounding rock is the main factor affecting the force of the cutter, which is related to the crack and fragment of rock. Under some high stress geological conditions, it is of great practical significance to develop the stress relationship between surrounding rock and TBM cutter. In order to study the efficiency of the cutter breaking different rocks and obtain the stress relationship between the cutter and different rocks, the rock breaking efficiency of TBM is evaluated by the specific energy formula. The calculation method is shown in equation 1.

$$E=\frac{F(p+l)}{V} \tag{1}$$

Where $E$ is breaking rocks energy, represents the external work required to break the rock, $p$ is penetration, $F$ is rolling load of cutter, $l$ is the spacing of cutter, $V$ is the volume of rock.

According to the structure distribution of TBM cutter and the calculation method of rock specific energy. The finite element model of cutter was established by ANSYS software. The finite element mesh is divided by Boolean calculation. After ANSYS post-processing, the stress changes of cutter shaft and outer ring of broken hard sandstone, shale and marble are shown in figure 1.

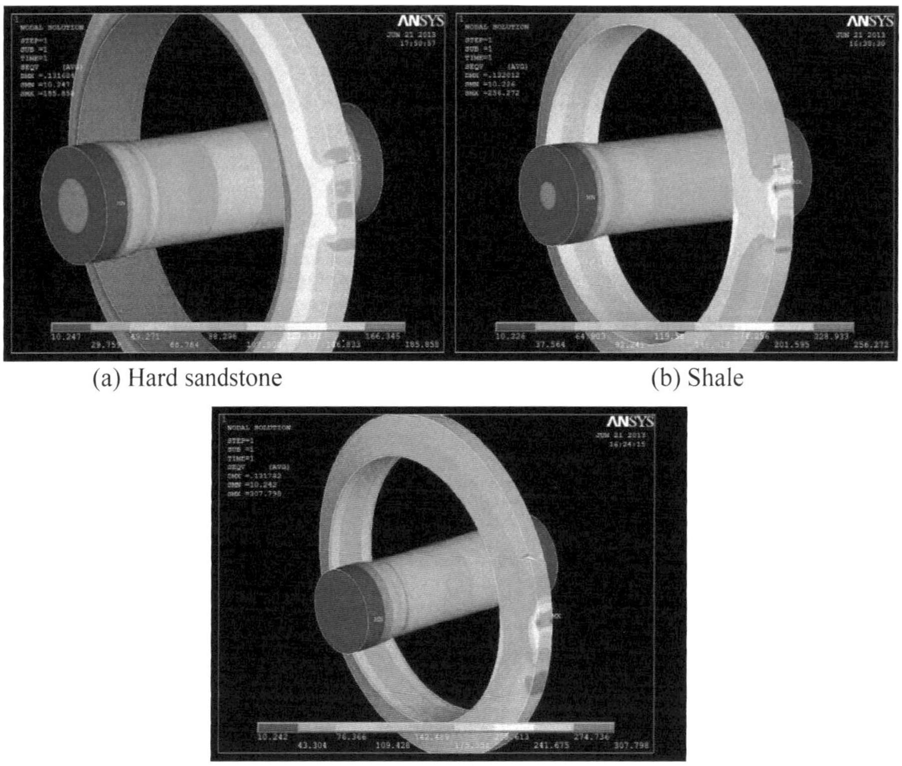


(a) Hard sandstone (b) Shale

(c) Marble

**Figure 1.** The stress changes of the cutter shaft and outer ring of broken hard sandstone, shale and marble.

It can be seen from figure 2 that the stress distribution of cutter shaft and cutter ring varies greatly with the rock. In the crushing condition of hard sandstone, the cutter ring and cutter axis are smaller than those of marble and hard sandstone. In the contact area between the cutter ring and rock, the maximum stress appears, and with the increase of the contact area of the cutter ring, the stress diffused. The stress value of the cutter ring is higher than that of the cutter shaft.

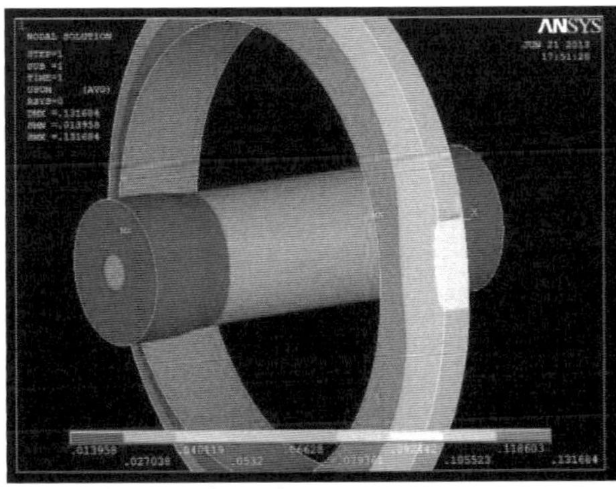


(a) Hard sandstone

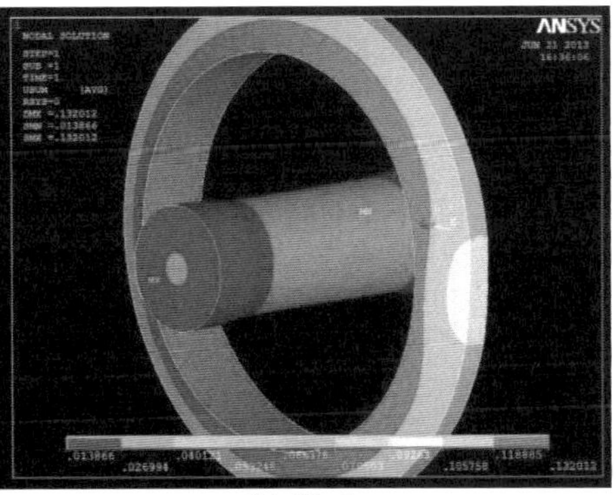


(b) Shale

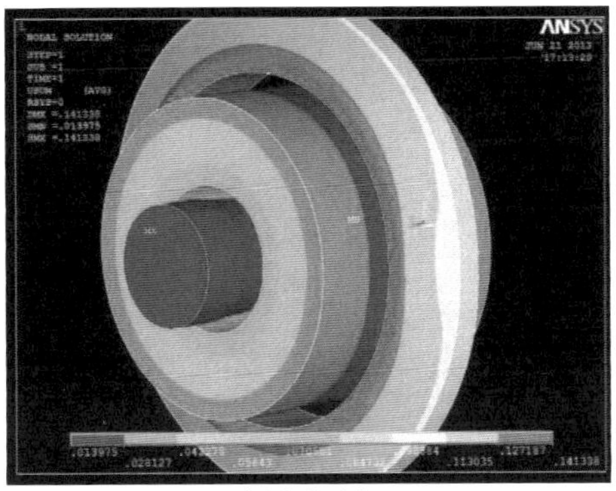


(c) Marble

**Figure 2.** The stress deformation changes of the cutter shaft and outer ring of broken hard sandstone, shale and marble.

## 3. Experimental Verification

According to the type of rock simulated, testing the force change of marble, hard sandstone and shale by different cutting rocks experiments. The rock crushing machine is shown in figure 3.

**Figure 3.** The rock crushing machine.

According to the experiment of the rock crushing machine, the stress experimental data of different rocks are obtained. The comparison of simulation experimental data is shown in figure 4. It can be seen that the experimental values are basically similar to the simulation, and the stress of marble is greater than that of other hard sandstone and shale.

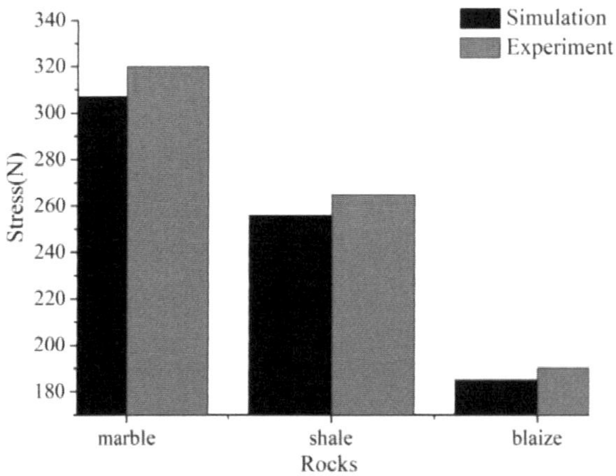


**Figure 4.** The comparison of simulation experimental data.

## 4. Conclusion

In this paper, ANSYS software is used to analyze the force state of cutter under different surrounding rock. In the state of hard rock, the cutter has a large force, and the crack propagation of rock is faster. With the change of rock type, the stress of the cutter is directly related to the hardness of the rock. In the process of crushing hard rock, the cutter with greater hardness should be used appropriately.

**Acknowledgements**

The project is funded by Youth Science and technology Innovation Special Project(2021-HJQZ-08).

## References

[1] Cao R L, Wang Y J, Chen C and Liu L P 2019 Development status and parameter analysis of TBM net tunneling speed prediction model *Water Resources and Hydropower Engineering* **50** 8
[2] Liu G Q, Liu B, Zhang Q M and Liu B L 2021 Analysis of Efficiency of Hob Crushing Jointed Rock Mass and Research on the Optimization Method of Knife Spacing *Chongqing Architecture* **20** 8
[3] Xia Y M, Wu C Z, Gu J J 2016 Mechanical characteristics of TBM disc cutter under the initial stress *Journal of Central South University (Science and Technology)* **47** 2
[4] Zhang M 2020 Analysis of Cutter Fracture in Shield Tunneling Machine in a Tunnel Project Construction Mechanization **41** 7
[5] Tan Q, Yang Y and Xia Y M 2018 Numerical and experiment study on rock breaking by cutter in rolling mode *Journal of Hunan University (Natural Science)* **45** 8
[6] Lu F, Zhang C, Sun J, Tian J X, Liu M and Wu Y H 2016 Experiment study on rock breaking simulation of double disc cutter of TBM *Chinese Journal of Engineering Design* **23** 1
[7] Zhang X H, Xia Y M, Tan Q, Lin L k, Lao T B and Liu J 2016 Study on the characteristics of breaking joint rock by tunnel boring machine single-point and double–point cutters *Harbin Engineering University* **37** 10

# Stiffness analysis of rolling bearing for concrete mixer

**Z Wang[1*], W Z He[1] and S Q Niu[1]**

[1] School of Mechanical Engineering, Shenyang Jianzhu University, Shenyang, China

* juven1126@163.com

**Abstract**. According to the rolling bearing geometry, the basic principle of kinematics and Hertz elastic contact theory, eccentric mass of mixing shaft, the centrifugal force of steel ball and gyroscopic moment are considered and the dynamic model of angular contact ball bearing is established in this paper. Using this model, the effects of preload, radial load and rotating speed on bearing stiffness and system dynamic characteristics are analyzed in detail. The results show that with the increase of radial load, the axial static stiffness decreases, while the radial stiffness increases gradually, and the increasing trend of radial static stiffness is more obvious than that of axial static stiffness. The increase of rotating speed will soften the stiffness of the bearing. The natural frequencies of each order of the system show a downward trend with the increase of rotating speed.

## 1. Introduction

Concrete is a mixture of cement, aggregate, additives and water in a certain proportion. It is the most widely used building material today. It is the most widely used building material and plays an important role in the national economy. The concrete mixer is a special machine for producing flow concrete that meets the quality requirements. Concrete mixer has the advantages of high productivity, good production quality and high degree of automation. It is an indispensable equipment in engineering construction.

With the worse and worse use environment of mixer in actual production, the requirements for key rotating parts (rolling bearings) in the equipment are increasing year by year. Therefore, the research on bearings with excellent performance has become a hot research field.

The system composed of propeller shaft bearing is referred to as propeller shaft system for short. Concrete mixer is a typical propeller shaft system. The installation position, geometric dimension and stiffness characteristics of bearing, and the structural dimension and material of coaxial itself have a great impact on the dynamic characteristics and operation stability of shafting. Wonglert et al. clarified and gained an insight into the impact of the length of the stiffened core and the strength of the deep cement mixing (DCM) socket on the behaviors of floating stiffened deep cement mixing (SDCM) columns [1]. Sheng et al. analyzed the relationship between load and deflection in the bearing dynamic model and proposed an analytical model based on implicit function differentiation to calculate the stiffness [2]. Hernot et al. established a 5-DOF model for solving the dynamic stiffness matrix of angular contact ball bearing [3]. Noel et al. further introduced the derivation process of dynamic stiffness matrix and applies it to spindle model [4]. Yang et al. analyzed the stiffness of main bearing under different bearing configurations [5]. EI-SAEIDY established a 5-degree-of-freedom rigid spindle rotor-angular contact bearing rigid matrix model to study the time-varying stiffness [6]. Cong et al. established a two degree of freedom bearing rotor system based on Jeffcott's rotor assumption to study the fault response of rolling bearings [7]. Van Ngoc et al. applied needle penetration tests, uniaxial compression tests and a thermogravimetric analysis to determine the extent of the deterioration of soil-cement columns exposed to synthetic sea water [8]. Zhang et al. presented an extended Jones–Harris stiffness model to ascertain the stiffness of the angular contact ball bearing considering five degrees of freedom [9]. Liu et al. proposed an analytical calculation method (SAMCA) for the internal load distribution and stiffness of a preloaded ACBB with combined loads including the radial and axial loads [10]. However, these works rarely consider eccentric mass of mixing shaft, the centrifugal force of steel ball and gyroscopic moment to establish the dynamic model of the bearing at the same time, and do not analyze the differential influence of radial load on the radial stiffness and axial stiffness of the bearing.

Based on the above research and considering the influence of unbalanced force, the centrifugal force of steel ball and gyroscopic moment, dynamic model of angular contact ball bearing is established in this paper. And the effects of radial load and rotating speed on the stiffness of angular contact ball bearing for mixing shaft are analyzed. The results provide a reference for the design and analysis of angular contact ball bearing and more accurate boundary conditions for the dynamic analysis of bearing rotor system in concrete mixer.

## 2. Ball motion model of angular contact ball bearing

When angular contact ball bearings are running, the curvature centers of the inner and outer race grooves of the bearings are not on the same line as the centers of the bearing balls due to the centrifugal force acting on the bearing balls. When the bearing is installed, the outer race is fixed on the bearing seat, so it is assumed that the curvature center of the outer race groove of the bearing is fixed, while the curvature center of the inner race groove displaces under the external load, and the bearing ball is subjected to centrifugal force, and its center will change. Figure 1 shows a simple model of angular contact ball bearing, $\alpha_0$ represents the initial contact angle of the bearing, $O$ represents the center of the ball, $A$ and $C$ represents the curvature center of inner and outer races respectively. The changes of curvature center of bearing inner race and bearing ball center before and after loading are shown in figure 2. In figure 2, $O'$ and $C'$ represent the center of the ball and the curvature center of the inner ring after the inner ring is stressed respectively. $\alpha_{ij}$and $\alpha_{ej}$ respectively represent the contact angle of the bearing ball with the inner and outer ring of the bearing under the function of forces.

According to the schematic diagram of the curvature center of bearing inner race and the change before and after the bearing center is stressed in figure 2, the deformation equation of bearing ball and groove of inner and outer race of bearing is established.

$$\begin{cases} \left(A_{1j} - X_{1j}\right)^2 + \left(A_{2j} - X_{2j}\right)^2 - \overline{O'C'}^2 = 0 \\ X_{1j}^2 + X_{2j}^2 - \overline{AO'}^2 = 0 \end{cases} \tag{1}$$

Where $A_{1j}$ and $A_{2j}$ respectively represent the axial and radial distance between the curvature centers of the inner and outer ring groove of the bearing at the $j$th ball after the bearing is stressed. $X_{1j}$ and $X_{2j}$ are the distance between the center of the bearing ball and the center of curvature of the bearing outer race groove in axial and radial directions after being stressed. $\overline{O'C'}$ and $\overline{AO'}$ are the distance between center of bearing ball and curvature center of inner and outer race groove after stress, respectively.

$A_{1j}$ and $A_{2j}$ can be expressed as in equation (2).

$$\begin{cases} A_{1j} = \overline{AC}\sin\alpha_0 + \overline{CC_z'} = \overline{AC}\sin\alpha_0 + \delta_z + R_i\theta\cos\psi_j \\ A_{2j} = \overline{AC}\cos\alpha_0 + \overline{CC_y'} = \overline{AC}\cos\alpha_0 + \delta_r\cos\psi_j \end{cases} \tag{2}$$

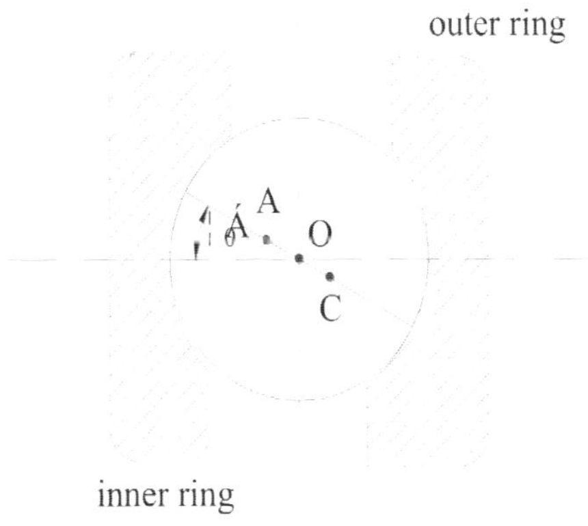


**Figure 1.** Angular contact ball model.

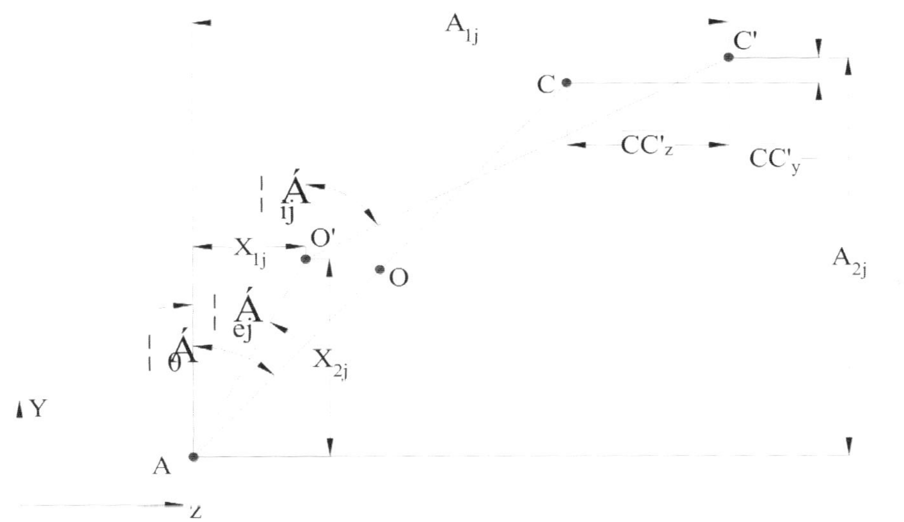


**Figure 2.** Variation diagram of curvature center of bearing inner ring.

Where $\delta_z$, $\delta_r$ and $\theta$ are the axial, radial and rotational displacement of the bearing inner ring during the high-speed operation of the bearing, respectively. $R_i$ represents the radius of the circle where the curvature center of the inner race is located. $\Psi_j = 2\pi(j-1)/N$ denotes angular position of the jth ball in the pitch circle. $N$ is the number of bearing balls. $\overline{AC}$ can be expressed as

$$\overline{AC} = (f_e + f_i - 1)D_w \tag{3}$$

Where $D_w$ is the bearing ball diameter, $f_i$ and $f_e$ respectively are the ratio of the curvature radius of the groove of the inner and outer races to the diameter of the ball.

$\overline{O'C'}$ and $\overline{AO'}$ can be expressed by equation (4). As in equation (4), $\delta_{ij}$ and $\delta_{ej}$ denote Hertz contact deformation of bearing ball with inner and outer race grooves, respectively.

$$\begin{cases} \overline{O'C'} = (f_i - 0.5D_w) + \delta_{ij} \\ \overline{AO'} = (f_e - 0.5D_w) + \delta_{ej} \end{cases} \tag{4}$$

## 3. Ball force model

During running, the bearing ball contacts with the inner and outer race grooves of the bearing, and is subjected to the force exerted by the inner and outer race grooves on the bearing ball as well as the centrifugal force caused by the movement of the bearing ball itself. In addition, it is also affected by the gyroscopic torque. The bearing ball force model is shown in figure 3. In this figure, $F_{cj}$ and $M_{gj}$ are centrifugal force and gyro moment respectively. $\lambda_{ij}M_{gj}/D_w$ and $\lambda_{ej}M_{gj}/D_w$ are the sliding friction between the ball rolling element and race. When bearing runs at high speed, the bearing can be approximated as "outer race control" state, that is, $\lambda_{ij} = 1$, $\lambda_{ej} = 1$. $Q_{ij}$ and $Q_{ej}$ represent the force exerted on the ball by the inner and outer race grooves of the bearing respectively, and the relationship between the force and the deformation can be expressed as

$$\begin{cases} Q_{ij} = K_{ij}\delta_{ij}^{1.5} \\ Q_{ej} = K_{ej}\delta_{ej}^{1.5} \end{cases} \tag{5}$$

Where $K_{ij}$ and $K_{ej}$ are related to the material properties of the bearings. According to the Hertz contact theory, it can be calculated by the following formula.

$$\begin{cases} K = \dfrac{\pi\kappa E}{3\tau}\sqrt{\dfrac{2\varepsilon R}{\tau}} \\ \kappa = 1.034\left(\dfrac{R_y}{R_x}\right)^{0.636} \\ \tau = 1.528 + 0.602\ln\left(\dfrac{R_x}{R_y}\right) \\ \varepsilon = 1.000 + 0.597\left(\dfrac{R_x}{R_y}\right) \end{cases} \tag{6}$$

Where $E = 2/(1-\upsilon_1^2)/E_1 + (1-\upsilon_2^2)/E_2$, $E_1$ and $E_2$ are the elastic modulus of two contactmaterials, $\upsilon_1$ and $\upsilon_2$ are the Poisson's ratio of ball and bearing groove material respectively, $R_x$ and $R_y$ are the effective contact radius of the bearing ball.

Figure 3 shows the force analysis of a ball of the bearing at high speed. The force balance equations in the horizontal and vertical directions of ball could be expressed as

$$\begin{cases} Q_{ij}\sin\alpha_{ij} - Q_{ej}\sin\alpha_{ej} - \dfrac{\lambda_{ij}M_{gj}}{D_w}\cos\alpha_{ij} + \dfrac{\lambda_{ej}M_{gj}}{D_w}\cos\alpha_{ej} = 0 \\ Q_{ij}\cos\alpha_{ij} - Q_{ej}\cos\alpha_{ej} - \dfrac{\lambda_{ij}M_{gj}}{D_w}\cos\alpha_{ij} + \dfrac{\lambda_{ej}M_{gj}}{D_w}\cos\alpha_{ej} + F_{cj} = 0 \end{cases} \tag{7}$$

According to the geometric relationship in figure 1, the contact angle between the ball and the inner and outer race grooves can be expressed as

$$\begin{cases} \sin\alpha_{ij} = \dfrac{A_{1j} - X_{1j}}{\overline{O'C'}} \\ \cos\alpha_{ij} = \dfrac{A_{2j} - X_{2j}}{\overline{O'C'}} \\ \sin\alpha_{ej} = \dfrac{X_{1j}}{\overline{AO'}} \\ \cos\alpha_{ej} = \dfrac{X_{2j}}{\overline{AO'}} \end{cases} \tag{8}$$

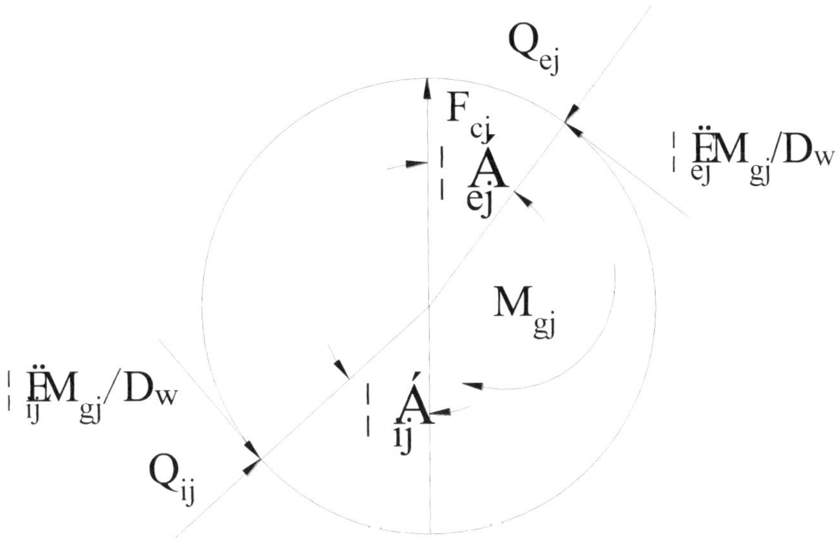


**Figure 3.** Load acting on the ball.

## 4. Bearing inner race force model

Bearing outer ring is fixed, the external load on the bearing is $F = \left[F_x\ F_y\ F_z\ M_x\ M_y\right]^T$. Then the force balance equation of the bearing inner ring is

$$\begin{cases} F_x + \overline{F_x} - \sum_{j=1}^{z}(Q_{ij}\cos\alpha_{ij} + \dfrac{\lambda_{ij}M_{gj}}{D}\sin\alpha_{ij})\sin\psi_j = 0 \\ F_y + \overline{F_y} - \sum_{j=1}^{z}(Q_{ij}\cos\alpha_{ij} + \dfrac{\lambda_{ij}M_{gj}}{D}\sin\alpha_{ij})\cos\psi_j = 0 \\ F_z - \sum_{j=1}^{z}(Q_{ij}\sin\alpha_{ij} - \dfrac{\lambda_{ij}M_{gj}}{D}\cos\alpha_{ij})\cos\psi_j = 0 \\ M_x + \overline{M_x} - \sum_{j=1}^{z}(Q_{ij}\sin\alpha_{ij} - \dfrac{\lambda_{ij}M_{gj}}{D}\cos\alpha_{ij})R_j\cos\psi_j = 0 \\ M_y + \overline{M_y} + \sum_{j=1}^{z}(Q_{ij}\sin\alpha_{ij} - \dfrac{\lambda_{ij}M_{gj}}{D}\cos\alpha_{ij})R_j\sin\psi_j = 0 \end{cases} \tag{9}$$

Where $\overline{F_x}$, $\overline{F_y}$, $\overline{M_x}$, $\overline{M_y}$ are the unbalanced force and torque caused by rotor eccentricity, $R_j$ is the force arm acting on the bearing inner race.

## 5. Stiffness characteristic analysis

Considering the centrifugal force of bearing ball and the expansion effect of bearing inner ring under the condition of high-speed rotation of angular contact ball bearing, according to the relationship between bearing deformation and force, the relationship between bearing deformation and force can be expressed as

$$K^b\delta = F \tag{10}$$

where $K^b$ is the 5×5 bearing stiffness matrix. Newton Raphson method has the characteristics of fast convergence and strong universality, so the above nonlinear equations are solved by this method. Newton Raphson method is used to solve equation (1), (7) and (9), and finally the stiffness matrix is obtained through equation (10). Matlab software was used for solving the nonlinear equations.

The 7008c angular contact ball bearing is selected as the research object to study the effects of bearing preload, radial force and speed on bearing stiffness. The bearings are made of GCr15. The specific parameters of ball bearing are shown in table 1.

**Table 1.** Structural parameters of bearing.

| parameter | Value |
|---|---|
| $D_w$/mm | 7.90 |
| $\alpha_0$/° | 15 |
| $f_i$ | 0.52 |
| $f_e$ | 0.52 |
| $E$/Gpa | 206 |
| $\upsilon$ | 0.3 |

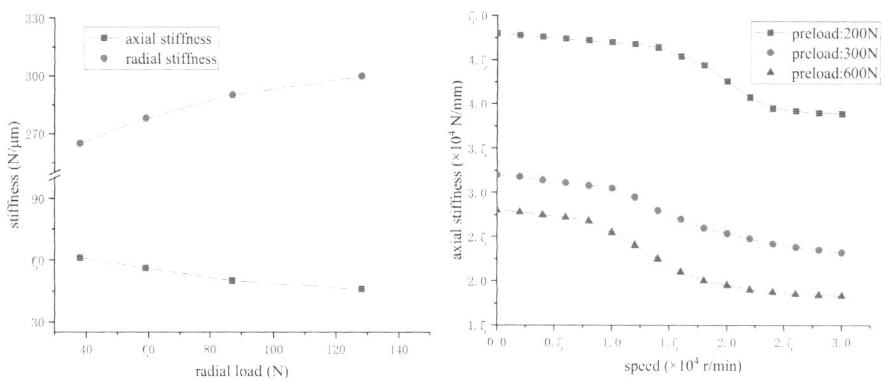


(a) (b)

**Figure 4.** (a) Variation of stiffness with radial load; (b) Variation curve of stiffness with speed under different preloads.

During the operation of the bearing, the bearing is mainly subjected to the radial load from the mixing shaft of concrete mixer. The variation curve of bearing stiffness with radial load is shown in figure 4(a). It can be seen from the figure 4(a) that with the increase of radial load, the axial stiffness of the bearing decreases, while the radial stiffness increases, and the increasing trend of radial stiffness is more obvious than the decreasing trend of axial stiffness.

The influence of the rotating speed of the mixing shaft of the concrete mixer on the bearing stiffness is shown in figure 4(b). It can be seen from figure 4(b) that the dynamic stiffness of the bearing decreases with the increase of the rotating speed, that is, the stiffness softening phenomenon occurs. Considering the softening phenomenon of angular contact ball bearing, as shown in figure 5, the natural frequency of each order of the system decreases with the increase of rotating speed, which is consistent with the decrease of bearing stiffness with rotating speed.

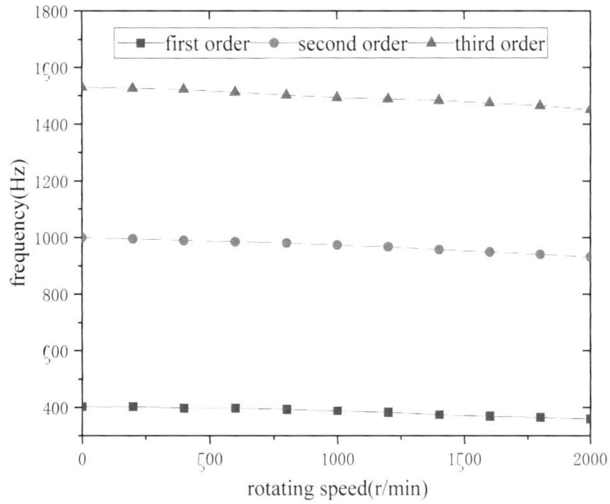


**Figure 5.** Influence of bearing stiffness on natural frequencies of the system.

## 6. Conclusion

In the bearing dynamic model in this paper, eccentric mass, the gyro moment and centrifugal force are considered to make the calculation results more accurate. The effects of preload, radial load and rotating speed on the dynamic characteristics of the bearing are analyzed. When the mixing shaft of concrete mixer is subjected to radial force, the bearing is subjected to radial load. With the increase of radial load, the axial static stiffness decreases, while the radial stiffness increases gradually, and the increasing trend of radial static stiffness is more obvious than that of axial static stiffness. The bearing stiffness will soften with the increase of rotating speed. And the greater the preload, the higher the softening point. Considering the softening phenomenon of angular contact ball bearing, the natural frequency of each order of the system decreases with the increase of rotating speed, which is consistent with the decrease of bearing stiffness with rotating speed. The work of this paper lays the foundation for the design and optimization of structure of concrete mixer.

## Acknowledgements

This work was supported by National Science Foundation of China: 51805337, National Science Foundation of China: 52175107, Key Laboratory of Vibration and Control of Aero-Propulsion System, Ministry of Education, Northeastern University: VCAME202008 and Scientific Research Project of Liaoning Provincial Education Department: lnjc202010.

## References

[1] Wonglert A, Jongpradist P, Jamsawang P and Larsson S 2018 Bearing capacity and failure behaviors of floating stiffened deep cement mixing columns under axial load *Soils Found.* **58** 2

[2] Sheng X, Li B, Wu Z and Li H 2014 Calculation of ball bearing speed-varying stiffness *Mech Mach Theory*. **81**
[3] Hernot X, Sartor M and Guillot J 2000 Calculation of the Stiffness Matrix of Angular Contact Ball Bearings by Using the Analytical Approach *J Mech Design*. **122** 1
[4] Noel D, Ritou M, Furet B and Le Loch S 2013 Complete Analytical Expression of the Stiffness Matrix of Angular Contact Ball Bearings *J Tribol*. **135** 4
[5] Yang Z H, Chen H and Yu T X 2018 Effects of rolling bearing configuration on stiffness of machine tool spindle *P I Mech Eng C-J Mec*. **232** 5
[6] El-Saeidy F 2011 Time-varying total stiffness matrix of a rigid machine spindle-angular contact ball bearings assembly: theory and analytical/experimental verifications *Shock Vib.* **18** 5
[7] Cong F, Chen J, Dong G and Pecht M 2013 Vibration model of rolling element bearings in a rotor-bearing system for fault diagnosis *J. Sound. Vib.* **332** 8
[8] Van Ngoc P, Turner B, Huang J and Kelly R 2017 Long-term strength of soil-cement columns in coastal areas *Soils Found.* **57** 4
[9] Zhang X N, Han Q K, Peng Z K and Chu F L 2013 Stability analysis of a rotor-bearing system with time-varying bearing stiffness due to finite number of balls and unbalanced force *J Sound Vib*. **332** 25
[10] Liu J, Tang C K, Wu H, Xu Z D and Wang L F 2019 An analytical calculation method of the load distribution and stiffness of an angular contact ball bearing *Mech Mach Theory*. **142**

IOP Conf. Series: Materials Science and Engineering 1242 (2022) 012040 doi:10.1088/1757-899X/1242/1/012040

# Research on basic principle and calibration experiment of fiber bragg grating sensor

**S Y Xing[1, 2], Q Wang[1*] and A X Liu[1, 3]**

[1] School of Mechanical Engineering, Shenyang Jianzhu University, Shenyang, China
[2] Shenyang Piotec Technology Co. Ltd, Shenyang, China
[3] North Huajin Chemical Industries Group Corporation, Panjin, China

* 591672678@qq.com

**Abstract**. Fiber bragg grating has the advantages of simple structure, light weight, small volume, low energy consumption, high sensitivity and measurement accuracy, and is widely used in the field of structural health monitoring. the structure and basic principle of fiber bragg grating sensor are introduced. the change of temperature, stress and other parameters will lead to the change of reflection center wavelength. The strain and temperature sensitivity coefficients of Fiber Bragg Grating sensors with different wavelengths are obtained. The fiber Bragg grating strain sensor is developed, and the calibration experiment of the sensor is carried out. The loading equipment is the loading tester. The linearity of the Fiber Bragg Grating sensor is good, and it has a good prospect of engineering civil application.

## 1. Introduction

Fiber bragg grating sensors is a kind of Fiber Sensor. Based on Fiber Grating, the sensing process can obtain the sensing information by external physical parameters, which is a wavelength modulation fiber sensor. Fiber bragg grating sensors are small in size and can realize distributed multi-point measurement. It has the advantages of electromagnetic interference resistance, high temperature resistance, radiation resistance, which can achieve easily remote real-time measurement and multi-signal measurement. And the key parameters of large structures can be monitored in real time through fiber bragg grating sensor. In 1978, K.O. Hill et al. [1] first discovered the light-induced grating effect of germanium-doped fiber by using the standing wave method in Canada. Using 1488 nm and 1514.5 nm argon ion laser to generate standing wave interference fringe in germanium doped fiber, the refractive index of optical fiber core changes periodically. The first Hill fiber Grating was successfully fabricated. In recent years, grating technology has been greatly developed due to the continuous research of scholars.

At present, the temperature compensation methods for fiber bragg grating sensors can be divided into two categories: one is automatic temperature compensation scheme, the other is passive temperature compensation scheme. In the single fiber bragg grating self-compensation method. Feng Liu et al. showed that the verification of radial pressure on grating sensor characteristics is analyzed by finite element software [2]. Hairuo Guo et al. analyzed a fiber mechanical-thermal composite multi-parameter sensing system, which is fed into the temperature and strain sensing unit and etalon composed of multi-channel fiber grating sensors, and the effectiveness of the design system is verified [3]. Qian Zheng designed a fiber grating demodulation system based on the working principle of fiber grating environmental monitoring sensor. The application of fiber Bragg grating sensor in environmental monitoring is analyzed [4]. Qiang Wang et al. proposed a calibration method for strain sensitivity coefficient of fiber bragg grating sensor under ultra-low temperature and wide temperature model, and the measurement results is verified through temperature experiments [5]. Yuegang Tan designed and implemented a kind of industrial fire warning fiber sensor with stainless steel capillary tube and reliable encapsulation, and greatly improved high temperature regeneration treatment [6]. Yuanze Xue et al. studied high temperature regenerative fiber Bragg grating (RFBG) temperature sensor and calibrated a high temperature regenerated fiber grating temperature sensor [7]. In addition, fiber bragg grating sensors are widely used in the engineering detection and monitoring of aircraft structures, composite cable structures and reinforced concrete structures.

According to the structure and basic principle of fiber Bragg grating sensor, the change of temperature, stress and other parameters will lead to the change of reflection center wavelength in the paper. The strain and temperature sensitivity coefficients of FBG sensors with different wavelength are calculated. The calibration experiment of the sensor was carried out. The research results have important technical support in practical application.

## 2. Basic principle of fiber grating sensing

According to the coupled mode theory, the central wavelength of the fiber grating $\lambda_B = 2n\Lambda$ (where $\lambda_B$ is the central wavelength of the fiber grating, $\Lambda$ is the grating period, and $n$ is the effective refractive index of the fiber core). It can be seen from the formula that the grating period $\Lambda$ and the effective refractive index $n$ of the fiber core have an impact on the reflected wavelength. Therefore, any parameter (such as temperature, stress, etc.) that can cause the change of the two values will lead to the change of the center wavelength of the reflected light. The schematic diagram of fiber bragg grating is shown in figure 1.

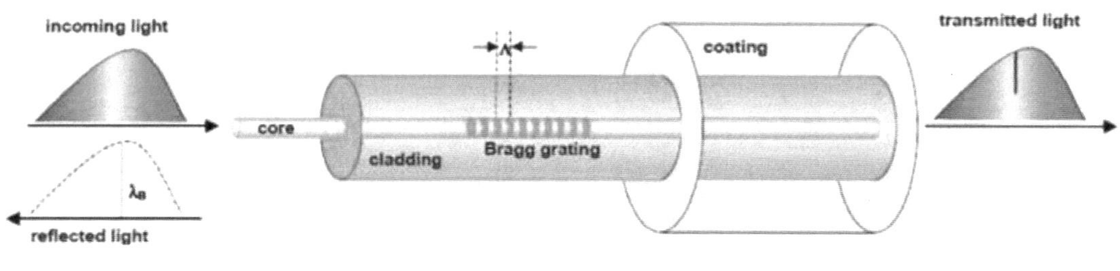


**Figure 1.** The schematic diagram of fiber bragg grating.

The sensing principle of fiber bragg grating sensor is that light waves with a certain bandwidth are shot into the fiber Bragg grating through a circulator using a broad spectrum light source (such as ASE or SLED). The qualified light waves are reflected. Because the wavelength of the fiber Bragg grating has a selective effect, the reflected light is sent to the wavelength measurement system again through the circulator, so that the change of the reflected wavelength of the fiber Bragg grating can be learned. When the fiber Bragg grating is used as the probe to measure the change of external temperature, stress or pressure, the grating spacing will change, resulting in the change of reflected wavelength. The change of external temperature, stress or pressure can be deduced by monitoring the change of wavelength.

## 3. Fiber grating calibration experiment

Loading equipment adopt load tester, models for CMT5000, is the combination of electronic technology and mechanical drive new material testing machine, with broad accurate loading speed and load range, measurement and control of load, deformation and displacement is of high accuracy and sensitivity. It can also be a constant loading, uniform deformation, the automatic control of constant displacement experiment. Calibration experiment of fiber Bragg grating sensor is shown in figure 2:

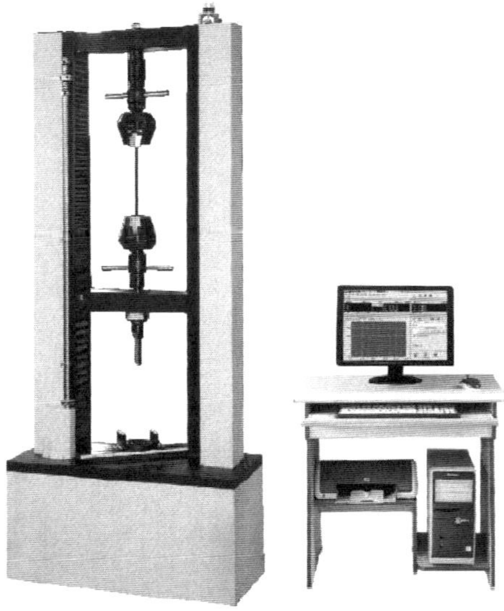

**Figure 2.** Calibration experiment of fiber bragg grating sensor.

Tensile experiment steps:

(1) The tensile steel plate is installed on the support of the new test machine, the support needs to clamp the steel plate, the fiber Bragg grating sensor is installed on the steel plate;

(2) The testing machine adopts force control, pre-loading 5 kN, so that the steel plate and fiber bragg grating sensor are in a state of stretching, loading 5 kN step by step, the testing machine loading force is 10 kN, 15 kN, 20 kN, 25 kN, 30 kN, until 35 kN;

(3) The fiber Bragg grating demodulation instrument saves and collects wavelength data instantly.

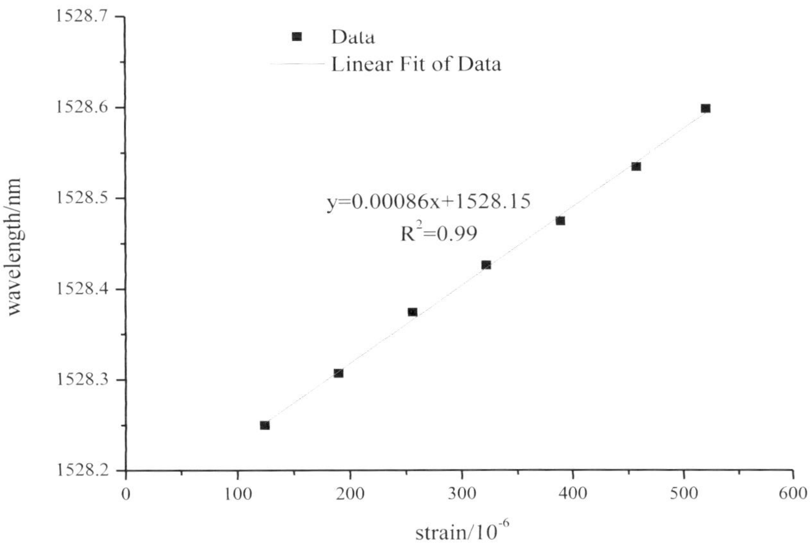


**Figure 3.** Calibration experiment of fiber grating sensor.

Figure 3 shows the calibration experiment curve of encapsulated fiber Bragg grating. The calibration curve equation is $Y = 0.00086x + 1528.15$, $R^2 = 0.99$, and the calibration coefficient is 0.00086 nm/με. The linearity of the curve is good, and the fiber Bragg grating sensor has a good prospect of engineering application.

## 4. Engineering applications

The sensor type is encapsulated fiber grating strain sensor. The reflection wavelength is 1555.4 nm, the sensor sensitivity coefficient is 5.8 pm/με, and the resolution is 1με. By pouring the concrete structure, fiber bragg grating strain sensor is embedded to obtain the internal force variation data of concrete structure under pouring, curing, mold removal and loading conditions, which can control and direct the on-site construction. The applicability and feasibility of fiber bragg grating sensor are verified. Concrete strain monitoring by fiber Bragg grating sensor is shown in figure 4.

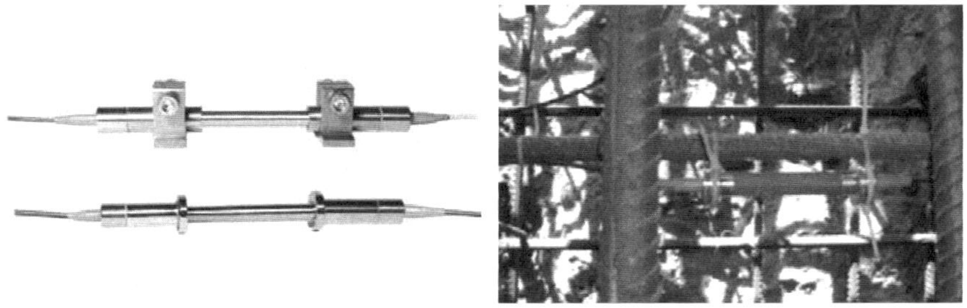

**Figure 4**. Concrete strain monitoring by fiber Bragg grating sensor.

## 5. Conclusion

In this paper, the coupling mode theory of fiber bragg grating sensor is studied, a reasonable fiber Bragg grating strain sensor model is established, a fiber Bragg grating strain sensor is developed, and sensor calibration experiments are carried out. The conclusions are:

(1) The change of environmental temperature, structural stress and other parameters will lead to the change of fiber grating period and core effective refractive index, and affect the change of reflection wavelength.

(2) The calibration coefficient of the developed fiber grating sensor is 0.00086 nm/με, and the linearity of the curve is good, $R^2 = 0.99$.

(3) Fiber bragg grating sensors have wide engineering application prospects due to their advantages of small volume, high sensitivity, corrosion resistance, electromagnetic interference resistance and high measurement accuracy.

**Acknowledgements**

The project is funded by Youth Science and technology Innovation Special Project(2021-HJQZ-08).

## References

[1] Hill K O, Fujii Y and Johnson D C 1978 Photosensitivity in optical fiber waveguides: Application to reflection filter fabrication *Appl. Phys. Let.* 32 10

[2] Liu F, Ma W L and Guo X 2015 Study on Radial Pressure Sensing Characteristics of EVA Sensitivity Enhanced Fiber Grating *Acta Optica Sinica.* **35** 10

[3] Guo H, Liu K, Jiang J F, Xu T H, Wang S, Sun Z S and et al 2021 Optical fiber high or low temperature mechanical and thermal demodulation system based on tunable laser Chinese Journal of Lasers. **48** 19

[4] Zheng Q 2021 Design and Application of Real-time Ambient Temperature Monitoring System of Fiber Grating *Technology and Industry* **2021** 7

[5] Wang Q, Ma G L, Gai W and Gu Z H 2021 Strain Calibration Method of FBG Sensor in Ultra-low Temperature and Wide Temperature Range *Journal of Vibration, Measurement &Diagnosis.* **41** 3

[6] Tan Y G, Wang X Fand Liu Y 2021 Industrial fire early warning optical fiber sensor and its sensing characteristics *Optical sensing.* **45** 3

[7] Xue Y Z, Wang X F and Tang C L 2021 Transducer and Microsystem Technologies *ACI Materials Journal.* **38** 5

# Study on the temperature field of stone processed by diamond tools

**Z Yang[1], Z Q Qu[1], R B Mu[1], K Y Ji[1] and M Zhao[2]**

[1] School of Intelligent Engineering, Shenyang City University, Shenyang, China
[2] School of Mechanical Engineering, College of Arts and Information Engineering, Dalian Polytechnic University, Dalian, China

* 174802854@qq.com

**Abstract.** The experiments were designed by testing methods and orthogonal single factor experimental design. XK5032C CNC milling machine, the equipment to achieve X-axis, Y-axis and 3-axis Z direction of the automatic feed, spindle cooling method using water-cooled. In different processing parameters (spindle speed, depth of cut and feed rate), measurement of the surface temperature of diamond tools. In the testing experiment result analysis, we conclude, the depth of cutting is the biggest influence on the surface temperature. From the orthogonal experiment data analysis, we found that, the temperature was reduced with spindle rotation but opposite with two other factors. By linear regression analysis. We get the best cutting results by choosing a large spindle rotation and a small feed rate.

## 1. Introduction

With the development of the world economy, people's living standards, natural stone, jade, artificial senior stone, ceramics, glass, and cement products have been widely used in housing, building decoration, making a variety of decorations, household items and road, bridge construction [1,2]. The processing of these materials all require a variety of different diamond tools. Affect the diamond milling cutter service life of many factors, in addition to its own composition and structure also includes diamond milling cutter performance, stone performance and processing process parameters. Stone processing process due to different processing methods, processing parameters are also different [3]. In the process of CNC cutting stone, cutting parameters are a major parameter affecting processing costs and processing efficiency. The main cutting parameters are spindle speed, feed rate and feed depth. By systematically studying the influence of different cutting parameters and stone characteristics on the surface temperature of diamond tools, we can optimize machining parameters and reasonably select diamond tools, reduce machining costs and improve production efficiency [4].

## 2. Experimental section

### *2.1. Equipment and materials*

XK5032C CNC milling machine, the equipment to achieve X-axis, Y-axis and 3-axis Z direction of the automatic feed, spindle cooling method using water-cooled.

Infrared thermometer, infrared thermometer consists of optical system, photoelectric detector, signal amplifier and signal processing, display output and other parts.

According to its basic characteristics or composition of the stone can be divided into granite, marble, and slate three categories (as shown in Table 1). The main finishing stones for building decoration are marble, granite and sandstone. The main processes are sawing, grinding and polishing, cutting finished panels, trimming and chamfering and slotting.

**Table 1.** Classification of stones according to their basic characteristics or composition.

| Category | Main components | Texture |
|---|---|---|
| Granite | A magmatic rock with aluminosilicate as the main component. Its main chemical composition is aluminum oxide and silicon oxide, and a small amount of calcium oxide, magnesium oxide | Harder |
| Marble | Chemical composition is calcium carbonate, about 50% or more, and magnesium carbonate, calcium oxide, manganese oxide and silicon dioxide, etc. | Softer |
| Sandstone | Quartz sandstone, feldspar sandstone, quartz chalk sandstone, sandy slate | Softer |

Diamond Thin Wall Drill, Diamond Thin Wall Drill is made of Cu-based metal powder, pressed by sinter molding method and welded to the HSS shank. The temperature generated on the tool surface is measured with an infrared measuring instrument in the range of -50 to 650. Direct measurement at room temperature without coolant.

Diamond circular saw, diamond circular saw is made of Cu-based metal powder, pressed into the bit by sinter modelling method and welded to the HSS shank.

### 2.2. *Experimental Analysis*

Single factor analysis and orthogonal experimental analysis were used to analyze the effect of diamond tools on their surface temperature when processing different parameters and different stones. Single factor mainly from the spindle rotation, feed speed and feed depth to study which parameter has the most influence on the tool surface temperature, and analyze what kind of stone cutting surface temperature of diamond tools is the highest. The orthogonal experiments were assigned using the orthogonal experimental table with the same main parameters as the single factor, and the most orthogonal experiments for marble and sandstone. For the orthogonal experiments, linear regression analysis was used to analyze each of these parameters, and finally, an empirical formula was derived.

## 3. Results and Discussion

Analysis of single-factor experimental results of drilling different stones with diamond tools.

**Table 2.** Temperature values for processing three types of stone.

| Stone types | Granite | Marble | Sandstone |
|---|---|---|---|
| 1 | 46 | 25.4 | 14 |
| 2 | 78 | 31.9 | 19.5 |
| 3 | 101 | 43.1 | 23 |
| 4 | 109 | 59.6 | 30 |
| 5 | 150 | 63 | 34 |
| 6 | 178 | 78 | 36 |
| 7 | 204 | 116 | 45 |
| 8 | 243 | 125 | 51 |
| AVERAGE | 138.63 | 67.75 | 31.56 |

In order to systematically study the effect of diamond tool processing stone on its surface temperature, granite, marble and sandstone were used to analyze the effect of different stones on diamond surface temperature under the same parameters. Table 2 shows the temperature values of three different stones at a spindle speed of 300 r/min and a feed rate of 15 mm/min at a feed depth of 3 mm.

Figure 1 variation curve of surface temperature of diamond thin-walled drill when processing three kinds of stone under the same processing parameters. It can be seen from the figure that Granite had the highest temperature with a rise rate of 28.1, while both marble and sandstone had 14.2 and 5.3. The temperature profile of sandstone is relatively flat with a relatively small range of temperature variation, while the granite profile is broader, with a maximum temperature of 243, which is five times higher than the minimum temperature. Marble, on the other hand, lies between granite and sandstone, and the temperature rise range is not very wide.

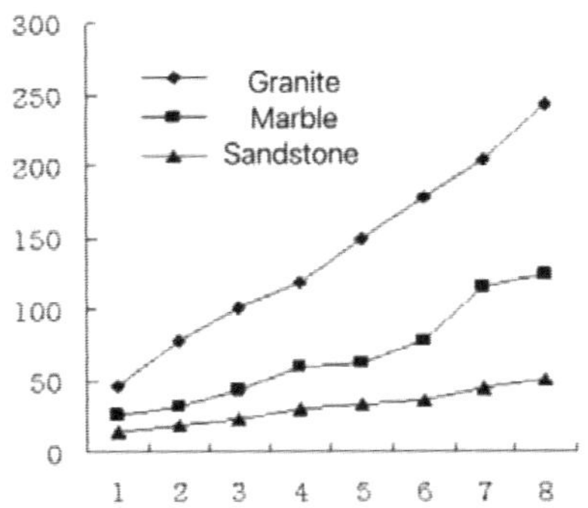


**Figure 1.** The change of surface temperature when three kinds of stone.

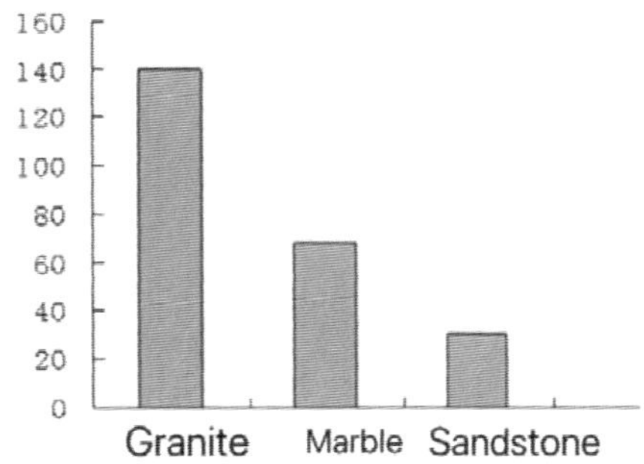


**Figure 2.** Average temperature profile of three types of stone.

Figure 2, temperature profiles and average temperature value profiles of three stones under the same processing parameters. It can be seen from the graph that the diamond surface temperature is maximum when drilling granite. Through the stone properties learned that the density of granite 2.5 ~ 2.7 kg/cm$^3$, compressive strength of 120 ~ 250 MPa, the density of marble 2.5 ~ 2.7 kg/cm$^3$, compressive strength of 47 ~ 120 MPa, sandstone density of 2.2 ~ 2.5 kg/cm$^3$, compressive strength of 47 ~ 120 MPa. The density of the three stones is basically the same, but the compressive strength of granite is much greater than that of marble and sandstone, almost twice that of marble and sandstone. The strength of the stone increases, the cutting force increases, and the diamond tool wears severely, leading to an increase in the surface temperature of the diamond drill bit. From figure 2. you can see that the surface temperature of the diamond tool when processing granite is 2 times that of processing marble and 5 times that of sandstone. From the machining experiment, we can see that when dry cutting granite, the surface temperature of the diamond tool rises very quickly while the temperature is very high, leading to the diamond bit falling off. The heat transfer coefficient of granite is 3.49 W/m K, while the heat transfer coefficient of copper is 370.49 W/m K. From this, we can see that most of the heat is transferred through the diamond bit for heat transfer, which leads to a sharp increase in the surface temperature of the diamond bit and serious diamond wear. The highest temperature is reached when processing granite, up to 243°C, while the lowest temperature is reached when processing sandstone, whose maximum value is 51°C.

### *3.1. The effect of diamond disc sawing marble on its surface temperature*

In order to study the characteristics of diamond tools, a diamond circular saw was used to cut marble to study the amount of change in the surface temperature of diamond tools. As shown in Figure 3, diamond circular saw cutting marble and measuring the diamond surface temperature of the schematic diagram.

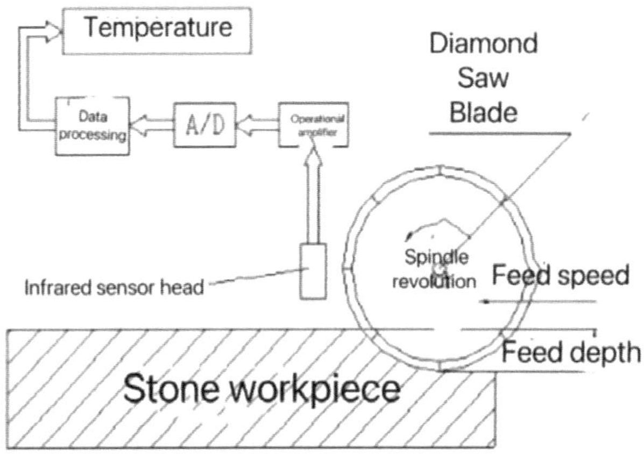


**Figure 3.** Diamond Isc Saw processing stone and temperature measurement diagram.

To obtain a more accurate picture of the factors affecting the temperature, we used the average temperature to synthesize the temperature profile.

**Table 3.** Combined analysis table data of different factors.

| Comprehensive Analysis (take the average value of temperature) | | | |
|---|---|---|---|
| Change spindle speed | 15.33 | 17.83 | 18.61 |
| Changing the feed depth | 17.83 | 20.1 | 23.52 |
| Changing the feed speed | 20.1 | 21.92 | 24.84 |

Figure 4, the experiment was conducted at a feed rate of 20 mm/min and a feed depth of 5 mm. From the average temperature graph in the figure 4, it can be seen that the temperature increases with the increase of spindle revolution, but the rise is not very large, the spindle revolution does not have a great influence on the temperature, and the temperature range is 15-19 degrees.

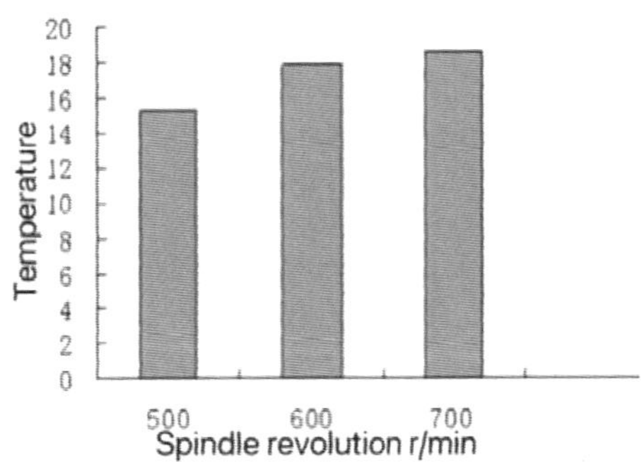


**Figure 4.** Spindle rotation and average temperature analysis.

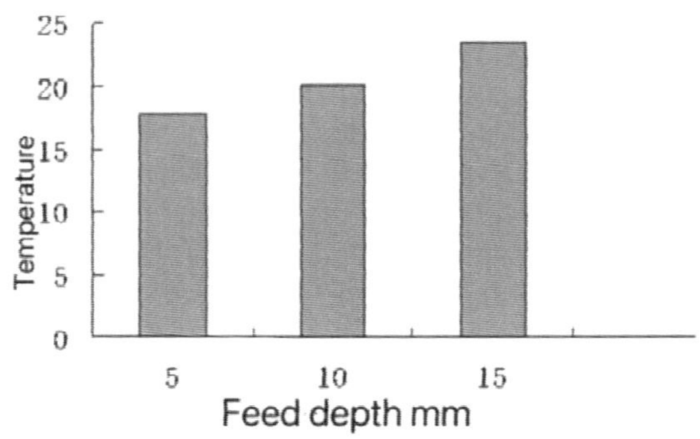


**Figure 5.** Feed depth average temperature analysis.

Figure 5, the spindle rotation is 600 r/min and the feed rate is 20 mm/min. From the graph, it can be out that the temperature increases with the increase of the feed depth, and the temperature is between 18 and 25.

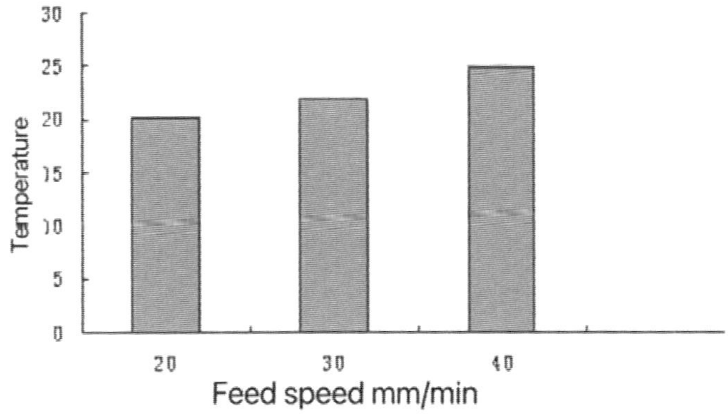


**Figure 6.** Feed rate average temperature analysis.

Figure 6, it was performed at a spindle speed of 600 r/min and a feed depth of 10 mm. The graph shows that the temperature increases with the increase of the feed rate.

### *3.2. Study on the orthogonality experimental method of cutting sandstone 1 by thin-walled diamond drilling*

**Table 4.** Results of sandstone orthogonality experiments.

| | | | | | | | | | | temperature variation | | | |
|---|---|---|---|---|---|---|---|---|---|---|---|---|---|
| Experimental number | 1 | 2 | 3 | 4 | 5 | 6 | 7 | 8 | 9 | I | II | III | R1 |
| A（Spindle speed）r/min | 300 | 300 | 300 | 400 | 400 | 400 | 500 | 500 | 500 | 83.13 | 87.79 | 117.20 | 34.07 |
| B（Feeding speed）mm/min | 10 | 15 | 20 | 10 | 15 | 20 | 10 | 15 | 20 | 84.36 | 97.76 | 106 | 21.64 |
| Temperature variation（°C） | 25.69 | 27.58 | 29.86 | 26.46 | 29.52 | 31.81 | 32.21 | 40.66 | 44.33 | | | | |

When processing stone with diamond thin-walled drill, there are many parameters that affect the surface temperature of thin-walled drill, including the physical and mechanical properties of the drill bit, physical and mechanical properties of the stone, processing parameters, cooling parameters, etc. The experiments focus on the influence of processing parameters and stone properties on the surface temperature of the drill bit. The orthogonal experimental analysis method was applied to analyze the diamond thin-walled drill. From the extreme difference R of the experimental results, the spindle speed has the greatest influence on the temperature variation, followed by the feed rate. From figure 7, it can be seen that the amount of temperature change of the diamond tool increases with the increase of the tool spindle speed and feed rate. In terms of the total amount of temperature change, the spindle speed has the greatest effect on the amount of temperature change.

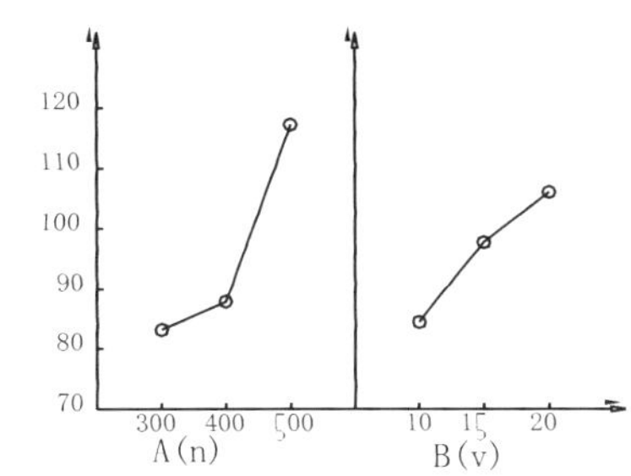


**Figure 7.** Orthogonal data plot.

*3.3. Study on the orthogonality experimental method of cutting sandstone 2 by diamond saw blade*

**Figure 8.** Sawing sandstone track.

**Table 5.** Sandstone orthogonal experimental data.

| Experiment | 1 | 2 | 3 | 4 | 5 | 6 | 7 | 8 | 9 | temperature variation | | | |
|---|---|---|---|---|---|---|---|---|---|---|---|---|---|
| | | | | | | | | | | I | II | III | R1 |
| A (Spindle revolutions) | 500 | 500 | 500 | 600 | 600 | 600 | 700 | 700 | 700 | 86.61 | 57.88 | 49.9 | 36.71 |
| B (Feeding speed) | 20 | 30 | 40 | 20 | 30 | 40 | 20 | 30 | 40 | 47.79 | 60.26 | 73.2 | 25.41 |
| C Cutting depth) | 5 | 10 | 15 | 10 | 15 | 5 | 15 | 5 | 10 | 42.01 | 57.72 | 81.52 | 39.51 |
| Experimental results | 13.14 | 21.59 | 38.74 | 17.14 | 25.27 | 15.47 | 17.51 | 13.40 | 18.99 | | | | |

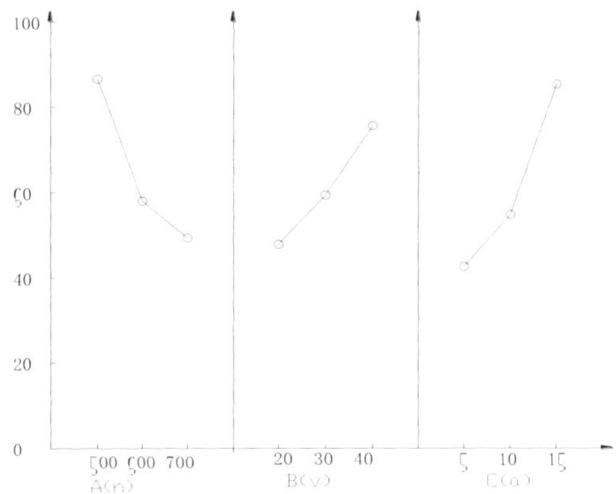


**Figure 9.** Orthogonal data plot.

The method of extreme difference analysis of orthogonal experimental results, the extreme difference R value of experimental results shows that the depth of cut has the greatest amount of influence on the tool surface temperature, followed by the feed depth, and the least influence is the spindle speed. From Figure 9. we can see that the surface temperature of the diamond tool decreases with the increase of the tool spindle speed, but increases with the increase of the tool feed rate and depth of cut.

The maximum amount of temperature variation in depth of cut is 2.5 times the minimum, and the temperature variation is relatively large.In terms of total temperature variation, the effect of feed depth on tool surface temperature is an order of magnitude greater than the other two parameters. If the minimum value of the tool surface temperature is the index parameter, it can be seen from the direct results of the experiment that the best group 1 experiment is used i.e. tool spindle speed of 500 r/min, feed rate of 20 mm/m in and cutting depth of 5 mm.

## 4. Linear Regression Analysis

When machining stone with diamond tools, the machining parameters (spindle revolution n, feed speed vf and feed depth ap) have a significant effect on the tool surface temperature. In most of the experiments, the feed depth ap is the main influencing factor. However, due to the combined effect of cutting parameters, spindle speed n, feed rate vf and feed depth ap, it is difficult to draw quantitative conclusions on the effect law of tool surface temperature. In this paper, the parameters of machining parameters that have a large impact on the tool surface temperature (spindle revolution n, feed speed vf and feed depth ap) are selected to establish a prediction model and apply the empirical formula.

$$T = AX^aY^bZ^c \tag{1}$$

Empirical formula for diamond surface temperature T :

$$T = 10^{2.4939} X^{-0.8949} Y^{0.5094} Z^{0.5471} \tag{2}$$

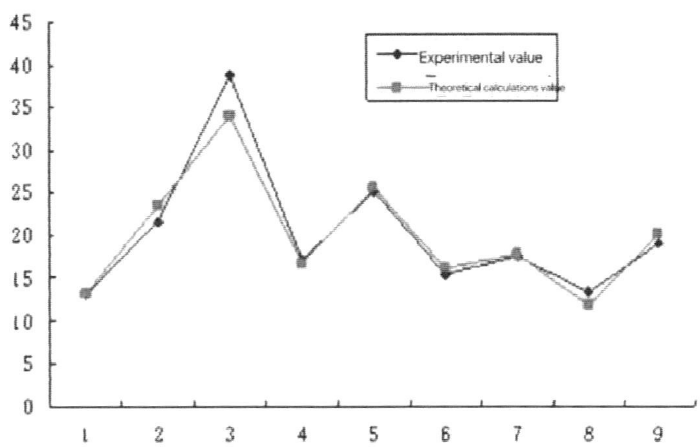


**Figure 10.** Comparison of experimental value and calculated value.

Apply equation (2) to calculate the diamond surface temperature T. The calculated value is compared with the actual measured value as shown in Figure 10. According to the above calculation results, the diamond surface temperature model fits well with the actual processing situation, and the prediction of diamond disc saw surface temperature with this model has great confidence.

## 5. Conclusion

To study the surface temperature of diamond tools used in stone production processing, the influence of processing parameters (spindle speed, feed rate, depth of cut) and stone properties on the surface temperature of diamond tools, and to find out the optimal processing parameters through experiments so as to improve the tool life.

The main parameters of stone processing are spindle rpm, feed depth, and feed speed, where feed depth has the greatest effect on tool surface temperature, followed by feed speed and spindle rpm. Diamond tool processing different stone, cutting granite when the highest tool surface temperature, followed by marble, sandstone is the smallest. This is due to the nature of the three types of stone, granite is harder and denser, cutting will generate a lot of heat, while sandstone is loose and generates less heat, marble is in between them.

In the orthogonal experiments, the temperature decreases with increasing spindle revolution and increases with increasing feed depth and feed speed. Applying linear regression analysis, the empirical equation (2) is derived. In order to get the best cutting effect, choosing a large spindle revolution and a small feed rate is the best choice.

In the case of maintaining production efficiency, should try to improve the speed of diamond tools, appropriate to reduce the feed depth and feed speed, so as to help reduce the surface temperature of the tool, thereby improving the life of the tool.

## References

[1] Zheng L X and Zhang C H **2001** Scientific and technological innovation - the road of diamond industry from big to strong [J]. *Stone*, **(3):25-28**

[2] Chen Y C and Kang Y **1996** Application of diamond tools in the construction materials industry. [J] *Journal of Southwest Engineering College*, **(4): 57-61**

[3] Zhao M, Lu D and Zhang X X **2008** Study on the effect of machining parameters on the wear amount of diamond milling cutters when milling stone [J]. *Stone*, **4(2):49-52**

[4] Cui M M **2016** Study on the wear reliability of diamond milling cutters [D]. *Shenyang: Shenyang University of Architecture*

# Experimental research of main characteristics in a complex multi-layer plate heat exchanger

**V S Yezhov[1], N E Semicheva[1*], A P Burtsev[1], A V Brezhnev[1] and N S Perepelitsa[1]**

[1] Southwest State University, Kursk, Russian Federation

* nsemicheva@yandex.ru

**Abstract.** The objective of this paper is research of main heat exchange characteristics in a complex multi-layer heat-exchanger with the purpose of determining heat exchanging process intensification dependencies and conditions during utilization of ventilation gas and emission low-grade heat with associated recovery of thermoelectricity.

## 1. Introduction

One of the main trends of increasing ventilation facility effectiveness is using low-grade heat utilization devices and equipment systems that allow to reduce harmful substance and low-grade heat emission aiming to protect the environment and improve ecological safety of the territory surrounding the building [1].

## 2. Methods

To reach the purposes mentioned above, a test unit was developed (figure 1a). It is comprised of two blocks: a complex multi-layer plate heat exchanger (CMPHE) and an absorber unit. CMPHE unit (figure 1b) is made in the form of ribbed heat transmitting sections (figure 2a) that have semiconducting thermoelectric Peltier elements installed inside them (figure 2b). Cylindrical vortex generators [1-5] are installed on the ribbing for heat transmission process intensification.

Air heated up in a calorifier was used as a heat-carrying agent. Air, collected by the vent from the laboratory, was used as an operating environment. Research experiments on heat exchange with counterflow in a thermoelectric section were conducted for a range of fixed speeds and air consumption rates at the test unit.

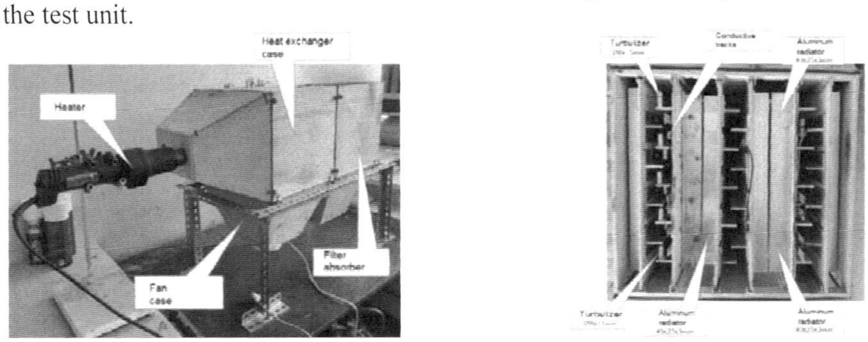


(a) (b)

**Figure 1.** Test unit: a – overall view; b – hot air flow channels

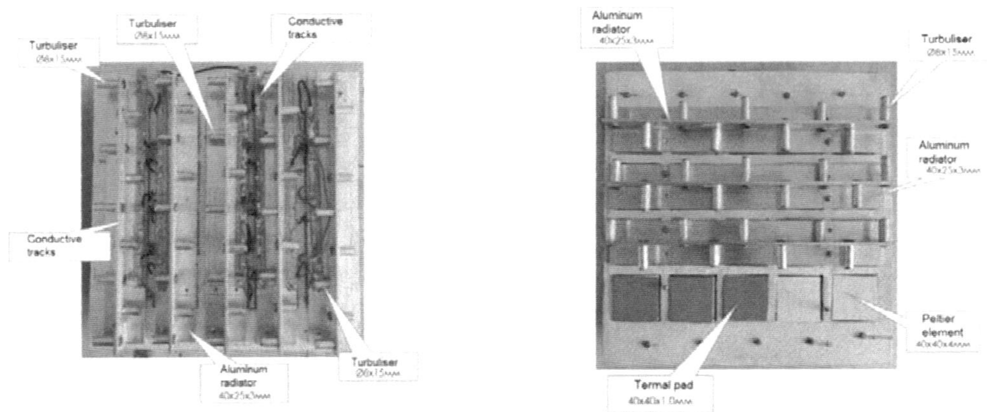


(a) (b)

**Figure 2.** Section of the test unit: a – cold side view; b – hot side view

Below are listed the main characteristics of the experimental equipment:
- thermal electromotive force quotient $\alpha = 12.97\cdot10^{-3}$ V/K;
- quality factor = $2.8\cdot10^{-3}$ $K^{-1}$;
- electric conductivity quotient $\sigma = 8\cdot10^{4}\,\Omega^{-1}\cdot m^{-1}$;
- element thermal conductivity quotient $\lambda = 146$ W/m$^{2}\cdot$°C;
- aluminum thermal conductivity quotient $\lambda = 220$ W/m$^{2}\cdot$°C;
- thermoelectric Peltier element size 40x40 mm;
- aluminum radiator, thickness - 3.0 mm, thermal conductivity coefficient - 220 W/(m·°C);
- thermal pad, width - 1.0 mm, thermal conductivity quotient – 6.0 W/(m·°C);
- Peltier element, width - 4.0 mm, thermal conductivity quotient 146.0 W/(m·°C).

Experiments were conducted under three calorifier operating modes. Experimental results are depicted in tables 1 and 2.

**Table 1.** Experimental results without vortex generators installed

| Operating mode | Hot air input temperature | Hot air output temperature | Cold air input temperature | Cold air output temperature | Voltage of 1 section, mV | Total voltage, mV | Electric current in 1 section, mA | Total electric current, mA | Hot flow speed, m/s | Cold flow speed, m/s |
|---|---|---|---|---|---|---|---|---|---|---|
| 1 | 50 | 42 | 20 | 21 | 27 | 60 | 2.4 | 2.4 | 2.4 | 1.9 |
| 2 | 70 | 58 | 20 | 21 | 63 | 150 | 6.9 | 6.9 | 2.7 | 1.9 |
| 3 | 90 | 75 | 20 | 22 | 144 | 357 | 13.8 | 13.8 | 3.0 | 1.9 |

**Table 2.** Experimental results with vortex generators installed

| Operating mode | Hot air input temperature | Hot air output temperature | Cold air input temperature | Cold air output temperature | Voltage of 1 section, mV | Total voltage, mV | Electric current in 1 section, mA | Total electric current mA | Hot flow speed m/s | Cold flow speed m/s |
|---|---|---|---|---|---|---|---|---|---|---|
| 1 | 50 | 29 | 20 | 23 | 90 | 200 | 8 | 8 | 1.9 | 1.2 |
| 2 | 70 | 41 | 20 | 23 | 210 | 500 | 23 | 23 | 2.4 | 1.2 |
| 3 | 90 | 60 | 20 | 24 | 480 | 1020 | 46 | 46 | 2.7 | 1.2 |

As seen in the experimental data, vortex generators installation has drastically increased electricity production as per unit of the thermoelectric element, as for the experiment setup in general. Introduce examples from the tables.

Besides, hot flow heat exchanger output temperatures have made a substantial decrease, consequentially increasing temperatures of the cold air entering the laboratory.

Moreover, as a result of increased aerodynamic resistance, a decrease in hot air flow speed has been observed, which is also depicted in the air pressure range oscillogram (figure 3).

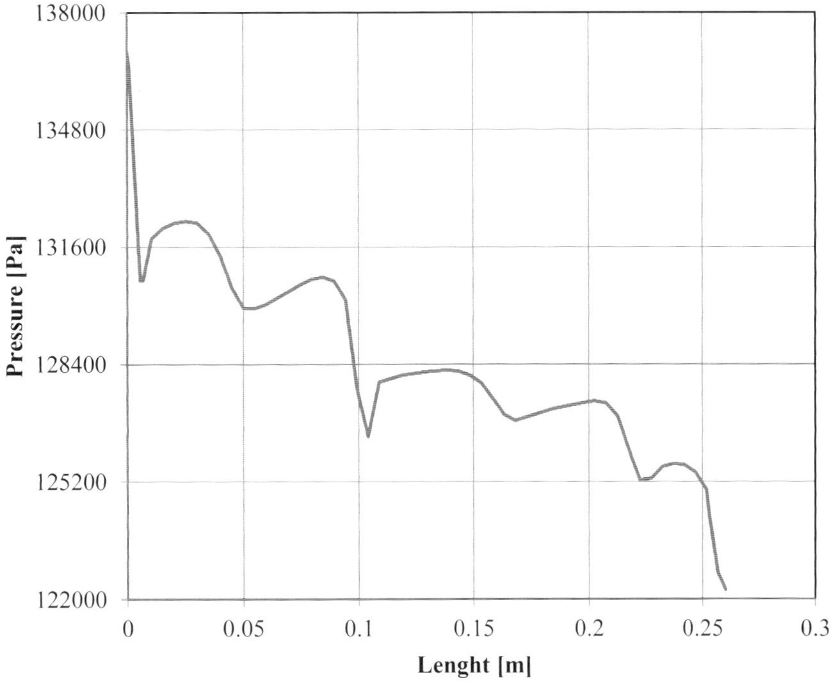


**Figure 3.** Oscillogram of hot air flow pressure range.

## 3. Results and Discussion

Based on the collected data, a calculation method of main thermo-technical characteristics of a complex multi-layer plate heat exchanger is presented below:

Average temperature in the heating channel: $\bar{t}'=\frac{50+29}{2}=39.5\ ^{o}C$

Average temperature in the cooling channel: $\bar{t}''=\frac{20+23}{2}=21.5\ ^{o}C$

Thermophysical properties of air at the pressure of $P=1.013\cdot10^{3}$ Pa are takenfrom reference data at the temperature of +39.5°C: $\lambda_1=0.0276$ W/m$^{2}\cdot$°C; $\nu_1=16.91\cdot10^{-6}$ [m$^2$/c], $Pr_1=0.699$.

Thermophysical properties of air at the pressure of $P=1.013\cdot10^{3}$ [Pa] are takenfrom reference data at the temperature of +21.5°C: $\lambda_1=0.026$ W/m$^{2}\cdot$°C; $\nu_1=15.2\cdot10^{-6}$[m$^2$/c], $Pr_1=0.703$.

Equivalent channel diameter calculation is done by the equation [m]:

$$d_i=\frac{4\cdot S}{P} \tag{1}$$

where: S – sectional area, m$^2$ and P – channel perimeter, m.

Equivalent heating channel diameter: $d_1=\frac{4\cdot(0.27\cdot0.04)}{2\cdot(0.27+0.04)}=0.07$ m.

Equivalent cooling channel diameter: $d_2=\frac{3\cdot(0.27\cdot0.04)}{2\cdot(0.27+0.04)}=0.05$ m.

The Reynolds number:

$$\mathrm{Re}_1=\frac{w_1 \cdot d_1}{\nu_1} \tag{2}$$

$$\mathrm{Re}_1=\frac{1.9 \cdot 0.07}{16.91 \cdot 10^{-6}}=7865.2$$

$$\mathrm{Re}_2=\frac{1.2 \cdot 0.05}{15.2 \cdot 10^{-6}}=3947.4$$

With a calculated Reynolds number, the operating mode is turbulent, therefore,Nusselt number criterial equation will look as follows:

$$Nu_{hc}=0.037 \cdot \mathrm{Re}_1^{\ 0.8} \cdot \mathrm{Pr}_1^{\ 0.43} \tag{3}$$

$$Nu_{hc}=0.037 \cdot 7865.2^{0,8} \cdot 0.699^{0,43}=41.5$$

Quotient of heat transfer from the heating medium to the radiator wall α1, [W/m²·°C]:

$$\alpha_1 = Nu_1 \cdot \frac{\lambda_1}{d_1} \tag{4}$$

$$\alpha_1=41.5 \cdot \frac{0,0276}{0,07}=16.4 \ [\mathrm{W/m^2 \cdot {}^{\circ}C}].$$

The Grashof number:

$$Gr = g \cdot d_2^{\ 3} \cdot \frac{\left[\left(\frac{1}{273+t_г}\right) \cdot (t_г - t_x)\right]}{\nu_2^{\ 2}} \tag{5}$$

$$G_r=9.81 \cdot 0.07^3 \cdot \frac{\left[\left(\frac{1}{273+39.5}\right) \cdot (39.5\text{-}21.5)\right]}{\left(16.91 \cdot 10^{-6}\right)^2}=6.77 \cdot 10^5$$

The Nusselt number criterial equation:

$$Nu_2 = 0,5 \cdot \left(\mathrm{Pr} \cdot Gr\right)_2^{0.25} \tag{6}$$

$$Nu_2=0.5 \cdot \left(0.699 \cdot 6.77 \cdot 10^5\right)^{0.25}=13.1$$

Quotient of heat transfer from radiator's outer surface to Peltier elements $\alpha_2$:

$$\alpha_2=13.1 \cdot \frac{0.026}{0.07}=4.87 \ \mathrm{W/m^2 \cdot {}^{\circ}C}.$$

Multi-layer wall heat transfer quotient:

$$K_I=\frac{1}{\left[\frac{1}{\alpha_1}+\Sigma\frac{\delta_i}{\lambda_i}+\frac{1}{\alpha_2}\right]} \tag{7}$$

$$K_I=\frac{1}{\left[\frac{1}{12.5}+2 \cdot \frac{0.003}{220}+2 \cdot \frac{0.001}{6.0}+\frac{0.004}{146.0}+\frac{1}{4.87}\right]}=3.55 \ \mathrm{W/(m \cdot {}^{\circ}C)}$$

Heat flow, W:

$$\mathrm{Q_h}=\mathrm{K_I} \cdot \mathrm{F_{ст}} \cdot \left(\bar{t}'\text{-}\bar{t}''\right) \tag{8}$$

$$\mathrm{Q_h}=(0.04 \cdot 0.04 \cdot 20) \cdot 3.55 \cdot (39.5\text{-}21.5)=2.04 \ \mathrm{W}$$

At the temperature of 39.5°C the main characteristics of produced electricity are:
- voltage, V: $U = 0.2\mathrm{V}$

- electric current, mA: $I = 8.0$mA

In accordance with the methodology given in [6], the auxiliary coefficient is determined by the formula:

$$m=\sqrt{1+0.5\cdot Z\cdot\left(t_1^{hc}+t_2^{cl}\right)} \tag{9}$$

$$m=\sqrt{1+0.5\cdot 2.8\cdot 10^{-3}\cdot(39.5+21.5)}=1.04$$

Thermoelectric converter resistance in accordance with reference data: R = 1.5 Ω.

Maximum electric power in accordance with reference data: P = 76 W. Efficiency [%]:

$$\eta=\frac{Q_h}{P}\times 100\% \tag{10}$$

$$\eta=\frac{2.04}{76}\times 100\%=2.68\%$$

Determining a system's cooling power requires calculating the cooling power of one Peltier element, W:

$$P_o=\frac{Q_o}{\varepsilon_{max}} \tag{11}$$

Refrigeration capacity of a thermal element with loss taken into account is determined by the equation [**Error! Reference source not found.**]:

$$Q_o=\alpha\cdot T_x\cdot I-0.5I^2R-\lambda(T_r-T_x) \tag{12}$$

$$Q_o=3.7\cdot 10^{-3}\cdot 39.5\cdot 0.008-0.5\cdot 0.008^2\cdot 1.5-3.54(39.5-21.5)=-63.7W$$

Maximum refrigerating quotient of a thermal element's reverse cycle, in which the working medium is represented by an electron gas and permanent loss is absent, is determined by the equation:

$$\varepsilon\frac{T_x}{T_r-T_x}\frac{m-\frac{T_r}{T_x}}{m+1}_{max} \tag{13}$$

$$\varepsilon\frac{21.5}{39.5-21.5}\frac{1.04-\frac{39.5}{21.5}}{1.04+1}_{max}$$

$$P_o=\frac{-63.7}{-0.47}=135.96W$$

Heat emitted by the hot junction outweighs the heat absorbed by the cold junction by the amount of electric energy expenditure:

$$Q_h^{'}=P_o+Q_o \tag{14}$$

$$Q_h^{'}=135.96+(-63.7)=72.06W$$

This given value corresponds with reference data of selected thermoelectric Peltier elements' semiconductors – maximum electric power in accordance withreference data is P = 76 W.

Dependence diagram of the multi-layer wall heat transfer quotient and the heatflow perceived by one Peltier element in relation to electric power generated by thesetup is demonstrated in figure 3.

Calculation results of main characteristics for the experimental setup's three operating modes are depicted in table 3.

**Table 3.** Results of main characteristics for three operatingmodes of the experimental setup.

| Parameter name | Setup operating mode | | |
|---|---|---|---|
| | 1 | 2 | 3 |
| Reynolds number | 7865.2 | 9076.2 | 9192.6 |
| Nusselt number criterial equation | 41.5 | 46.5 | 46.8 |
| Quotient of heat transfer from the heating medium to the radiator wall, W/m²·°C | 16.4 | 19.1 | 20.1 |
| The Reynolds number | 3947.4 | 3947.4 | 3888.5 |
| Nusselt number criterial equation | 24 | 24 | 23.7 |
| Quotient of heat transfer from the heating medium to the radiator wall, W/m²·°C | 12.5 | 12.5 | 12.4 |
| The Grashof number | 677795.1 | 1016468.3 | 1212309.6 |
| Nusselt number criterial equation | 13.1 | 14.5 | 15.1 |
| Quotient of heat transfer from the radiator's outer surface to Peltier elements | 4.87 | 5.39 | 565 |
| Multi-layer wall heat transferquotient, W/m²·°C | 3.55 | 3.92 | 4.11 |
| Heat flow received by 1 Peltierelement, W | 2.04 | 4.26 | 6.97 |
| Auxiliary quotient | 1.04 | 1.05 | 1.07 |
| Energy conversion efficiency, % | 2.68% | 5.61% | 9.17% |
| Thermal element's refrigeration capacity with losses taken into account, W | -63.9 | -133.3 | -217.8 |
| Maximum refrigerating quotient of athermal element's reverse cycle | -0.47 | -47 | -0.47 |
| Refrigerating power, W | 135.96 | 283.62 | 463.4 |
| Heat per unit of thermoelectricityproduction, W | 72.06 | 150.32 | 245.6 |

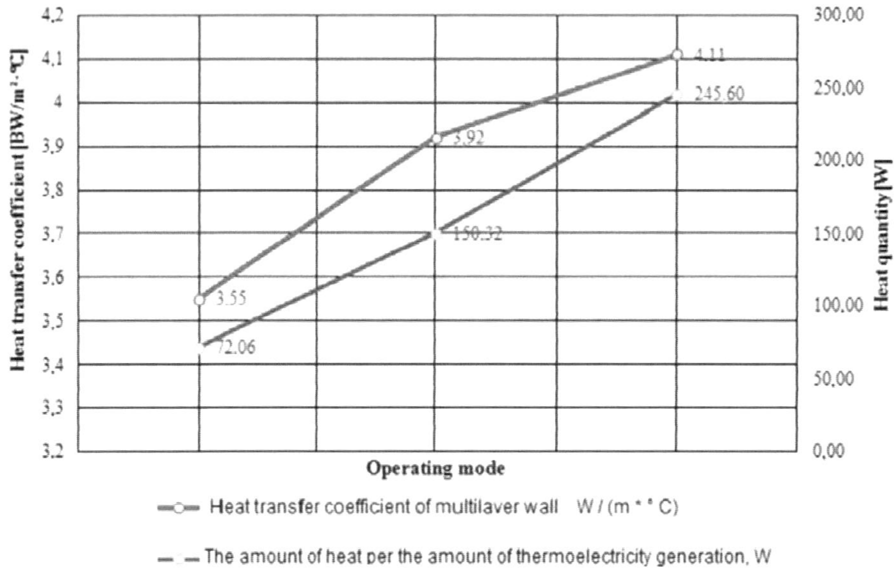


**Figure 4.** Relationship between multi-layer wall heat transfer quotient and electric power produced.

## 4. Conclusions

Developed a complex multi-layer plate heat exchanger (CMPHE) experimental setup comprised of a thermoelectric source of EMF that allows disposal of low grade ventilation emission heat with temperature ranging from 30°C to 90°C of residential, public and industrial buildings with associated partial conversion of heat to electricity;

Proposed a method of calculating multi-layer wall heat transfer quotient with CMPHE operating at three different modes;

Determined a direct correlation between a multi-layer wall heat transfer quotient, heat consumed by a Peltier element, and the amount of electricity produced.

The resulting electricity can be used for additional heating of the supply air or to ensure the autonomous operation of the automation and control system of the ventilation unit

**References**

[1] Ezhov, V. S., Evgenievna, S. N., Nikolaevich, T. D., Petrovich, B. A., & Sergeevich, P. N. 2021 Version of a mathematical model of purge ventilation system with a complex recuperative heat exchanger, *Journal of Applied Engineering Science*, **19**(1), 246-251

[2] Ezhov V S, Semicheva N E, Burtsev A P, Perepelitsa N S 2020 Complex heat exchanger made of multilayer plates *Patent 2737574 RF* dated 01.12.2020 11 p

[3] Ezhov V S, Burtsev A P 2020 Plate heat exchanger *Patent 2736316 RF* dated 08/13/2020 11 p.

[4] Ezhov V S, Semicheva N E, Burtsev A P 2020 Innovative technologies for improving the environmental and energy-saving characteristics of heat generators / monograph // *YuZGU Kursk: Publishing house of JSC Universitetskaya kniga* 157 p

[5] Halperin L G 2011 Modeling of thermal processes M .: UrFU 112 c

IOP Conf. Series: Materials Science and Engineering 1242 (2022) 012043 doi:10.1088/1757-899X/1242/1/012043

# Cement materials obtained by partial replacement of cement powder with brick powder

**A Zaharie[1*], M L Țințișan[1], A C Siomin[1], D L Manea[1] and M L Pleșa[1]**

[1]Technical University of Cluj-Napoca, Faculty of Civil Engineering, Romania

* anamaria.zaharie@ccm.utcluj.ro

**Abstract**. The current evolution of the construction materials manufacturing industry requires an increased focus on the quantities of raw materials used, this can be adjusted by obtaining new sustainable materials with low energy and raw material consumption. The negative impact of the exploitation of natural resources is felt both from the point of view of the population's health and of the economic branch. Urban development and the progress of society have led to an uncontrolled increase in the amount of waste resulting from construction and demolition. Over the years it has been shown that waste can be reintroduced into the circuit for reuse. Waste reuse positively influences the regeneration of the environment and the improvement of economic resources. The waste used in this study is the brick powder type waste, this waste being part of the category of recyclable waste resulting from building, rehabilitation modernization and demolition of different constructions This article studies the mechanical characteristics at 3, 7 and 28 days, of a series of CSIV type millet recipes with brick powder waste. Brick powder can be used in building materials without depleting the consumption of raw materials and without degrading the environment.

## 1. Introduction

At the present time, one of the most used materials around the worldwide are cement-based materials [1]. According to recent studies it has been observed that, to produce a 1000 kg of Portland cement 927 kg of $CO_2$ is emitted [2].

Currently, the production of the materials we use every day is responsible for 45% of $CO_2$ emissions [3]. Carbon neutrality is a goal that must be achieved for our safety and health. The reduction of carbon dioxide in the atmosphere can once be achieved by reducing the amount of cement produced annually and by adopting a circular economy solution. The circular economy involves the reuse of waste. A waste with recycling potential is the brick resulting from the process of construction and demolition of buildings.

The amount of waste resulting from constructions and demolitions in the period 2015-2020 is on average 3398.83 thousand tons/ year [4]. According to some studies, replacing cement with lime powder and brick powder in concrete recipes can reduce costs by 7 to 12.5% [5].

The aim of this study is to reuse the brick waste powder as a partial substitute of cement, in different proportions, taking also in considering not affecting the properties of recipes. According to the results obtained on fresh and hardened plastering mortars, the brick powder wastes can be used in percentage of 5%, 10% and 15% of the amount of cement. The flexural strength and compression strength values are much higher than standard mortars and, also higher than the minimum value according to the available norms, which is 6 $N/mm^2$.

## 2. Materials and Recipes

### *2.1. Materials*

The recipes were made with classic materials such as cement, sand and water in which binder component was replaced in proportion of 5%, 10% and 15% with brick powder.

The specific characteristics of the powdered cement were determined. The Blaine method obtained a specific cement surface area of 5568,8 $cm^2/g$.

Recipes of CSIV plaster mortars involve the use of aggregates on different types, ranging from 0.00-4.00mm. The experimental program used river sand as an aggregate sorted into five sizes (figure 1).

**Figure 1.** Sorted river sand.

The brick powder waste, results from constructions and demolition works, was used in the production of ecological plaster mortars. The resulting powder is similar in size to the cement used in the recipes. The fineness of the brick powder studied is 3397.5 $cm^2/g$. (figure 2).

**Figure 2.** Brick powder waste.

To prevent the hardening of mortar recipes, the water used in experimental program has to be clear water, without impurities or odour. [6]

### *2.2. Composition of ceramic waste mortars*
The compositions of ceramic waste recipes involve the use of brick powder waste in proportion of 5%, 10% and 15% as a replacement of cement. The quantities of the series of recipes for CS IV type plaster mortars being presented in table 1.

**Table 1.** Mortars recipes studied in the experimental program

| Recipe type | CSIV | BP 5% | BP 10% | BP 15% |
|---|---|---|---|---|
| Portland cement [kg] | 385 | 365.75 | 346.5 | 327.25 |
| River Sand [kg] | | | | |
| 0÷0.25 mm | 77.5 | 77.5 | 77.5 | 77.5 |
| 0.25÷0.50 mm | 232.5 | 232.5 | 232.5 | 232.5 |
| 0.50÷1 mm | 378.5 | 378.5 | 378.5 | 378.5 |
| 1÷2 mm | 232.5 | 232.5 | 232.5 | 232.5 |
| 2÷4 mm | 620 | 620 | 620 | 620 |
| Brick Powder [kg] | - | 19.25 | 38.5 | 57.75 |
| Water/Cement | 0.69 | 0.58 | 0.59 | 0.61 |

The steel moulds in which the mortar recipes were cast have the size of 40x40x160 mm. Until the test maturity of the recipes, all specimens have been kept in accordance with SR EN 1015-11 [7].

## 3. Methods
All the specimens were subjected to determinations according to the standards SR EN 1015 1 ÷ 21/2020). The methods used to determine the physical and mechanical characteristics were specific surface, apparent density, mortar consistency, tensile strength and compressive strength.

The value obtained represent the arithmetic average of three specimens that were subjected.

*Apparent density.* To determine the apparent density, the mortar must be in a homogeneous state. All the tests were made according to the standards [7-8].

*Consistency of the mortar.* The consistency of all types of mortars was made on the fresh mortars using spreading mass apparatus [9].

*Determination of flexural strength.* To study the evolution in time of hardened mortars, the flexural strength test was performed in time at 3, 7 and 28 days on prism sized 40x40x160mm (figure 3) [10].

**Figure 3.** Determination of flexural strength.

*Determination of compressive strength.* According to the rules in force (SR EN 1015-11: 2002 [10], see figure 4), the determination of compressive strength was made on the rest of the prisms after the flexural strength was performed. The determinations were made at 3, 7 and 28 days with the hydraulic press apparatus.

**Figure 4**. Determination of compressive strength at 28 days.

## 4. Results and discussion

The apparent density was performed in two states: apparent density for fresh mortars and, also for the hardened mortars. The results are according to table 2 and table 3.

**Table 2**. Values of Apparent density on fresh mortars.

| Recipe | CS IV | BP 5% | BP10% | BP 15% |
|---|---|---|---|---|
| Apparent density fresh mortars [kg/m$^3$] | 2274 | 2217.4 | 2209.8 | 2201.5 |

The values of apparent density of fresh mortars is between 2201.5 - 2217.4 kg /m$^3$ (see table 3). The recipe with the highest value of apparent density is BP 5% brick powder waste. The values of the apparent density of fresh mortar show a proportional decrease with the increase in the amount of brick powder waste.

**Table 3**. Values of apparent density of hardened mortars.

| | Mortar age | | | | | |
|---|---|---|---|---|---|---|
| | 3 days | | 7 days | | 28 days | |
| Recipes types | Weight [kg] | Apparent density [kg/m$^3$] | Weight [kg] | Apparent density [kg/m$^3$] | Weight [kg] | Apparent density [kg/m$^3$] |
| S | 0.575 | 2250 | 0.563 | 2200 | 0.567 | 2210 |
| BP 5% | 0.561 | 2191 | 0.547 | 2137 | 0.545 | 2129 |
| BP 10% | 0.571 | 2230 | 0.569 | 2223 | 0.561 | 2191 |
| BP 15 % | 0.585 | 2285 | 0.573 | 2238 | 0.569 | 2223 |

According to table 3, increasing the amount of brick powder in the plasters mortars recipes decreases the value of the apparent density (table 3).

**Table 4.** Consistency average values of tested mortars.

| Recipes | CS IV | BP 5% | BP 10% | BP 15% |
|---|---|---|---|---|
| Consistency [mm] | 182.5 | 189.5 | 197.0 | 202.0 |

The mortar with the highest value of consistency is the mortar in which the cement has been replaced in proportion of 15%. As for the average values obtained for the other types of mortars, they decrease in value as the percentages of brick powder in the proposed recipes increase (see table 4).

**Table 5.** Values of Bending tensile Strength [N/mm$^2$].

| Recipes / Mortar age | CS IV | BP 5% | BP 10% | BP 15% |
|---|---|---|---|---|
| 3 days | 4.47 | 3.86 | 4.76 | 4.78 |
| 7 days | 5.22 | 4.20 | 4.77 | 4.82 |
| 28 days | 6.08 | 7.15 | 7.06 | 6.84 |

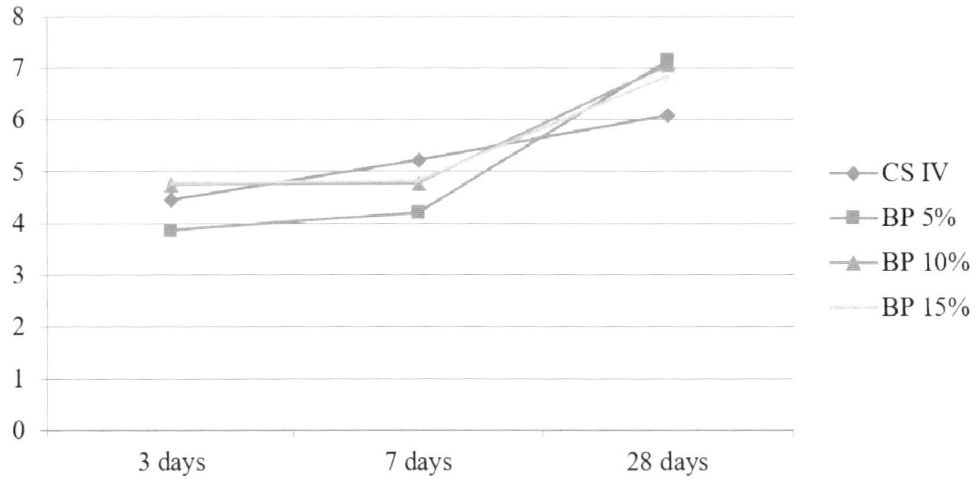


**Figure 5.** Representation influence of brick powder on Flexural strength determination [N/mm$^2$].

The higher value of tensile strength is the recipe with a proportion of 5% brick powder. All proportions replaced with brick powder have values above the value of standard CS IV mortar (figure 5 and table 5).

The higher value of compressive strength is the recipe with a proportion of 10% brick powder (figure 5). All the values according to table 6 and figure 7 are above 6N/mm$^2$ the standard limit of compressive strength).

The use of brick powder waste positively influences the test results, regardless of its proportion. The recipe with the highest values being the recipe with a substitute is 5% brick powder).

**Figure 6.** Tests of specimens with hydraulic press.

**Table 6.** Results of compressive strength [N/mm$^2$].

| Recipes type | CS IV | BP 5% | BP 10% | BP 15% |
|---|---|---|---|---|
| Mortar age | | | | |
| 3 days | 34.86 | 28.56 | 26.11 | 21.10 |
| 7 days | 36.71 | 35.24 | 32.33 | 30.77 |
| 28 days | 39.11 | 45.52 | 41.90 | 41.84 |

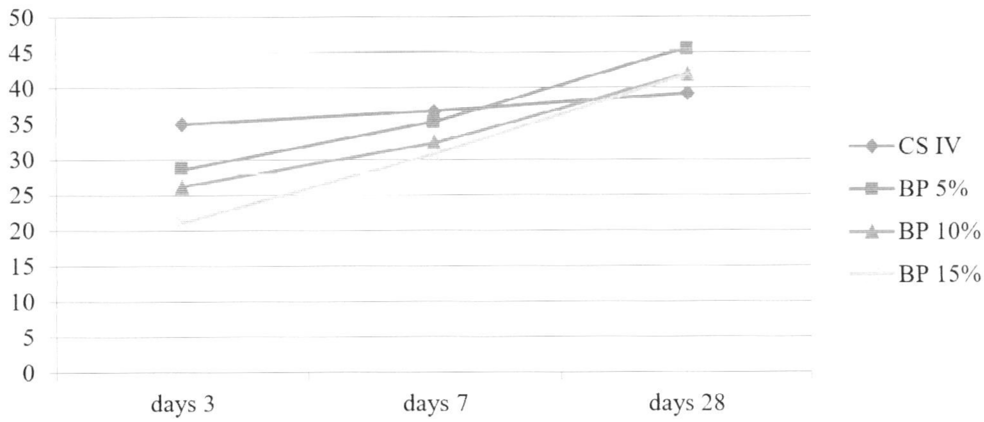


**Figure 7.** Representation of compressive strength value [N/mm$^2$].

## 5. Conclusions

The paper studies the plastering mortars where the cement was replaced in 5%, 10% and 15% with brick powder waste. The physical properties of brick powder were also studied, and the specific surface of cement is higher than brick powder wastes.

According to European Standards the brick powder can be used in the recipes of plaster mortars.

Replacing the cement with brick powder waste in the recipes of plaster mortars reduces the amount of $CO_2$ emissions.

The mechanical characteristics (tensile strength and compressive strength) of mortars in hardened state with brick powder substitute, presents higher values of the CS IV recipe studied.

The highest value in terms of determinations made in the experimental program is the recipe with 5% proportion of the amount of cement.

Air pollution is currently the biggest health risk in the world; therefore it is necessary to recycle and reuse the wastes in order to reduce the degree of pollution.

**References**

[1] Olivier J Janssens-Maenhout G Muntean M and Peters J 2016 Trends in global $CO_2$ emissions: 2016 *Report* Haga: PBL Netherlands Environmental Assessment Agency
[2] Marceau Medgar L Nisbet Michael A and VanGeem Martha G 2006 Life Cycle Inventory of Portland Cement Manufacture *SN2095b* Portland Cement Association Skokie IL p 69
[3] Commitee on the Environment public Health and Food Safety 2020 *Darft Report on the new circular economy action plan* 2077 INI p 9
[4] Romanian Association for waste maagement 2021 Construction and demolation waste management in Romania p 10
[5] Basit A, Aftab M, Khan I, Ahmed M Nasir Ayaz Khan and M Umar 2018 Effect of brick dust and lime powder on the performance of plain cement concrete p 1
[6] SR EN 1008-2003 Concrete preparation water. Specifications for sampling ,testing and assessment of water use capabilities, including water recovered from concrete industry processes, as water preparation for concrete
[7] SR EN 1015-6:2001/A1:2007 Methods of testing masonry mortar Part 6: Determination of apparent density of fresh mortar.
[8] SR EN 1015-10:2002/A1:2007 Methods of testing masonry mortar Part 10: Determination of apparent density of hardened mortar.
[9] SR EN 1015-3:2001 Methods of testing masonry mortar Part 3: Determination of consistency of fresh mortar (with spreading table)
[10] SR EN 1015-11:2020 Methods of testing masonry mortar Part 11: Determination of bending and compressive strength of hardened mortar

# Analysis of the openings of aircraft composite laminated wing beam web under shear load

**X L Zhang[1,2*], J C Wang[2], Q Xu[2], Y Y Wang[2] and C Y Li[1]**

[1] School of Mechanical Engineering and Automation, Northeastern University, Shenyang, China
[2] School of Intelligent Systems Engineering, Shenyang City University, Shenyang, China

* zhangxiulu1987@126.com

**Abstract**. In this paper, experiments and finite element methods are used to study the openings of the composite wing spar web of a civil aircraft. First, according to the design principles of composite laminates, determine the layup sequence and layup ratio of the spar, and then lay the beam webs and the beam edge strips integrally; then, calculate the equivalent of the spar by using the classic laminate theory Material properties, the method to calculate the equivalent elastic modulus of composite laminates of different thicknesses is derived; then, the influence of opening shape and opening angle on the strength of composite wing spar web under shear load is analyzed. Finally, the reinforcement design is carried out on the opening position. This research method not only solves specific engineering design problems, but also proposes a set of practical, cost-effective and efficient design methods for composite material opening shapes, which has a certain generality. Its design ideas and methods provide technical support for civil aircraft structural design.

## 1. Introduction

Compared with metal materials, composite materials have excellent mechanical properties. In general, composite materials have the characteristics of high specific strength, greater specific stiffness, simple manufacturing process, and strong design-ability of the material. At present, due to the needs of the aviation industry and other industries, especially in recent years, the aviation industry has made great progress in the design and manufacture of high-performance composite materials in terms of aircraft weight reduction and cost considerations. Currently, most advanced civil aircraft use composite materials for the center wing [1].

Fiber-reinforced resin-based composites have the characteristics of high specific strength, high specific stiffness, excellent fatigue performance, and can be cut. It has been widely used in the design and manufacture of civil aircraft structures, and has been completed from the fairing, front and rear edges, disturbances, etc. The transition from non-force-bearing structures such as fairings, front and rear edges, and spoilers to main bearing structures such as vertical stabilizers and wings has been completed [2-4]. However, in the actual structure of the aircraft, due to the passage of hydraulic pipes, wiring harnesses, control cables, etc., as well as inspection and maintenance, equipment installation, etc., it is inevitable to provide various types and sizes of openings. However, the opening will cut the fiber, which will cause the rigidity and strength of the laminate to decrease, and it will also cause the stress concentration at the edge of the hole. Therefore, accurate stress analysis and accurate strength prediction for composite open structures are an indispensable link in the design and analysis of composite structures, and it is also the key to improving the efficiency of composite structures [5-8].

At present, domestic and foreign research on composite material openings mainly focuses on the exploration of opening reinforcement methods, and certain progress has also been made. However, the interlayer shear strength and rigidity of composite laminates are relatively low, the openings are not easy to reinforce, and the reinforcement range will be larger than that of metal structures. On the contrary, if the appropriate opening shape can be selected at the beginning of the conceptual design, it will greatly reduce the requirements for reinforcement of the opening in the later stage [9].

In order to better perform finite element simulation modeling and meshing of composite spar webs, solid elements, shell elements and beam elements are integrated in the finite element software, which greatly improves the calculation accuracy [10]. By analyzing the SAC values of the finite element modal results and the experimental modal results, the rationality and feasibility of the constructed composite spar web finite element calculation simulation model is proved. Then, perform strength analysis in the finite element software to find the optimal opening plan.

## 2. Design of composite laminates

For composite laminates, only a reasonable ply design can maximize the use of the fiber's high strength and high stiffness characteristics, and can ensure the strength, stiffness and stability requirements under various design loads.

(1) The standard laying angles of $0^0$, $\pm45^0$, $90^0$ are generally used, and only a limited number of non-standard angles will be used in very special circumstances. Because the fiber in the $0^0$ direction mainly bears the tension and compression load in the layer, the fiber in the $\pm45^0$ direction mainly bears the shear load in the layer, and the fiber in the $90^0$ direction mainly plays the role of improving the transverse strength and adjusting the Poisson's ratio. Therefore, these four laying angles can generally meet the needs of the design load. At the same time, each of these four laying angles should account for at least 10% to prevent the substrate in a certain direction from being directly loaded [11].

(2) There must be enough plies, and the fiber axis is consistent with the direction of tension and compression load; avoid overlapping of plies in the same direction, let alone stacking more than 5 plies in the same direction to reduce edge delamination.

(3) When the load is mainly in the $0^0$ direction, the $90^0$-direction layer group should be used as little as possible, and it can be separated by layers in the $0^0$ direction or $\pm45^0$ direction to reduce the shear stress and normal stress; in addition, in the layer The outer layer of the pressure plate uses $\pm45^0$ direction paving, which can effectively improve the impact resistance of the structure [12].

(4) The symmetrical and balanced layering design can avoid the warpage caused by the tension-bending-torsion coupling, and it is also conducive to process processing. At the same time, each $+45^0$ layer should be equipped with $-45^0$ layers balance.

The spar mainly bears shear load and bending load. According to the paving design principle, the ratio of ±450 direction and 00 direction plies should be increased, and the ±450 direction plies should be placed as far away from the center plane as possible to improve the structure Critical load; At the same time, a balanced and symmetrical layering is adopted, and the number of layers is appropriately increased, which can effectively utilize the high strength and high rigidity characteristics of the fiber. The upper and lower skins are mainly subjected to tension, compression, bending and shearing, and the ply ratio is the same as that of the spar.

For composite laminates, the order of plies, the total number of plies and the direction of the material coordinate system need to be specified. Take the spar of the spar as the X direction of the material coordinate system (that is, the 0° layering direction), the normal direction of the shell element is the Z direction of the material coordinate system, and the Y direction of the material coordinate system is determined by the right-hand rule. See table 1 for the layup sequence of the spar structure. In order for the beam web and the beam edge strip to be laid integrally, the plies of the beam web and the beam edge strip are designed to be exactly the same, both are 30 layers, and the thickness is 4.5 mm; the number of layers of the skin is 36. The thickness is 5.4 mm.

**Table 1.** Laying sequence of spar structure.

| Region | Laying sequence |
|---|---|
| web | [45/-45/0/90/45/-45/0/45/-45/0/45/-45/0/0/90] s |
| cap | [45/-45/0/90/45/-45/0/45/-45/0/45/-45/0/0/90] s |
| skin | [45/-45/0/90/45/-45/0/-45/0/45/-45/0/0/45/-45/0/45/-45] s |

## 3. Specimen conditioning for

### *3.1. Dry conditioning*

Dry Specimen conditioning: Exposed specimens to the laboratory environment (21°C ± 5°C, 50% ± 10% RH) more than 24 hours.

### *3.2. Wet conditioning*

Measure the weight $W_b$ of specimens and travelers to be used in moisture absorption testing. Then dry them at 70°C. Weigh and record travelers once per 24 hours. The average desorption content of the travelers coupon changes by less than 0.02% for the consecutive readings at an interval of 7 days. Calculate the average desorption percent of the travelers change by Eq1.

$$\left| \frac{W_i - W_b}{W_b} - \frac{W_{i-1} - W_b}{W_b} \right| \times 100 < 0.02\% \tag{1}$$

where:

$W_i$——value at current time, lb;

$W_{i-1}$——value at previous time, lb, and;

$W_b$——traveler mass, lb.

Put the dried specimen and travelers into the conditioning chamber SH050A (see figure 1) at 70°C and relative humidity of 85%. Weigh and record travelers once per 24 hours. The average moisture content of the travelers coupon changes by less than 0.02% for the consecutive readings at an interval of 7 days. Calculate the percent change in mass for moisture absorption by Eq2.

$$\left| \frac{W_i - W_{od}}{W_{od}} - \frac{W_{i-1} - W_{od}}{W_{od}} \right| \times 100 < 0.02\% \tag{2}$$

where:

$W_i$——value at current time, lb;

$W_{i-1}$——value at previous time, lb, and;

$W_{od}$——oven-dry specimen mass, lb.

Formulation 3 shows how to calculate the moisture equilibrium content $M_m$ (%) of all the specimen.

$$M_m = \frac{W_m - W_{od}}{W_{od}} \times 100 \tag{3}$$

where:

$M_m$——moisture equilibrium content, % and

$W_m$——moisture equilibrium content.

Put specimens into the sealed bag after moisture equilibrium, store all materials at the standard laboratory atmosphere until required subsequent material test. The specimen shall be out of the sealed bag for more than one hour.

**Figure 1**. The Conditioning Chamber (SH050A).

## 4. Establishment and verification of FEM model

### *4.1. Structure introduction*

Figure 2 shows a composite wing spar of a civil aircraft, which is mainly composed of beam webs, beam edge strips and skin. The laminates are all made of medium-mold high-strength carbon fiber toughened epoxy resin prepreg, and their thickness is 4.5 mm, 4.5 mm and 5.4 mm, the overall size is 1500 mm (length) × 380 mm (width) × 400 mm (height). The beam web openings are mainly in three forms: triangle, diamond and circle.

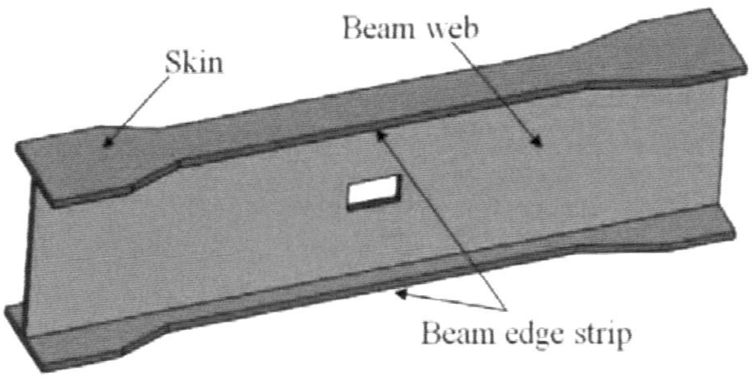


**Figure 2**. Schematic diagram of composite wing spar.

### *4.2. Establishment of simulation model*

Both the skin and the spar are simulated using shell elements (QUAD4), and the mesh is divided according to a density of approximately 5 mm. The plane where the shell elements are located is the midplane of the solid part, and the overall finite element model of the spar contains about 60,000 elements. The meshes near the openings are refined.

For the composite laminate structure, as long as it does not fail under load (that is, the laminate is not buckled), the strain value of each layer in the laminate is the same, but the stress of each layer is not the same. Therefore, the allowable strain is used as the design limit value in the design of composite structures. The strain field at the edge of the composite material opening is complex, the stress concentration is high, and the free edge deformation is large, so the hole edge strain analysis is carried out.

In the modeling of the spar web, except for some components which are three-dimensional solid models and beam elements, the rest are mainly two-dimensional shell elements. Compared with the thin plate, the thin shell increases the internal force in the plane, and then the internal force of the thin film neglected by the thin plate can be taken into account in the case of a small deflection, so the model will be closer to the real situation of the model. Therefore, the model simulated by the two-dimensional shell element is more accurate to approximate the actual structure of the structure than the model by the plate element, that is, it has higher accuracy [13].

### 4.3. Boundary conditions

The spar is mainly subjected to bending and shear loads. The restraint and loading conditions are shown in figure 3. At the constraint end on the left side of the fixture, all the degrees of freedom of the model nodes are constrained for fixed support; at the same time, in order to prevent torsional deformation during the analysis process, the degrees of freedom in the two directions outside the loading end are constrained. Finally, a 3-direction load is applied on the reference point.

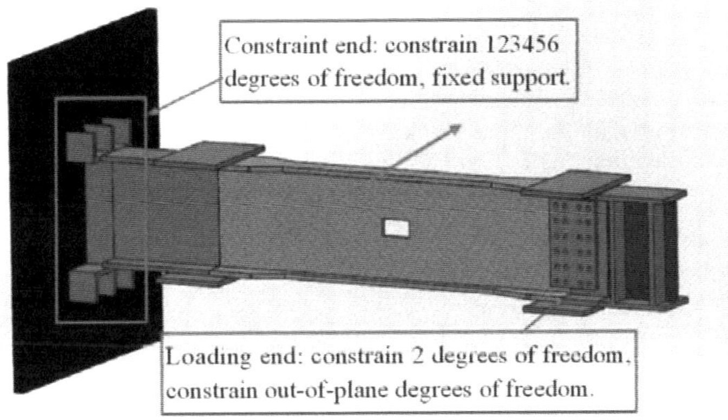


**Figure 3**. Boundary conditions of the wing beam shear test.

### 4.4. Model verification

The rationality and feasibility of the model are verified by comparing the values of the finite element free mode and the experimental free mode. The hammering method used in the experiment is a typical single-point excitation method. Since the excitation device is not directly connected to the structure being tested, it will not add additional mass to the structure, so it will not affect the structure of the structure [14].

The beam web is suspended on the support frame with springs so that it is in an approximate "free-free" state. According to the structure of the beam web, a number of test points and an excitation point are selected, the force hammer is used for excitation, the data acquisition system picks up the vibration signal of the test point, and the modal parameters of the beam web are identified by PolyMAX of Belgium LMS [15].

According to the classical laminate theory, the rationality and feasibility of the simulation model could be verified by the strain assurance criterion (SAC). The SAC matrix was used for the comparison of the strain data identified from the measured data (i.e. experimental analysis) and the strain data computed from the analytical model (i.e. FEM analysis). The measured strain of the composite stiffened panel was extracted from the experimental method, as shown in figure 4.

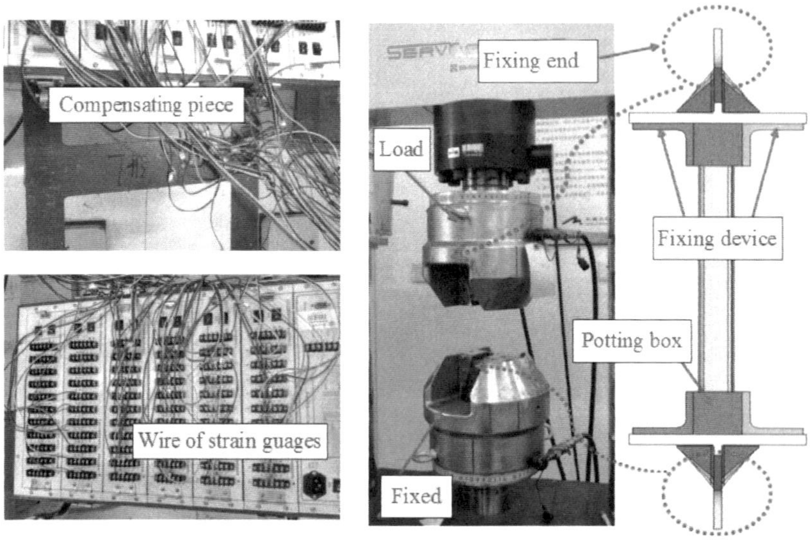


**Figure 4**. Experiment to extract the strain value of specimen.

The maximum error of strain value between the FEM and the experiment did not exceed 2% at the linear range, representing that the results were in good agreement. Most of the SAC values were greater than 0.7, denoting that the finite element simulation model of the composite stiffened panel structure had a high approximation to the real situation. Consequently, it proved that the calculation and analysis model was reasonable and feasible, providing a foundation for further research and analysis.

## 5. Simulation analysis of open beam web

### *5.1. The influence of different opening shapes*

Using the controlled variable method, the three finite element models are controlled to be equal in size. The base materials are all 2D orthotropic materials, and the basic parameters are the same. The number of layers, thickness and direction are exactly the same, the boundary conditions are the same, and the applied loads are all the cutting force of 100 N/mm, the opening area is the same, both are 10000 mm$^2$. The web models of triangular, rhombic and circular open beams were established for simulation analysis [16].

It can be seen from the strain cloud diagram in figure 5 that the strain distribution of beam web models with different opening shapes is different. The relationship between the area of the high strain area (1760-3530 $\mu\varepsilon$ ) at the edge of the opening is: rhombus>circle>triangle. It can be concluded that the larger the area of the high-strain region, the smaller the strain concentration, and the less prone to failure of the model. At the same time, the maximum strain on the edge of the triangular opening is 3530 $\mu\varepsilon$ , the diamond opening is 1740 $\mu\varepsilon$ , and the circular opening is 2370 $\mu\varepsilon$ . Therefore, when the opening area is the same, the beam web model of the diamond-shaped opening scheme is stronger and is the least prone to shear failure, followed by the circular opening scheme, and the triangular opening scheme is the most prone to shear failure.

As the shear load continues to increase, the maximum strain values of the edges of the openings of different shapes gradually increase. The maximum strain value at the edge of the diamond-shaped opening is the smallest, and as the shear load increases, its rise is the slowest.

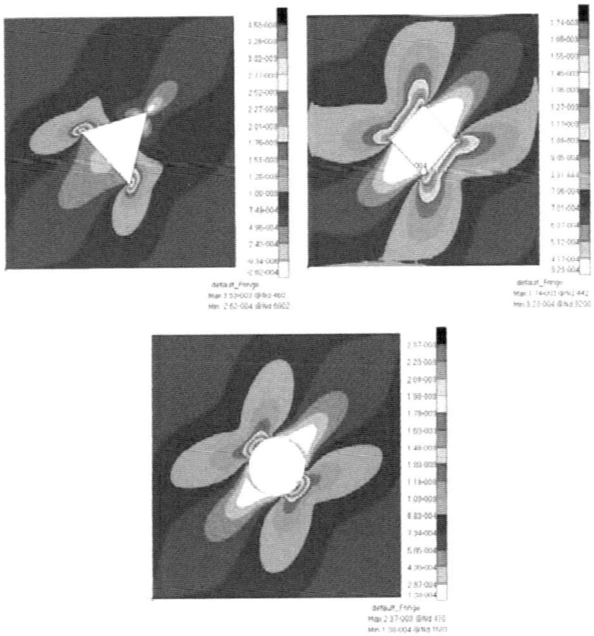

**Figure 5**. Strain cloud diagram of spars with different opening shapes under shear load.

The opening areas of the three shapes are the same, but the number of fibers in the ±45° direction of the cut composite laminate is different. Under the action of shear load, the main bearing function is the fiber in the direction of ±45°. Under the same opening area, the effective length of the triangular opening cutting the fiber in the ±45° direction is 152.0 mm, the effective length of the diamond opening is 100.0 mm, and the effective length of the circular opening is 112.8 mm. For beam webs under shear load, the model with fewer fibers in the ±45° direction of the opening cut is less likely to fail. Therefore, beam webs with diamond-shaped openings are the least likely to undergo shear failure

*5.2 Influence of different opening angles*

In order to further verify the relationship between the strength of the model and the number of cut fibers in the opening, the beam web models with square openings of 0°, 11.25°, 22.5°, 33.75° and 45° under the same conditions were established for simulation analysis. Because the spars are all laid out symmetrically and evenly, there is no need to build models with square openings of 0°, -11.25°, -22.5°, -33.75°, and -45°.

It can be seen from the strain cloud diagram in figure 6 that the strain distribution under different angles of the square opening is different. The relationship between the area of the high strain area (1760-3530 $\mu\varepsilon$ ) at the edge of the opening is: 45°> 22.5°> 0°. It can be concluded that the larger the area of the high-strain region, the smaller the strain concentration, and the less prone to failure of the model. At the same time, the maximum strain at the edge of a 0° square opening is 3170 $\mu\varepsilon$ , a 22.5° square opening is 2960 $\mu\varepsilon$ , and a 45° square opening is 1740 $\mu\varepsilon$ . Therefore, the beam web model of the 45° (-45°) square opening scheme is the least prone to shear failure, followed by the 22.5° square opening scheme, and the 0° square opening scheme is the most prone to shear failure.

With the continuous increase of the shear load, the maximum strain values of the edges of the square openings with different angles gradually increase. The maximum strain at the edge of the 45° square opening is the smallest, and as the shear load increases, it rises the slowest.

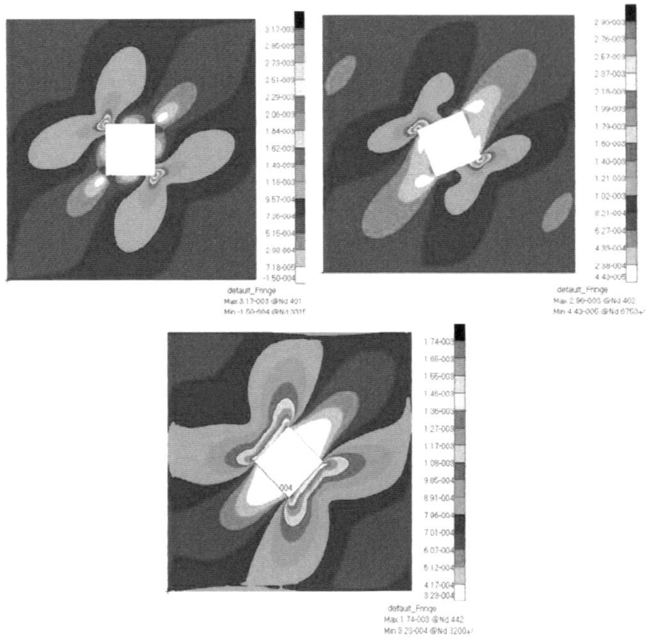

**Figure 6**. Strain cloud diagram of spar with different opening angles under shear load.

The effective length of the 0° square opening to cut the fiber in the ±45° direction is 141.4 mm, the 11.25° opening is 133.6 mm, the 22.5° opening is 123.5 mm, the 33.7 5° opening is 114.2 mm, and the 45° opening is 100.0 mm. For beam webs under shear load, the model with fewer fibers in the ±45° direction cut off the opening is less likely to fail. Therefore, beam webs with diamond-shaped ±45° openings are the least likely to fail.

## 6. Opening reinforcement design

The edge of the opening will produce stress concentration and edge effects, which will cause the edge of the opening to be partially damaged in advance. Therefore, the opening of the laminate generally needs to be reinforced. Reinforcement measures should be determined according to the specific conditions, including patch reinforcement, recessed mouth frame reinforcement, hole edge edging, hole edge softening, etc. After the opening is reinforced, its load-bearing capacity should be restored to meet the requirements of the structure's load-bearing capacity. Reinforcement measures should be feasible and reliable in quality, and the increased quality should be as small as possible. Among them, the method based on patch reinforcement is worth recommending.

Since the fibers of the opening are cut, a good reinforcement effect can be obtained by using the patch reinforcement around the opening, which can be preferentially used when permitted. Taking a circular opening as an example, the patch diameter is generally 1.5 to 2 times the diameter of the opening, as shown in figure 7. The material of the reinforcing piece is the same as that of the mother board, and the area of the patch has sufficient interlayer shear strength when patching. Due to the small area of the web, the patch is directly connected to the vertical flange of the edge strip, which not only maintains the continuity of the fiber, but also the web layup at the reinforcement is balanced and symmetrical.

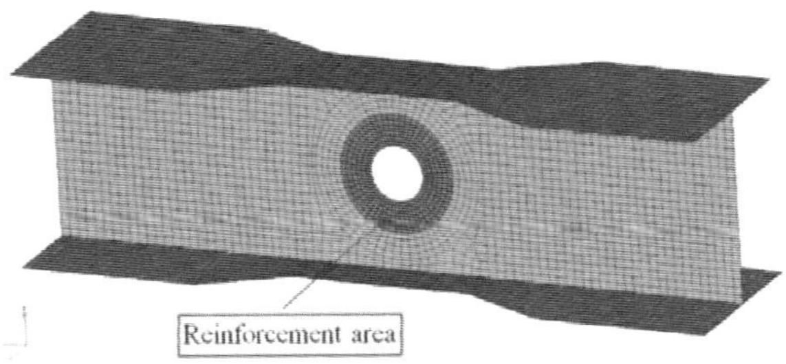


**Figure 7**. Schematic diagram of opening reinforcement design.

Calculating and analyzing the reinforced circular opening spar, it is found that the main feature of opening deformation is still the spatial bending along the diagonal opening edge, but the maximum strain value is effectively controlled, and the distribution is more even than before reinforcement Some meet the requirements of strain allowable value, as shown in figure 8. Therefore, this reinforcement method is effective and feasible for improving the strength and rigidity of the opening. The focus is on the width of the reinforcement piece, the width of the transition section and the design of the ply.

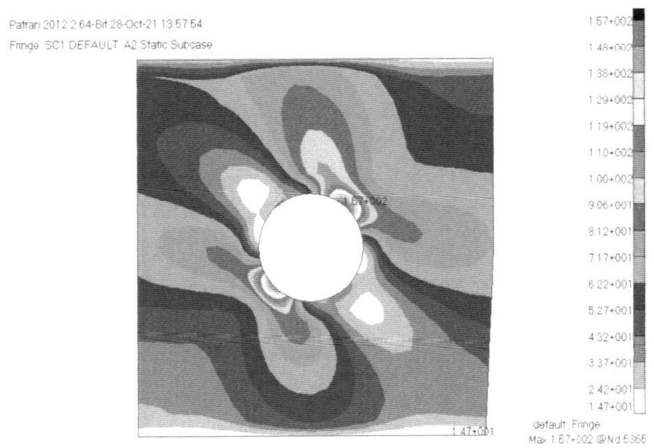

**Figure 8**. Strain cloud diagram of the spar after the circular opening is reinforced.

## 7. Conclusion

The design of composite material openings has always been a difficult point in the design of aircraft structures. This article focuses on the research of composite material wing beam web opening structure, and gives finite element modeling, material property definition and boundary condition definition methods, and then carries out the wing beam shearing Analysis of the maximum strain at the edge of the opening under shear load. The comparative analysis results show that the beam web model with the diamond-shaped opening is stronger when the opening area is the same and the shape is different, and shear failure is the least likely to occur in comparison. When the opening area and shape are the same, the opening the less the number of fibers cut in the ±45° direction, the less prone to shear failure. Considering the analysis results comprehensively, it is more reasonable to design the beam web with a diamond-shaped ±45° angle opening under the action of shear load. In addition, this paper also analysis the force of the opening and surrounding reinforcement structure, analysis the reinforcement form of the circular opening structure, and verified the rationality of the design of the opening reinforcement structure through experiments. The design ideas and methods have a strong reference value for the reinforcement design of composite material opening structures, simplifying the design process, shortening the development cycle, reducing the number of tests, reducing R&D costs, and improving the performance of composite beam webs. Has an important guiding and reference role for the initial research and development of civil aircraft.

**Acknowledgements**
The work was supported by National Natural Science Foundation of China (Grant No. 52075087), the Fundamental Research Funds for the Central Universities (Grant No. N2003006), National Natural Science Foundation of China (Grant No. U1708254). Thanks to AVIC SAC Commercial Aircraft Co. Ltd for their supports in the manufacture of specimens and NDT inspection.

**References**
[1] Nsengiyumva W, Zhong S, Lin J, Zhang Q, Zhong J, and Huang Y 2021 *Compos. Struct.* **256** 112951
[2] Hassan M H, Othman A R and Kamaruddin S 2017 *Int. J. Adv. Manuf. Tech.* **91** 4081
[3] Zimmermann N and Wang P H 2020 *Eng. Fail. Anal.* **115**
[4] Chen B 1987 *J. Rock Drilling Mach. & Pneu. Tools* **47** 47
[5] Skalka P, Slámečka K, Pokluda J and Čelko L 2015 *J. Sur. & Coat. Tech.* **274** 26
[6] Slámečka K, Skalka P, Pokluda J and Čelko L 2016 *J. Sur. & Coat. Tech.* **304** 574
[7] Yin A, Feng Z, Hui R and Guo H 2018 *J. Foundry Tech. Tools* **39** 2854
[8] Su X, Feng Z, Gao C, and Hao W 2017 *J. Foundry Tech. Tools* **42** 156
[9] Xiao H, Duan Z, Li N, Xiong C, Zhou R, Lu D and Jiang Y 2019 *J. Rare Metal Mat. and Eng.* **48** 235
[10] Tian R and Wang Z 2002 *Copper alloy and its processing manual* (Chang Sha: Central South University Press)
[11] Xu B and Liu S 2002 *New technology in surface engineering* (Bei Jing: national defence industrial press)
[12] Abramovich H and Weller T 2009 *J. Mech. Mater. Struct.* **4** 1187
[13] Liu N, Beata P A and Jeffers A E 2019 *Numer. Heat Tr. B-Fund* **75** 347
[14] Liew K M, Pan Z Z and Zhang L W 2019 *Comput. Struct.* **216** 240
[15] Liu N, Ren X and Lua J 2020 *Compos. Struct.* **237** 111893
[16] Liu N, Johnson E L, Rajanna M R, Lua J, Phan N and Hsu M C 2021 *Comput. Method Appl.* **385** 114005

# AUTHOR INDEX